Springer Undergraduate Mathematics Series

Other books in this series

A First Course in Discrete Mathematics *I. Anderson*
Analytic Methods for Partial Differential Equations *G. Evans, J. Blackledge, P. Yardley*
Basic Linear Algebra, Second Edition *T.S. Blyth and E.F. Robertson*
Basic Stochastic Processes *Z. Brzeźniak and T. Zastawniak*
Complex Analysis *J.M. Howie*
Elementary Differential Geometry *A. Pressley*
Elementary Number Theory *G.A. Jones and J.M. Jones*
Elements of Abstract Analysis *M. Ó Searcóid*
Elements of Logic via Numbers and Sets *D.L. Johnson*
Essential Mathematical Biology *N.F. Britton*
Fields, Flows and Waves: An Introduction to Continuum Models *D.F. Parker*
Further Linear Algebra *T.S. Blyth and E.F. Robertson*
Geometry *R. Fenn*
Groups, Rings and Fields *D.A.R. Wallace*
Hyperbolic Geometry *J.W. Anderson*
Information and Coding Theory *G.A. Jones and J.M. Jones*
Introduction to Laplace Transforms and Fourier Series *P.P.G. Dyke*
Introduction to Ring Theory *P.M. Cohn*
Introductory Mathematics: Algebra and Analysis *G. Smith*
Linear Functional Analysis *B.P. Rynne and M.A. Youngson*
Mathematics for Finance: An Introduction to Financial Engineering *M. Capiński and T. Zastawniak*
Matrix Groups: An Introduction to Lie Group Theory *A. Baker*
Measure, Integral and Probability, Second Edition *M. Capiński and E. Kopp*
Multivariate Calculus and Geometry, Second Edition *S. Dineen*
Numerical Methods for Partial Differential Equations *G. Evans, J. Blackledge, P. Yardley*
Probability Models *J. Haigh*
Real Analysis *J.M. Howie*
Sets, Logic and Categories *P. Cameron*
Special Relativity *N.M.J. Woodhouse*
Symmetries *D.L. Johnson*
Topics in Group Theory *G. Smith and O. Tabachnikova*
Vector Calculus *P.C. Matthews*

Duncan Marsh

Applied Geometry for Computer Graphics and CAD

Second Edition

With 127 Figures

Cover ilustration elements reproduced by kind permission of:
Aptech Systems, Inc., Publishers of the GAUSS Mathematical and Statistical System, 23804 S.E. Kent-Kangley Road, Maple Valley, WA 98038, USA. Tel: (206) 432-7855 Fax (206) 432-7832 email: info@aptech.com URL: www.aptech.com.
American Statistical Association: Chance Vol 8 No 1, 1995 article by KS and KW Heiner 'Tree Rings of the Northern Shawangunks' page 32 fig 2.
Springer-Verlag: Mathematica in Education and Research Vol 4 Issue 3 1995 article by Roman E Maeder, Beatrice Amrhein and Oliver Gloor 'Illustrated Mathematics: Visualization of Mathematical Objects' page 9 fig 11, originally published as a CD Rom 'Illustrated Mathematics' by TELOS: ISBN 0-387-14222-3, German edition by Birkhauser: ISBN 3-7643-5100-4.
Mathematica in Education and Research Vol 4 Issue 3 1995 article by Richard J Gaylord and Kazume Nishidate 'Traffic Engineering with Cellular Automata' page 35 fig 2. Mathematica in Education and Research Vol 5 Issue 2 1996 article by Michael Trott 'The Implicitization of a Trefoil Knot' page 14.
Mathematica in Education and Research Vol 5 Issue 2 1996 article by Lee de Cola 'Coins, Trees, Bars and Bells: Simulation of the Binomial Process' page 19 fig 3. Mathematica in Education and Research Vol 5 Issue 2 1996 article by Richard Gaylord and Kazume nishidate 'Contagious Spreading' page 33 fig 1. Mathematica in Education and Research Vol 5 Issue 2 1996 article by Joe Buhler and Stan Wagon 'Secrets of the Madelung Constant' page 50 fig 1.

British Library Cataloguing in Publication Data
Marsh, Duncan
Applied geometry for computer graphics and CAD. — 2nd ed. —
(Springer undergraduate mathematics series)
1. Geometry — Data processing 2. Computer graphics —
Mathematics 3. Computer-aided design — Mathematics
I. Title
516′.0028566
ISBN 1852338016

Library of Congress Cataloging-in-Publication Data
Marsh, Duncan
Applied geometry for computer graphics and CAD / Duncan Marsh.—2nd ed.
p. cm. — (Springer undergraduate mathematics series)
Includes bibliographical references and index.
ISBN 1-85233-801-6 (alk. paper)
1. Computer graphics. 2. Computer-aided design. 3. Geometry—Data processing. I. Title. II. Series.
T385.M3648 2004
516—dc22 2004054958

Springer Undergraduate Mathematics Series ISSN 1615-2085
ISBN 1-85233-801-6 2nd edition Springer-Verlag London Berlin Heidelberg
ISBN 1-85233-080-1 1st edition Springer-Verlag London Berlin Heidelberg
Springer Science+Business Media
springeronline.com

Printed and bound in the United States of America

First published 1999
Second edition 2005

Typesetting: Camera ready by the author
12/3830-543210 Printed on acid-free paper SPIN 10946442

To Tine and Emma

Preface to the Second Edition

The second edition of *Applied Geometry for Computer Graphics and CAD* features three substantial new sections and an additional chapter. The new topics, which include discussions of quaternions, surfaces, solid modelling and rendering, give further insight into the applications of geometry in computer graphics and CAD. The text has been revised throughout, and supplemented with further examples and exercises: the second addition contains more than 300 exercises and over 120 illustrations.

In Chapter 3, a new section introduces quaternions, an important method of representing orientation that is used in computer graphics animation.

Chapter 9 has been expanded to provide two new sections that focus on the applications of surfaces in CAD: Section 9.6 describes skin and loft surfaces (including Gordon–Coons surfaces), and Section 9.7 discusses geometric modelling. The chapter also benefits from additional examples of applications of surfaces; for example, offset and blend surfaces, and shelling and thickening operations.

A new final chapter addresses rendering methods in computer graphics and CAD, and presents an introduction to silhouettes and shadows.

There is a web site for the book which contains additional information and further web links:

www.springeronline.com/1-85233-801-6/

Cambridgeshire, UK — Duncan Marsh

Preface to the First Edition

Applied Geometry for Computer Graphics and CAD explores the application of geometry to computer graphics and computer-aided design (CAD). The textbook considers two aspects: the *manipulation* and the *representation* of geometric objects. The first three chapters describe how points and lines can be represented by Cartesian (affine) and homogeneous coordinates. Planar and spatial transformations are introduced to construct objects from geometric primitives, and to manipulate existing objects. Chapter 4 describes the method of rendering three-dimensional objects on a computer screen by application of a linear projection, and the construction of the *complete viewing pipeline*. The material then develops into a study of planar and spatial curves. Conics are described in some detail, but the main emphasis is a discussion of the two main curve representations used in CAD packages and in computer graphics, namely, Bézier and B-spline curves. The techniques of the earlier chapters are applied to these curves in order to manipulate and view them. The important de Casteljau and de Boor algorithms, for (integral and rational) Bézier and B-spline curves respectively, are derived and applied. The representations of curves lead naturally into surface representations, namely Bézier, B-spline and NURBS surfaces. The transition is relatively painless since many properties of the curve representations correspond to similar surface properties. The final chapter introduces curvature for curves and surfaces.

The book includes more than 250 exercises. Some exercises encourage the reader to implement a number of the techniques and algorithms which are discussed. These exercises can be carried out using a computer algebra package in order to avoid the complexity of computer programming. Certainly this is the most accessible route to obtaining quality graphics. Alternatively, the algorithms can be implemented using the reader's favourite programming language together with a library of graphics routines (e.g. PHIGS, OpenGL, or

GKS). The two approaches can be mixed as some computer algebra packages can make use of procedures written in programming languages such as C and FORTRAN. A number of exercises indicate investigations which would be suitable for coursework, labs or projects.

The book assumes a knowledge of vectors, matrices, and calculus. However, the course has been taught to engineering and computing students with only a little knowledge of these topics; with some additional material, these topics can be taught on a need to know basis. Indeed, the material in the book provides a source of motivation for teaching elementary algebra and calculus to non-mathematics students. Prerequisite reading on vectors, matrices and continuity of functions can be found in Chapters 4 and 7 of the SUMS series text *Introductory Mathematics: Algebra and Analysis* by Geoff Smith.

The author would like to thank a number of people. First, the mathematics, computing and engineering students at Napier University who took the modules on which this book is based. Second, my colleagues at Napier University; in particular, Dr. Winston Sweatman who shares an office with me (need I say more!). Finally, my wife Tine and daughter Emma for their continuing love and support.

Edinburgh, UK Duncan Marsh

Contents

1
Transformations of the Plane

1.1 Introduction

The two main areas of application which are considered in this textbook are computer graphics and computer-aided design (CAD). In computer graphics applications, geometric objects are defined in terms of a number of basic building blocks called *graphical primitives*. There are primitives which correspond to points, lines, curves, and surfaces. For example, a rectangle can be defined by its four sides. Each side is constructed from a line segment primitive by applying a number of geometric operations, called transformations, which position, orientate or scale the line primitive. Five types of transformation are particularly relevant in applications, namely, translations, scalings, reflections, rotations, and shears. These are introduced in Sections 1.2–1.6. Applications of transformations are considered in Section 1.8. In particular, Section 1.8.1 exemplifies, in more detail, how objects can be defined by applying transformations to graphical primitives by a process called *instancing*. Each primitive has a mathematical representation which can be expressed as a data or type structure for storage and manipulation by a computer. The mathematical representation of primitives is discussed in Chapters 5–9.

Given a fixed unit of length, and two perpendicular lines of reference called the x-axis and the y-axis, each point $\mathbf{P}$ of the plane is represented by an ordered pair of real numbers (x, y) such that the perpendicular distance of $\mathbf{P}$ from the y-axis is x units and the distance of $\mathbf{P}$ from the x-axis is y units. The ordered pair (x, y) is called the *Cartesian* or *affine coordinates* of $\mathbf{P}$, and the set of all

ordered pairs of real numbers (x, y) is called the *Cartesian* or *affine plane* and denoted $\mathbb{R}^2$. The axes intersect in a point $\mathbf{O}$, with coordinates $(0, 0)$, called the *origin*. The point $\mathbf{P}$ with coordinates (x, y) will be denoted $\mathbf{P}(x, y)$. For the purposes of computation the point may also be represented by the row vector (x, y) or the row matrix $\begin{pmatrix} x & y \end{pmatrix}$.

For constants A, B, C (A and B not both zero) the set of points (x, y) satisfying the equation

$$Ax + By + C = 0$$

is a *line* which is said to be defined in *implicit form*. The line through a point (p_1, p_2) in the direction of the vector (v_1, v_2) can be defined *parametrically* by

$$(x(t), y(t)) = (p_1 + v_1 t, p_2 + v_2 t) \ .$$

Each value of the parameter t corresponds to a point on the line. For instance, evaluating $x(t)$ and $y(t)$ at $t = 0$ yields the point (p_1, p_2), and evaluating at $t = 1$ yields the point $(p_1 + v_1, p_2 + v_2)$. Any parametrically defined line can be expressed in implicit form by eliminating t from $x = p_1 + v_1 t$ and $y = p_2 + v_2 t$, to give

$$v_2 x - v_1 y + (p_2 v_1 - p_1 v_2) = 0 \ .$$

It also follows that the line with equation $Ax + By + C = 0$ has the direction of the vector $\pm(-B, A)$ and normal direction (the direction perpendicular to the line) $\pm(A, B)$.

The line through the two points $\mathbf{P}$ and $\mathbf{Q}$ is denoted $\overline{\mathbf{PQ}}$. The line *segment* $\mathbf{PQ}$ (with *endpoints* $\mathbf{P}$ and $\mathbf{Q}$) is the portion of the line $\overline{\mathbf{PQ}}$ between the points $\mathbf{P}$ and $\mathbf{Q}$.

Example 1.1

Consider the line passing through the point (a, b), and making an angle α with the x-axis. By elementary trigonometry, a point (x, y) on the line satisfies $\tan(\alpha) = (y - b)/(x - a)$. Hence the line is given in implicit form by $\tan(\alpha)x - y + b - \tan(\alpha)a = 0$.

Example 1.2

Consider two lines $A_1 x + B_1 y + C_1 = 0$ and $A_2 x + B_2 y + C_2 = 0$ with directions $\mathbf{v} = (-B_1, A_1)$ and $\mathbf{w} = (-B_2, A_2)$ respectively. Suppose θ is the angle between the lines. Then the vector identity $\mathbf{v} \cdot \mathbf{w} = |\mathbf{v}|\,|\mathbf{w}| \cos\theta$ and the trigonometric identity $\cos^2\theta + \sin^2\theta = 1$ give

$$\cos\theta = \frac{A_1 A_2 + B_1 B_2}{\left(A_1^2 + B_1^2\right)^{1/2} \left(A_2^2 + B_2^2\right)^{1/2}}, \qquad \sin\theta = \frac{A_1 B_2 - B_1 A_2}{\left(A_1^2 + B_1^2\right)^{1/2} \left(A_2^2 + B_2^2\right)^{1/2}} \ .$$

Hence

$$\tan\theta = \frac{A_1B_2 - A_2B_1}{A_1A_2 + B_1B_2} \, .$$

It follows that the two lines are parallel if and only if $\theta = 0$, that is, if and only if $A_1B_2 = A_2B_1$.

EXERCISES

1.1. Show that the angle α that the line $Ax + By + C = 0$ makes with the x-axis is given by $\tan(\alpha) = -A/B$.

1.2. Determine an implicit equation for the line $(2+3t, 5-4t)$. Determine the angle that the line makes with the x-axis.

1.3. Show that, for points $\mathbf{P}(p_1, p_2)$ and $\mathbf{Q}(q_1, q_2)$, the line $\overline{\mathbf{PQ}}$ has the parametric form $(1-t)(p_1, p_2) + t(q_1, q_2)$, that is, $(x(t), y(t)) = (p_1 - tp_1 + tq_1, p_2 - tp_2 + tq_2)$ for $t \in \mathbb{R}$. Show also that the segment $\mathbf{PQ}$ is given by the same equation for $t \in [0, 1]$.

1.4. Show that $A_1x + B_1y + C_1 = 0$ and $A_2x + B_2y + C_2 = 0$ are perpendicular if and only if $A_1A_2 + B_1B_2 = 0$.

Definition 1.3

A (linear) *transformation* of the plane is a mapping $L : \mathbb{R}^2 \to \mathbb{R}^2$ of the plane to itself of the form

$$L(x, y) = (ax + by + c, dx + ey + f) \, , \tag{1.1}$$

for some constant real numbers a, b, c, d, e, f. The point $\mathbf{P}' = L(\mathbf{P})$ is called the *image* of $\mathbf{P}$. If S is a subset of $\mathbb{R}^2$, then the set of all points $L(x, y)$, for $(x, y) \in S$, is called the *image* of S and denoted $L(S)$.

Example 1.4

Let $L(x, y) = (2x + 3y + 4, 5x + 6y + 7)$. The images of the points $(4, 2)$, $(2, 1)$, and $(0, 0)$ are $L(4, 2) = (18, 39)$, $L(2, 1) = (11, 23)$, and $L(0, 0) = (4, 7)$.

Lemma 1.5

If $aB - bA$ and $dB - eA$ are not both zero, then the transformation L given by (1.1) maps the line $Ax + By + C = 0$ (A and B not both zero) to the line

$$(eA - dB)\,x + (aB - bA)\,y + (bf - ce)\,A - (af - cd)\,B + (ae - bd)\,C = 0 \, . \tag{1.2}$$

If $aB - bA = 0$ and $eA - dB = 0$, then $ae - bd = 0$ and L maps every point on the line to the point $((cB - bC)/B, (fB - eC)/B)$.

Proof

Let L be the transformation given by (1.1). Consider the line $Ax+By+C=0$, and suppose $B \neq 0$. (The case $B = 0$ is left as an exercise to the reader.) Then each point on the line has the form $\left(t, -\frac{A}{B}t - \frac{C}{B}\right)$. So $L\left(t, -\frac{A}{B}t - \frac{C}{B}\right) = (x, y)$ where

$$x = \frac{(aB - bA)\,t - bC + cB}{B} \text{ and } y = \frac{(dB - eA)\,t - eC + fB}{B}. \tag{1.3}$$

If $aB - bA \neq 0$ or $dB - eA \neq 0$, then t can be eliminated from equations (1.3) to give (1.2) and the first part of the lemma is proved.

Suppose $aB - bA = 0$ and $eA - dB = 0$. Since A and B are not both zero, it follows that $ae - bd = 0$. Every point on the line maps to the point $(X, Y) = ((cB - bC)/B, (fB - eC)/B)$. □

Definition 1.6

A transformation L given by (1.1) is said to be *singular* whenever

$$\begin{vmatrix} a & b \\ d & e \end{vmatrix} = ae - bd = 0\,, \tag{1.4}$$

and *non-singular* otherwise.

EXERCISES

1.5. The proof of Lemma 1.5 shows that whenever a linear transformation L given by (1.1) maps a line to a point, then $aB - bA = dB - eA = 0$. Hence $ae - bd = 0$, and L is singular. Show the converse, that if L is singular (so that $ae - bd = 0$), then there exists a line $Ax+By+C=0$ which is mapped by L to a point.

1.6. Suppose L is a non-singular transformation. Show that the line segment with endpoints $\mathbf{P}(p_1, p_2)$ and $\mathbf{Q}(q_1, q_2)$ maps to the line segment with endpoints $L(\mathbf{P})$ and $L(\mathbf{Q})$.

Remark 1.7

Throughout the book the term *object* is used rather vaguely. A planar object is a subset of $\mathbb{R}^2$, and a spatial object is a subset of $\mathbb{R}^3$. In most applications

an object has a geometrical structure such as that of being a “point”, a “line”, a “curve”, a “collection of curves”, or a “region of points”.

1.2 Translations

A *translation* is a transformation which maps a point $\mathbf{P}(x, y)$ to a point $\mathbf{P}'(x', y')$ by adding a constant amount to each coordinate so that

$$x' = x + h, \quad y' = y + k \ ,$$

for some constants h and k. The translation has the effect of moving $\mathbf{P}$ in the direction of the x-axis by h units, and in the direction of the y-axis by k units. If $\mathbf{P}$ and $\mathbf{P}'$ are written as row vectors, then

$$(x', y') = (x, y) + (h, k) \ .$$

To translate an object it is necessary to add the vector (h, k) to every point of that object. The translation is denoted $\mathsf{T}\,(h, k)$. A translation can also be executed using matrix addition if (x, y) is represented as the row matrix $\begin{pmatrix} x & y \end{pmatrix}$.

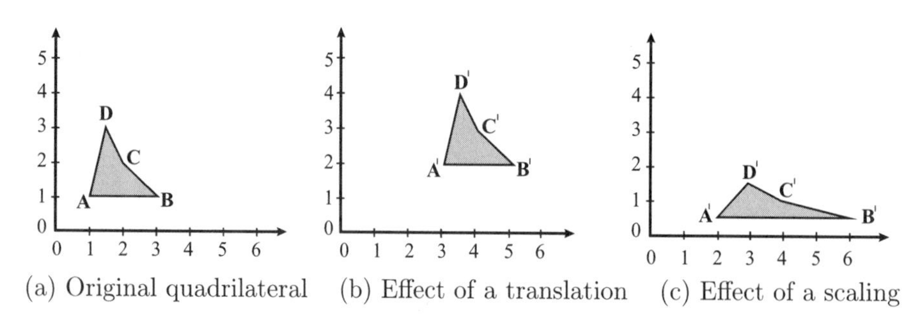

(a) Original quadrilateral (b) Effect of a translation (c) Effect of a scaling

Figure 1.1

Example 1.8

Consider a quadrilateral with vertices $\mathbf{A}(1,1)$, $\mathbf{B}(3,1)$, $\mathbf{C}(2,2)$, and $\mathbf{D}(1.5,3)$. Applying the translation $\mathsf{T}\,(2,1)$, the images of the vertices are

$$\begin{aligned}
\mathbf{A}' &= (1,1) + (2,1) = (3,2) \ , \\
\mathbf{B}' &= (3,1) + (2,1) = (5,2) \ , \\
\mathbf{C}' &= (2,2) + (2,1) = (4,3) \ , \text{ and} \\
\mathbf{D}' &= (1.5,3) + (2,1) = (3.5,4) \ .
\end{aligned}$$

Figure 1.1 shows (a) the original, and (b) the translated quadrilateral.

Definition 1.9

The transformation which leaves all points of the plane unchanged is called the *identity transformation* and denoted I. The *inverse transformation* of L, denoted L^{-1}, is the transformation such that (i) L^{-1} maps every image point $L(\mathbf{P})$ back to its original position $\mathbf{P}$, and (ii) L maps every image point $L^{-1}(\mathbf{P})$ to $\mathbf{P}$. Inverse transformations will be discussed further in Section 2.5.1.

Example 1.10

Consider the translation $\mathsf{T}(h,k)$ which maps a point $\mathbf{P}(x,y)$ to $\mathbf{P}'(x+h,y+k)$. The transformation T^{-1} required to map $\mathbf{P}'$ back to $\mathbf{P}$ is the inverse translation $\mathsf{T}(-h,-k)$. For instance, applying $\mathsf{T}(-2,-1)$ to the point $\mathbf{A}'$ of Example 1.8 gives $(3,2)+(-2,-1)=(1,1)$, and hence maps $\mathbf{A}'$ back to $\mathbf{A}$. The reader can check that the same translation returns the other images to their original locations.

Exercise 1.7

(a) Apply the translation $\mathsf{T}(3,-2)$ to the quadrilateral of Example 1.8, and make a sketch of the transformed quadrilateral.

(b) Determine the inverse transformation of $\mathsf{T}(3,-2)$. Apply the inverse to the transformed quadrilateral to verify that the inverse returns the quadrilateral to its original position.

1.3 Scaling about the Origin

A *scaling about the origin* is a transformation which maps a point $\mathbf{P}(x,y)$ to a point $\mathbf{P}'(x',y')$ by multiplying the x and y coordinates by non-zero constant *scaling factors* s_x and s_y, respectively, to give

$$x' = s_x x \quad \text{and} \quad y' = s_y y \ .$$

A scaling factor s is said to be an *enlargement* if $|s|>1$, and a *contraction* if $|s|<1$. A scaling transformation is said to be *uniform* whenever $s_x = s_y$. By representing a point (x,y) as a row matrix $\begin{pmatrix} x & y \end{pmatrix}$, the scaling transformation can be performed by a matrix multiplication

$$\mathbf{P}' = \begin{pmatrix} x & y \end{pmatrix} \begin{pmatrix} s_x & 0 \\ 0 & s_y \end{pmatrix} = \begin{pmatrix} s_x x & s_y y \end{pmatrix} .$$

The matrix

$$\mathsf{S}(s_x, s_y) = \begin{pmatrix} s_x & 0 \\ 0 & s_y \end{pmatrix}$$

is called the *scaling transformation matrix*.

Example 1.11

To apply the scaling transformation $\mathsf{S}(2, 0.5)$ to the quadrilateral of Example 1.8, the coordinates of the four vertices of the quadrilateral are represented by the rows of the 4×2 matrix

$$\begin{pmatrix} \mathbf{A} \\ \mathbf{B} \\ \mathbf{C} \\ \mathbf{D} \end{pmatrix} = \begin{pmatrix} 1 & 1 \\ 3 & 1 \\ 2 & 2 \\ 1.5 & 3 \end{pmatrix},$$

and multiplied by the scaling transformation matrix

$$\begin{pmatrix} \mathbf{A}' \\ \mathbf{B}' \\ \mathbf{C}' \\ \mathbf{D}' \end{pmatrix} = \begin{pmatrix} 1 & 1 \\ 3 & 1 \\ 2 & 2 \\ 1.5 & 3 \end{pmatrix} \begin{pmatrix} 2 & 0 \\ 0 & 0.5 \end{pmatrix} = \begin{pmatrix} 2 & 0.5 \\ 6 & 0.5 \\ 4 & 1 \\ 3 & 1.5 \end{pmatrix}.$$

The rows of the resulting matrix are the coordinates of the images of the vertices. The original quadrilateral and its scaled image are shown in Figures 1.1(a) and (c). The quadrilateral is scaled by a factor 2 in the x-direction and by a factor 0.5 in the y-direction.

Remark 1.12

The quadrilateral of Example 1.11 has experienced a translation due to the fact that scaling transformations are performed about the origin $\mathbf{O}$. (Scalings about an arbitrary point are considered in Section 2.4.2.) The true effect of a scaling about the origin is to scale the position vectors $\overrightarrow{\mathbf{OP}}$ of each point $\mathbf{P}$ in the plane. For instance, in Example 1.11 vectors $\overrightarrow{\mathbf{OA}}$, $\overrightarrow{\mathbf{OB}}$, $\overrightarrow{\mathbf{OC}}$, and $\overrightarrow{\mathbf{OD}}$ have been scaled by the factors 2 and 0.5 in the x- and y-directions as shown in Figure 1.2. Since the positions of all four points $\mathbf{A}$, $\mathbf{B}$, $\mathbf{C}$, and $\mathbf{D}$ have changed, there is a combined effect of scaling and translating of the object. The origin is the only point unaffected by a scaling about the origin.

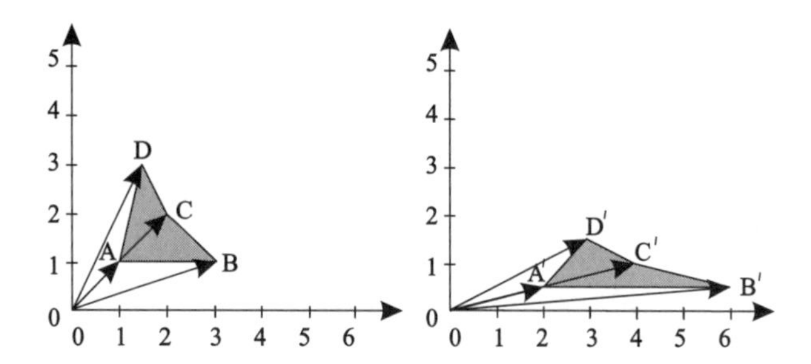

Figure 1.2 Effect of scaling on position vectors

EXERCISES

1.8. Apply the scaling transformation $\mathsf{S}(-1,1)$ to the quadrilateral of Example 1.8. Describe the effect of the transformation.

1.9. Show that the inverse transformation $\mathsf{S}(s_x, s_y)^{-1}$ of a scaling $\mathsf{S}(s_x, s_y)$ (with $s_x \neq 0$ and $s_y \neq 0$) is the scaling $\mathsf{S}(1/s_x, 1/s_y)$.

1.4 Reflections

Two effects which are commonly used in CAD or computer drawing packages are the horizontal and vertical "flip" or "mirror" effects. Pictures which have undergone a horizontal or vertical flip are shown in Figure 1.3(a). A flip of an object is obtained by applying a transformation known as a *reflection*. Consider a fixed line ℓ in the plane. The reflected image of a point $\mathbf{P}$, a distance d from ℓ, is determined as follows. If $d = 0$ then $\mathbf{P}$ is a point on ℓ and the image is $\mathbf{P}$. Otherwise, take the unique line ℓ_1 through $\mathbf{P}$ and perpendicular to ℓ. Then, as showed in Figure 1.3(b), there are two distinct points on ℓ_1, $\mathbf{P}$ and $\mathbf{P}'$, which are a distance d away from ℓ. The point $\mathbf{P}'$ is the required image of $\mathbf{P}$.

It is easily verified that the reflection R_x in the x-axis is the transformation $L(x,y) = (x,-y)$, and the reflection R_y in the y-axis is $L(x,y) = (-x,y)$. The reflection R_x can be computed by the matrix multiplication

$$\mathsf{R}_x \begin{pmatrix} x & y \end{pmatrix} = \begin{pmatrix} x & y \end{pmatrix} \begin{pmatrix} 1 & 0 \\ 0 & -1 \end{pmatrix} = \begin{pmatrix} x & -y \end{pmatrix},$$

and R_y by

$$\mathsf{R}_y \begin{pmatrix} x & y \end{pmatrix} = \begin{pmatrix} x & y \end{pmatrix} \begin{pmatrix} -1 & 0 \\ 0 & 1 \end{pmatrix} = \begin{pmatrix} -x & y \end{pmatrix}.$$

The reflection R_y was encountered in Exercise 1.8. Reflections in arbitrary lines are discussed in Section 2.5.3.

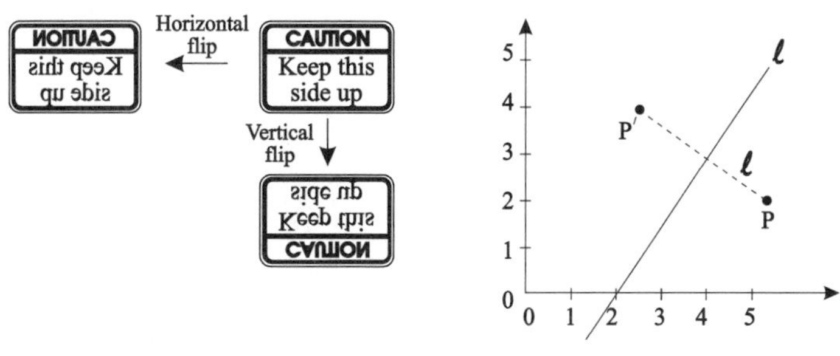

(a) Horizontal and vertical flips (b) Reflection in the line ℓ

Figure 1.3

EXERCISES

1.10. Apply the reflection R_x to the quadrilateral of Example 1.8.

1.11. Verify that $\mathsf{R}_x = \mathsf{S}(1, -1)$ and $\mathsf{R}_y = \mathsf{S}(-1, 1)$.

1.12. Show that the inverse of R_x is R_x, that is, $\mathsf{R}_x^{-1} = \mathsf{R}_x$. Similarly, show that $\mathsf{R}_y^{-1} = \mathsf{R}_y$.

1.5 Rotation about the Origin

A *rotation* about the origin through an angle θ has the effect that a point $\mathbf{P}(x, y)$ is mapped to a point $\mathbf{P}'(x', y')$ so that the initial point $\mathbf{P}$ and its image point $\mathbf{P}'$ are the same distance from the origin, and the angle between lines $\overline{\mathbf{OP}}$ and $\overline{\mathbf{OP}'}$ is θ. There are two possible image points which satisfy these properties depending on whether the rotation is carried out in a clockwise or anticlockwise direction. It is the convention that a positive angle θ represents an *anticlockwise* direction so that a $\pi/2$ rotation about the origin maps points on the x-axis to points on the y-axis.

Referring to Figure 1.4, let $\mathbf{P}'(x', y')$ denote the image of a point $\mathbf{P}(x, y)$ following a rotation about the origin through an angle θ (in an anticlockwise direction). Suppose the line $\overline{\mathbf{OP}}$ makes an angle ϕ with the x-axis, and that $\mathbf{P}$ is a distance r from the origin. Then $(x, y) = (r\cos\phi, r\sin\phi)$. $\mathbf{P}'$ makes an angle $\theta + \phi$ with the x-axis, and therefore $(x', y') = (r\cos(\theta + \phi), r\sin(\theta + \phi))$. The addition formulae for trigonometric functions yield

$$
\begin{aligned}
x' &= r\cos(\theta + \phi) = r\cos\theta\cos\phi - r\sin\theta\sin\phi = x\cos\theta - y\sin\theta \ , \text{ and} \\
y' &= r\sin(\theta + \phi) = r\sin\theta\cos\phi + r\cos\theta\sin\phi = x\sin\theta + y\cos\theta \ .
\end{aligned}
$$

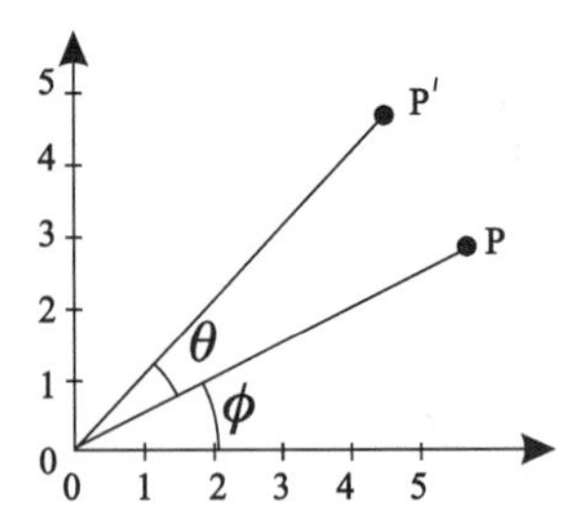

Figure 1.4 Rotation of a point **P** about the origin

The coordinates (x', y') can be obtained from (x, y) by the matrix multiplication

$$\mathbf{P}' = \begin{pmatrix} x & y \end{pmatrix} \begin{pmatrix} \cos\theta & \sin\theta \\ -\sin\theta & \cos\theta \end{pmatrix} = \begin{pmatrix} x\cos\theta - y\sin\theta & x\sin\theta + y\cos\theta \end{pmatrix} .$$

The matrix

$$\mathsf{Rot}\,(\theta) = \begin{pmatrix} \cos\theta & \sin\theta \\ -\sin\theta & \cos\theta \end{pmatrix}$$

is called the *rotation matrix.*

Example 1.13

The rotation matrices of rotations about the origin through $\pi/2$, π, and $3\pi/2$ radians are

$$\mathsf{Rot}\,(\pi/2) = \begin{pmatrix} 0 & 1 \\ -1 & 0 \end{pmatrix}, \mathsf{Rot}\,(\pi) = \begin{pmatrix} -1 & 0 \\ 0 & -1 \end{pmatrix}, \mathsf{Rot}\,(3\pi/2) = \begin{pmatrix} 0 & -1 \\ 1 & 0 \end{pmatrix} .$$

Example 1.14

Applying the rotation $\mathsf{Rot}\,(\pi/2)$ to the quadrilateral of Example 1.8, gives the points

$$\begin{pmatrix} \mathbf{A}' \\ \mathbf{B}' \\ \mathbf{C}' \\ \mathbf{D}' \end{pmatrix} = \begin{pmatrix} 1 & 1 \\ 3 & 1 \\ 2 & 2 \\ 1.5 & 3 \end{pmatrix} \begin{pmatrix} 0 & 1 \\ -1 & 0 \end{pmatrix} = \begin{pmatrix} -1 & 1 \\ -1 & 3 \\ -2 & 2 \\ -3 & 1.5 \end{pmatrix} .$$

The image of the quadrilateral is shown in Figure 1.5.

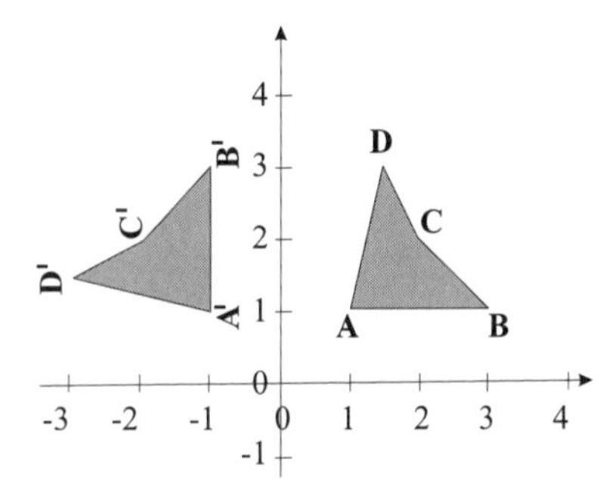

Figure 1.5 Rotation of the quadrilateral about the origin through $\pi/2$

EXERCISES

1.13. Apply rotations about the origin through the angles $\pi/3$, $2\pi/3$, and $\pi/4$ to the triangle with vertices $\mathbf{P}(1,1)$, $\mathbf{Q}(3,1)$, and $\mathbf{R}(2,2)$. Sketch the resulting triangles.

1.14. Show that $\mathsf{Rot}\,(\theta)^{-1} = \mathsf{Rot}\,(-\theta)$.

1.15. Do the transformations $\mathsf{Rot}\,(\pi/2)$ and R_y have the same effect?

1.6 Shears

Given a fixed direction in the plane specified by a unit vector $\mathbf{v} = (v_1, v_2)$, consider the lines ℓ_d with direction $\mathbf{v}$ and a distance d from the origin as shown in Fig.ure 1.6. A *shear about the origin of factor* r *in the direction* $\mathbf{v}$ is defined to be the transformation which maps a point $\mathbf{P}$ on ℓ_d to the point $\mathbf{P}' = \mathbf{P} + rd\mathbf{v}$. Thus the points on ℓ_d are translated along ℓ_d (that is, in the direction of $\mathbf{v}$) through a distance of rd. Shears can be used to obtain italic fonts from normal fonts (see Section 8.1.3).

Example 1.15

To determine a shear in the direction of the x-axis with factor r, let $\mathbf{v} = (1,0)$. The line in the direction of $\mathbf{v}$ through an arbitrary point $\mathbf{P}(x_0, y_0)$ has the equation $y = y_0$. The line is a distance y_0 from the origin. Thus $\mathbf{P}$ is mapped to $\mathbf{P}'(x_0 + ry_0, y_0)$ and hence

$$\begin{pmatrix} x' & y' \end{pmatrix} = \begin{pmatrix} x_0 + ry_0 & y_0 \end{pmatrix} = \begin{pmatrix} x & y \end{pmatrix} \begin{pmatrix} 1 & 0 \\ r & 1 \end{pmatrix}.$$

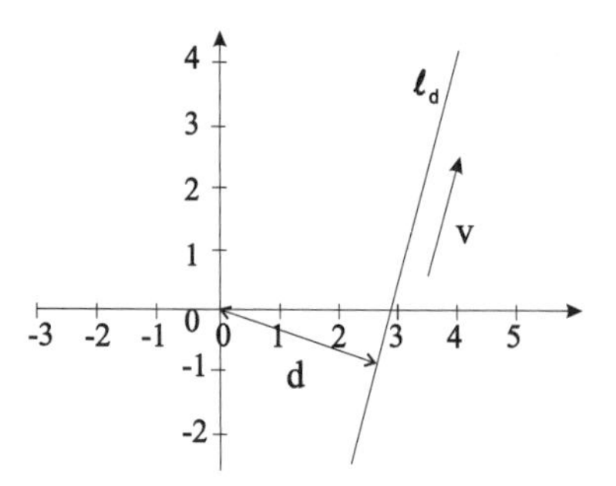

Figure 1.6 Shear in the direction $\mathbf{v}$

The general shear transformation matrix is determined as follows. The line through $\mathbf{P}(x_0, y_0)$ with direction $\mathbf{v} = (v_1, v_2)$ has the equation

$$v_2 x - v_1 y + (v_1 y_0 - v_2 x_0) = 0 \ .$$

Since $\mathbf{v}$ is a unit vector, the distance from this line to the origin is

$$d = v_1 y_0 - v_2 x_0 \ .$$

There are two lines a given distance away from the origin with a specified direction, and the lines on either side of ℓ_0 (denoting the line through the origin with direction $\mathbf{v}$) are distinguished by the sign of $v_1 y_0 - v_2 x_0$. It follows that the shear transformation maps $\mathbf{P}(x_0, y_0)$ to

$$\mathbf{P}' = \mathbf{P} + rd\mathbf{v} = (x_0 + r(v_1 y_0 - v_2 x_0)v_1, y_0 + r(v_1 y_0 - v_2 x_0)v_2) \ .$$

Thus the shear has transformation matrix

$$\mathsf{Sh}(\mathbf{v}, r) = \begin{pmatrix} 1 - rv_1 v_2 & -rv_2^2 \\ rv_1^2 & 1 + rv_1 v_2 \end{pmatrix} .$$

In particular,

$$\mathsf{Sh}((1,0), r) = \begin{pmatrix} 1 & 0 \\ r & 1 \end{pmatrix}$$

verifying the result of Example 1.15.

Example 1.16

The shear in the direction $\mathbf{v} = \left(2 \big/ \sqrt{5}\,, 1 \big/ \sqrt{5}\right)$ with a factor $r = 1.5$ has transformation matrix

$$\begin{aligned} \mathsf{Sh}\left(\left(2\big/\sqrt{5}\,, 1\big/\sqrt{5}\right), 1.5\right) &= \begin{pmatrix} 1 - 1.5\left(\frac{2}{\sqrt{5}}\right)\left(\frac{1}{\sqrt{5}}\right) & -1.5\left(\frac{1}{\sqrt{5}}\right)^2 \\ 1.5\left(\frac{2}{\sqrt{5}}\right)^2 & 1 + 1.5\left(\frac{2}{\sqrt{5}}\right)\left(\frac{1}{\sqrt{5}}\right) \end{pmatrix} \\ &= \begin{pmatrix} 0.4 & -0.3 \\ 1.2 & 1.6 \end{pmatrix} . \end{aligned}$$

Applying the shear to the quadrilateral of Example 1.8,

$$\begin{pmatrix} 1 & 1 \\ 3 & 1 \\ 2 & 2 \\ 1.5 & 3 \end{pmatrix} \begin{pmatrix} 0.4 & -0.3 \\ 1.2 & 1.6 \end{pmatrix} = \begin{pmatrix} 1.6 & 1.3 \\ 2.4 & 0.7 \\ 3.2 & 2.6 \\ 4.2 & 4.35 \end{pmatrix}.$$

The effect of the shear is shown in Figure 1.7.

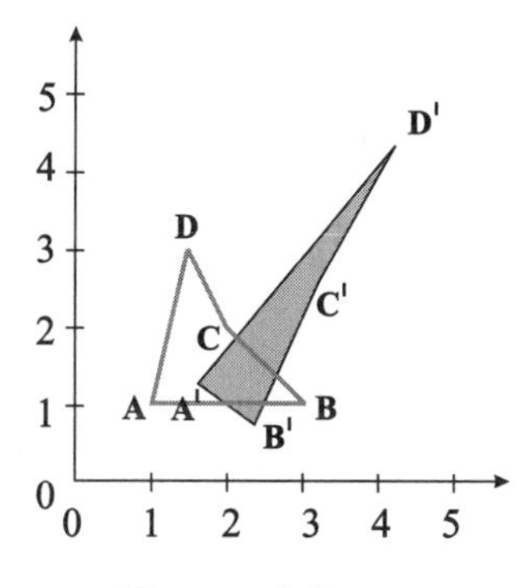

Figure 1.7

Exercise 1.16

Determine the transformation matrix for a shear with (a) direction $(3, -4)$ and factor $r = 4$, and (b) direction $(8, 6)$ and factor $r = -1$.

1.7 Concatenation of Transformations

In many applications it is desirable to apply more than one transformation to an object. For instance, a translation and a rotation may be required to position and orientate an object. The process of following one transformation by another to form a new transformation with a combined effect is called *concatenation* or *composition* of transformations. The term *concatenation* is the most commonly used in computer graphics. All of the transformations described in the earlier sections can be concatenated to obtain new transformations.

Example 1.17

A rotation about the origin through an angle $\pi/3$ is obtained by applying the matrix $\mathsf{Rot}\,(\pi/3)$

$$\begin{aligned}\begin{pmatrix} x' & y' \end{pmatrix} &= \begin{pmatrix} x & y \end{pmatrix}\begin{pmatrix} \cos(\pi/3) & \sin(\pi/3) \\ -\sin(\pi/3) & \cos(\pi/3) \end{pmatrix} \\ &= \begin{pmatrix} x & y \end{pmatrix}\begin{pmatrix} 1/2 & \sqrt{3}/2 \\ -\sqrt{3}/2 & 1/2 \end{pmatrix}.\end{aligned}$$

Next apply a scaling by a factor of 6 in the x-direction and 2 in the y-direction

$$\begin{aligned}\begin{pmatrix} x'' & y'' \end{pmatrix} &= \begin{pmatrix} x' & y' \end{pmatrix}\begin{pmatrix} 6 & 0 \\ 0 & 2 \end{pmatrix} \\ &= \begin{pmatrix} x & y \end{pmatrix}\begin{pmatrix} 1/2 & \sqrt{3}/2 \\ -\sqrt{3}/2 & 1/2 \end{pmatrix}\begin{pmatrix} 6 & 0 \\ 0 & 2 \end{pmatrix}.\end{aligned}$$

Hence, the concatenated transformation has transformation matrix

$$\begin{pmatrix} 1/2 & \sqrt{3}/2 \\ -\sqrt{3}/2 & 1/2 \end{pmatrix}\begin{pmatrix} 6 & 0 \\ 0 & 2 \end{pmatrix} = \begin{pmatrix} 3 & \sqrt{3} \\ -3\sqrt{3} & 1 \end{pmatrix}.$$

A problem is encountered whenever translations are concatenated with other types of transformation since it is necessary to combine a matrix (or vector) addition for the translation with a matrix multiplication for the other transformations. This is an awkward procedure remedied only by the introduction of homogeneous coordinates, as discussed in Chapter 2. Thus concatenation will not be discussed any further, and the approach of using 2×2 matrix multiplications will be abandoned. The homogeneous coordinate system offers the following advantages for the execution of transformations.

1. All transformations can be represented by matrices, and performed by matrix multiplication.
2. Concatenation of transformations is performed by matrix multiplication of the transformation matrices.
3. Inverse transformations are obtained by taking a matrix inverse.

The effort expended has not been in vain since the "homogeneous" transformation matrices are closely related to those described in this chapter.

1.8 Applications

1.8.1 Instancing

A geometric object is created by defining the different parts which make up the object. For example, the front of a house in Figure 1.8 consists of a number of rectangles, or rather scaled squares, which form the walls, windows, and door of the house. The square is an example of a *picture element.* For convenience, picture elements are defined in their own local coordinate system called the *modelling coordinate system*, and are constructed from *graphical primitives* which are the basic building blocks. Picture elements are defined once, but may be used many times in the construction of objects. The number and type of graphical primitives available depends on the computer graphics system.

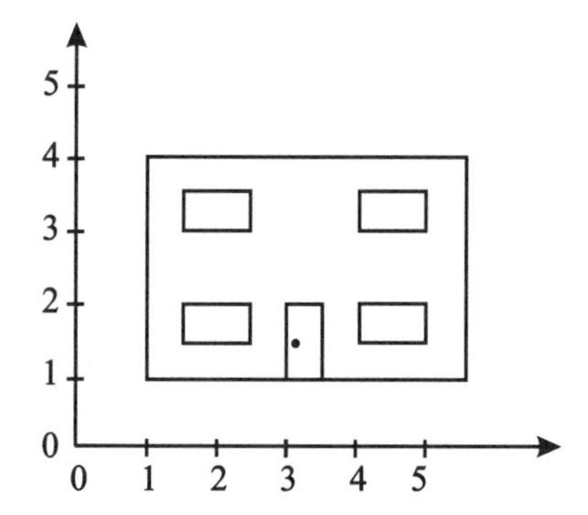

Figure 1.8 Front of a house obtained from instances of **Square** and **Point**

For example, a square with vertices $(0,0)$, $(1,0)$, $(1,1)$, and $(0,1)$ can be obtained using the graphical primitive for the line segment, denoted **Line**, which joins the points $(0,0)$ and $(1,0)$. One possible construction of the square is obtained in the following manner.

1. Draw **Line**. This produces the horizontal base of the square.
2. Apply a rotation about the origin through an angle $\pi/2$ to a copy of **Line**, and then apply a translation of 1 unit in the x-direction. This gives the right vertical edge of the square.
3. Apply a translation of 1 unit in the y-direction to a copy of **Line**. This gives the top of the square.
4. Apply a rotation about the origin through an angle $\pi/2$ to a copy of **Line**. This gives the left vertical edge of the square.

A transformed copy of a graphical primitive or picture element is called an *instance*. The square, denoted **Square**, is defined by four instances of **Line** as depicted in Figure 1.9.

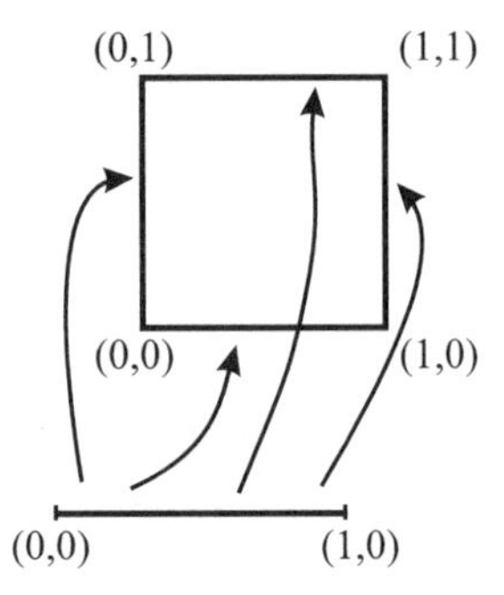

Figure 1.9 **Square** obtained from four instances of **Line**

The completed "real" object is defined in *world coordinates* by applying a *modelling coordinate transformation* to each picture element. The house of Figure 1.8 is defined by six instances of the picture element **Square,** and one instance of the primitive **Point** (for the door handle). In particular, the front door is obtained by applying a scaling of 0.5 unit in the x-direction, followed by a translation of 3 units in the x-direction and 1 unit in the y-direction.

The line primitive of most graphics systems will be more sophisticated than the one described above: the line primitive might be defined by two arbitrary points, or the system might have a *polyline* primitive consisting of a chain of line segments connecting a sequence of user specified points.

In the above discussion, instancing has been described in words since without homogeneous coordinates the concatenation of transformations is awkward. In the proposed homogeneous coordinate system described in the next chapter each instance of a picture element or object can be represented by a single modelling transformation matrix.

EXERCISES

1.17. Each window and the outline of the house is obtained by instances of **Square**. Describe in words the sequence of transformations used for each instance.

1.18. Investigate the graphical primitives available in graphics systems such as PHIGS, GKS, and OpenGL. See for example [14] and the web page for the book.

1.8.2 Robotics

The Denavit–Hartenberg notation, a standard representation used to define a robotic mechanism, describes how each rigid link of a mechanism is related to the neighbouring link (or links) by means of transformations. To exemplify this, consider a planar 2R robot manipulator arm (Figure 1.10) consisting of two links. The first link is attached to the base by a revolute joint J_1. A revolute joint permits the link to rotate about a point. The second link is attached to the first link by a second revolute joint J_2. The robot hand or end effector is attached to the second link. The position and orientation of the robot hand is controlled by turning the links about the two joints.

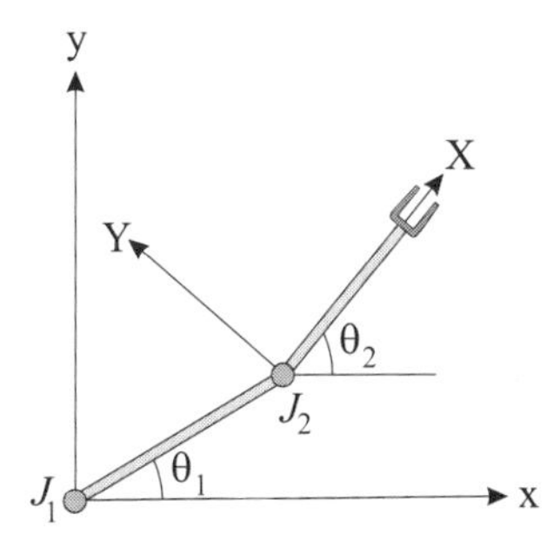

Figure 1.10 2R robot manipulator

Define an (x, y)-coordinate system with J_1 as the origin as shown in Figure 1.10. The second link is given its own (X, Y)-coordinate system with J_2 as the origin. Let d be the distance between J_1 and J_2, and let the angles between links 1 and 2 and the x-axis be θ_1 and θ_2 respectively. The position and orientation of the second link is obtained by applying a rotation $\mathsf{Rot}(\theta_2)$ followed by a translation $\mathsf{T}(d\cos\theta_1, d\sin\theta_1)$. Given the (X, Y) coordinates of a point $\mathbf{P}$, the (x, y)-coordinates of $\mathbf{P}$ are obtained by the transformation

$$\begin{aligned}\begin{pmatrix} x & y \end{pmatrix} &= \begin{pmatrix} X & Y \end{pmatrix}\begin{pmatrix} \cos\theta_2 & \sin\theta_2 \\ -\sin\theta_2 & \cos\theta_2 \end{pmatrix} + \begin{pmatrix} d\cos\theta_1 & d\sin\theta_1 \end{pmatrix} \\ &= \begin{pmatrix} X\cos\theta_2 - Y\sin\theta_2 + d\cos\theta_1 & X\sin\theta_2 + Y\cos\theta_2 + d\sin\theta_1 \end{pmatrix}.\end{aligned}$$

The ultimate aim is to express such concatenations with one matrix multiplication with the assistance of homogeneous coordinates.

EXERCISES

1.19. Suppose an affine transformation $L(x, y) = (ax + by + c, dx + ey + f)$ is applied to a triangle $\mathcal{T}$ with vertices **A**, **B**, **C** and area $\mathcal{A}$. Show that the area of $L(\mathcal{T})$ is $(ad - bc) \cdot \mathcal{A}$.

1.20. Prove that a transformation maps the midpoint of a line segment to the midpoint of the image.

1.21. Write a computer program or use a computer package to implement the various types of transformation. Apply the program to the examples of the chapter.

2 Homogeneous Coordinates and Transformations of the Plane

2.1 Introduction

In Chapter 1 planar objects were manipulated by applying one or more transformations. Section 1.7 identified the problem that the concatenation of a translation with a rotation, scaling or shear requires an awkward combination of a matrix addition and a matrix multiplication. The problem can be avoided by using an alternative coordinate system for which computations are performed by 3×3 matrix multiplications. Since

$$\begin{aligned}\begin{pmatrix} x' & y' & 1 \end{pmatrix} &= \begin{pmatrix} x & y & 1 \end{pmatrix} \begin{pmatrix} a & d & 0 \\ b & e & 0 \\ c & f & 1 \end{pmatrix} \\ &= \begin{pmatrix} ax+by+c & dx+ey+f & 1 \end{pmatrix} \end{aligned} \tag{2.1}$$

it follows that

$$x' = ax + by + c \quad \text{and} \quad y' = dx + ey + f \,.$$

To this end a new coordinate system is defined in which the point with Cartesian coordinates (x, y) is represented by the *homogeneous* or *projective* coordinates $(x, y, 1)$, or any multiple (rx, ry, r) with $r \neq 0$. The set of all homogeneous coordinates (x, y, w) is called the *projective plane* and denoted $\mathbb{P}^2$. In order to carry out transformations using matrix computations the homogeneous coordinates (x, y, w) are represented by the row matrix $(x \ y \ w)$. Equation (2.1) implies that any planar transformation can be performed by a 3×3 matrix multiplication and using homogeneous coordinates. Sometimes homogeneous coordinates will be denoted by capitals (X, Y, W) in order to distinguish them from the affine coordinates (x, y).

Example 2.1

1. $(1, 2, 3)$, $(2, 4, 6)$, and $(-1, -2, -3)$ are all homogeneous coordinates of the point $(1/3, 2/3)$ since

$$(1/3, 2/3, 1) = \frac{1}{3}(1, 2, 3) = \frac{1}{6}(2, 4, 6) = (-1)(-1, -2, -3) .$$

2. The Cartesian coordinates of the point with homogeneous coordinates $(X, Y, W) = (6, 4, 2)$ are obtained by dividing the coordinates through by $W = 2$ to give alternative homogeneous coordinates $(3, 2, 1)$. Thus the Cartesian coordinates of the point are $(x, y) = (3, 2)$.

EXERCISES

2.1. Which of the following homogeneous coordinates $(2, 6, 2)$, $(2, 6, 4)$, $(1, 3, 1)$, $(-1, -3, -2)$, $(1, 3, 2)$, and $(4, 12, 8)$ represent the point $(1/2, 3/2)$?

2.2. Write down two sets of homogeneous coordinates of $(2, -3)$.

2.3. A point has Cartesian coordinates $(5, -20)$ and homogeneous coordinates $(-5, ?, -1)$ and $(10, -40, ?)$. Fill in the missing entries indicated by a "?".

Definition 2.2

A *(projective) transformation* of the projective plane is a mapping $L : \mathbb{P}^2 \to \mathbb{P}^2$ of the form

$$\begin{aligned} L(x, y, w) &= \begin{pmatrix} x & y & w \end{pmatrix} \begin{pmatrix} a & d & g \\ b & e & h \\ c & f & k \end{pmatrix} && (2.2) \\ &= (ax + by + cw, dx + ey + fw, gx + hy + kw) , && (2.3) \end{aligned}$$

for some constant real numbers a, b, c, d, e, f, g, h, k. A matrix which represents a linear transformation of the projective plane is called a *homogeneous transformation matrix*. When $g = h = 0$ and $k \neq 0$, L is said to be an *affine transformation.* Affine transformations correspond to transformations of the Cartesian plane.

Remark 2.3

If alternative homogeneous coordinates (rx, ry, rw) are taken in (2.2) then

$$L(rx, ry, rw) = (arx + bry + crw, drx + ery + frw, grx + hry + krw) ,$$

and dividing through by r gives the homogeneous coordinates (2.3). Thus $L(rx, ry, rw)$ and $L(x, y, w)$ map to the same point, and therefore the definition of a transformation does not depend on the choice of homogeneous coordinates for a given point.

2.1.1 Homogeneous Coordinates

A more formal definition of homogeneous coordinates is obtained in terms of an equivalence relation.

Definition 2.4

A *relation* $\sim$ on a set S is a rule which determines whether two members of the set S are considered related or not. If s_1 is related to s_2, then this is expressed by writing $s_1 \sim s_2$.

Example 2.5

"Greater than", with its usual meaning, is a relation on $\mathbb{R}$. The relationship "3 is greater than 2" is written $3 \sim 2$. The relation "greater than" is generally written $3 > 2$ where the symbol $\sim$ is substituted by $>$. The number 2 is not related to 3 since *it is not true* that $2 > 3$.

Definition 2.6

A relation $\sim$ on a set S is said to be

1. *reflexive* if $s \sim s$ for all s in S;
2. *symmetric* if whenever $s_1 \sim s_2$, then $s_2 \sim s_1$;
3. *transitive* if whenever $s_1 \sim s_2$ and $s_2 \sim s_3$, then $s_1 \sim s_3$;
4. an *equivalence relation* if $\sim$ is reflexive, symmetric, and transitive.

Example 2.7

The relation $>$ on $\mathbb{R}$ is transitive, but not reflexive or symmetric. The relations $\geq$ and $\leq$ are both reflexive and transitive, but not symmetric. The most familiar equivalence relation on $\mathbb{R}$ is $=$.

Definition 2.8

Let s_1 be a member of S. The subset of S, consisting of every s in S which is related to s_1, is called the *equivalence class* of s_1 and denoted by $[s_1]$. A member of an equivalence class $[s_1]$ is called a *representative* of $[s_1]$. Clearly, if s is a representative of $[s_1]$ then $s \sim s_1$.

Homogeneous coordinates arise as equivalence classes determined by the following lemma which defines an equivalence relation on $S = \mathbb{R}^3 \backslash \{(0,0,0)\}$ (that is, S consists of all $\mathbb{R}^3$ excluding the origin).

Lemma 2.9

The relation $\sim$ on the set $S = \mathbb{R}^3 \backslash \{(0,0,0)\}$ defined by

$$(x_0, y_0, w_0) \sim (x_1, y_1, w_1) \Leftrightarrow (x_1, y_1, w_1) = r(x_0, y_0, w_0) \text{ for some } r \neq 0$$

is an equivalence relation.

Proof

1. The relation $\sim$ is reflexive since $(x_0, y_0, w_0) = 1(x_0, y_0, w_0)$.
2. The relation $\sim$ is symmetric since if $(x_0, y_0, w_0) \sim (x_1, y_1, w_1)$, then $(x_1, y_1, w_1) = r(x_0, y_0, w_0)$ for some $r \neq 0$. Thus $(x_0, y_0, w_0) = \frac{1}{r}(x_1, y_1, w_1)$, and hence $(x_1, y_1, w_1) \sim (x_0, y_0, w_0)$.
3. Suppose $(x_0, y_0, w_0) \sim (x_1, y_1, w_1)$, and $(x_1, y_1, w_1) \sim (x_2, y_2, w_2)$. Then $(x_1, y_1, w_1) = r_1(x_0, y_0, w_0)$ for some $r_1 \neq 0$, and $(x_2, y_2, w_2) = r_2(x_1, y_1, w_1)$ for some $r_2 \neq 0$. So

   $$(x_2, y_2, w_2) = r_2(x_1, y_1, w_1) = r_2 r_1 (x_0, y_0, w_0), \text{ for } r_2 r_1 \neq 0 ,$$

   and hence $(x_2, y_2, w_2) \sim (x_0, y_0, w_0)$. Hence $\sim$ is transitive.

□

The equivalence classes $[(x, y, w)]$ are the sets

$$[(x, y, w)] = \{ \ r(x, y, w) \mid r \in \mathbb{R}, r \neq 0 \ \} .$$

The projective plane $\mathbb{P}^2$ is defined to be the set of all equivalence classes. An equivalence class is referred to as a *point* of the projective plane.

In practice, operations of the projective plane are carried out by taking a representative for each equivalence class. Homogeneous coordinates (X, Y, W) with $W \neq 0$ have a representative of the form $(x, y, 1)$ where $x = X/W$, and $y = Y/W$. Thus there is a $1 - 1$ correspondence between points (x, y) of the Cartesian plane and points (X, Y, W) in the projective plane with $W \neq 0$. Points with $W = 0$ are discussed in Section 2.2. Then, a transformation is a mapping of equivalence classes, that is, a mapping of points in the projective plane. Remark 2.3 states that the definition of a transformation does not depend on the choice of the representative of an equivalence class.

Exercise 2.4

Define a relation $\sim$ on non-singular 3×3 matrices by $M_1 \sim M_2$ if and only if $M_1 = \mu M_2$ for some $\mu \neq 0$. Show that $\sim$ is an equivalence relation.

2.2 Points at Infinity

Homogeneous coordinates of the form $(x, y, 0)$ do not correspond to a point in the Cartesian plane, but represent the unique *point at infinity* in the direction $(x\ y)$. To justify this remark, consider the line $(x(t), y(t)) = (tx + a, ty + b)$ through the point (a, b) with direction $(x\ y)$. The point $(tx + a, ty + b)$ has homogeneous coordinates $(tx + a, ty + b, 1)$ and multiplying through by $1/t$ (for $t \neq 0$) gives alternative homogeneous coordinates $(x + a/t, y + b/t, 1/t)$. Points on the line an infinite distance away from the origin in the Cartesian plane may be obtained by letting t tend to infinity. The limiting point of $(x + a/t, y + b/t, 1/t)$ as $t \to \infty$ is $(x, y, 0)$. Therefore, it is natural to interpret the homogeneous coordinates $(x, y, 0)$ as the point at infinity in the direction (x, y). The projective plane may be interpreted as the Cartesian plane together with all the points at infinity.

The projective plane also makes sense of the intuitive notion that two parallel lines intersect at infinity. For instance, consider the parallel lines

$$x + 2y = 1\,, \text{ and} \tag{2.4}$$

$$x + 2y = 2\,. \tag{2.5}$$

Let (X, Y, W) be homogeneous coordinates of a point (x, y) on the line (2.4). Then $(x, y) = (X/W, Y/W)$ and hence

$$(X/W) + 2\,(Y/W) = 1\,.$$

Multiplying through by W, yields the *homogeneous equation* of the line

$$X + 2Y = W \ . \tag{2.6}$$

Similarly, the homogeneous equation of (2.5) is

$$X + 2Y = 2W \ . \tag{2.7}$$

Equations (2.6) and (2.7) have common solutions of the form $(-2r, r, 0)$. The solutions are all homogeneous coordinates of the point $(-2, 1, 0)$ which is the unique point of intersection of the parallel lines. It is easily verified that $(-2, 1, 0)$ is the point at infinity in the direction of the lines. A similar argument yields that all parallel lines intersect in a unique point at infinity.

EXERCISES

2.5. Find the point at infinity in the direction of the vector $(6, -3)$.

2.6. Find the point at infinity on the line $4x - 3y + 1 = 0$.

2.7. Determine the homogeneous equation of the line $3x + 4y = 5$.

2.8. Determine the homogeneous coordinates of the point at infinity which is the intersection of the lines $2x - 9y = 5$ and $2x - 9y = 7$. Verify that the intersection is the point at infinity in the direction of the lines.

2.9. Determine the point at infinity on the line $ax + by + c = 0$. Conclude that all lines in the direction $(-b, a)$ intersect in a unique point at infinity.

2.3 Visualization of the Projective Plane

There are two models that interpret homogeneous coordinates geometrically, and hence enable the projective plane to be visualized.

2.3.1 Line Model of the Projective Plane

The line model of the projective plane is obtained by representing the point with homogeneous coordinates $\mu(X, Y, W)$, $\mu \neq 0$, by the line through the origin with direction (X, Y, W) in (X, Y, W)-space. Since the point with Cartesian coordinates (x, y) has homogeneous coordinates of the form $(X, Y, W) = r(x, y, 1)$

for $r \neq 0$, there is a $1-1$ correspondence between points (x, y) of the Cartesian plane and the lines

$$\{\ r(x, y, 1) \mid r \in \mathbb{R}\ \} \tag{2.8}$$

as illustrated in Figure 2.1. There is also a $1-1$ correspondence between the points (x, y) and the points $(x, y, 1)$ of the $W = 1$ plane.

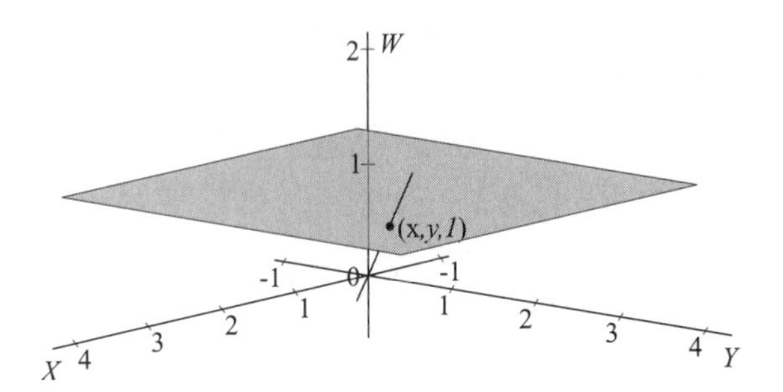

Figure 2.1 The line model of the projective plane

The $W = 1$ plane is inadequate for studying the projective plane since points at infinity do not correspond to points in the $W = 1$ plane, nor to lines of the form (2.8). Instead, points at infinity correspond to lines in the $W = 0$ plane. For example, the parallel lines (2.4) and (2.5) correspond to the planes in (X, Y, W)-space defined by Equations (2.6) and (2.7). The planes intersect in a line through the origin in the $W = 0$ plane as shown in Figure 2.2. The line is parametrized by $(-2t, t, 0)$ and corresponds to the point at infinity $(-2, 1, 0)$ which is the intersection of the two parallel lines. The difficulty with the line model is that lines in the projective plane correspond to planes in the model, and more generally, curves in the projective plane correspond to surfaces. To visualize curves in the projective plane the spherical model is introduced.

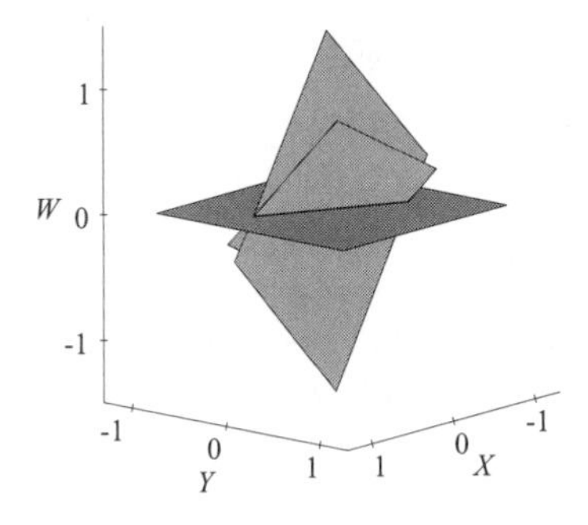

Figure 2.2 Intersection of planes corresponding to parallel lines in the Cartesian plane

2.3.2 Spherical Model of the Projective Plane

The spherical model of the projective plane is obtained by representing the point with homogeneous coordinates $\mu(X, Y, W)$, $\mu \neq 0$, by the points of intersection of the line through the origin with direction (X, Y, W) and the unit sphere centred at the origin $X^2 + Y^2 + W^2 = 1$ as illustrated in Figure 2.3. The intersections are the antipodal points

$$\pm \left(\frac{X}{X^2 + Y^2 + W^2}, \frac{Y}{X^2 + Y^2 + W^2}, \frac{W}{X^2 + Y^2 + W^2} \right) .$$

Since antipodal points on the sphere correspond to the same point in the projective plane, it suffices to consider the upper half-sphere together with (half of) the equator. (The equator is the circle of intersection of the sphere with the $W = 0$ plane.) Points at infinity $(X, Y, 0)$ correspond to points on the equator.

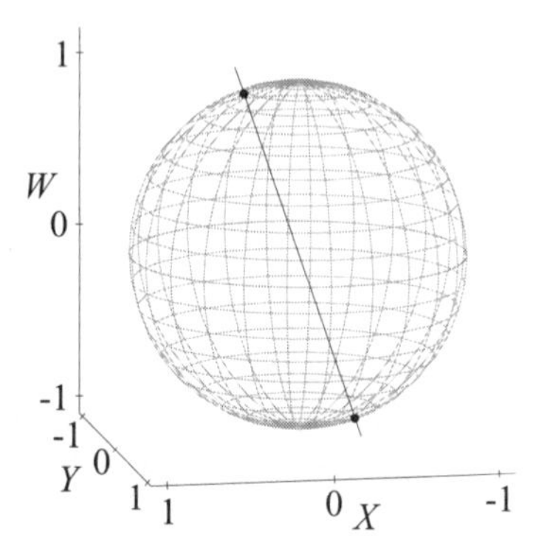

Figure 2.3 Spherical model of the projective plane. Antipodal points represent the same homogeneous point.

Thus the sphere provides a way of visualizing *all* homogeneous coordinates. For instance, the intersection of parallel lines can be visualized in the spherical model. Lines in the Cartesian plane correspond to planes which intersect the sphere in a great circle. The intersection of two parallel lines corresponds to the intersection of the two great circles on the sphere, namely, two antipodal points at infinity on the equator. Figure 2.4 shows how two great circles, representing the lines (2.4) and (2.5), intersect in the antipodal points $(-2/\sqrt{5}\,, 1/\sqrt{5}\,, 0)$ and $(2/\sqrt{5}\,, -1/\sqrt{5}\,, 0)$ on the equator.

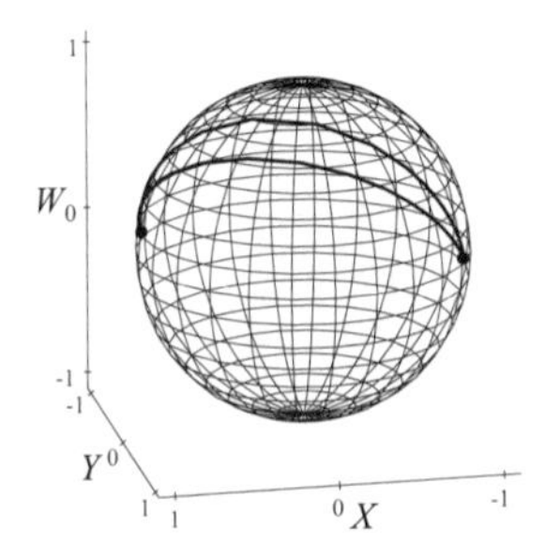

Figure 2.4 Intersection of parallel lines on the spherical model of the projective plane

2.4 Transformations in Homogeneous Coordinates

In the following sections the homogeneous transformation matrices for translations, scalings, and rotations are described. In order to minimize on notation, a transformation and its homogeneous transformation matrix will be given the same notation. For instance, a translation and its translation matrix are both denoted $\mathsf{T}(h,k)$.

2.4.1 Translations

The homogeneous translation matrix for the translation $\mathsf{T}(h,k)$ is

$$\mathsf{T}(h,k) = \begin{pmatrix} 1 & 0 & 0 \\ 0 & 1 & 0 \\ h & k & 1 \end{pmatrix}.$$

Then

$$\begin{pmatrix} x & y & 1 \end{pmatrix} \begin{pmatrix} 1 & 0 & 0 \\ 0 & 1 & 0 \\ h & k & 1 \end{pmatrix} = \begin{pmatrix} x+h & y+k & 1 \end{pmatrix},$$

verifying that the point (x, y) is translated to $(x+h, y+k)$.

Example 2.10

In Example 1.8 the translation $\mathsf{T}(2,1)$ was applied to the quadrilateral with vertices $\mathbf{A}(1,1)$, $\mathbf{B}(3,1)$, $\mathbf{C}(2,2)$, and $\mathbf{D}(1.5,3)$. Let the homogeneous coordinates of the 4 vertices be expressed as the rows of a 4×3 matrix. The translation is applied by multiplying the matrix of vertices by the translation matrix. The rows of the resulting matrix are the homogeneous coordinates of images of the vertices.

$$\begin{pmatrix} 1 & 1 & 1 \\ 3 & 1 & 1 \\ 2 & 2 & 1 \\ 1.5 & 3 & 1 \end{pmatrix} \begin{pmatrix} 1 & 0 & 0 \\ 0 & 1 & 0 \\ 2 & 1 & 1 \end{pmatrix} = \begin{pmatrix} 3 & 2 & 1 \\ 5 & 2 & 1 \\ 4 & 3 & 1 \\ 3.5 & 4 & 1 \end{pmatrix} .$$

$$\text{vertices} \times \text{translation} = \text{images of vertices}$$

The images have Cartesian coordinates $\mathbf{A}'(3,2)$, $\mathbf{B}'(5,2)$, $\mathbf{C}'(4,3)$, and $\mathbf{D}'(3.5,4)$.

2.4.2 Scaling about the Origin

The homogeneous scaling matrix is

$$\mathsf{S}(s_x, s_y) = \begin{pmatrix} s_x & 0 & 0 \\ 0 & s_y & 0 \\ 0 & 0 & 1 \end{pmatrix} .$$

Then

$$\begin{pmatrix} x & y & 1 \end{pmatrix} \begin{pmatrix} s_x & 0 & 0 \\ 0 & s_y & 0 \\ 0 & 0 & 1 \end{pmatrix} = \begin{pmatrix} s_x x & s_y y & 1 \end{pmatrix} ,$$

verifying that the point $(x, y, 1)$ is mapped to $(s_x x, s_y y, 1)$. The scaling can also be performed by the scaling matrix

$$\mathsf{S}(s_x, s_y; s_w) = \begin{pmatrix} s_x & 0 & 0 \\ 0 & s_y & 0 \\ 0 & 0 & s_w \end{pmatrix}$$

for $s_w \neq 0$. The transformation $\mathsf{S}(s_x, s_y; s_w)$ represents a scaling about the origin by a factor of s_x/s_w in the x-direction, and by a factor of s_y/s_w in the y-direction. The semicolon before the s_w is used to distinguish the planar scaling from the spatial scaling which is introduced in Chapter 3.

Example 2.11

A scaling about the origin by a factor of 4 in the x-direction, and by a factor of 2 in the y-direction, of the unit square with vertices $(1,1)$, $(2,1)$, $(2,2)$, and $(1,2)$ is determined by

$$\begin{pmatrix} 1 & 1 & 1 \\ 2 & 1 & 1 \\ 2 & 2 & 1 \\ 1 & 2 & 1 \end{pmatrix} \begin{pmatrix} 4 & 0 & 0 \\ 0 & 2 & 0 \\ 0 & 0 & 1 \end{pmatrix} = \begin{pmatrix} 4 & 2 & 1 \\ 8 & 2 & 1 \\ 8 & 4 & 1 \\ 4 & 4 & 1 \end{pmatrix} .$$

The image is a square with vertices $(4,2)$, $(8,2)$, $(8,4)$, and $(4,4)$.

2.4.3 Rotation about the Origin

In homogeneous coordinates the transformation matrix for a rotation $\mathsf{Rot}\,(\theta)$ about the origin through an angle θ is

$$\mathsf{Rot}\,(\theta) = \begin{pmatrix} \cos\theta & \sin\theta & 0 \\ -\sin\theta & \cos\theta & 0 \\ 0 & 0 & 1 \end{pmatrix},$$

where a positive angle denotes an anticlockwise rotation. Hence

$$\begin{pmatrix} x & y & 1 \end{pmatrix} \begin{pmatrix} \cos\theta & \sin\theta & 0 \\ -\sin\theta & \cos\theta & 0 \\ 0 & 0 & 1 \end{pmatrix} = \begin{pmatrix} x\cos\theta - y\sin\theta & x\sin\theta + y\cos\theta & 1 \end{pmatrix}.$$

Example 2.12

An anticlockwise rotation about the origin through an angle $\pi/3$ of the unit square with vertices $(1,1)$, $(2,1)$, $(2,2)$, and $(1,2)$ is determined by

$$\begin{pmatrix} 1 & 1 & 1 \\ 2 & 1 & 1 \\ 2 & 2 & 1 \\ 1 & 2 & 1 \end{pmatrix} \begin{pmatrix} 0.5 & 0.866 & 0 \\ -0.866 & 0.5 & 0 \\ 0 & 0 & 1.0 \end{pmatrix} = \begin{pmatrix} -0.366 & 1.366 & 1.0 \\ 0.134 & 2.232 & 1.0 \\ -0.732 & 2.732 & 1.0 \\ -1.232 & 1.866 & 1.0 \end{pmatrix}.$$

The image is a square with vertices $(-0.366, 1.366)$, $(0.134, 2.232)$, $(-0.732, 2.732)$, and $(-1.232, 1.866)$.

EXERCISES

2.10. Apply the translation $\mathsf{T}\,(-2,-1)$ to the quadrilateral, obtained in Example 2.10, with vertices $\mathbf{A}'(3,2)$, $\mathbf{B}'(5,2)$, $\mathbf{C}'(4,3)$, and $\mathbf{D}'(3.5,4)$.

2.11. Write down the transformation matrix which has the effect of a scaling by a factor of 2 in the x-direction and by a factor of 1.5 in the y-direction. Apply the transformation to the quadrilateral of Example 2.10. Compare the result with Example 1.11.

2.12. Write down the transformation matrix which has the effect of an anticlockwise rotation about the origin through an angle $\pi/2$. Apply the transformation to the quadrilateral of Example 2.10.

2.13. Determine the matrix for the inverse scaling transformation of Exercise 2.11.

2.14. Determine the homogeneous transformation matrix of $\mathsf{Rot}\,(\theta)^{-1}$.

2.15. Determine the homogeneous transformation matrices for reflections in the x- and y-axes.

2.5 Concatenation of Transformations

In homogeneous coordinates, the concatenation of transformations T_1 and T_2, denoted $T_1 \circ T_2$, can be performed with matrix multiplications alone. For example, a rotation $\mathsf{Rot}\,(\theta)$ about the origin followed by a translation $\mathsf{T}\,(h,k)$ is denoted $\mathsf{Rot}\,(\theta) \circ \mathsf{T}\,(h,k)$, and has the homogeneous transformation matrix

$$\begin{aligned}\mathsf{Rot}\,(\theta)\,\mathsf{T}\,(h,k) &= \begin{pmatrix} \cos\theta & \sin\theta & 0 \\ -\sin\theta & \cos\theta & 0 \\ 0 & 0 & 1 \end{pmatrix} \begin{pmatrix} 1 & 0 & 0 \\ 0 & 1 & 0 \\ h & k & 1 \end{pmatrix} \\ &= \begin{pmatrix} \cos\theta & \sin\theta & 0 \\ -\sin\theta & \cos\theta & 0 \\ h & k & 1 \end{pmatrix}.\end{aligned}$$

Example 2.13

The transformation matrix which represents an anticlockwise rotation of $3\pi/2$ about the origin followed by a scaling by a factor of 3 units in the x-direction and 2 units in the y-direction is

$$\mathsf{Rot}\,(3\pi/2)\,\mathsf{S}\,(3,2) = \begin{pmatrix} 0 & -1 & 0 \\ 1 & 0 & 0 \\ 0 & 0 & 1 \end{pmatrix} \begin{pmatrix} 3 & 0 & 0 \\ 0 & 2 & 0 \\ 0 & 0 & 1 \end{pmatrix} = \begin{pmatrix} 0 & -2 & 0 \\ 3 & 0 & 0 \\ 0 & 0 & 1 \end{pmatrix}.$$

EXERCISES

2.16. Determine the matrix which represents the operations of Example 2.13 performed in *reverse order*. What can be deduced about the order in which transformations are performed?

2.17. Determine the matrix which represents an anticlockwise rotation about the origin through an angle π followed by a scaling by a factor of 4 in the x-direction and by a factor of 0.5 in the y-direction.

2.18. Determine the matrix which represents a translation of 4 units in the x-direction followed by a rotation about the point $(2, 3)$ through an angle $\pi/2$ in a *clockwise* direction.

2.5.1 Inverse Transformations

The *identity* and *inverse* transformations were introduced in Section 1.2. The *identity transformation* I is the transformation which has the effect of leaving all points of the plane unchanged. The *inverse* of a transformation L, denoted L^{-1}, has the effect of mapping images of the transformation L back to their original points. These transformations can be given a more precise definition in terms of the concatenation of transformations.

Definition 2.14

The *identity* transformation of the plane, denoted I, is the transformation for which $I \circ L = L \circ I = L$, for all planar transformations L. The transformation matrix of the identity transformation is the 3×3 identity matrix I_3 (that is, the matrix with values of 1's on the leading diagonal and 0's elsewhere).

Definition 2.15

The inverse L^{-1} of a transformation L is the transformation such that $L \circ L^{-1} = I$ and $L^{-1} \circ L = I$.

Lemma 2.16

Let the homogeneous transformation matrix of L be T. A necessary and sufficient condition for the inverse L^{-1} to exist is that T^{-1} exists and is the transformation matrix of L^{-1}.

Proof

Suppose L has an inverse L^{-1} with transformation matrix T_1. The concatenation $L \circ L^{-1} = I$ has transformation matrix $\mathrm{T}\mathrm{T}_1 = I_3$. Similarly, $L^{-1} \circ L = I$ has transformation matrix $\mathrm{T}_1\mathrm{T} = I_3$. Thus by the definition of a matrix inverse $\mathrm{T}_1 = \mathrm{T}^{-1}$.

Conversely, suppose T has an inverse T^{-1}, and let L_1 be the transformation defined by T^{-1}. Since $\mathrm{T}\mathrm{T}^{-1} = I_3$ and $\mathrm{T}^{-1}\mathrm{T} = I_3$ it follows that $L \circ L_1 = I$ and $L_1 \circ L = I$. Hence L_1 is the inverse transformation of L. □

Definition 2.17

A transformation $L : \mathbb{P}^2 \to \mathbb{P}^2$ which has an inverse L^{-1} is called a *non-singular transformation*. Lemma 2.16 shows that a transformation is a non-singular transformation if and only if its transformation matrix is non-singular.

Example 2.18

Non-singular matrices A and B satisfy $(AB)^{-1} = B^{-1}A^{-1}$. Further, $\mathsf{S}(s_1, s_2)^{-1} = \mathsf{S}(1/s_1, 1/s_2)$ and $\mathsf{Rot}(\theta)^{-1} = \mathsf{Rot}(-\theta)$ (Exercises 1.9 and 1.14). This gives a straightforward way of determining the inverse transformation matrix of the concatenated transformation $\mathsf{Rot}(3\pi/2) \circ \mathsf{S}(3, 2)$:

$$
\begin{aligned}
(\mathsf{Rot}(3\pi/2)\,\mathsf{S}(3,2))^{-1} &= \mathsf{S}(3,2)^{-1}\,\mathsf{Rot}(3\pi/2)^{-1} \\
&= \mathsf{S}(1/3,1/2)\,\mathsf{Rot}(-3\pi/2) \\
&= \begin{pmatrix} \frac{1}{3} & 0 & 0 \\ 0 & \frac{1}{2} & 0 \\ 0 & 0 & 1 \end{pmatrix} \begin{pmatrix} \cos\left(-\frac{3\pi}{2}\right) & \sin\left(-\frac{3\pi}{2}\right) & 0 \\ -\sin\left(-\frac{3\pi}{2}\right) & \cos\left(-\frac{3\pi}{2}\right) & 0 \\ 0 & 0 & 1 \end{pmatrix} \\
&= \begin{pmatrix} 0 & \frac{1}{3} & 0 \\ -\frac{1}{2} & 0 & 0 \\ 0 & 0 & 1 \end{pmatrix}.
\end{aligned}
$$

Alternatively, using Example 2.13

$$
(\mathsf{Rot}(3\pi/2)\,\mathsf{S}(3,2))^{-1} = \begin{pmatrix} 0 & -2 & 0 \\ 3 & 0 & 0 \\ 0 & 0 & 1 \end{pmatrix}^{-1} = \begin{pmatrix} 0 & \frac{1}{3} & 0 \\ -\frac{1}{2} & 0 & 0 \\ 0 & 0 & 1 \end{pmatrix}.
$$

EXERCISES

2.19. Determine the transformation matrix of the inverse of the concatenation $\mathsf{T}(-2, 5) \circ \mathsf{Rot}(-\pi/3)$.

2.20. Use a graphics calculator or mathematics computer package to compute the inverse of the transformation with matrix

$$
\begin{pmatrix} 1.0 & 0.5 & 0.0 \\ 0.8 & -1.2 & 0.0 \\ 4.0 & -2.0 & 1.0 \end{pmatrix}.
$$

2.21. Consider a (rectangular) Cartesian coordinate system with origin $\mathbf{O}$ and coordinates (x, y), and a second system with origin $\mathbf{O}'(x_0, y_0)$

and coordinates (x', y'). The origin and axes of the first system can be mapped to those of the second by applying a rotation $\mathsf{Rot}(\theta)$ followed by the translation $\mathsf{T}(x_0, y_0)$. The (x, y)-coordinates of a point given in (x', y')-coordinates is obtained by applying the *orthogonal change of coordinates* transformation

$$\begin{aligned} x &= x' \cos\theta - y' \sin\theta + x_0 \\ y &= x' \sin\theta + y' \cos\theta + y_0 \,. \end{aligned}$$

a) Determine the homogeneous transformation matrix A of the change of coordinates and show that $\det(\mathsf{A}) = 1$.

b) Determine the inverse change of coordinates transformation which determines the (x', y')-coordinates of a point (x, y).

c) Show that a change of coordinates preserves the angle between a pair of lines.

d) Show that the x'- and y'-axes, expressed in (x, y)-coordinates, are given by the equations

$$\begin{aligned} (x - x_0)\sin\theta - (y - y_0)\cos\theta &= 0\,, \quad \text{and} \\ (x - x_0)\cos\theta + (y - y_0)\sin\theta &= 0\,. \end{aligned}$$

2.5.2 Rotation about an Arbitrary Point

A rotation through an angle θ about an arbitrary point (x_0, y_0) is obtained by performing a translation which maps (x_0, y_0) to the origin, followed by a rotation through an angle θ about the origin, and followed by a translation which maps the origin to (x_0, y_0). The rotation matrix is

$$\begin{aligned} \mathsf{Rot}_{(x_0,y_0)}(\theta) &= \mathsf{T}(-x_0, -y_0)\,\mathsf{Rot}(\theta)\,\mathsf{T}(x_0, y_0) \\ &= \begin{pmatrix} 1 & 0 & 0 \\ 0 & 1 & 0 \\ -x_0 & -y_0 & 1 \end{pmatrix} \begin{pmatrix} \cos\theta & \sin\theta & 0 \\ -\sin\theta & \cos\theta & 0 \\ 0 & 0 & 1 \end{pmatrix} \begin{pmatrix} 1 & 0 & 0 \\ 0 & 1 & 0 \\ x_0 & y_0 & 1 \end{pmatrix} \\ &= \begin{pmatrix} \cos\theta & \sin\theta & 0 \\ -\sin\theta & \cos\theta & 0 \\ \begin{matrix}(-x_0\cos\theta \\ +y_0\sin\theta + x_0)\end{matrix} & \begin{matrix}(-x_0\sin\theta \\ -y_0\cos\theta + y_0)\end{matrix} & 1 \end{pmatrix}. \end{aligned}$$

Example 2.19

A square has vertices $\mathbf{A}(1,1)$, $\mathbf{B}(2,1)$, $\mathbf{C}(2,2)$, and $\mathbf{D}(1,2)$. Calculate the coordinates of the vertices when the rectangle is rotated about $\mathbf{B}$ through an angle $\pi/4$. The required transformation is

$$\begin{aligned}
&\mathsf{T}(-2,-1)\,\mathsf{Rot}(\pi/4)\,\mathsf{T}(2,1) \\
&= \begin{pmatrix} 1 & 0 & 0 \\ 0 & 1 & 0 \\ -2 & -1 & 1 \end{pmatrix} \begin{pmatrix} 0.7071 & 0.7071 & 0 \\ -0.7071 & 0.7071 & 0 \\ 0 & 0 & 1 \end{pmatrix} \begin{pmatrix} 1 & 0 & 0 \\ 0 & 1 & 0 \\ 2 & 1 & 1 \end{pmatrix} \\
&= \begin{pmatrix} 0.7071 & 0.7071 & 0 \\ -0.7071 & 0.7071 & 0 \\ 1.2929 & -1.1213 & 1 \end{pmatrix}.
\end{aligned}$$

Applying the transformation to the vertices,

$$\begin{pmatrix} 1 & 1 & 1 \\ 2 & 1 & 1 \\ 2 & 2 & 1 \\ 1 & 2 & 1 \end{pmatrix} \begin{pmatrix} 0.7071 & 0.7071 & 0 \\ -0.7071 & 0.7071 & 0 \\ 1.2929 & -1.1213 & 1 \end{pmatrix} = \begin{pmatrix} 1.2929 & 0.2929 & 1 \\ 2 & 1 & 1 \\ 1.2929 & 1.7071 & 1 \\ 0.5858 & 1 & 1 \end{pmatrix}$$

gives $\mathbf{A}'(1.2929, 0.2929)$, $\mathbf{B}'(2,1)$, $\mathbf{C}'(1.2929, 1.7071)$, and $\mathbf{D}'(0.5858, 1.0)$. The rotated square is illustrated in Figure 2.5.

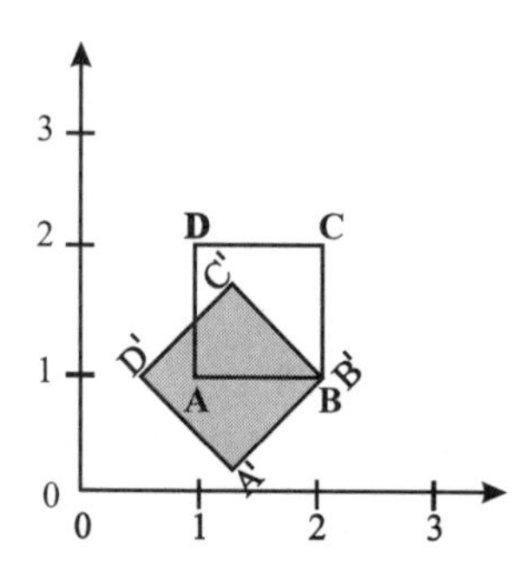

Figure 2.5

2.5.3 Reflection in an Arbitrary Line

Reflections in the x- and y-axes were derived in Exercise 2.15.A reflection in an arbitrary line ℓ with equation $ax + by + c = 0$ is obtained by transforming the line to one of the axes, reflecting in that axis, and then applying the inverse of the first transformation. Suppose $b \neq 0$.

1. The line ℓ intersects the y-axis in the point $(0, -c/b)$.
2. Apply a translation mapping $(0, -c/b)$ to the origin, and thus mapping ℓ to a line ℓ' through the origin with an identical gradient to ℓ.
3. The gradient of ℓ' is $\tan\theta = -a/b$, where θ is the angle that ℓ makes with the x-axis. Rotate ℓ' about the origin through an angle $-\theta$. The line is now mapped to the x-axis.
4. Apply a reflection in the x-axis.
5. Apply the inverse of the rotation of step 3, followed by the inverse of the translation of step 2.

The concatenation of the above transformations is

$$\begin{pmatrix} 1 & 0 & 0 \\ 0 & 1 & 0 \\ 0 & c/b & 1 \end{pmatrix} \begin{pmatrix} \cos\theta & -\sin\theta & 0 \\ \sin\theta & \cos\theta & 0 \\ 0 & 0 & 1 \end{pmatrix} \begin{pmatrix} 1 & 0 & 0 \\ 0 & -1 & 0 \\ 0 & 0 & 1 \end{pmatrix}$$

$$\times \begin{pmatrix} \cos\theta & \sin\theta & 0 \\ -\sin\theta & \cos\theta & 0 \\ 0 & 0 & 1 \end{pmatrix} \begin{pmatrix} 1 & 0 & 0 \\ 0 & 1 & 0 \\ 0 & -c/b & 1 \end{pmatrix} \tag{2.9}$$

$$= \begin{pmatrix} \cos^2\theta - \sin^2\theta & 2\cos\theta\sin\theta & 0 \\ 2\cos\theta\sin\theta & \sin^2\theta - \cos^2\theta & 0 \\ 2\frac{c}{b}\sin\theta\cos\theta & \frac{c}{b}\left(\sin^2\theta - \cos^2\theta - 1\right) & 1 \end{pmatrix}. \tag{2.10}$$

Since $\tan\theta = \sin\theta/\cos\theta = -a/b$, it follows that $\sin\theta = a/\left(a^2+b^2\right)^{1/2}$ and $\cos\theta = -b/\left(a^2+b^2\right)^{1/2}$ (Exercise 2.25). Hence, $\cos^2\theta = b^2/\left(a^2+b^2\right)$, $\sin^2\theta = a^2/\left(a^2+b^2\right)$, $\sin\theta\cos\theta = -ab/\left(a^2+b^2\right)$, and $\cos^2\theta - \sin^2\theta = \left(b^2-a^2\right)/\left(a^2+b^2\right)$. Finally, substitution for the trigonometric functions in (2.10) yields

$$\begin{pmatrix} \frac{b^2-a^2}{a^2+b^2} & -\frac{2ab}{a^2+b^2} & 0 \\ -\frac{2ab}{a^2+b^2} & -\frac{b^2-a^2}{a^2+b^2} & 0 \\ -\frac{2ac}{a^2+b^2} & -\frac{2bc}{a^2+b^2} & 1 \end{pmatrix}.$$

Since in homogeneous coordinates multiplication by a factor does not affect the result, the above matrix can be multiplied by a factor $\left(a^2+b^2\right)$ to give the general reflection matrix

$$\mathsf{R}_{(a,b,c)} = \begin{pmatrix} b^2-a^2 & -2ab & 0 \\ -2ab & -b^2+a^2 & 0 \\ -2ac & -2bc & a^2+b^2 \end{pmatrix}. \tag{2.11}$$

EXERCISES

2.22. Show that the concatenation of two rotations, the first through an angle θ about a point $\mathbf{P}(x_0, y_0)$ and the second about a point $\mathbf{Q}(x_1, y_1)$ (distinct from $\mathbf{P}$) through an angle $-\theta$, is equivalent to a translation.

2.23. Determine the transformation matrix of a reflection in the line $5x - 2y + 8 = 0$. Express the reflection first using (2.11) and then as a concatenation of transformations (2.9).

2.24. Demonstrate that if the coordinates of points are expressed by rational numbers (whole numbers and fractions), then a reflection in a line defined by rational coefficients a, b, c can be computed using integer arithmetic.

2.25. Use trigonometry to verify the result used in the derivation of (2.11) that if $\tan\theta = -a/b$, then $\sin\theta = a/\left(a^2+b^2\right)^{1/2}$ and $\cos\theta = -b/\left(a^2+b^2\right)^{1/2}$.

2.6 Applications

2.6.1 Instancing

In Section 1.8.1 the model of the front of a house was defined by instancing the picture element **Square** with vertices $(0,0)$, $(1,0)$, $(1,1)$, and $(0,1)$. The front door was obtained by applying a scaling of 0.5 units in the x-direction, followed by a translation of 3 units in the x-direction and 1 unit in the y-direction. Transformations applied to picture elements and primitives to obtain instances are called *modelling transformations*. The front door is obtained from **Square** by applying the modelling transformation $\mathsf{S}(1,3) \circ \mathsf{T}(4,0)$ which has the *modelling transformation matrix*

$$\mathsf{S}(1,3)\mathsf{T}(4,0) = \begin{pmatrix} 0.5 & 0 & 0 \\ 0 & 1 & 0 \\ 0 & 0 & 1 \end{pmatrix} \begin{pmatrix} 1 & 0 & 0 \\ 0 & 1 & 0 \\ 3 & 1 & 1 \end{pmatrix} = \begin{pmatrix} 0.5 & 0 & 0 \\ 0 & 1 & 0 \\ 3 & 1 & 1 \end{pmatrix}.$$

The vertices of the door are obtained by applying the modelling transformation matrix to the vertices of the **Square** primitive, giving

$$\begin{pmatrix} 0 & 0 & 1 \\ 1 & 0 & 1 \\ 1 & 1 & 1 \\ 0 & 1 & 1 \end{pmatrix} \begin{pmatrix} 1 & 0 & 0 \\ 0 & 1 & 0 \\ 3 & 1 & 1 \end{pmatrix} = \begin{pmatrix} 3 & 1 & 1 \\ 4 & 1 & 1 \\ 4 & 2 & 1 \\ 3 & 2 & 1 \end{pmatrix}.$$

So in world coordinates the vertices are $(3,1)$, $(4,1)$, $(4,2)$ and $(3,2)$.

Exercise 2.26

Determine the modelling transformation matrices of the four instances of **Square** which define the windows of the front of the house in Figure 1.8. Complete the picture element **House** by determining the modelling transformation matrix of the primitive **Point** which is a small circle centred at the point $(0, 0)$. Now create a modern housing estate by instancing **House**!

2.6.2 Device Coordinate Transformation

Sections 1.8.1 and 2.6.1 discuss how the model of an object is obtained by instancing a number of picture elements and graphical primitives. The object (the front of a house) is defined in a two-dimensional world coordinate system. The object is displayed in a device window, such as a computer screen, by applying a *device coordinate transformation.* The process of viewing an object defined in a *three*-dimensional world coordinate system is discussed later in Chapter 4.

Suppose the world coordinate system is the (x, y)-plane. The region of the plane to be displayed by the device is specified by a rectangular *window* with lower left corner $(x_{\min}, y_{\min})$ and upper right corner $(x_{\max}, y_{\max})$. Any part of the object lying outside this region is "clipped" and is not displayed. The coordinate system of a display device is determined by its resolution. For example, a computer screen consists of a rectangular array of pixels. The number of pixels in the horizontal (h) and vertical (v) directions is written $h \times v$ and called the screen resolution. The origin is assumed to be the lower left corner of the screen, and the pixels are labelled with coordinates (h, v) where h and v are non-negative integers. Figure 2.6 illustrates a screen with a resolution of 1280×1024 pixels, and a window given by $(x_{\min}, y_{\min}) = (-10, -5)$ and $(x_{\max}, y_{\max}) = (10, 5)$. The window is mapped onto the screen by the device coordinate transformation which is the concatenation of (i) the translation $\mathsf{T}(10, 5)$ which maps the point $(-10, -5)$ to the origin, and (ii) a scaling $\mathsf{S}(1280/20, 1024/10)$ which makes the rectangle the same size as the screen. Therefore, the device coordinate transformation is

$$\begin{aligned} \mathsf{T}(10,5)\,\mathsf{S}(1280/20, 1024/10) &= \begin{pmatrix} 1 & 0 & 0 \\ 0 & 1 & 0 \\ 10 & 5 & 1 \end{pmatrix} \begin{pmatrix} 64 & 0 & 0 \\ 0 & 102.4 & 0 \\ 0 & 0 & 1 \end{pmatrix} \\ &= \begin{pmatrix} 64 & 0 & 0 \\ 0 & 102.4 & 0 \\ 640 & 512 & 1 \end{pmatrix}. \end{aligned}$$

Hence the Cartesian coordinates of the point on the screen corresponding to the point (x, y) in the window are $(64x + 640, 102.4y + 512)$.

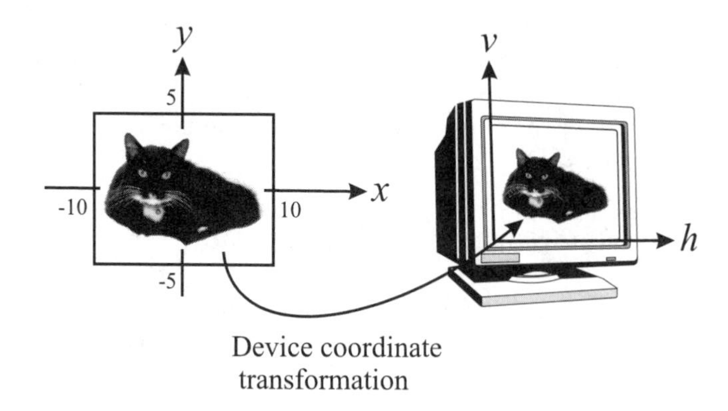

Figure 2.6 A device coordinate transformation makes Max a film star

Exercise 2.27

Suppose the window specified above is to be mapped onto a rectangular *device window* of the computer screen with lower left corner $(200, 200)$ and upper right corner $(600, 400)$. Determine the device coordinate transformation matrix.

2.7 Point and Line Geometry in Homogeneous Coordinates

The general equation of a line in the Cartesian plane is $ax + by + c = 0$. Suppose (X, Y, W) are the homogeneous coordinates of the point (x, y), so that $x = X/W$ and $y = Y/W$. Substituting for x and y in the equation of the line, and multiplying through by W, yields the condition for (X, Y, W) to be a point on the line

$$aX + bY + cW = 0 \; . \tag{2.12}$$

The equation is known as the *homogeneous line equation*. The line is uniquely defined by the coefficients a, b, and c, or any non-zero multiple ra, rb, and rc of them. Therefore, it is natural to specify the line by the homogeneous *line coordinates*

$$\ell = (a, b, c) \; .$$

It is also useful to consider ℓ to be a vector known as the *line vector*. Since any non-zero multiple of ℓ defines the same line, only the direction of ℓ is of importance. Let $\mathbf{P}(X,Y,W)$ be a point on the line. By permitting the homogeneous coordinates (X,Y,W) to be treated as a vector, Equation (2.12) may be expressed as the dot product

$$\ell \cdot \mathbf{P} = aX + bY + cW = 0\,. \tag{2.13}$$

The identity (2.13) leads to two useful operations: (i) determining the line through two distinct points, and (ii) determining the point of intersection of two lines.

To Find the Equation of the Line Through Two Points

Suppose ℓ is the line vector of a line containing two distinct points $\mathbf{P}_1(X_1,Y_1,W_1)$ and $\mathbf{P}_2(X_2,Y_2,W_2)$. Then (2.13) yields

$$\ell \cdot \mathbf{P}_1 = 0 \quad \text{and} \quad \ell \cdot \mathbf{P}_2 = 0\,.$$

For any two vectors, the condition $\mathbf{a} \cdot \mathbf{b} = 0$ implies that $\mathbf{a}$ and $\mathbf{b}$ are perpendicular. Hence, ℓ is a vector perpendicular to both $\mathbf{P}_1$ and $\mathbf{P}_2$. To determine ℓ it is sufficient to determine any vector perpendicular to $\mathbf{P}_1$ and $\mathbf{P}_2$. In particular, the cross product gives a vector perpendicular to two given vectors, thus $\ell = \mathbf{P}_1 \times \mathbf{P}_2$ (or any multiple of $\mathbf{P}_1 \times \mathbf{P}_2$). Hence, the equation of the line through two points can be determined by taking the "cross product" of the homogeneous coordinates of the points.

Example 2.20

The line ℓ passing through $(0,5)$ and $(6,-7)$ satisfies

$$\ell \cdot (0,5,1) = 0 \quad \text{and} \quad \ell \cdot (6,-7,1) = 0\,.$$

Hence

$$\ell = (0,5,1) \times (6,-7,1) = (12,6,-30)$$

giving the line $12x + 6y - 30 = 0$.

To Determine the Point of Intersection of Two Lines

Suppose $\mathbf{P}$ is the point of intersection of two lines ℓ_1 and ℓ_2. Then $\mathbf{P}$ is a point on both lines and (2.13) yields

$$\ell_1 \cdot \mathbf{P} = 0 \quad \text{and} \quad \ell_2 \cdot \mathbf{P} = 0\,.$$

Hence $\mathbf{P}$ is a vector perpendicular to both ℓ_1 and ℓ_2, and hence it is sufficient to take $\mathbf{P} = \ell_1 \times \ell_2$ (or any multiple of it). The cross product yields the homogeneous coordinates of the point of intersection.

Example 2.21

The point $\mathbf{P}$ of intersection of the lines $x - 7y + 8 = 0$ and $3x - 4y + 1 = 0$ satisfies

$$(1, -7, 8) \cdot \mathbf{P} = 0 \quad \text{and} \quad (3, -4, 1) \cdot \mathbf{P} = 0 \, .$$

Hence

$$\mathbf{P} = (1, -7, 8) \times (3, -4, 1) = (25, 23, 17) \; .$$

The Cartesian coordinates of the intersection point are $(25/17, 23/17)$.

Example 2.22

The point $\mathbf{P}$ of intersection of the lines $2x - 5y = 0$ and $2x - 5y + 3 = 0$ has homogeneous coordinates

$$\mathbf{P} = (2, -5, 0) \times (2, -5, 3) = (-15, -6, 0) \; .$$

The point of intersection $(-15, -6, 0)$ is a point at infinity since the lines are parallel.

EXERCISES

2.28. Determine the line passing through $(1, 3)$ and $(4, -2)$.

2.29. Determine the point of intersection of the lines $x - 3y + 7 = 0$ and $4x + 3y - 5 = 0$.

2.30. The methods used to determine the line through two distinct points and the point of intersection of two lines both involve the cross product. This is due to the duality between points and lines in the plane which relates results about points and lines to a dual result about lines and points. For example, the property "points r_1, r_2, and r_3 are *collinear* if and only if $r_1 \cdot (r_2 \times r_3) = 0$" has the *dual* property "lines ℓ_1, ℓ_2, and ℓ_3 are *concurrent* if and only if $\ell_1 \cdot (\ell_2 \times \ell_3) = 0$". Investigate further the property of duality [24, pp78–80].

3 Homogeneous Coordinates and Transformations of Space

3.1 Homogeneous Coordinates

Homogeneous coordinates in three-dimensional space are derived in a similar manner as homogeneous coordinates of the plane. A point (x, y, z) in three-dimensional Cartesian space $\mathbb{R}^3$ is represented in the four-dimensional space $\mathbb{R}^4$ by the vector $(x, y, z, 1)$, or by any multiple (rx, ry, rz, r) (with $r \neq 0$). When $W \neq 0$, the homogeneous coordinates (X, Y, Z, W) represent the Cartesian point $(x, y, z) = (X/W, Y/W, Z/W)$. A point of the form $(X, Y, Z, 0)$ does not correspond to a Cartesian point, but represents the *point at infinity* in the direction of the three-dimensional vector (X, Y, Z). The set of all homogeneous coordinates (X, Y, Z, W) is called *(three-dimensional) projective space* and denoted $\mathbb{P}^3$. Homogeneous coordinates (x, y, z, w) are frequently represented by the row matrix $(x\ y\ z\ w)$ for matrix computations.

Example 3.1

The homogeneous coordinates $(2, 3, 4, 5)$, $(-4, -6, -8, -10)$, and $(6, 9, 12, 15)$ all represent the point with Cartesian coordinates $(2/5, 3/5, 4/5)$.

Definition 3.2

A (projective) *transformation* of projective space is a mapping $L : \mathbb{P}^3 \to \mathbb{P}^3$ of the form

$$L(x, y, z, w) = \begin{pmatrix} x & y & z & w \end{pmatrix} \begin{pmatrix} m_{11} & m_{12} & m_{13} & m_{14} \\ m_{21} & m_{22} & m_{23} & m_{24} \\ m_{31} & m_{32} & m_{33} & m_{34} \\ m_{41} & m_{42} & m_{43} & m_{44} \end{pmatrix}.$$

The 4×4 matrix M is called the *homogeneous transformation matrix* of L. If M is a non-singular matrix then L is called a *non-singular* transformation. If $m_{14} = m_{24} = m_{34} = 0$ and $m_{44} \neq 0$, then L is said to be an *affine transformation.* (Affine transformations correspond to translations, scalings, rotations etc. of three-dimensional Cartesian space.)

3.2 Transformations of Space

A number of transformations of space are considered, namely, translations, scalings, reflections, rotations, and the composition of these transformations. As in the planar case, compositions of three-dimensional transformations are performed by multiplication of the transformation matrices.

3.2.1 Translations

The transformation matrix of a translation by x_0, y_0, and z_0 units in the x-, y-, and z-directions respectively, is

$$\mathsf{T}(x_0, y_0, z_0) = \begin{pmatrix} 1 & 0 & 0 & 0 \\ 0 & 1 & 0 & 0 \\ 0 & 0 & 1 & 0 \\ x_0 & y_0 & z_0 & 1 \end{pmatrix}.$$

The point with homogeneous coordinates $\mathbf{P}(x, y, z, 1)$ is translated to the point $\mathbf{P}'$ given by

$$\begin{pmatrix} x+x_0 & y+y_0 & z+z_0 & 1 \end{pmatrix} = \begin{pmatrix} x & y & z & 1 \end{pmatrix} \begin{pmatrix} 1 & 0 & 0 & 0 \\ 0 & 1 & 0 & 0 \\ 0 & 0 & 1 & 0 \\ x_0 & y_0 & z_0 & 1 \end{pmatrix}.$$

Hence, $\mathbf{P}(x, y, z)$ is transformed to $\mathbf{P}'(x + x_0, y + y_0, z + z_0)$ as required.

3.2.2 Scalings and Reflections

A scaling about the origin by a factor s_x/s_w, s_y/s_w, and s_z/s_w in the x-, y-, and z-directions respectively, is obtained by the following transformation matrix

$$\mathsf{S}\left(s_x, s_y, s_z, s_w\right) = \begin{pmatrix} s_x & 0 & 0 & 0 \\ 0 & s_y & 0 & 0 \\ 0 & 0 & s_z & 0 \\ 0 & 0 & 0 & s_w \end{pmatrix}.$$

Frequently, s_w is taken to be 1.

The transformation matrices of the reflections R_{yz} in the $x = 0$ plane, R_{xz} in the $y = 0$ plane, and R_{xy} in the $z = 0$ plane, are obtained by taking a scaling of -1 in one of the coordinate directions,

$$\begin{aligned} \mathsf{R}_{yz} &= \begin{pmatrix} -1 & 0 & 0 & 0 \\ 0 & 1 & 0 & 0 \\ 0 & 0 & 1 & 0 \\ 0 & 0 & 0 & 1 \end{pmatrix}, \\ \mathsf{R}_{xz} &= \begin{pmatrix} 1 & 0 & 0 & 0 \\ 0 & -1 & 0 & 0 \\ 0 & 0 & 1 & 0 \\ 0 & 0 & 0 & 1 \end{pmatrix}, \\ \mathsf{R}_{xy} &= \begin{pmatrix} 1 & 0 & 0 & 0 \\ 0 & 1 & 0 & 0 \\ 0 & 0 & -1 & 0 \\ 0 & 0 & 0 & 1 \end{pmatrix}. \end{aligned}$$

3.2.3 Rotations about the Coordinate Axes

Rotations in space take place about a line called the *rotation axis*. The rotations about the three coordinate axes are called the *primary rotations*.

1. Rotation about the x-axis through an angle θ_x

$$\mathsf{Rot}_x\left(\theta_x\right) = \begin{pmatrix} 1 & 0 & 0 & 0 \\ 0 & \cos\theta_x & \sin\theta_x & 0 \\ 0 & -\sin\theta_x & \cos\theta_x & 0 \\ 0 & 0 & 0 & 1 \end{pmatrix}.$$

2. Rotation about the y-axis through an angle θ_y

$$\mathsf{Rot}_y(\theta_y) = \begin{pmatrix} \cos\theta_y & 0 & -\sin\theta_y & 0 \\ 0 & 1 & 0 & 0 \\ \sin\theta_y & 0 & \cos\theta_y & 0 \\ 0 & 0 & 0 & 1 \end{pmatrix}.$$

3. Rotation about the z-axis through an angle θ_z

$$\mathsf{Rot}_z(\theta_z) = \begin{pmatrix} \cos\theta_z & \sin\theta_z & 0 & 0 \\ -\sin\theta_z & \cos\theta_z & 0 & 0 \\ 0 & 0 & 1 & 0 \\ 0 & 0 & 0 & 1 \end{pmatrix}.$$

Figure 3.1 shows the directions which the primary rotations take when the rotation angle is *positive.* The directions are easily remembered by the mnemonic

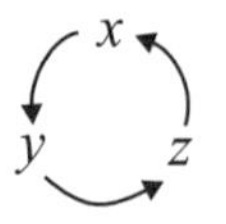

For instance, to determine the positive sense of a rotation about the y-axis, cover up the "y" to reveal $z \to x$. The arrow indicates that a positive angle of rotation has the effect of moving points on the z-axis towards the x-axis. A two-dimensional rotation in the xy-plane about the origin yields the same

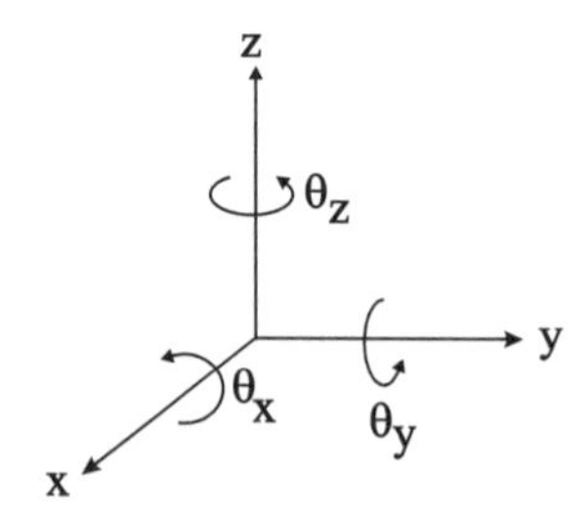

Figure 3.1 Definition of positive rotation angles

transformation of points in the plane as a three-dimensional rotation of the plane about the z-axis. Rotations about an arbitrary line are obtained in Section 3.2.4 by a composition of translations and primary rotations.

Example 3.3

The transformation matrix M which represents a rotation of an angle $\pi/6$ about the y-axis followed by a translation $\mathsf{T}(1,-1,2)$ is

$$\begin{aligned}\mathsf{Rot}_y(\pi/6)\,\mathsf{T}(1,-1,2) &= \begin{pmatrix} 0.866 & 0 & -0.5 & 0 \\ 0 & 1.0 & 0 & 0 \\ 0.5 & 0 & 0.866 & 0 \\ 0 & 0 & 0 & 1.0 \end{pmatrix}\begin{pmatrix} 1 & 0 & 0 & 0 \\ 0 & 1 & 0 & 0 \\ 0 & 0 & 1 & 0 \\ 1 & -1 & 2 & 1 \end{pmatrix} \\ &= \begin{pmatrix} 0.866 & 0 & -0.5 & 0 \\ 0 & 1.0 & 0 & 0 \\ 0.5 & 0 & 0.866 & 0 \\ 1.0 & -1.0 & 2.0 & 1.0 \end{pmatrix}.\end{aligned}$$

3.2.4 Rotation about an Arbitrary Line

Rotation through an angle θ about an arbitrary rotation axis is obtained by transforming the rotation axis to one of the coordinate axes, applying a primary rotation through an angle θ about the coordinate axis, and applying the transformation which maps the coordinate axis back to the rotation axis. Let the rotation axis be the line through the points $\mathbf{P}(p_1, p_2, p_3)$ and $\mathbf{Q}(q_1, q_2, q_3)$. Let $\mathbf{R}(r_1, r_2, r_3)$ be the unit vector in the direction $\mathbf{Q} - \mathbf{P}$. Then the rotation can be performed as follows.

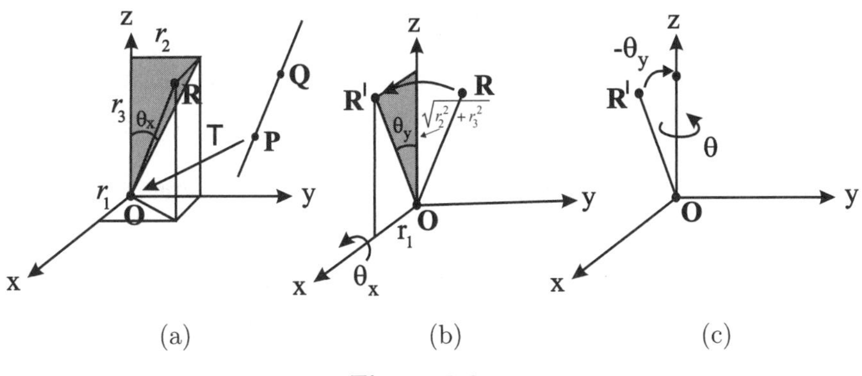

Figure 3.2

1. Apply the translation $\mathsf{T}(-p_1, -p_2, -p_3)$ which maps $\mathbf{P}$ to the origin and the rotation axis to the line $\overline{\mathbf{OR}}$ as shown in Figure 3.2(a). If $\mathbf{R}$ is parallel

to the x-axis (when $r_2 = r_3 = 0$) then the required rotation matrix is

$$\mathsf{T}(-p_1, -p_2, -p_3)\,\mathsf{Rot}_x(\theta)\,\mathsf{T}(p_1, p_2, p_3)\ .$$

Likewise, if $\mathbf{R}$ is parallel to the y-axis (when $r_1 = r_3 = 0$) or the z-axis (when $r_1 = r_2 = 0$) then the required rotation matrices are

$$\mathsf{T}(-p_1, -p_2, -p_3)\,\mathsf{Rot}_y(\theta)\,\mathsf{T}(p_1, p_2, p_3)$$

or

$$\mathsf{T}(-p_1, -p_2, -p_3)\,\mathsf{Rot}_z(\theta)\,\mathsf{T}(p_1, p_2, p_3)$$

respectively.

2. Suppose r_2 and r_3 are not both zero. Apply a rotation through an angle θ_x about the x-axis so that the line $\overline{\mathbf{OR}}$ is mapped into the xz-plane. Referring to Figure 3.2(a), an application of trigonometry to the shaded triangle yields that the line $\overline{\mathbf{OR}}$ makes an angle θ_x with the xz-plane where

$$\sin\theta_x = r_2/\sqrt{r_2^2 + r_3^2}, \quad \text{and} \quad \cos\theta_x = r_3/\sqrt{r_2^2 + r_3^2}\ .$$

The desired rotation $\mathsf{Rot}_x(\theta_x)$ maps $\mathbf{R}$ to the point $\mathbf{R}'(r_1, 0, \sqrt{r_2^2 + r_3^2})$ as depicted in Figure 3.2(b).

3. Apply a rotation about the y-axis so that the line $\overline{\mathbf{OR}'}$ is mapped to the z-axis. Applying trigonometry to the shaded triangle of Figure 3.2(b) the required angle is found to be $-\theta_y$ where

$$\sin\theta_y = r_1, \quad \text{and} \quad \cos\theta_y = \sqrt{r_2^2 + r_3^2}\ .$$

4. Apply a rotation through an angle θ about the z-axis (Figure 3.2(c)).

5. Apply the inverses of the transformations 1–3 in reverse order.

Thus the general rotation through an angle θ about the line through the points $\mathbf{P}(p_1, p_2, p_3)$ and $\mathbf{Q}(q_1, q_2, q_3)$ has transformation matrix

$$\begin{aligned}\mathsf{T}(-p_1, -p_2, -p_3)\,\mathsf{Rot}_x(\theta_x)\,\mathsf{Rot}_y(-\theta_y)\,\mathsf{Rot}_z(\theta) \times \\ \mathsf{Rot}_y(\theta_y)\,\mathsf{Rot}_x(-\theta_x)\mathsf{T}(p_1, p_2, p_3)\ ,\end{aligned}$$

where $\mathbf{R}(r_1, r_2, r_3)$ is the unit vector in the direction $\mathbf{Q} - \mathbf{P}$, and

$$\begin{aligned}\sin\theta_x &= r_2/\sqrt{r_2^2 + r_3^2}, \quad \cos\theta_x = r_3/\sqrt{r_2^2 + r_3^2}\,, \\ \sin\theta_y &= r_1, \quad \text{and} \quad \cos\theta_y = \sqrt{r_2^2 + r_3^2}\ .\end{aligned}$$

Example 3.4

Compute the transformation matrix of the rotation through an angle θ about the line through the points $\mathbf{P}(2,1,5)$ and $\mathbf{Q}(4,7,2)$. Then

$$\mathbf{Q}-\mathbf{P}=(4,7,2)-(2,1,5)=(2,6,-3)\ ,$$

$|(2,6,-3)|=7$, and hence $\mathbf{R}=(2/7,6/7,-3/7)$. Then $\sqrt{r_2^2+r_3^2}=\frac{3}{7}\sqrt{5}$, and $\sin\theta_x=2\,/\sqrt{5}$, $\cos\theta_x=-1\,/\sqrt{5}$, $\sin\theta_y=2/7$, and $\cos\theta_y=\frac{3}{7}\sqrt{5}$. The rotation matrix is

$$\begin{pmatrix}1&0&0&0\\0&1&0&0\\0&0&1&0\\-2&-1&-5&1\end{pmatrix}\begin{pmatrix}1&0&0&0\\0&-\frac{1}{\sqrt{5}}&\frac{2}{\sqrt{5}}&0\\0&-\frac{2}{\sqrt{5}}&-\frac{1}{\sqrt{5}}&0\\0&0&0&1\end{pmatrix}$$

$$\times\begin{pmatrix}\frac{3}{7\sqrt{5}}&0&\frac{2}{7}&0\\0&1&0&0\\-\frac{2}{7}&0&\frac{3}{7\sqrt{5}}&0\\0&0&0&1\end{pmatrix}\begin{pmatrix}\cos\theta&\sin\theta&0&0\\-\sin\theta&\cos\theta&0&0\\0&0&1&0\\0&0&0&1\end{pmatrix}$$

$$\times\begin{pmatrix}\frac{3}{7\sqrt{5}}&0&-\frac{2}{7}&0\\0&1&0&0\\\frac{2}{7}&0&\frac{3}{7\sqrt{5}}&0\\0&0&0&1\end{pmatrix}\begin{pmatrix}1&0&0&0\\0&-\frac{1}{\sqrt{5}}&-\frac{2}{\sqrt{5}}&0\\0&\frac{2}{\sqrt{5}}&-\frac{1}{\sqrt{5}}&0\\0&0&0&1\end{pmatrix}\begin{pmatrix}1&0&0&0\\0&1&0&0\\0&0&1&0\\2&1&5&1\end{pmatrix}=$$

$$\begin{pmatrix}\frac{45}{49}\cos\theta+\frac{4}{49} & -\frac{3}{7}\sin\theta-\frac{12}{49}\cos\theta+\frac{12}{49} & -\frac{6}{7}\sin\theta+\frac{6}{49}\cos\theta-\frac{6}{49} & 0\\ -\frac{12}{49}\cos\theta+\frac{3}{7}\sin\theta+\frac{12}{49} & \frac{13}{49}\cos\theta+\frac{36}{49} & \frac{2}{7}\sin\theta+\frac{18}{49}\cos\theta-\frac{18}{49} & 0\\ \frac{6}{49}\cos\theta+\frac{6}{7}\sin\theta-\frac{6}{49} & -\frac{2}{7}\sin\theta+\frac{18}{49}\cos\theta-\frac{18}{49} & \frac{40}{49}\cos\theta+\frac{9}{49} & 0\\ -\frac{108}{49}\cos\theta-\frac{33}{7}\sin\theta+\frac{108}{49} & \frac{16}{7}\sin\theta-\frac{79}{49}\cos\theta+\frac{79}{49} & \frac{10}{7}\sin\theta-\frac{230}{49}\cos\theta+\frac{230}{49} & 1\end{pmatrix}.$$

3.2.5 Reflection in an Arbitrary Plane

Reflection in an arbitrary *reflection plane* $ax+by+cz+d=0$ is obtained by making a transformation which maps the plane to one of the xy-, xz- or yz-planes, followed by a primary reflection in the chosen plane, and followed by the transformation which maps the plane back to the reflection plane. The transformation is obtained as follows.

1. Determine a point $\mathbf{P}(p_1, p_2, p_3)$ on the plane (for example, the intersection of the plane with one of the axes). Apply the translation $\mathsf{T}(-p_1, -p_2, -p_3)$ to map $\mathbf{P}$ to the origin, and to map the reflection plane to the plane through the origin with normal direction (a, b, c). Let $\mathbf{R} = (r_1, r_2, r_3)$ denote the unit vector in the direction of (a, b, c). If $\mathbf{R}$ is parallel to the x-axis (when $r_2 = r_3 = 0$) then the required reflection matrix is

$$\mathsf{T}(-p_1, -p_2, -p_3)\, \mathsf{R}_{yz} \mathsf{T}(p_1, p_2, p_3) \ .$$

Likewise, if $\mathbf{R}$ is parallel to the y-axis (when $r_1 = r_3 = 0$) or the z-axis (when $r_1 = r_2 = 0$) then the required reflection matrices are

$$\mathsf{T}(-p_1, -p_2, -p_3)\, \mathsf{R}_{xz}(\theta)\, \mathsf{T}(p_1, p_2, p_3)$$

and

$$\mathsf{T}(-p_1, -p_2, -p_3)\, \mathsf{R}_{xy}(\theta)\, \mathsf{T}(p_1, p_2, p_3)$$

respectively.

2. Suppose r_2 and r_3 are not both zero. Following step 2 of the method of the general rotation, there is a composition of rotations $\mathsf{Rot}_x(\theta_x) \circ \mathsf{Rot}_y(-\theta_y)$, such that $\sin\theta_x = r_2/\sqrt{r_2^2 + r_3^2}$, $\cos\theta_x = r_3/\sqrt{r_2^2 + r_3^2}$, $\sin\theta_y = r_1$, $\cos\theta_y = \sqrt{r_2^2 + r_3^2}$, which maps the line $\overline{\mathbf{OR}}$ to the z-axis, and the translated reflection plane to the xy-plane.

3. Apply the reflection in the xy-plane.

4. Apply the inverses of the transformations 1–2 in reverse order.

The general reflection is

$$\mathsf{T}(-p_1, -p_2, -p_3)\, \mathsf{Rot}_x(\theta_x)\, \mathsf{Rot}_y(-\theta_y)\, \mathsf{R}_{xy} \times \\ \mathsf{Rot}_y(\theta_y)\, \mathsf{Rot}_x(-\theta_x)\, \mathsf{T}(p_1, p_2, p_3) \ .$$

Example 3.5

The transformation matrix for a reflection in the plane $2x - y + 2z - 2 = 0$ is obtained as follows. Translate the point $(1, 0, 0)$ of the plane to the origin. The translated plane is $2x - y + 2z = 0$ which has unit normal direction $\mathbf{R} = (2/3, -1/3, 2/3)$. Then the composition of rotations $\mathsf{Rot}_x(\theta_x) \circ \mathsf{Rot}_y(-\theta_y)$, such that $\sin\theta_x = -1/\sqrt{5}$, $\cos\theta_x = 2/\sqrt{5}$, $\sin(-\theta_y) = -\sin\theta_y = -2/3$, $\cos(-\theta_y) = \cos\theta_y = \sqrt{5}/3$, maps the plane to the xy-plane. The reflection matrix is

$$\begin{pmatrix} 1 & 0 & 0 & 0 \\ 0 & 1 & 0 & 0 \\ 0 & 0 & 1 & 0 \\ -1 & 0 & 0 & 1 \end{pmatrix} \begin{pmatrix} 1 & 0 & 0 & 0 \\ 0 & 2/\sqrt{5} & -1/\sqrt{5} & 0 \\ 0 & 1/\sqrt{5} & 2/\sqrt{5} & 0 \\ 0 & 0 & 0 & 1 \end{pmatrix} \begin{pmatrix} \sqrt{5}/3 & 0 & 2/3 & 0 \\ 0 & 1 & 0 & 0 \\ -2/3 & 0 & \sqrt{5}/3 & 0 \\ 0 & 0 & 0 & 1 \end{pmatrix}$$

$$\times \begin{pmatrix} 1 & 0 & 0 & 0 \\ 0 & 1 & 0 & 0 \\ 0 & 0 & -1 & 0 \\ 0 & 0 & 0 & 1 \end{pmatrix} \begin{pmatrix} \sqrt{5}/3 & 0 & -2/3 & 0 \\ 0 & 1 & 0 & 0 \\ 2/3 & 0 & \sqrt{5}/3 & 0 \\ 0 & 0 & 0 & 1 \end{pmatrix} \begin{pmatrix} 1 & 0 & 0 & 0 \\ 0 & 2/\sqrt{5} & 1/\sqrt{5} & 0 \\ 0 & -1/\sqrt{5} & 2/\sqrt{5} & 0 \\ 0 & 0 & 0 & 1 \end{pmatrix}$$

$$\times \begin{pmatrix} 1 & 0 & 0 & 0 \\ 0 & 1 & 0 & 0 \\ 0 & 0 & 1 & 0 \\ 1 & 0 & 0 & 1 \end{pmatrix} = \frac{1}{9} \begin{pmatrix} 1 & 4 & -8 & 0 \\ 4 & 7 & 4 & 0 \\ -8 & 4 & 1 & 0 \\ 8 & -4 & 8 & 9 \end{pmatrix}.$$

3.3 Applications

3.3.1 Computer-aided Design

In the design of parts for manufacture a common construction is that of a surface of revolution. A surface of revolution is obtained as the locus of a curve in the xz-plane which is rotated about the z-axis through 2π radians (or possibly a smaller angle). More general surfaces of revolution can be obtained by rotating a curve around an arbitrary line. As a simple illustration, consider forming a surface by rotating the curve consisting of the consecutive linear segments joining the points $(1,0,2)$, $(2,0,1)$, $(3,0,-1)$, and $(1,0,-2)$. The surface is approximated by rotating the curve through $2\pi j/n$ radians for $j = 0,\ldots,n$ to give $n+1$ instances of the curve (so that the last instance equals the first). The result is a rectangular mesh of points and these can be filled in by quadrilateral patches to give a reasonable impression of the surface. For instance, let $n = 20$ then the instances of the curve are obtained by applying the $\mathsf{Rot}_z\,(2\pi j/10)$

rotation matrix

$$\begin{pmatrix} 1 & 0 & 2 & 1 \\ 2 & 0 & 1 & 1 \\ 3 & 0 & -1 & 1 \\ 1 & 0 & -2 & 1 \end{pmatrix} \begin{pmatrix} \cos\frac{j\pi}{5} & \sin\frac{j\pi}{5} & 0 & 0 \\ -\sin\frac{j\pi}{5} & \cos\frac{j\pi}{5} & 0 & 0 \\ 0 & 0 & 1 & 0 \\ 0 & 0 & 0 & 1 \end{pmatrix}$$
$$= \begin{pmatrix} \cos\frac{j\pi}{5} & \sin\frac{j\pi}{5} & 2 & 1 \\ 2\cos\frac{j\pi}{5} & 2\sin\frac{j\pi}{5} & 1 & 1 \\ 3\cos\frac{j\pi}{5} & 3\sin\frac{j\pi}{5} & -1 & 1 \\ \cos\frac{j\pi}{5} & \sin\frac{j\pi}{5} & -2 & 1 \end{pmatrix}.$$

Evaluating the points for $j = 0, \ldots, 20$ and plotting yields an approximation to the surface of revolution illustrated in Figure 3.3. Surfaces of revolution will be developed further in Section 9.4.4.

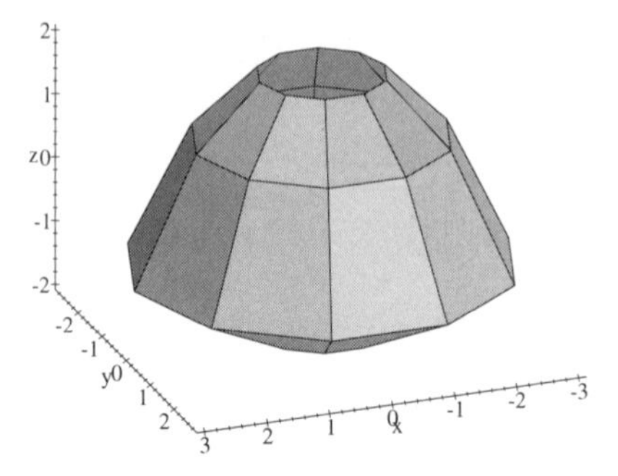

Figure 3.3 Approximate surface of revolution

3.3.2 Orientation of a Rigid Body

The *orientation* of a rigid body is determined by the angles subtended by a frame on the body relative to a fixed reference frame. A body can be positioned with any desired orientation by applying a rotation about each of the axes. For instance, $\mathsf{Rot}_x(\theta_x)\,\mathsf{Rot}_y(\theta_y)\,\mathsf{Rot}_z(\theta_z)$, which has transformation matrix

$$\begin{pmatrix} 1 & 0 & 0 & 0 \\ 0 & \cos\theta_x & \sin\theta_x & 0 \\ 0 & -\sin\theta_x & \cos\theta_x & 0 \\ 0 & 0 & 0 & 1 \end{pmatrix} \begin{pmatrix} \cos\theta_y & 0 & -\sin\theta_y & 0 \\ 0 & 1 & 0 & 0 \\ \sin\theta_y & 0 & \cos\theta_y & 0 \\ 0 & 0 & 0 & 1 \end{pmatrix}$$

$$\times \begin{pmatrix} \cos\theta_z & \sin\theta_z & 0 & 0 \\ -\sin\theta_z & \cos\theta_z & 0 & 0 \\ 0 & 0 & 1 & 0 \\ 0 & 0 & 0 & 1 \end{pmatrix} =$$

$$\begin{pmatrix} \cos\theta_y\cos\theta_z & \cos\theta_y\sin\theta_z & -\sin\theta_y & 0 \\ \begin{matrix}(\sin\theta_x\sin\theta_y\cos\theta_z \\ -\cos\theta_x\sin\theta_z)\end{matrix} & \begin{matrix}(\sin\theta_x\sin\theta_y\sin\theta_z \\ +\cos\theta_x\cos\theta_z)\end{matrix} & \sin\theta_x\cos\theta_y & 0 \\ \begin{matrix}(\cos\theta_x\sin\theta_y\cos\theta_z \\ +\sin\theta_x\sin\theta_z)\end{matrix} & \begin{matrix}(\cos\theta_x\sin\theta_y\sin\theta_z \\ -\sin\theta_x\cos\theta_z)\end{matrix} & \cos\theta_x\cos\theta_y & 0 \\ 0 & 0 & 0 & 1 \end{pmatrix}.$$

The angles $\theta_x, \theta_y, \theta_z$ are known as the *Euler angles*. Rotations about the x, y and z axes are referred to as *pitch*, *yaw* and *roll* respectively. The orientation $(\theta_x, \theta_y, \theta_z) = (\pi/6, 7\pi/6, \pi/3)$ is illustrated in Figure 3.4. The rotations can be taken in any order yielding a number of ways of expressing an orientation. When $\theta_y = \pi/2$ the above transformations simplifies to

$$\begin{pmatrix} 0 & 0 & -1 & 0 \\ \sin\theta_x\cos\theta_z - \cos\theta_x\sin\theta_z & \sin\theta_x\sin\theta_z + \cos\theta_x\cos\theta_z & 0 & 0 \\ \cos\theta_x\cos\theta_z + \sin\theta_x\sin\theta_z & \cos\theta_x\sin\theta_z - \sin\theta_x\cos\theta_z & 0 & 0 \\ 0 & 0 & 0 & 1 \end{pmatrix}$$

$$= \begin{pmatrix} 0 & 0 & -1 & 0 \\ \sin(\theta_x - \theta_z) & \cos(\theta_x - \theta_z) & 0 & 0 \\ \cos(\theta_x - \theta_z) & -\sin(\theta_x - \theta_z) & 0 & 0 \\ 0 & 0 & 0 & 1 \end{pmatrix}.$$

Therefore, the angles θ_x and θ_z are not independent. This loss of one degree of freedom is referred to as *gimbal lock* . Similar problems arise if the primary rotations are taken in a different order. Despite this deficiency the method is popular in animation due to its simplicity. The deficiency can be overcome by the use of quaternions introduced in Section 3.5.

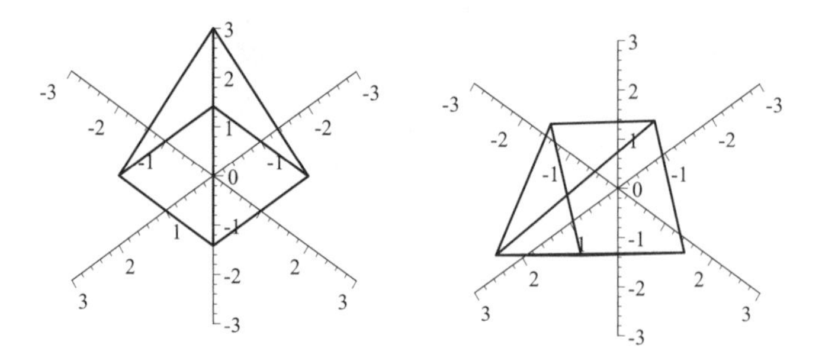

Figure 3.4

3.4 Geometric Methods for Lines and Planes in Space

In Cartesian coordinates a plane is given by an equation of the form $ax + by + cz + d = 0$. The equation in homogeneous coordinates is obtained by substituting $x = X/W$, $y = Y/W$, $z = Z/W$ in the equation and multiplying by W. Hence the homogeneous coordinates (X, Y, Z, W) of points on the plane satisfy

$$aX + bY + cZ + dW = 0.$$

In Section 2.7 lines in the plane were represented by a line vector. Likewise, planes in three-dimensional space are specified by a *plane vector* or *plane coordinates*

$$\mathbf{n} = \begin{pmatrix} a & b & c & d \end{pmatrix}.$$

The condition that a point $\mathbf{R}(X, Y, Z, W)$ lies on a plane with plane vector $\mathbf{n}$ can be expressed as a dot product $\mathbf{n} \cdot \mathbf{R} = 0$.

Plane Through Three Distinct Points

In Section 2.7 the unique line in the plane through two distinct points was obtained by performing a cross product. The analogous problem in space is to determine the unique plane $\mathbf{n}$ through three distinct points $\mathbf{P}_i(X_i, Y_i, Z_i, W_i)$ $(i = 1, 2, 3)$. The plane vector $\mathbf{n}$ satisfies

$$\mathbf{n} \cdot \mathbf{P}_1 = 0, \quad \mathbf{n} \cdot \mathbf{P}_2 = 0, \quad \mathbf{n} \cdot \mathbf{P}_3 = 0\,.$$

Thus $\mathbf{n}$ is perpendicular to three vectors $\mathbf{P}_1$, $\mathbf{P}_2$, and $\mathbf{P}_3$. The condition for this to occur is that

$$\mathbf{n} = \begin{vmatrix} \mathbf{e}_1 & \mathbf{e}_2 & \mathbf{e}_3 & \mathbf{e}_4 \\ X_1 & Y_1 & Z_1 & W_1 \\ X_2 & Y_2 & Z_2 & W_2 \\ X_3 & Y_3 & Z_3 & W_3 \end{vmatrix} \tag{3.1}$$

(or $\mathbf{n}$ is any multiple of the vector determinant (3.1)) where $\mathbf{e}_1 = (1,0,0,0)$, $\mathbf{e}_2 = (0,1,0,0)$, $\mathbf{e}_3 = (0,0,1,0)$, and $\mathbf{e}_4 = (0,0,0,1)$, the unit vectors in the coordinate directions of $\mathbb{R}^4$. The vector obtained from the determinant (3.1) is denoted by $\operatorname{orth}(\mathbf{P}_1, \mathbf{P}_2, \mathbf{P}_3)$.

Intersection of Three Planes

Analogously, the point of intersection $\mathbf{P}$ of three planes $\mathbf{n}_1$, $\mathbf{n}_2$, and $\mathbf{n}_3$ satisfies

$$\mathbf{n}_1 \cdot \mathbf{P} = 0, \quad \mathbf{n}_2 \cdot \mathbf{P} = 0, \quad \mathbf{n}_3 \cdot \mathbf{P} = 0.$$

Hence the homogeneous coordinates of the intersection are given by $\mathbf{P} = \operatorname{orth}(\mathbf{P}_1, \mathbf{P}_2, \mathbf{P}_3)$.

Example 3.6

The plane through the points $(5,4,2)$, $(-1,7,3)$, and $(2,-2,9)$ is given by the determinant

$$\begin{aligned}
\begin{vmatrix} \mathbf{e}_1 & \mathbf{e}_2 & \mathbf{e}_3 & \mathbf{e}_4 \\ 5 & 4 & 2 & 1 \\ -1 & 7 & 3 & 1 \\ 2 & -2 & 9 & 1 \end{vmatrix}
&= \mathbf{e}_1 \begin{vmatrix} 4 & 2 & 1 \\ 7 & 3 & 1 \\ -2 & 9 & 1 \end{vmatrix} - \mathbf{e}_2 \begin{vmatrix} 5 & 2 & 1 \\ -1 & 3 & 1 \\ 2 & 9 & 1 \end{vmatrix} \\
&\quad + \mathbf{e}_3 \begin{vmatrix} 5 & 4 & 1 \\ -1 & 7 & 1 \\ 2 & -2 & 1 \end{vmatrix} - \mathbf{e}_4 \begin{vmatrix} 5 & 4 & 2 \\ -1 & 7 & 3 \\ 2 & -2 & 9 \end{vmatrix} \\
&= 27\mathbf{e}_1 + 39\mathbf{e}_2 + 45\mathbf{e}_3 - 381\mathbf{e}_4 \,.
\end{aligned}$$

Thus giving the line $27x + 39y + 45z - 381 = 0$.

Example 3.7

The point of intersection of the three planes

$$3x + 5y + z = 2, \quad 7x - 4z = -1, \quad 2y + 5z + 8 = 0$$

is obtained by computing the determinant

$$\begin{vmatrix} \mathbf{e}_1 & \mathbf{e}_2 & \mathbf{e}_3 & \mathbf{e}_4 \\ 3 & 5 & 1 & -2 \\ 7 & 0 & -4 & 1 \\ 0 & 2 & 5 & 8 \end{vmatrix} = -199\mathbf{e}_1 + 237\mathbf{e}_2 - 314\mathbf{e}_3 + 137\mathbf{e}_4 \,.$$

Thus giving the point with homogeneous coordinates $(-199, 237, -314, 137)$ and Cartesian coordinates $\left(-\frac{199}{137}, \frac{237}{137}, -\frac{314}{137}\right)$.

Points and planes in three-dimensional space may be assigned unique homogeneous coordinates. A line in space may be specified by any two distinct points on the line or by two distinct planes which contain the line. To yield unique coordinates for a line, take any two points with homogeneous coordinates $\mathbf{P}(x_0, x_1, x_2, x_3)$ and $\mathbf{Q}(y_0, y_1, y_2, y_3)$ on the line and let

$$p_{ij} = x_i y_j - x_j y_i$$

for $i = 0, \ldots, 3$ and $j = 0, \ldots, 3$. Then $p = (p_{12}, p_{20}, p_{01}, p_{03}, p_{13}, p_{23})$ are uniquely defined *homogeneous line coordinates*, also known as *Plücker* or *Grassmann* coordinates. The line coordinates are independent of the choice of points. For, suppose that $\mathbf{U}(u_0, u_1, u_2, u_3) = a\mathbf{P} + b\mathbf{Q}$ and $\mathbf{V}(v_0, v_1, v_2, v_3) = c\mathbf{P} + d\mathbf{Q}$ are another choice of distinct points on the line. Then

$$\begin{aligned} u_i v_j - u_j v_i &= (ax_i + by_i)(cx_j + dy_j) - (ax_j + by_j)(cx_i + dy_i) \\ &= (ad - bc)(x_i y_j - x_j y_i) = (ad - bc)\, p_{ij} \,. \end{aligned}$$

Thus any distinct pair of points used to define the line results in a scalar multiple of p_{ij}. It follows that the line coordinates, considered as homogeneous coordinates (in a five-dimensional projective space) uniquely represent lines in three-dimensional space.

Alternatively, a line can be defined by the intersection of two planes. If the planes have plane vectors $\mathbf{L}(\ell_0, \ell_1, \ell_2, \ell_3)$ and $\mathbf{M}(m_0, m_1, m_2, m_3)$, then let

$$\rho_{ij} = \ell_i m_j - \ell_j m_i$$

for $i = 0, \ldots, 3$ and $j = 0, \ldots, 3$. Then $\rho = (\rho_{12}, \rho_{20}, \rho_{01}, \rho_{03}, \rho_{13}, \rho_{23})$ are called the *dual* line coordinates. The line and dual line coordinates are related by

$$\frac{\rho_{12}}{p_{03}} = \frac{\rho_{20}}{p_{13}} = \frac{\rho_{01}}{p_{23}} = \frac{\rho_{03}}{p_{12}} = \frac{\rho_{13}}{p_{20}} = \frac{\rho_{23}}{p_{01}} \,,$$

that is,

$$(p_{12}, p_{20}, p_{01}, p_{03}, p_{13}, p_{23}) = \mu\,(\rho_{03}, \rho_{13}, \rho_{23}, \rho_{12}, \rho_{20}, \rho_{01}) \,,$$

for some $\mu \neq 0$.

Lemma 3.8

Two lines with line coordinates $p = (p_{12}, p_{20}, p_{01}, p_{03}, p_{13}, p_{23})$ and $q = (\rho_{12}, \rho_{20}, \rho_{01}, \rho_{03}, \rho_{13}, \rho_{23})$ intersect if and only if

$$p_{12}q_{03} + p_{20}q_{13} + p_{01}q_{23} + p_{03}q_{12} + p_{13}q_{20} + p_{23}q_{01} = 0 \,.$$

Proof

If p is the line through points with homogeneous coordinates $\mathbf{P}_1(x_0, x_1, x_2, x_3)$ and $\mathbf{P}_2(y_0, y_1, y_2, y_3)$, and q is the line through points $\mathbf{Q}_1(X_0, X_1, X_2, X_3)$ and $\mathbf{Q}_2(Y_0, Y_1, Y_2, Y_3)$, then the lines have a common point if and only if $\alpha\mathbf{P}_1 + \beta\mathbf{P}_2 = \gamma\mathbf{Q}_1 + \delta\mathbf{Q}_2$ for some $\alpha, \beta, \gamma, \delta$. Thus $\mathbf{P}_1, \mathbf{P}_2, \mathbf{Q}_1, \mathbf{Q}_2$ are linearly dependent vectors implying

$$\begin{vmatrix} x_0 & x_1 & x_2 & x_3 \\ y_0 & y_1 & y_2 & y_3 \\ X_0 & X_1 & X_2 & X_3 \\ Y_0 & Y_1 & Y_2 & Y_3 \end{vmatrix} = 0 .$$

Expansion of the determinant (using Laplace's expansion is the most succinct method) gives

$$p_{12}q_{03} + p_{20}q_{13} + p_{01}q_{23} + p_{03}q_{12} + p_{13}q_{20} + p_{23}q_{01} = 0 .$$

□

EXERCISES

3.1. Determine the Cartesian coordinates of the following points $(3, 6, 5, 2)$, $(2, 4, 6, 4)$, $(0, 0, 2, 1)$, $(2, 0, 0, 2)$.

3.2. Determine the point at infinity in the directions of the following vectors $(3, 4, 1)$ and $(7, 2, 0)$.

3.3. Determine the homogeneous transformation matrices for the following.

(a) A rotation about the z-axis through an angle of $\pi/4$.

(b) A scaling by a factor of 3 units in the y-direction, followed by a translation of 2 units in the x-direction and 5 units in the z-direction, followed by a rotation about the x-axis through an angle $7\pi/6$.

(c) A reflection in the plane $6x - 6y + 3z - 5 = 0$.

(d) A rotation about the line through the points $(2, 1, 2)$ and $(8, 3, 5)$ through an angle $5\pi/6$.

3.4. Determine the Cartesian coordinates of the point of intersection of the three planes $2x - y + z = 0$, $-4x + 3y - 2z - 5 = 0$, and $x + y - 6 = 0$.

3.5. Determine the plane through $(1, 1, -1)$, $(-9, 7, 3)$, and $(2, 0, 5)$.

3.6. Show that every point on the line through the points with homogeneous coordinates $\mathbf{P}$ and $\mathbf{Q}$ has homogeneous coordinates of the form $\alpha\mathbf{P} + \beta\mathbf{Q}$.

3.7. Prove that three points $\mathbf{P}, \mathbf{Q}$, and $\mathbf{R}$ are collinear if and only if $\mathrm{orth}(\mathbf{P}, \mathbf{Q}, \mathbf{R}) = 0$. What does $\mathrm{orth}(\mathbf{n}_1, \mathbf{n}_2, \mathbf{n}_3) = 0$ imply about three planes $\mathbf{n}_1$, $\mathbf{n}_2$, and $\mathbf{n}_3$?

3.8. (a) What does $\mathrm{orth}(\mathbf{P}, \mathbf{Q}, \mathbf{R}) \cdot \mathbf{e}_4 = 0$ imply about points $\mathbf{P}$, $\mathbf{Q}$, $\mathbf{R}$?

(b) What does $\mathrm{orth}(\mathbf{n}_1, \mathbf{n}_2, \mathbf{n}_3) \cdot \mathbf{e}_4 = 0$ say about the lines $\mathbf{n}_1$, $\mathbf{n}_2$, $\mathbf{n}_3$?

3.9. Let $\mathbf{P}(x_0, x_1, x_2, 1)$ and $\mathbf{Q}(y_0, y_1, y_2, 1)$ and let $p_{ij} = x_i y_j - x_j y_i$. Show that $\omega = (p_{03}, p_{13}, p_{23})$ and $\mathbf{v} = (p_{12}, p_{20}, p_{01})$ and that the Cartesian coordinates of $\mathbf{P}$ and $\mathbf{Q}$ satisfy $\omega = \mathbf{P} - \mathbf{Q}$, and $\mathbf{v} = \mathbf{P} \times \mathbf{Q}$. Deduce that ω determines the direction of the line, and $\mathbf{v}$ is normal to the plane containing the line and the origin.

3.10. Show that $\omega \cdot \mathbf{v} = 0$, and hence deduce that

$$(p_{03}, p_{13}, p_{23}) \cdot (p_{12}, p_{20}, p_{01}) = p_{03}p_{12} + p_{13}p_{20} + p_{23}p_{01} = 0 \,. \quad (3.2)$$

Thus $p = (p_{12}, p_{20}, p_{01}, p_{03}, p_{13}, p_{23})$ corresponds to a line in three-dimensional space if and only if p lies on the quadric defined by Equation (3.2).

3.11. Determine the matrix for a rotation through an angle θ about an axis that passes through the origin and has direction given by the unit vector (r_1, r_2, r_3).

3.5 Quaternions

Quaternions provide an alternative to matrices as a way of representing orientations in three-dimensional space. They are used to apply rotations to objects in computer graphics animation. Discovered by William Hamilton in 1843, quaternions may be considered a generalisation of complex numbers. Complex numbers are represented in the form $a + b\mathbf{i}$ where $\mathbf{i}^2 = -1$, and a and b are real numbers. The operations of addition and multiplication are defined as

$$\begin{aligned}
(a_1 + b_1\mathbf{i}) + (a_2 + b_2\mathbf{i}) &= (a_1 + a_2) + (b_1 + b_2)\mathbf{i}\,, \quad \text{and} \\
(a_1 + b_1\mathbf{i})(a_2 + b_2\mathbf{i}) &= a_1a_2 + a_1b_2\mathbf{i} + b_1a_2\mathbf{i} + b_1b_2\mathbf{i}^2 \\
&= (a_1a_2 - b_1b_2) + (a_1b_2 + b_1a_2)\mathbf{i}\,.
\end{aligned}$$

A quaternion is an extended complex number $s+x\mathbf{i}+y\mathbf{j}+z\mathbf{k}$, where s, x, y, z are real numbers, and $\mathbf{i}^2 = -1$, $\mathbf{j}^2 = -1$, $\mathbf{k}^2 = -1$, and $\mathbf{ijk} = -1$. These identities imply a further six identities: $\mathbf{ij} = -\mathbf{ji} = \mathbf{k}$, $\mathbf{jk} = -\mathbf{kj} = \mathbf{i}$, and $\mathbf{ik} = -\mathbf{ki} = -\mathbf{j}$. For instance, $\mathbf{ijk} = -1$ implies that $\mathbf{ijkk} = -\mathbf{k}$, and since $\mathbf{k}^2 = -1$ it follows that $\mathbf{ij} = \mathbf{k}$. The remaining identities are left as an exercise (Exercise 3.14). Addition of quaternions is similar to that of complex numbers:

$$\begin{aligned}&(s_1 + x_1\mathbf{i} + y_1\mathbf{j} + z_1\mathbf{k}) + (s_2 + x_2\mathbf{i} + y_2\mathbf{j} + z_2\mathbf{k})\\ &\quad = (s_1 + s_2) + (x_1 + x_2)\mathbf{i} + (y_1 + y_2)\mathbf{j} + (z_1 + z_2)\mathbf{k}\,.\end{aligned}$$

Multiplication is given by

$$\begin{aligned}&(s_1 + x_1\mathbf{i} + y_1\mathbf{j} + z_1\mathbf{k})(s_2 + x_2\mathbf{i} + y_2\mathbf{j} + z_2\mathbf{k})\\
&= s_1s_2 + s_1x_2\mathbf{i} + s_1y_2\mathbf{j} + s_1z_2\mathbf{k} + s_2x_1\mathbf{i} + x_1x_2\mathbf{i}^2 + x_1y_2\mathbf{ij} + x_1z_2\mathbf{ik}\\
&\quad + s_2y_1\mathbf{j} + x_2y_1\mathbf{ji} + y_1y_2\mathbf{j}^2 + y_1z_2\mathbf{jk} + s_2z_1\mathbf{k} + x_2z_1\mathbf{ki} + y_2z_1\mathbf{kj} + z_1z_2\mathbf{k}^2\\
&= s_1s_2 + s_1x_2\mathbf{i} + s_1y_2\mathbf{j} + s_1z_2\mathbf{k} + s_2x_1\mathbf{i} - x_1x_2 + x_1y_2\mathbf{k} - x_1z_2\mathbf{j}\\
&\quad + s_2y_1\mathbf{j} - x_2y_1\mathbf{k} - y_1y_2 + y_1z_2\mathbf{i} + s_2z_1\mathbf{k} + x_2z_1\mathbf{j} - y_2z_1\mathbf{i} - z_1z_2\\
&= (s_1s_2 - x_1x_2 - y_1y_2 - z_1z_2) + (s_1x_2 + s_2x_1 + y_1z_2 - y_2z_1)\mathbf{i}\\
&\quad + (s_1y_2 + s_2y_1 - x_1z_2 + x_2z_1)\mathbf{j} + (s_1z_2 + s_2z_1 + x_1y_2 - x_2y_1)\mathbf{k}\,.\end{aligned}$$

Alternatively, a quaternion may be written in the form $(s, \mathbf{v})$ where $\mathbf{v} = (x, y, z)$. The operations of addition and multiplication are

$$\begin{aligned}&(s_1, \mathbf{v_1}) + (s_2, \mathbf{v_2}) = (s_1 + s_2, \mathbf{v}_1 + \mathbf{v}_2)\,, \quad \text{and}\\
&(s_1, \mathbf{v_1})(s_2, \mathbf{v_2}) = (s_1s_2 - \mathbf{v_1}\cdot\mathbf{v_2}, s_1\mathbf{v}_2 + s_2\mathbf{v}_1 + (\mathbf{v_1}\times\mathbf{v_2}))\,.\end{aligned}$$

Since $\mathbf{v_1}\times\mathbf{v_2} = -\mathbf{v_2}\times\mathbf{v_1}$, the multiplication of quaternions is *non-commutative*: in general, $(s_1, \mathbf{v_1})(s_2, \mathbf{v_2}) \neq (s_2, \mathbf{v_2})(s_1, \mathbf{v_1})$.

Example 3.9

(a)

$$\begin{aligned}(3 + 5\mathbf{i} - 2\mathbf{j} + 7\mathbf{j}) + (1 - 4\mathbf{j} - 3\mathbf{k}) &= (3 + 1) + 5\mathbf{i} + (-2 - 4)\mathbf{j} + (7 - 3)\mathbf{k}\\ &= 4 + 5\mathbf{i} - 6\mathbf{j} + 4\mathbf{k}\,.\end{aligned}$$

(b)

$$\begin{aligned}&(2 + 3\mathbf{i} + 5\mathbf{k})(2 - 5\mathbf{j} + 2\mathbf{k})\\
&= 4 - 10\mathbf{j} + 4\mathbf{k} + 6\mathbf{i} - 15\mathbf{ij} + 6\mathbf{ik} + 10\mathbf{k} - 25\mathbf{kj} + 10\mathbf{k}^2\\
&= 4 - 10\mathbf{j} + 4\mathbf{k} + 6\mathbf{i} - 15\mathbf{k} - 6\mathbf{j} + 10\mathbf{k} + 25\mathbf{i} - 10\\
&= -6 + 31\mathbf{i} - 16\mathbf{j} - \mathbf{k}\,.\end{aligned}$$

(c)

$$\begin{aligned}
&(-3,(4,1,2))(2,(-1,0,3)) \\
&\quad = (-6-((4,1,2)\cdot(-1,0,3)), \\
&\qquad -3(-1,0,3)+2(4,1,2)+((4,1,2)\times(-1,0,3))) \\
&\quad = (-6-2,(3,0,-9)+(8,2,4)+(3,-14,1)) \\
&\quad = (-8,(14,-12,-4)) .
\end{aligned}$$

Quaternions of the form $(s, \mathbf{0})$ are identified with real numbers s and it common to write the quaternion as s. In particular, $(0, \mathbf{0})$ is denoted 0, and $(1, \mathbf{0})$ is denoted 1. Quaternions of the form $(0, \mathbf{v})$ are called *pure imaginary* quaternions and are identified with three-dimensional vectors $\mathbf{v}$.

The following algebraic properties are satisfied by all quaternions $\mathbf{p} = (s_1, \mathbf{v}_1)$, $\mathbf{q} = (s_2, \mathbf{v}_2)$ and $\mathbf{r} = (s_3, \mathbf{v}_3)$.

Additive identity: $\mathbf{p} + 0 = 0 + \mathbf{p}$.

Multiplicative identity: $1\mathbf{p} = \mathbf{p}1 = \mathbf{p}$.

Commutative addition: $\mathbf{p} + \mathbf{q} = \mathbf{q} + \mathbf{p}$.

Associative addition: $(\mathbf{p} + \mathbf{q}) + \mathbf{r} = \mathbf{p} + (\mathbf{q} + \mathbf{r})$.

Associative multiplication: $(\mathbf{pq})\mathbf{r} = \mathbf{p}(\mathbf{qr})$.

Distributive: $\mathbf{p}(\mathbf{q} + \mathbf{r}) = \mathbf{pq} + \mathbf{pr}$ and $(\mathbf{p} + \mathbf{q})\mathbf{r} = \mathbf{pr} + \mathbf{qr}$.

No zero divisors: If $\mathbf{pq} = 0$ then $\mathbf{p} = 0$ or $\mathbf{q} = 0$.

Most of the properties can be obtained directly from the definitions of addition and multiplication (Exercise 3.17).

The property of no zero divisors is proved as follows. Suppose $\mathbf{pq} = 0$. Then $(s_1 s_2 - \mathbf{v}_1 \cdot \mathbf{v}_2, s_1 \mathbf{v}_2 + s_2 \mathbf{v}_1 + (\mathbf{v}_1 \times \mathbf{v}_2)) = 0$ which implies that $\mathbf{v}_1 \cdot \mathbf{v}_2 = s_1 s_2$ and

$$s_1 \mathbf{v}_2 + s_2 \mathbf{v}_1 + (\mathbf{v}_1 \times \mathbf{v}_2) = \mathbf{0} . \tag{3.3}$$

Applying the dot product of $\mathbf{v}_1$ to both sides of Equation (3.3) gives

$$s_1 \mathbf{v}_2 \cdot \mathbf{v}_1 + s_2 \mathbf{v}_1 \cdot \mathbf{v}_1 + (\mathbf{v}_1 \times \mathbf{v}_2) \cdot \mathbf{v}_1 = \mathbf{0} .$$

Then, since $\mathbf{v}_1 \cdot \mathbf{v}_2 = s_1 s_2$ and $(\mathbf{v}_1 \times \mathbf{v}_2) \cdot \mathbf{v}_1 = 0$, it follows that

$$s_1^2 s_2 + s_2 \mathbf{v}_1 \cdot \mathbf{v}_1 = s_2(s_1^2 + \mathbf{v}_1 \cdot \mathbf{v}_1) = \mathbf{0} .$$

Similarly, the following condition is also satisfied:

$$s_1(s_2^2 + \mathbf{v}_2 \cdot \mathbf{v}_2) = \mathbf{0} .$$

There are three cases to consider: (i) $(s_1^2 + \mathbf{v}_1 \cdot \mathbf{v}_1) = \mathbf{0}$, (ii) $(s_2^2 + \mathbf{v}_2 \cdot \mathbf{v}_2) = \mathbf{0}$, and (iii) $s_1 = s_2 = 0$. When $s_1^2 + \mathbf{v}_1 \cdot \mathbf{v}_1 = 0$, then $s_1 = 0$ and $\mathbf{v}_1 = \mathbf{0}$, and therefore $\mathbf{p} = 0$. Likewise, when $s_2^2 + \mathbf{v}_2 \cdot \mathbf{v}_2 = 0$, then $\mathbf{q} = 0$. Finally, when $s_1 = s_2 = 0$, then $-\mathbf{v}_1 \cdot \mathbf{v}_2 = \mathbf{v}_1 \times \mathbf{v}_2 = 0$. Therefore, either $\mathbf{v}_1 = \mathbf{0}$ and hence $\mathbf{p} = 0$, or $\mathbf{v}_2 = \mathbf{0}$ and hence $\mathbf{q} = 0$.

Let $\mathbf{q} = (s, \mathbf{v}) = s + x\mathbf{i} + y\mathbf{j} + z\mathbf{k}$ be any quaternion, then the *conjugate* quaternion, denoted $\overline{\mathbf{q}}$, is defined to be $(s, -\mathbf{v}) = s - x\mathbf{i} - y\mathbf{j} - z\mathbf{k}$. Then

$$\begin{aligned} \mathbf{q}\overline{\mathbf{q}} &= (s^2 + \mathbf{v} \cdot \mathbf{v}, -s\mathbf{v} + s\mathbf{v} - (\mathbf{v} \times \mathbf{v})) \\ &= (s^2 + \mathbf{v} \cdot \mathbf{v}, \mathbf{0}) = (s^2 + |\mathbf{v}|^2, \mathbf{0}) \\ &= s^2 + |\mathbf{v}|^2 . \end{aligned}$$

The *modulus* of $\mathbf{q}$, denoted $|\mathbf{q}|$, is defined to be

$$|\mathbf{q}| = (\mathbf{q}\overline{\mathbf{q}})^{1/2} = (s^2 + |\mathbf{v}|^2)^{1/2} .$$

A quaternion $\mathbf{q}$ satisfying $|\mathbf{q}| = 1$ is said to be a *unit* quaternion. Every non-zero quaternion $\mathbf{q}$ has a multiplicative *inverse* quaternion, denoted $\mathbf{q}^{-1}$, satisfying $\mathbf{q}\mathbf{q}^{-1} = \mathbf{q}^{-1}\mathbf{q} = 1$ (see Exercise 3.15). The inverse is

$$\mathbf{q}^{-1} = \frac{\overline{\mathbf{q}}}{|\mathbf{q}|^2} . \qquad (3.4)$$

Readers with a knowledge of algebraic structures may conclude that the algebraic properties described earlier, together with the existence of additive and multiplicative inverses (Exercises 3.15 and 3.16), imply that the quaternions are a *non-commutative division ring.*

Example 3.10

Let $\mathbf{q} = (2, (-1, 0, 3))$. Then $\overline{\mathbf{q}} = (2, (1, 0, -3))$, and $|\mathbf{q}| = (2^2 + (-1, 0, 3) \cdot (-1, 0, 3))^{1/2} = \sqrt{14}$. Hence $\mathbf{q}^{-1} = \overline{\mathbf{q}}/|\mathbf{q}|^2 = \frac{1}{14}(2, (1, 0, -3)) = (\frac{1}{7}, (\frac{1}{14}, 0, -\frac{3}{14}))$.

EXERCISES

3.12. Determine the following sums and products of quaternions

(a) $(7 + 3\mathbf{i} + 5\mathbf{j} - 3\mathbf{k}) + (-2 + 3\mathbf{i} + 6\mathbf{j} - 4\mathbf{k})$,

(b) $(9, (2, -1, 3)) + (-7, (1, 0, -2))$,

(c) $(2 + 4\mathbf{i} - 9\mathbf{j} + 5\mathbf{k})(5 + 3\mathbf{i} - 2\mathbf{k})$,

(d) $(-2, (3, 2, -5))(7, (0, 1, 4))$, and

(e) $(7,(0,1,4))(-2,(3,2,-5))$.

3.13. Find the (multiplicative) inverses of (a) $(2,(5,-3,4))$, and (b) $(-3,(4,0,-1))$.

3.14. Show that $\mathbf{ij} = -\mathbf{ji} = \mathbf{k}$, $\mathbf{jk} = -\mathbf{kj} = \mathbf{i}$, and $\mathbf{ik} = -\mathbf{ki} = -\mathbf{j}$.

3.15. Show that the (multiplicative) inverse quaternion given by (3.4) satisfies $\mathbf{qq}^{-1} = \mathbf{q}^{-1}\mathbf{q} = 1$.

3.16. Show that every quaternion $\mathbf{q}$ has an additive inverse, denoted $-\mathbf{q}$ satisfying $\mathbf{q} + (-\mathbf{q}) = (-\mathbf{q}) + \mathbf{q} = 0$.

3.17. Use the definitions of addition and multiplication to prove the algebraic properties of quaternions given on page 58.

3.18. Show that (a) $\overline{\mathbf{q_1 q_2}} = \overline{\mathbf{q_2}}\,\overline{\mathbf{q_1}}$, and (b) $|\mathbf{q_1 q_2}| = |\mathbf{q_1}||\mathbf{q_2}|$.

3.19. Let $\mathbf{I} = (0, \mathbf{v})$. Show that $\mathbf{I}^2 = -|\mathbf{v}|^2$.

3.20. Write a computer program to perform quaternion addition, multiplication, conjugation, and to find multiplicative inverses. (Alternatively, do the same using a computer algebra package.)

Lemma 3.11

Any unit quaternion $\mathbf{q} = (s, \mathbf{v})$ has the form

$$\mathbf{q} = (\cos\theta, \sin\theta\mathbf{I})$$

for some angle θ and unit vector $\mathbf{I}$.

Proof

Since $|\mathbf{q}|^2 = s^2 + |\mathbf{v}|^2 = 1$, it follows that $-1 \le s \le 1$. So $s = \cos\theta$ for some $0 \le \theta \le \pi$. Then

$$|\mathbf{v}|^2 = 1 - s^2 = 1 - \cos^2\theta = \sin^2\theta\ .$$

Therefore,

$$\mathbf{v} = |\mathbf{v}|\frac{\mathbf{v}}{|\mathbf{v}|} = \sin\theta\mathbf{I}\ ,$$

where $\mathbf{I} = \mathbf{v}/|\mathbf{v}|$, and so $\mathbf{q} = (\cos\theta, \sin\theta\mathbf{I})$. □

Let $e^{\theta\mathbf{I}}$ denote $(\cos\theta, \sin\theta\mathbf{I})$. A modification to the proof of Lemma 3.11 yields that any quaternion $\mathbf{q}$ can be expressed in the *polar form* $\mathbf{q} = re^{\theta\mathbf{I}} = r(\cos\theta, \sin\theta\mathbf{I})$. The set of quaternions of the form $(a, b\mathbf{I})$, for some fixed unit vector $\mathbf{I}$, has very similar properties to the complex numbers $a + b\mathbf{i}$. Indeed, there is a result for quaternions corresponding to the Theorem of de Moivres for complex numbers:

$$(r(\cos\theta, \sin\theta\mathbf{I}))^n = r^n(\cos n\theta, \sin n\theta\mathbf{I}) . \qquad (3.5)$$

Example 3.12

To express $\mathbf{q} = (4, (1, 2, -2))$ in polar form, let $r = |\mathbf{q}| = 5$. Then, $\frac{1}{5}\mathbf{q} = \left(\frac{4}{5}, \left(\frac{1}{5}, \frac{2}{5}, -\frac{2}{5}\right)\right)$ is a unit quaternion. Following the proof of Lemma 3.11 gives $\cos\theta = \frac{4}{5}$, $\sin\theta = \frac{3}{5}$, and $\theta = 0.6435$ radians. Further, $|(1, 2, -2)| = 3$ and so $\mathbf{I} = \frac{1}{3}(1, 2, -2) = \left(\frac{1}{3}, \frac{2}{3}, -\frac{2}{3}\right)$. Hence $\mathbf{q} = 5\left(\cos\theta + \sin\theta\mathbf{I}\right)$.

Let $\mathbf{q} = s + x\mathbf{i} + y\mathbf{j} + z\mathbf{k}$ and $\mathbf{p} = w + p_1\mathbf{i} + p_2\mathbf{j} + p_3\mathbf{k}$. The left and right quaternion multiplications $\mathbf{qp}$ and $\mathbf{pq}$ can be written as the matrix multiplications

$$\mathbf{qp} = \mathbf{pL_q} = \begin{pmatrix} w & p_1 & p_2 & p_3 \end{pmatrix} \begin{pmatrix} s & x & y & z \\ -x & s & z & -y \\ -y & -z & s & x \\ -z & y & -x & s \end{pmatrix}, \quad \text{and}$$

$$\mathbf{pq} = \mathbf{pR_q} = \begin{pmatrix} w & p_1 & p_2 & p_3 \end{pmatrix} \begin{pmatrix} s & x & y & z \\ -x & s & -z & y \\ -y & z & s & -x \\ -z & -y & x & s \end{pmatrix}.$$

Suppose that $\mathbf{q}$ is a non-zero quaternion, and let $C_\mathbf{q}(\mathbf{p}) = \mathbf{qpq}^{-1}$. (Note that when $\mathbf{q}$ is a unit quaternion $C_\mathbf{q}(\mathbf{p}) = \mathbf{qp\overline{q}}$.) Then

$$C_\mathbf{q}(\mathbf{p}) = (\mathbf{qp})\,\mathbf{q}^{-1} = (\mathbf{pL_q})\,\mathbf{q}^{-1} = (\mathbf{pL_q})\,\mathbf{R}_{\mathbf{q}^{-1}} = \mathbf{p}\left(\mathbf{L_q}\mathbf{R}_{\mathbf{q}^{-1}}\right),$$

or, alternatively,

$$C_\mathbf{q}(\mathbf{p}) = \mathbf{q}\left(\mathbf{pq}^{-1}\right) = \mathbf{q}\left(\mathbf{pR}_{\mathbf{q}^{-1}}\right) = \left(\mathbf{pR}_{\mathbf{q}^{-1}}\right)\mathbf{L_q} = \mathbf{p}\left(\mathbf{R}_{\mathbf{q}^{-1}}\mathbf{L_q}\right).$$

Then

$$\mathsf{C}_\mathbf{q} = \mathbf{L_q}\mathbf{R}_{\mathbf{q}^{-1}} = \mathbf{R}_{\mathbf{q}^{-1}}\mathbf{L_q} =$$

$$\begin{pmatrix} s^2+x^2+y^2+z^2 & 0 & 0 & 0 \\ 0 & s^2+x^2-y^2-z^2 & 2xy+2sz & 2xz-2sy \\ 0 & 2xy-2sz & s^2-x^2+y^2-z^2 & 2yz+2sx \\ 0 & 2xz+2sy & 2yz-2sx & s^2-x^2-y^2+z^2 \end{pmatrix}.$$

Now suppose that $\mathbf{q} = (\cos\frac{\theta}{2}, \sin\frac{\theta}{2}\mathbf{I})$ where $\mathbf{I} = (r_1, r_2, r_3)$. Then, $s = \cos\frac{\theta}{2}$, $x = r_1\sin\frac{\theta}{2}$, $y = r_2\sin\frac{\theta}{2}$ and $z = r_3\sin\frac{\theta}{2}$, and substitution into $\mathsf{C}_\mathbf{q}$ yields (after some algebraic manipulation and row and column swapping) the matrix of Exercise 3.11. Thus $C_{\mathbf{q}}$ is the matrix for a rotation about $\mathbf{I}$ through an angle θ. If the point with homogeneous coordinates $\mathbf{p} = (p_1, p_2, p_3, w)$ is identified with the quaternion $\mathbf{p} = w + p_1\mathbf{i} + p_2\mathbf{j} + p_3\mathbf{k}$, then $C_{\mathbf{q}}(\mathbf{p}) = \mathbf{q}\mathbf{p}\mathbf{q}^{-1}$ yields the rotation of $\mathbf{p}$ about $\mathbf{I}$, and thus proving the following key theorem linking quaternions to rotations. (The converse, that any rotation is given by $C_{\mathbf{q}}$ for some unit quaternion $\mathbf{q}$, is left as an exercise.)

Theorem 3.13

Let $\mathbf{q} = (\cos\frac{\theta}{2}, \sin\frac{\theta}{2}\mathbf{I}))$ be a unit quaternion, and $\mathbf{p}$ any quaternion. Then $C_{\mathbf{q}}(\mathbf{p}) = \mathbf{q}\mathbf{p}\mathbf{q}^{-1}$ yields a rotation of $\mathbf{p}$ about the axis $\mathbf{I}$ through an angle θ. Conversely, any rotation is given by $C_{\mathbf{q}}$ for some unit quaternion $\mathbf{q}$.

The next lemma provides an alternative way of computing $C_{\mathbf{q}}(\mathbf{p})$.

Lemma 3.14

Let $\mathbf{q} = (s, \mathbf{v})$ and $\mathbf{p} = (w, \mathbf{x})$. Then

$$C_{\mathbf{q}}(\mathbf{p}) = \mathbf{q}\mathbf{p}\mathbf{q}^{-1} = \left(w, \frac{1}{|\mathbf{q}|^2}\left((s^2 - \mathbf{v}\cdot\mathbf{v})\mathbf{x} + 2(\mathbf{x}\cdot\mathbf{v})\mathbf{v} - 2s(\mathbf{x}\times\mathbf{v})\right)\right).$$

Proof

$$\begin{aligned}\mathbf{q}\mathbf{p}\mathbf{q}^{-1} &= \mathbf{q}\left((w,\mathbf{0}) + (0,\mathbf{x})\right)\mathbf{q}^{-1} \\ &= \mathbf{q}(w,\mathbf{0})\mathbf{q}^{-1} + \mathbf{q}(0,\mathbf{x})\mathbf{q}^{-1}.\end{aligned} \tag{3.6}$$

But

$$\mathbf{q}(w,\mathbf{0})\mathbf{q}^{-1} = (w,\mathbf{0})\mathbf{q}\mathbf{q}^{-1} = (w,\mathbf{0}), \tag{3.7}$$

and

$$\begin{aligned}\mathbf{q}(0,\mathbf{x})\mathbf{q}^{-1} &= \frac{1}{|\mathbf{q}|^2}(s,\mathbf{v})(0,\mathbf{x})(s,-\mathbf{v}) = \frac{1}{|\mathbf{q}|^2}(s,\mathbf{v})(\mathbf{x}\cdot\mathbf{v}, s\mathbf{x} - (\mathbf{x}\times\mathbf{v})) \\ &= \frac{1}{|\mathbf{q}|^2}\,(s(\mathbf{x}\cdot\mathbf{v}) - \mathbf{v}\cdot(s\mathbf{x} - (\mathbf{x}\times\mathbf{v})), \\ &\qquad s^2\mathbf{x} - s(\mathbf{x}\times\mathbf{v}) + (\mathbf{x}\cdot\mathbf{v})\mathbf{v} + s(\mathbf{v}\times\mathbf{x}) - \mathbf{v}\times(\mathbf{x}\times\mathbf{v})) \\ &= \left(0, \frac{1}{|\mathbf{q}|^2}\left((s^2 - \mathbf{v}\cdot\mathbf{v})\mathbf{x} + 2(\mathbf{x}\cdot\mathbf{v})\mathbf{v} - 2s(\mathbf{x}\times\mathbf{v})\right)\right)\end{aligned} \tag{3.8}$$

using the vector identity $\mathbf{a}\times(\mathbf{b}\times\mathbf{c}) = (\mathbf{a}\cdot\mathbf{c})\mathbf{b} - (\mathbf{a}\cdot\mathbf{b})\mathbf{c}$. Then (3.6), (3.7) and (3.8) give

$$\mathbf{q}\mathbf{p}\mathbf{q}^{-1} = \left(w, \frac{1}{|\mathbf{q}|^2}\left((s^2 - \mathbf{v}\cdot\mathbf{v})\mathbf{x} + 2(\mathbf{x}\cdot\mathbf{v})\mathbf{v} - 2s(\mathbf{x}\times\mathbf{v})\right)\right).$$

□

Note that, since the w-coordinate remains unchanged in the calculation of $C_{\mathbf{q}}(\mathbf{p})$, it is common practice to identify the point with affine coordinates (p_1, p_2, p_3) with the pure imaginary quaternion $(0, (p_1, p_2, p_3))$ rather than with $(1, (p_1, p_2, p_3))$.

Example 3.15

A rotation about an axis with direction $(-4, 2, 4)$ through an angle $\pi/3$ is obtained as follows. Normalize $(-4, 2, 4)$ to give $\mathbf{I} = (-2/3, 1/3, 2/3)$. Then

$$\mathbf{q} = \left(\cos(\pi/3), \sin(\pi/3)\left(-\frac{2}{3}, \frac{1}{3}, \frac{2}{3}\right)\right) = \left(\frac{1}{2}, \left(-\frac{\sqrt{3}}{3}, \frac{\sqrt{3}}{6}, \frac{\sqrt{3}}{3}\right)\right).$$

The rotation is applied to the point $(3, 6, -5)$, say, by letting $\mathbf{p} = (0, (3, 6, -5))$ (or $\mathbf{p} = (1, (3, 6, -5)))$ and computing

$$\begin{aligned}
C_{\mathbf{q}}(\mathbf{p}) &= \mathbf{q}\mathbf{p}\mathbf{q}^{-1} \\
&= \left(\frac{1}{2}, \left(-\frac{\sqrt{3}}{3}, \frac{\sqrt{3}}{6}, \frac{\sqrt{3}}{3}\right)\right)(0, (3, 6, -5))\left(\frac{1}{2}, \left(\frac{\sqrt{3}}{3}, -\frac{\sqrt{3}}{6}, -\frac{\sqrt{3}}{3}\right)\right) \\
&= \left(\frac{5\sqrt{3}}{3}, \left(\frac{3}{2} - \frac{17\sqrt{3}}{6}, 3 - \frac{2\sqrt{3}}{3}, -\frac{5}{2} - \frac{5\sqrt{3}}{2}\right)\right)\left(\frac{1}{2}, \left(\frac{\sqrt{3}}{3}, -\frac{\sqrt{3}}{6}, -\frac{\sqrt{3}}{3}\right)\right) \\
&= \left(0, \left(\frac{11}{6} - \frac{17\sqrt{3}}{6}, -\frac{14}{3} - \frac{2\sqrt{3}}{3}, -\frac{5}{6} - \frac{5\sqrt{3}}{2}\right)\right).
\end{aligned}$$

The rotated point is $\left(\frac{11}{6} - \frac{17\sqrt{3}}{6}, -\frac{14}{3} - \frac{2\sqrt{3}}{3}, -\frac{5}{6} - \frac{5\sqrt{3}}{2}\right)$ which is approximately $(-3.074, -5.821, -5.163)$.

EXERCISES

3.21. Find the polar forms for (a) $(2, (1, 2, 4))$ and (b) $(5, (-2, 2, 4))$.

3.22. Show that two quaternions $\mathbf{q}_1 = r_1 e^{\theta_1 \mathbf{I}}$ and $\mathbf{q}_2 = r_2 e^{\theta_2 \mathbf{I}}$ satisfy (a) $\mathbf{q}_1\mathbf{q}_2 = r_1 r_2 e^{(\theta_1+\theta_2)\mathbf{I}}$, and (b) $\mathbf{q}_1{}^{-1} = r_1 e^{-\theta_1 \mathbf{I}}$.

3.23. Show that $C_{\mathbf{q}_1}$ and $C_{\mathbf{q}_2}$ yield the same rotation if and only if $\mathbf{q}_1 = \lambda \mathbf{q}_2$, for some real number $\lambda \neq 0$.

3.24. Show that the rotation $C_{\mathbf{q}_1\mathbf{q}_2}$ gives the rotation $C_{\mathbf{q}_1}$ followed by the rotation $C_{\mathbf{q}_2}$.

3.25. Determine the quaternion $\mathbf{q}$ that represents a rotation about the axis $(-1, 2, 2)$ through an angle $\pi/4$. Apply the rotation to the point $(5, 6, 7)$.

3.26. Show that, for any quaternion $\mathbf{q}$, $|C_{\mathbf{q}}(\mathbf{x})| = |\mathbf{x}|$. ($C_{\mathbf{q}}$ is said to be an *isometry*.)

3.27. Prove the converse to Theorem 3.13, that any three-dimensional rotation is given by $C_{\mathbf{q}}$ for some unit quaternion $\mathbf{q}$

3.28. Write a computer program (or use a computer algebra package) to perform rotations using the operation $C_{\mathbf{q}}$.

Example 3.16

The rotations about the coordinate axis are $\mathsf{Rot}_x(\theta_x) = \left(\cos\frac{1}{2}\theta_x, \left(\sin\frac{1}{2}\theta_x, 0, 0\right)\right)$, $\mathsf{Rot}_y(\theta_y) = \left(\cos\frac{1}{2}\theta_y, \left(0, \sin\frac{1}{2}\theta_y, 0\right)\right)$, and $\mathsf{Rot}_z(\theta_z) = \left(\cos\frac{1}{2}\theta_z, \left(0, 0, \sin\frac{1}{2}\theta_z\right)\right)$. Then the representation of orientation given by Euler angles $(\theta_x, \theta_y, \theta_z)$ (described in Section 3.3.2) has the form $\mathbf{q}\mathbf{p}\mathbf{q}^{-1}$ where

$$\mathbf{q} = \mathsf{Rot}_z(\theta_z)\mathsf{Rot}_y(\theta_y)\mathsf{Rot}_x(\theta_x)\ .$$

Suppose $\theta_y = \pi/2$. Then

$$\begin{aligned}
\mathbf{q} &= \mathsf{Rot}_z(\theta_z)\mathsf{Rot}_y(\pi/2)\mathsf{Rot}_x(\theta_x) \\
&= \left(\cos\frac{\theta_z}{2}, \left(0, 0, \sin\frac{\theta_z}{2}\right)\right)\left(\frac{1}{\sqrt{2}}, \left(0, \frac{1}{\sqrt{2}}, 0\right)\right)\left(\cos\frac{\theta_x}{2}, \left(\sin\frac{\theta_x}{2}, 0, 0\right)\right) \\
&= \frac{1}{\sqrt{2}}\left(\cos\frac{\theta_z}{2}, \left(0, 0, \sin\frac{\theta_z}{2}\right)\right)\left(\cos\frac{\theta_x}{2}, \left(\sin\frac{\theta_x}{2}, \cos\frac{\theta_x}{2}, -\sin\frac{\theta_x}{2}\right)\right) \\
&= \frac{1}{\sqrt{2}}\left(\cos\frac{\theta_x}{2}\cos\frac{\theta_z}{2} + \sin\frac{\theta_x}{2}\sin\frac{\theta_z}{2}, \left(\sin\frac{\theta_x}{2}\cos\frac{\theta_z}{2} - \sin\frac{\theta_z}{2}\cos\frac{\theta_x}{2},\right.\right. \\
&\qquad \left.\left.\cos\frac{\theta_x}{2}\cos\frac{\theta_z}{2} + \sin\frac{\theta_x}{2}\sin\frac{\theta_z}{2}, \sin\frac{\theta_x}{2}\cos\frac{\theta_z}{2} - \sin\frac{\theta_z}{2}\cos\frac{\theta_x}{2}\right)\right) \\
&= \frac{1}{\sqrt{2}}\left(\cos\frac{\theta_x - \theta_z}{2}, \left(\sin\frac{\theta_x - \theta_z}{2}, \cos\frac{\theta_x - \theta_z}{2}, \sin\frac{\theta_x - \theta_z}{2}\right)\right)\ .
\end{aligned}$$

As remarked in Section 3.3.2, the Euler angle representation has the problem of *gimbal lock* caused by the loss of one degree of freedom. In Example 3.16 the Euler parameters θ_x and θ_z are not independent and account for the loss of one freedom. Quaternions overcome the gimbal lock problem since any orientation can be expressed by a unit quaternion $(\cos\theta, \sin\theta\mathbf{I})$.

Unit quaternions $\mathbf{q} = (s, (x, y, z))$ can be represented geometrically by a point on the unit sphere $|\mathbf{q}|^2 = s^2 + x^2 + y^2 + z^2 = 1$ in four-dimensional space $\mathbb{R}^4$. Note that, as a consequence of Exercise 3.23, antipodal points on the sphere represent the same rotation or orientation. An animation of an object between a start orientation $\mathbf{q_s}$ and an end orientation $\mathbf{q_e}$ can be performed by determining a curve (contained in the sphere) that interpolates the two points. Each point on the curve corresponds to a quaternion that specifies an intermediate orientation of the object. An interpolating curve can be obtained by considering great arcs on a sphere in $\mathbb{R}^3$ and extending to the sphere of unit quaternions in $\mathbb{R}^4$ by analogy.

Consider two points on the unit sphere in $\mathbb{R}^3$ with (unit) position vectors $\mathbf{a}$ and $\mathbf{b}$. A point $\mathbf{p}$ on the great arc through $\mathbf{a}$ and $\mathbf{b}$ lies in the plane through the origin containing the directions $\mathbf{a}$ and $\mathbf{b}$. Hence $\mathbf{p} = \alpha\mathbf{a} + \beta\mathbf{b}$, for some real numbers α and β. Suppose that $\mathbf{a}$ and $\mathbf{b}$ make an angle ϕ, and $\mathbf{a}$ and $\mathbf{p}$ make an angle θ. Then the following conditions are satisfied:

$$\begin{aligned} \mathbf{a}\cdot\mathbf{a} &= \mathbf{b}\cdot\mathbf{b} = \mathbf{p}\cdot\mathbf{p} = 1\,, \\ \mathbf{a}\cdot\mathbf{b} &= \cos\phi\,, \quad \mathbf{a}\cdot\mathbf{p} = \cos\theta\,, \quad \mathbf{b}\cdot\mathbf{p} = \cos(\phi-\theta)\,. \end{aligned}$$

Then

$$\begin{aligned} \mathbf{a}\cdot\mathbf{p} &= \mathbf{a}\cdot(\alpha\mathbf{a}+\beta\mathbf{b}) = \alpha\mathbf{a}\cdot\mathbf{a}+\beta\mathbf{a}\cdot\mathbf{b} = \alpha+\beta\cos\phi\,, \\ \mathbf{b}\cdot\mathbf{p} &= \mathbf{b}\cdot(\alpha\mathbf{a}+\beta\mathbf{b}) = \alpha\mathbf{a}\cdot\mathbf{b}+\beta\mathbf{b}\cdot\mathbf{b} = \alpha\cos\phi+\beta\,, \end{aligned}$$

giving

$$\cos\theta = \alpha+\beta\cos\phi\,, \tag{3.9}$$

$$\cos(\phi-\theta) = \alpha\cos\phi+\beta\,. \tag{3.10}$$

Solving (3.9) and (3.10) for α and β, and applying trigonometric formulae gives

$$\begin{aligned} \alpha &= \frac{\cos\theta-\cos\phi\cos(\phi-\theta)}{1-\cos^2\phi} = \frac{\sin(\phi-\theta)}{\sin\phi}\,, \quad \text{and} \\ \beta &= \frac{\cos(\phi-\theta)-\cos\phi\cos\theta}{1-\cos^2\phi} = \frac{\sin\theta}{\sin\phi}\,. \end{aligned}$$

The arc is parametrized on the interval $0 \le t \le 1$ by setting $\theta = t\phi$ to give

$$\mathbf{p}(t) = \alpha\mathbf{a}+\beta\mathbf{b} = \frac{\sin((1-t)\phi)}{\sin\phi}\mathbf{a} + \frac{\sin(t\phi)}{\sin\phi}\mathbf{b}\,.$$

By analogy, the formula is extended to give an arc $\mathbf{q}(t)$ interpolating two quaternions $\mathbf{q}_s$ and $\mathbf{q}_e$:

$$\mathbf{q}(t) = \frac{\sin((1-t)\phi)}{\sin\phi}\mathbf{q}_s + \frac{\sin(t\phi)}{\sin\phi}\mathbf{q}_e\,. \tag{3.11}$$

The arc can also be expressed as the quaternion multiplication

$$\mathbf{q}(t) = (\mathbf{q_e}\mathbf{q_s}^{-1})^t\mathbf{q_s} \quad \text{or} \quad \mathbf{q}(t) = \mathbf{q_s}(\mathbf{q_s}^{-1}\mathbf{q_e})^t\,,$$

for $0 \le t \le 1$, but this formulation is less useful for applications.

Example 3.17

Let $\mathbf{q_s} = \left(\frac{1}{5}, \left(\frac{2}{5}, \frac{4}{5}, \frac{2}{5}\right)\right)$ and $\mathbf{q_e} = \left(\frac{2}{3}, \left(0, \frac{1}{3}, \frac{2}{3}\right)\right)$. Then $\cos\phi = \left(\frac{1}{5}, \frac{2}{5}, \frac{4}{5}, \frac{2}{5}\right) \cdot \left(\frac{2}{3}, 0, \frac{1}{3}, \frac{2}{3}\right) = \frac{2}{3}$, giving $\sin\phi = \frac{\sqrt{5}}{3}$ and $\phi = 0.8411$. Then

$$\begin{aligned}\mathbf{q}(t) &= \sin(0.8411(1-t))\left(\frac{3}{5\sqrt{5}}, \left(\frac{6}{5\sqrt{5}}, \frac{12}{5\sqrt{5}}, \frac{6}{5\sqrt{5}}\right)\right) \\ &\quad + \sin(0.8411t)\left(\frac{2}{\sqrt{5}}, \left(0, \frac{1}{\sqrt{5}}, \frac{2}{\sqrt{5}}\right)\right).\end{aligned}$$

Intermediate orientations of the animation are obtained by evaluating $\mathbf{q}(t)$ for $0 \le t \le 1$. For instance, the intermediate quaternion corresponding to $t = 0.4$ is obtained by evaluating $\mathbf{q}(0.4) = (0.4250, (0.2595, 0.6666, 0.5547))$.

Figure 3.5 illustrates the result of applying the animation to the triangle with vertices $(5, 5, 0)$, $(10, 5, 0)$ and $(7, 10, 0)$. The location at which the triangle is defined is different to the start location because the start orientation is not specified by identity quaternion. The figure shows the start, end and an intermediate position of the triangle. The locus of a point $\mathbf{p}$ is (in general) a curve parametrized by $\mathbf{C}(t) = \mathbf{q}(t)\mathbf{p}\mathbf{q}(t)^{-1}$. The figure shows three curves that are the loci of the triangle vertices.

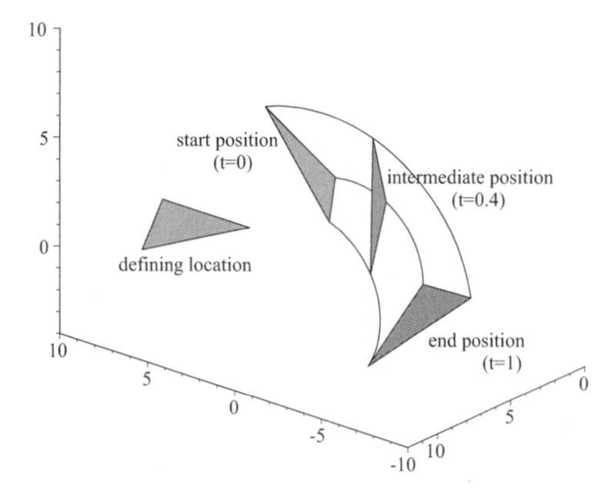

Figure 3.5 Animation of a triangle generated by quaternions

Exercise 3.29

Let $\mathbf{q_s} = \left(\frac{2}{7}, \left(0, \frac{6}{7}, -\frac{3}{7}\right)\right)$ and $\mathbf{q_e} = \left(\frac{2}{3}, \left(\frac{2}{3}, 0, \frac{1}{3}\right)\right)$. Apply formula (3.11) to obtain a motion with initial orientation $\mathbf{q_s}$ and final orientation $\mathbf{q_e}$. Suppose the motion is applied to a triangular body with vertices $(10, 0, 0)$, $(15, 0, 0)$ and $(12, 10, 0)$. Determine parametric expressions for the curves that are the loci of the vertices.

4 Projections and the Viewing Pipeline

4.1 Introduction

This chapter describes the process of visualizing three-dimensional objects. Current display devices such as computer monitors and printers are two-dimensional, and therefore it is necessary to obtain a planar view of the object which gives the impression of the omitted third dimension. Visualization of an object is achieved by a sequence of operations called the viewing pipeline (see Figure 4.1). Firstly, a projection is applied which maps the object to a new "flat" object in a specified plane known as the viewplane. The "flat" object represents a planar view of the object expressed in three-dimensional world coordinates. Secondly, a coordinate system in the viewplane is defined by specifying a point as origin, and two perpendicular vectors which give the directions of the coordinate axes. A viewplane coordinate mapping is applied to express the "flat" object in terms of the chosen two-dimensional viewplane coordinate system. Finally, the "flat" object is mapped to the computer screen by means of a two-dimensional device coordinate transformation.

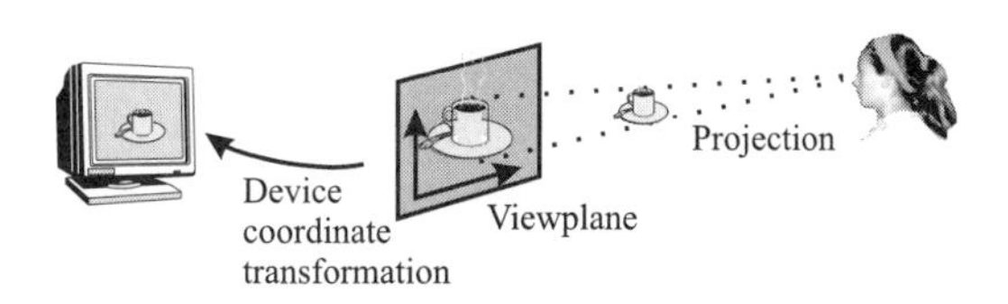

Figure 4.1 The viewing pipeline

The discussion begins in Section 4.2 with projections of the plane onto a line, and is followed by projections of three-dimensional space onto a plane in Section 4.3. Section 4.4 introduces the viewplane coordinate mapping which converts the three-dimensional world coordinate definition of the view to two-dimensional coordinates. The final step of mapping the view to the display device is discussed in Section 4.5. In Section 11.6, projections are used to create shadows which arise when light sources illuminate objects in a scene.

4.2 Projections of the Plane

A view of a spatial object is obtained by a mapping or projection of three-dimensional space onto a plane. Consider first the simpler problem of projecting the plane onto a line contained in the plane. Let ℓ be a line in the plane, and let $\mathbf{V}$ be a point not on the line. The *perspective projection* from $\mathbf{V}$ onto ℓ is the transformation which maps any point $\mathbf{P}$, distinct from $\mathbf{V}$, onto the point $\mathbf{P}'$ which is the intersection of the lines $\overline{\mathbf{VP}}$ and ℓ, as illustrated in Figure 4.2. The point $\mathbf{V}$ is called the *viewpoint* or *centre of perspectivity*, and the line ℓ is called the *viewline*. The next theorem shows that this mapping is indeed a transformation.

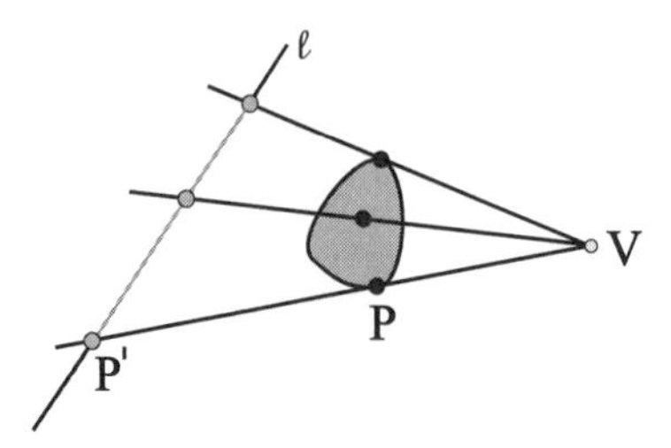

Figure 4.2 Perspective projection from the viewpoint $\mathbf{V}$ onto the line ℓ

Theorem 4.1

The perspective projection from the viewpoint $\mathbf{V}$ (expressed in homogeneous coordinates) onto the viewline with line vector ℓ is the two-dimensional transformation given by the matrix

$$\mathsf{M} = \ell^T \mathbf{V} - (\ell \cdot \mathbf{V}) I_3 ,$$

where I_3 denotes the 3×3 identity matrix.

Proof

Referring to Figure 4.2, the image $\mathbf{P}'$ of a point $\mathbf{P}$ is obtained as the intersection of the viewline ℓ with the line through $\mathbf{V}$ and $\mathbf{P}$. The techniques of Section 2.7 imply that the line through $\mathbf{V}$ and $\mathbf{P}$ has the line vector $\mathbf{V} \times \mathbf{P}$, and therefore intersects ℓ in the point with homogeneous coordinates given by $\ell \times (\mathbf{V} \times \mathbf{P})$. Applying the vector identity $\mathbf{A} \times (\mathbf{B} \times \mathbf{C}) = (\mathbf{C} \cdot \mathbf{A})\,\mathbf{B} - (\mathbf{A} \cdot \mathbf{B})\,\mathbf{C}$ yields

$$\mathbf{P}' = \ell \times (\mathbf{V} \times \mathbf{P}) = (\mathbf{P} \cdot \ell)\,\mathbf{V} - (\ell \cdot \mathbf{V})\,\mathbf{P}\,.$$

Replacing vectors by row matrices, and the dot product by a matrix multiplication, yields

$$\mathbf{P}' = \mathbf{P}\ell^T\mathbf{V} - \mathbf{P}\,(\ell \cdot \mathbf{V})I_3 = \mathbf{P}\left(\ell^T\mathbf{V} - (\ell \cdot \mathbf{V})I_3\right)\,.$$

Thus $\mathbf{P}' = \mathbf{P}\mathsf{M}$, where $\mathsf{M} = \ell^T\mathbf{V} - (\ell \cdot \mathbf{V})I_3$ as required. □

Definition 4.2

The matrix M is called the *projection matrix* of the perspective projection from $\mathbf{V}$ onto ℓ. Lines through the viewpoint are called *projectors*. The viewpoint $\mathbf{V}$ can be a point at infinity in which case the projection is called a *parallel projection*. It is common practice to use the term "perspective projection" to mean a non-parallel projection.

For a parallel projection with viewpoint $\mathbf{V}\,(v_1, v_2, 0)$ the projectors correspond to parallel lines in the Cartesian plane with direction (v_1, v_2) as shown in Figure 4.3.

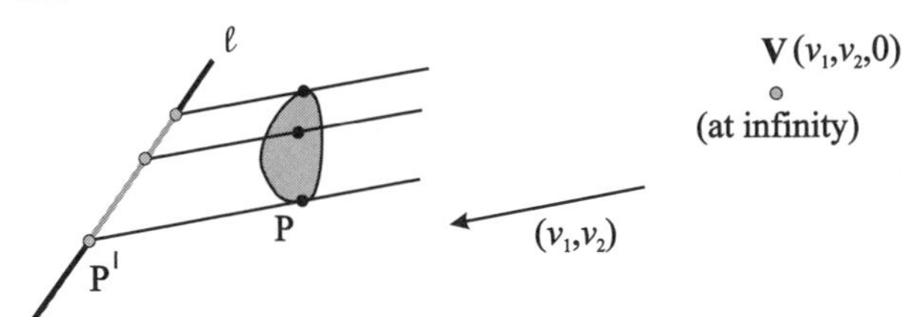

Figure 4.3 Parallel projection in the direction (v_1, v_2) onto the line ℓ

Example 4.3

The perspective projection of the triangle with vertices $\mathbf{A}(2,3)$, $\mathbf{B}(4,4)$, and $\mathbf{C}(3,-1)$ onto the line $5x + y - 4 = 0$ from the viewpoint with Cartesian coordinates $(10,2)$ is illustrated in Figure 4.4. The homogeneous viewpoint is

$\mathbf{V}(10,2,1)$, the line vector is $\ell = (5,1,-4)$, and $\ell \cdot \mathbf{V} = (5,1,-4)\cdot(10,2,1) = 48$. Hence

$$\mathsf{M} = \begin{pmatrix} 5 \\ 1 \\ -4 \end{pmatrix} \begin{pmatrix} 10 & 2 & 1 \end{pmatrix} - 48 \begin{pmatrix} 1 & 0 & 0 \\ 0 & 1 & 0 \\ 0 & 0 & 1 \end{pmatrix}$$

$$= \begin{pmatrix} 50 & 10 & 5 \\ 10 & 2 & 1 \\ -40 & -8 & -4 \end{pmatrix} - \begin{pmatrix} 48 & 0 & 0 \\ 0 & 48 & 0 \\ 0 & 0 & 48 \end{pmatrix} = \begin{pmatrix} 2 & 10 & 5 \\ 10 & -46 & 1 \\ -40 & -8 & -52 \end{pmatrix}.$$

The images of the vertices are obtained by multiplying the homogeneous coordinates of $\mathbf{A}$, $\mathbf{B}$, and $\mathbf{C}$ by M. Then

$$\begin{pmatrix} \mathbf{A}' \\ \mathbf{B}' \\ \mathbf{C}' \end{pmatrix} = \begin{pmatrix} \mathbf{A} \\ \mathbf{B} \\ \mathbf{C} \end{pmatrix} \mathsf{M} = \begin{pmatrix} 2 & 3 & 1 \\ 4 & 4 & 1 \\ 3 & -1 & 1 \end{pmatrix} \mathsf{M} = \begin{pmatrix} -6 & -126 & -39 \\ 8 & -152 & -28 \\ -44 & 68 & -38 \end{pmatrix}.$$

The Cartesian coordinates of the vertex images are $\mathbf{A}'(6/39, 126/39)$, $\mathbf{B}'(-8/28, 152/28)$, and $\mathbf{C}'(44/38, -68/38)$.

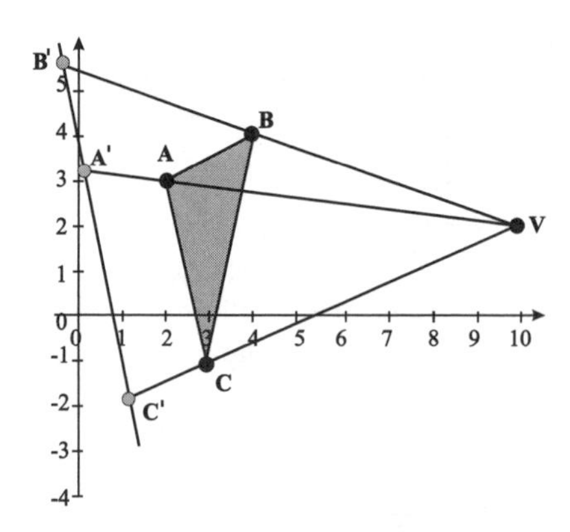

Figure 4.4 Perspective projection of Example 4.3

Example 4.4

The parallel projection of the triangle with vertices $\mathbf{A}(2,3)$, $\mathbf{B}(4,4)$, and $\mathbf{C}(3,-1)$ onto the line $3x + 2y - 4 = 0$ in the direction of the y-axis is shown in Figure 4.5. The viewpoint is $\mathbf{V}(0,1,0)$, the point at infinity in the direction of the y-axis. Then $\ell = (3,2,-4)$, and $\ell \cdot \mathbf{V} = (3,2,-4)\cdot(0,1,0) = 2$. Thus

$$\mathsf{M} = \begin{pmatrix} 3 \\ 2 \\ -4 \end{pmatrix} \begin{pmatrix} 0 & 1 & 0 \end{pmatrix} - 2 \begin{pmatrix} 1 & 0 & 0 \\ 0 & 1 & 0 \\ 0 & 0 & 1 \end{pmatrix} = \begin{pmatrix} -2 & 3 & 0 \\ 0 & 0 & 0 \\ 0 & -4 & -2 \end{pmatrix},$$

and

$$\begin{pmatrix} \mathbf{A}' \\ \mathbf{B}' \\ \mathbf{C}' \end{pmatrix} = \begin{pmatrix} 2 & 3 & 1 \\ 4 & 4 & 1 \\ 3 & -1 & 1 \end{pmatrix} \begin{pmatrix} -2 & 3 & 0 \\ 0 & 0 & 0 \\ 0 & -4 & -2 \end{pmatrix} \mathsf{M} = \begin{pmatrix} -4 & 2 & -2 \\ -8 & 8 & -2 \\ -6 & 5 & -2 \end{pmatrix}.$$

Thus the Cartesian coordinates of the images are $\mathbf{A}'(2,-1)$, $\mathbf{B}'(4,-4)$, and $\mathbf{C}'(3,-5/2)$.

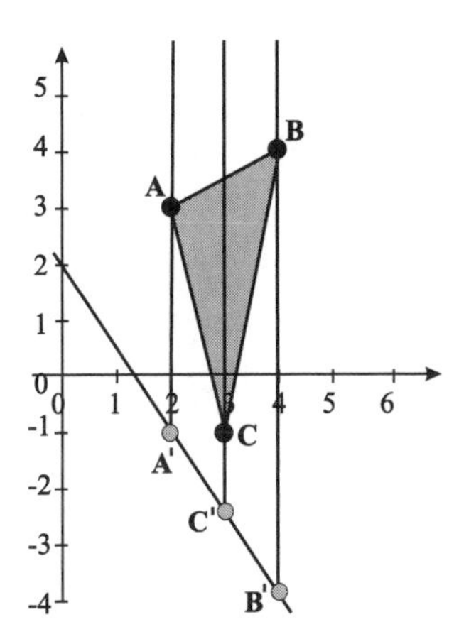

Figure 4.5 Parallel projection of Example 4.4

EXERCISES

4.1. Determine the projection matrix for a perspective projection with viewpoint $(2,11)$ and viewline $-3x+12y-5=0$.

4.2. Determine the projection matrix for a parallel projection in the direction $(3,-2)$ and viewline $7x-5y-2=0$.

4.3. Determine the projection matrix for a perspective projection with viewpoint $(7,-3)$ and viewline $x-y+9=0$. Apply the projection to the triangle with vertices $\mathbf{A}(2,2)$, $\mathbf{B}(4,3)$, and $\mathbf{C}(3,5)$. Make a sketch showing the projection of the triangle onto the line.

4.4. Determine the projection matrix for a parallel projection in the direction $(-1,4)$ and viewline $2x-y+8=0$. Apply the projection to the triangle with vertices $\mathbf{A}(2,2)$, $\mathbf{B}(4,3)$, and $\mathbf{C}(3,5)$. Make a sketch showing the projection of the triangle onto the line.

4.5. Let $\mathbf{P}=(p_1,p_2,p_3)$, $\mathbf{V}=(v_1,v_2,v_3)$, and $\ell=(a,b,c)$. Write out the proof of Theorem 4.1 in full.

4.3 Projections of Three-dimensional Space

Projections of three-dimensional space follow a similar line of development to projections of the plane. Let $\mathbf{n}$ be the plane vector of a viewplane, and let $\mathbf{V}$ be a point not on the viewplane. The *perspective projection* from $\mathbf{V}$ onto $\mathbf{n}$ is the transformation which maps any point $\mathbf{P}$, distinct from $\mathbf{V}$, onto the point $\mathbf{P}'$ which is the intersection of the line $\overline{\mathbf{VP}}$ and the plane $\mathbf{n}$, as illustrated in Figure 4.6(a). If $\mathbf{V}$ is a point at infinity then the projection is called a *parallel projection*, as illustrated in Figure 4.6(b). The term "perspective projection" is generally used to mean a projection which is not parallel.

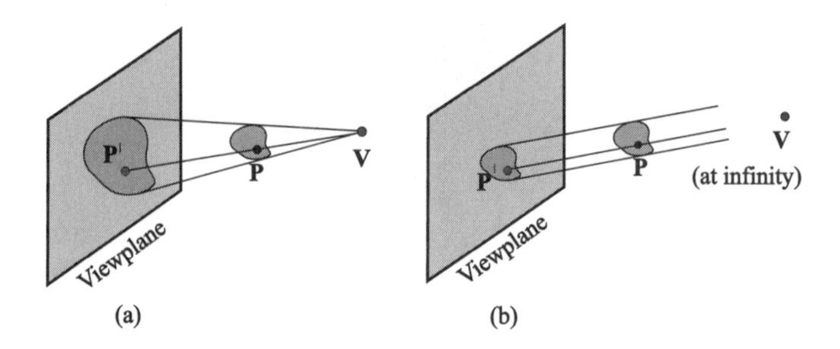

Figure 4.6 Perspective and parallel three-dimensional projections

Theorem 4.5

The projection with homogeneous viewpoint $\mathbf{V}$ and viewplane with plane vector $\mathbf{n}$ is the three-dimensional transformation given by the matrix

$$\mathsf{M} = \mathbf{n}^T\mathbf{V} - (\mathbf{n}\cdot\mathbf{V})I_4\,,$$

where I_4 denotes the 4×4 identity matrix,

$$\mathsf{M}=\begin{pmatrix} (-n_2v_2 - n_3v_3 - n_4v_4) & n_1v_2 & n_1v_3 & n_1v_4 \\ n_2v_1 & (-n_1v_1 - n_3v_3 - n_4v_4) & n_2v_3 & n_2v_4 \\ n_3v_1 & n_3v_2 & (-n_1v_1 - n_2v_2 - n_4v_4) & n_3v_4 \\ n_4v_1 & n_4v_2 & n_4v_3 & (-n_1v_1 - n_2v_2 - n_3v_3) \end{pmatrix}.$$

Proof

Let $\mathbf{P}$ be an arbitrary point to be projected ($\mathbf{P} \neq \mathbf{V}$). If $\mathbf{n} \cdot \mathbf{P} = 0$, then $\mathbf{P}$ is a point on the viewplane and its projected image is $\mathbf{P}$. To verify that $\mathbf{P}\mathsf{M} = \mathbf{P}$, note that $\mathbf{n} \cdot \mathbf{P} = \mathbf{P}\mathbf{n}^T = 0$ (representing vectors by row matrices), and hence

$$\mathbf{P}\mathsf{M} = \mathbf{P}\mathbf{n}^T\mathbf{V} - \mathbf{P}(\mathbf{n} \cdot \mathbf{V})I_4 = -\mathbf{P}(\mathbf{n} \cdot \mathbf{V})I_4 .$$

Since $-\mathbf{P}(\mathbf{n} \cdot \mathbf{V})I_4$ is a multiple of $\mathbf{P}$, $\mathbf{P}\mathsf{M}$ are homogeneous coordinates of $\mathbf{P}$, that is, $\mathbf{P}' = \mathbf{P}\mathsf{M}$.

Suppose $\mathbf{n} \cdot \mathbf{P} \neq 0$. Then every point on the line through $\mathbf{P}$ and $\mathbf{V}$ has homogeneous coordinates of the form $\alpha\mathbf{P} + \beta\mathbf{V}$ for some α and β (Exercise 3.6). The line intersects the viewplane when $\mathbf{n} \cdot (\alpha\mathbf{P} + \beta\mathbf{V}) = 0$. Then $\alpha\,(\mathbf{n} \cdot \mathbf{P}) + \beta\,(\mathbf{n} \cdot \mathbf{V}) = 0$, giving $\alpha = -\beta\,(\mathbf{n} \cdot \mathbf{V})\,/(\mathbf{n} \cdot \mathbf{P}) = 0$. Substituting for α, the point of intersection is found to have homogeneous coordinates

$$\mathbf{P}' = \alpha\mathbf{P} + \beta\mathbf{V} = (-\beta\,(\mathbf{n} \cdot \mathbf{V})\,/(\mathbf{n} \cdot \mathbf{P}))\,\mathbf{P} + \beta\mathbf{V} .$$

Multiplying the coordinates by the scalar $(\mathbf{n} \cdot \mathbf{P})$ gives the alternative homogeneous coordinates

$$\mathbf{P}' = (\mathbf{n} \cdot \mathbf{P})\,\mathbf{V} - (\mathbf{n} \cdot \mathbf{V})\,\mathbf{P} .$$

Then, in matrix form,

$$\mathbf{P}' = \left(\mathbf{P}\mathbf{n}^T\right)\mathbf{V} - \mathbf{P}\,(\mathbf{n} \cdot \mathbf{V})\,I_4 = \mathbf{P}\left(\mathbf{n}^T\mathbf{V} - (\mathbf{n} \cdot \mathbf{V})I_4\right) .$$

Hence

$$\mathsf{M} = \mathbf{n}^T\mathbf{V} - (\mathbf{n} \cdot \mathbf{V})I_4 .$$

□

Example 4.6

Consider a parallel projection of the prism shown in Figure 4.7 onto the plane $z = 0$ in a direction parallel to the z-axis. The viewpoint is $\mathbf{V}\,(0, 0, 1, 0)$, the point at infinity in the direction of the z-axis, and the viewplane has the equation $0x + 0y + 1z + 0 = 0$, so $\mathbf{n} = (0, 0, 1, 0)$. Thus

$$\mathsf{M} = \begin{pmatrix} 0 \\ 0 \\ 1 \\ 0 \end{pmatrix} \begin{pmatrix} 0 & 0 & 1 & 0 \end{pmatrix} - 1 \begin{pmatrix} 1 & 0 & 0 & 0 \\ 0 & 1 & 0 & 0 \\ 0 & 0 & 1 & 0 \\ 0 & 0 & 0 & 1 \end{pmatrix} = \begin{pmatrix} -1 & 0 & 0 & 0 \\ 0 & -1 & 0 & 0 \\ 0 & 0 & 0 & 0 \\ 0 & 0 & 0 & -1 \end{pmatrix} .$$

The prism has vertices $\mathbf{A}(0,0,0)$, $\mathbf{B}(2,0,0)$, $\mathbf{C}(2,3,0)$, $\mathbf{D}(0,3,0)$, $\mathbf{E}(1,2,1)$, $\mathbf{F}(1,1,1)$. Applying the projection to the vertices of the prism gives

$$\begin{pmatrix} \mathbf{A}' \\ \mathbf{B}' \\ \mathbf{C}' \\ \mathbf{D}' \\ \mathbf{E}' \\ \mathbf{F}' \end{pmatrix} = \begin{pmatrix} 0 & 0 & 0 & 1 \\ 2 & 0 & 0 & 1 \\ 2 & 3 & 0 & 1 \\ 0 & 3 & 0 & 1 \\ 1 & 2 & 1 & 1 \\ 1 & 1 & 1 & 1 \end{pmatrix} \begin{pmatrix} -1 & 0 & 0 & 0 \\ 0 & -1 & 0 & 0 \\ 0 & 0 & 0 & 0 \\ 0 & 0 & 0 & -1 \end{pmatrix}$$

$$= \begin{pmatrix} 0 & 0 & 0 & -1 \\ -2 & 0 & 0 & -1 \\ -2 & -3 & 0 & -1 \\ 0 & -3 & 0 & -1 \\ -1 & -2 & 0 & -1 \\ -1 & -1 & 0 & -1 \end{pmatrix}.$$

Following the usual procedure of dividing each point by its fourth coordinate yields the Cartesian coordinates $\mathbf{A}'(0,0,0)$, $\mathbf{B}'(2,0,0)$, $\mathbf{C}'(2,3,0)$, $\mathbf{D}'(0,3,0)$, $\mathbf{E}'(1,2,0)$, $\mathbf{F}'(1,1,0)$. The image of the prism is shown in Figure 4.8(a).

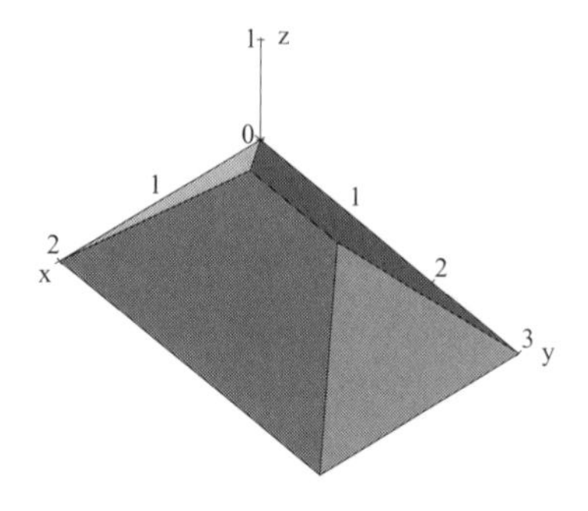

Figure 4.7 Prism of Example 4.6

Example 4.7

Consider a perspective projection onto the plane $z = 0$ from the viewpoint $(1,5,3)$. The viewpoint has homogeneous coordinates $\mathbf{V}\,(1,5,3,1)$ and the view-plane vector is $\mathbf{n} = (0,0,1,0)$. Thus

$$\mathsf{M} = \begin{pmatrix} 0 \\ 0 \\ 1 \\ 0 \end{pmatrix} \begin{pmatrix} 1 & 5 & 3 & 1 \end{pmatrix} - 3 \begin{pmatrix} 1 & 0 & 0 & 0 \\ 0 & 1 & 0 & 0 \\ 0 & 0 & 1 & 0 \\ 0 & 0 & 0 & 1 \end{pmatrix} = \begin{pmatrix} -3 & 0 & 0 & 0 \\ 0 & -3 & 0 & 0 \\ 1 & 5 & 0 & 1 \\ 0 & 0 & 0 & -3 \end{pmatrix}.$$

Applying the projection to the vertices of the prism of Example 4.6 yields

$$\begin{pmatrix} \mathbf{A}' \\ \mathbf{B}' \\ \mathbf{C}' \\ \mathbf{D}' \\ \mathbf{E}' \\ \mathbf{F}' \end{pmatrix} = \begin{pmatrix} 0 & 0 & 0 & 1 \\ 2 & 0 & 0 & 1 \\ 2 & 3 & 0 & 1 \\ 0 & 3 & 0 & 1 \\ 1 & 2 & 1 & 1 \\ 1 & 1 & 1 & 1 \end{pmatrix} \begin{pmatrix} -3 & 0 & 0 & 0 \\ 0 & -3 & 0 & 0 \\ 1 & 5 & 0 & 1 \\ 0 & 0 & 0 & -3 \end{pmatrix}$$

$$= \begin{pmatrix} 0 & 0 & 0 & -3 \\ -6 & 0 & 0 & -3 \\ -6 & -9 & 0 & -3 \\ 0 & -9 & 0 & -3 \\ -2 & -1 & 0 & -2 \\ -2 & 2 & 0 & -2 \end{pmatrix}.$$

The images have Cartesian coordinates $\mathbf{A}'(0,0,0)$, $\mathbf{B}'(2,0,0)$, $\mathbf{C}'(2,3,0)$, $\mathbf{D}'(0,3,0)$, $\mathbf{E}'(1,0.5,0)$, $\mathbf{F}'(1,-1,0)$. The image of the prism is shown in Figure 4.8(b).

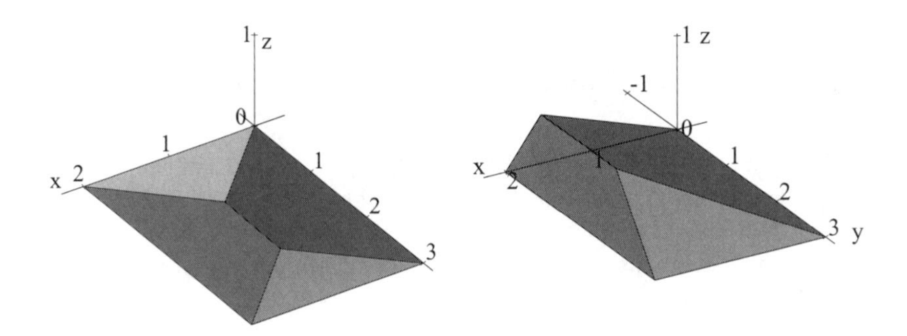

Figure 4.8 Images of the prism after the application of (a) the parallel projection of Example 4.6, and (b) the perspective projection of Example 4.7

EXERCISES

Determine the projection matrix M for the following.

4.6. Perspective projection onto the viewplane $-x+3y+2z-4=0$ from the viewpoint $(2,-1,1)$.

4.7. Perspective projection onto the viewplane $5x - 3z + 2 = 0$ from the viewpoint $(1, 4, -1)$.

4.8. Parallel projection onto the viewplane $2y+3z+4 = 0$ in the direction of the vector $(1, -2, 3)$.

4.9. Parallel projection onto the viewplane $7x-8y+5 = 0$ in the direction of the vector $(0, 4, 9)$.

4.10. Let a tetrahedron have vertices $\mathbf{A}(0, 0, 0)$, $\mathbf{B}(1, 0, 0)$, $\mathbf{C}(0, 1, 0)$, and $\mathbf{D}(1, 1, 1)$. Apply each of the projections of Exercises 4.6–4.9 to the tetrahedron.

4.11. Implement the three-dimensional projection procedure with the following specification. A viewpoint and viewplane are input by the user, and the computed projection matrix is obtained as output. In addition, the projected images of a number of input data points are determined. Computer algebra packages have a procedure for multiplying matrices; but, if you are writing a computer program, then you will need to devise your own algorithm to do this.

4.4 The Viewplane Coordinate Mapping

In the previous section three-dimensional projections were applied to give a planar representation of a view of an object. At this stage of the viewing process the view of the object is expressed in homogeneous three-dimensional world coordinates. The next stage is to define a two-dimensional *viewplane coordinate system* on the viewplane, and to represent the object view in terms of these coordinates. The viewing pipeline will be completed by specifying a rectangular *viewplane window* which identifies the region of the viewplane to be viewed. The viewplane window is mapped onto a rectangular *device* or *viewport window* of the display device. Any part of the view lying inside the viewplane window is mapped to the device window and displayed, but any part of the view lying outside the rectangle is *clipped*, and does not appear as part of the displayed image.

The viewplane (X, Y)-coordinate system is specified in terms of the world coordinate system by an origin $\mathcal{O}(q_1, q_2, q_3)$, and two *unit* vectors $\mathbf{r} = (r_1, r_2, r_3)$ and $\mathbf{s} = (s_1, s_2, s_3)$ which indicate the directions of the X- and Y-axes, respectively. Consider a point on the viewplane with homogeneous world coordinates $\mathbf{P}'(x, y, z, w)$, and homogeneous viewplane coordinates $\mathbf{P}''(X, Y, W)$. The aim is to obtain $\mathbf{P}''$ from $\mathbf{P}'$ by a mapping of the form $\mathbf{P}'' = \mathbf{P}' \cdot \mathsf{VC}$, where VC is a 4×3 matrix. Rather than compute VC directly, the strategy is to determine a 3×4 matrix K such that $\mathbf{P}''\mathsf{K} = \mathbf{P}'$, and then to express VC is terms of K. The reason for this is that K has a simple derivation.

The matrix K can be determined if the homogeneous world coordinates, and corresponding viewplane coordinates, of four points on the viewplane are known. Consider the following four points (expressed in homogeneous world coordinates):

(1) the origin $\mathcal{O}(q_1, q_2, q_3, 1)$,

(2) the point at infinity $\mathbf{R}(r_1, r_2, r_3, 0)$ in the direction of the X-axis,

(3) the point at infinity $\mathbf{S}(s_1, s_2, s_3, 0)$ in the direction of the Y-axis, and

(4) the point $\mathbf{T}(t_1, t_2, t_3, 1) = (q_1 + r_1 + s_1, q_2 + r_2 + s_2, q_3 + r_3 + s_3, 1)$ which is one unit in the X-direction and one unit in the Y-direction from the origin.

The homogeneous viewplane coordinates of the four points are $(0, 0, 1)$, $(1, 0, 0)$, $(0, 1, 0)$, and $(1, 1, 1)$ respectively. Then

$$\begin{pmatrix} \mathcal{O} \\ \mathbf{R} \\ \mathbf{S} \\ \mathbf{T} \end{pmatrix} = \begin{pmatrix} q_1 & q_2 & q_3 & 1 \\ r_1 & r_2 & r_3 & 0 \\ s_1 & s_2 & s_3 & 0 \\ t_1 & t_2 & t_3 & 1 \end{pmatrix} = \begin{pmatrix} 0 & 0 & 1 \\ 1 & 0 & 0 \\ 0 & 1 & 0 \\ 1 & 1 & 1 \end{pmatrix} \begin{pmatrix} r_1 & r_2 & r_3 & 0 \\ s_1 & s_2 & s_3 & 0 \\ q_1 & q_2 & q_3 & 1 \end{pmatrix},$$

and so corresponding points are correctly mapped to each other. Hence the required matrix is

$$\mathsf{K} = \begin{pmatrix} r_1 & r_2 & r_3 & 0 \\ s_1 & s_2 & s_3 & 0 \\ q_1 & q_2 & q_3 & 1 \end{pmatrix}. \tag{4.1}$$

The viewplane coordinate mapping is an inverse of the mapping determined by the matrix K. Since K is not a square matrix there is no matrix inverse K^{-1}, but there is a *right inverse*, a 4×3 matrix denoted K^R, for which $\mathsf{K}\mathsf{K}^R = I_3$. Since $\mathsf{K}\mathsf{K}^T(\mathsf{K}\mathsf{K}^T)^{-1} = I_3$, a right inverse of K is $\mathsf{K}^R = \mathsf{K}^T(\mathsf{K}\mathsf{K}^T)^{-1}$. Further, since $\mathbf{P}' = \mathbf{P}''\mathsf{K}$ it follows that

$$\begin{aligned} \mathbf{P}'\mathsf{K}^R &= \mathbf{P}'\mathsf{K}^T(\mathsf{K}\mathsf{K}^T)^{-1} = (\mathbf{P}''\mathsf{K})\,\mathsf{K}^T(\mathsf{K}\mathsf{K}^T)^{-1} \\ &= \mathbf{P}''\left(\mathsf{K}\mathsf{K}^T\right)(\mathsf{K}\mathsf{K}^T)^{-1} = \mathbf{P}''. \end{aligned}$$

Hence the viewplane coordinate mapping $\mathbf{P}'' = \mathbf{P}'\mathsf{VC}$ is given by the matrix

$$\mathsf{VC} = \mathsf{K}^R = \mathsf{K}^T(\mathsf{K}\mathsf{K}^T)^{-1}.$$

Observe that the viewplane coordinate matrix VC does not depend on a particular projection but is determined by the choice of origin and the directions of the X- and Y-axes. Thus VC can be applied to any view of an object.

Example 4.8

Consider the perspective projection of the prism onto the plane $z = 0$ from the viewpoint $\mathbf{V}(1,5,3)$ determined in Example 4.7. Let a coordinate system on the viewplane be given by origin $\mathcal{O}(1,2,0)$, X-axis direction $(3,4,0)$, and Y-axis direction $(-4,3,0)$. Then the unit vectors in the directions of the axes are $\mathbf{r} = (3/5, 4/5, 0)$ and $\mathbf{s} = (-4/5, 3/5, 0)$. The viewplane coordinate matrix $\mathsf{VC} = \mathsf{K}^T(\mathsf{KK}^T)^{-1}$ is obtained in several steps:

$$\begin{aligned}
\mathsf{KK}^T &= \begin{pmatrix} 3/5 & 4/5 & 0 & 0 \\ -4/5 & 3/5 & 0 & 0 \\ 1 & 2 & 0 & 1 \end{pmatrix} \begin{pmatrix} 3/5 & -4/5 & 1 \\ 4/5 & 3/5 & 2 \\ 0 & 0 & 0 \\ 0 & 0 & 1 \end{pmatrix} \\
&= \begin{pmatrix} 1 & 0 & 11/5 \\ 0 & 1 & 2/5 \\ 11/5 & 2/5 & 6 \end{pmatrix}, \\
(\mathsf{KK}^T)^{-1} &= \begin{pmatrix} 146/25 & 22/25 & -11/5 \\ 22/25 & 29/25 & -2/5 \\ -11/5 & -2/5 & 1 \end{pmatrix}, \\
\mathsf{VC} &= \begin{pmatrix} 3/5 & -4/5 & 1 \\ 4/5 & 3/5 & 2 \\ 0 & 0 & 0 \\ 0 & 0 & 1 \end{pmatrix} \begin{pmatrix} 146/25 & 22/25 & -11/5 \\ 22/25 & 29/25 & -2/5 \\ -11/5 & -2/5 & 1 \end{pmatrix} \\
&= \begin{pmatrix} 3/5 & -4/5 & 0 \\ 4/5 & 3/5 & 0 \\ 0 & 0 & 0 \\ -11/5 & -2/5 & 1 \end{pmatrix}.
\end{aligned}$$

The homogeneous viewplane coordinates of the prism vertices are determined by applying VC to the projected vertices obtained in Example 4.7:

$$\begin{pmatrix} \mathbf{A}'' \\ \mathbf{B}'' \\ \mathbf{C}'' \\ \mathbf{D}'' \\ \mathbf{E}'' \\ \mathbf{F}'' \end{pmatrix} = \begin{pmatrix} 0 & 0 & 0 & -3 \\ -6 & 0 & 0 & -3 \\ -6 & -9 & 0 & -3 \\ 0 & -9 & 0 & -3 \\ -2 & -1 & 0 & -2 \\ -2 & 2 & 0 & -2 \end{pmatrix} \begin{pmatrix} 3/5 & -4/5 & 0 \\ 4/5 & 3/5 & 0 \\ 0 & 0 & 0 \\ -11/5 & -2/5 & 1 \end{pmatrix}$$

$$
= \begin{pmatrix} 33/5 & 6/5 & -3 \\ 3 & 6 & -3 \\ -21/5 & 3/5 & -3 \\ -3/5 & -21/5 & -3 \\ 12/5 & 9/5 & -2 \\ 24/5 & 18/5 & -2 \end{pmatrix}.
$$

Hence, the Cartesian viewplane coordinates of the vertices are $\mathbf{A}''(-11/5, -2/5)$, $\mathbf{B}''(-1, -2)$, $\mathbf{C}''(7/5, -1/5)$, $\mathbf{D}''(1/5, 7/5)$, $\mathbf{E}''(-6/5, -9/10)$, $\mathbf{F}''(-12/5, -9/5)$. The projected prism in viewplane coordinates is illustrated in Figure 4.9.

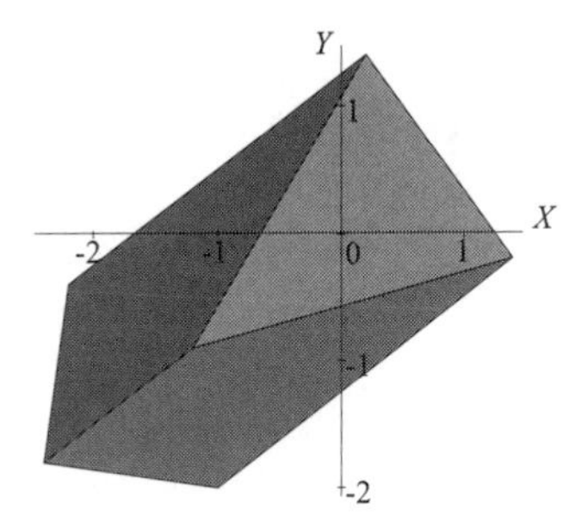

Figure 4.9 Projected prism of Example 4.8 in viewplane coordinates

EXERCISES

4.12. Consider the perspective projection of the prism onto the $z = 0$ plane from the viewpoint $\mathbf{V}(1, 5, 3)$ determined in Example 4.7. Let a coordinate system on the viewplane be given by origin $\mathcal{O}(4, 3, 0)$, X-axis direction $(12, 5, 0)$, and Y-axis direction $(-5, 12, 0)$. Determine the viewplane coordinate matrix VC. (Remember $\mathbf{r}$ and $\mathbf{s}$ must be unit vectors.) Apply VC to the vertices of the prism and make a sketch of the image.

4.13. Consider the projection of the tetrahedron with vertices $(0, 1, 0)$, $(3, 1, 1)$, $(-1, -1, 1)$, $(0, -2, -1)$ onto the viewplane $5x - 3z + 2 = 0$ from the viewpoint $(1, 4, -1)$ determined in Exercises 4.7 and 4.10. Let a viewplane coordinate system be defined by origin $\mathcal{O}(-1, 1, 1)$, X-axis direction $(3, 0, 5)$, and Y-axis direction $(0, -1, 0)$. Determine the viewplane coordinate matrix VC. Apply the matrix to the vertices of the projected tetrahedron, and make a sketch of the image.

4.14. Implement the viewplane coordinate mapping using a computer package, or by writing a computer program, with the following specification. The viewplane vector, origin and axes directions are given as input. The following checks on the input are carried out: that the origin is a point on the plane, that the axes directions are perpendicular, and that the axes directions are perpendicular to the plane vector. The matrix VC is determined and applied to input data points, or concatenated with a projection matrix. Most computer algebra packages have a procedure for determining matrix inverses. A subroutine to obtain a 3×3 matrix inverse is available on the book website (see the Preface to the Second Edition).

4.5 The Viewing Pipeline

So far two stages of the viewing pipeline have been considered in detail. First, the projection onto a viewplane $\mathbf{P}' = \mathbf{P}\mathsf{M}$ derived in Section 4.3, and second, the viewplane coordinate mapping $\mathbf{P}'' = \mathbf{P}'\mathsf{VC}$ derived in Section 4.4. The final stage to be considered is the device coordinate transformation $\mathbf{P}''' = \mathbf{P}''\mathsf{DC}$ which was introduced in Section 2.6.2. The concatenation of these three transformations yields the *viewing pipeline* which has the form $\mathbf{P}''' = \mathbf{P}\mathsf{VP}$ where $\mathsf{VP} = \mathsf{M} \cdot \mathsf{VC} \cdot \mathsf{DC}$.

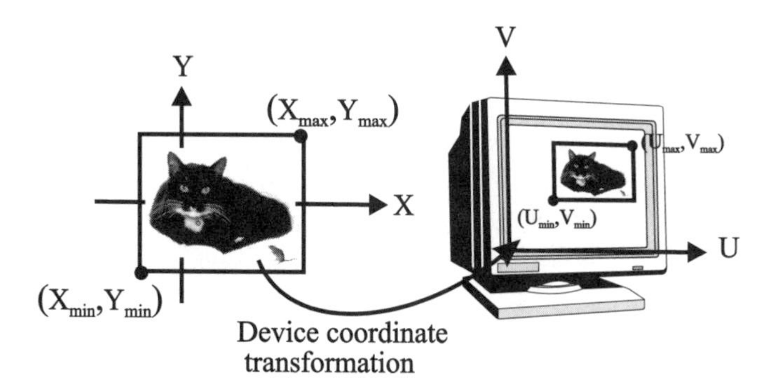

Figure 4.10

Referring to Figure 4.10, the region of the viewplane to be displayed is specified by a rectangular viewplane window with lower left corner $(X_{\min}, Y_{\min})$ and upper right corner $(X_{\max}, Y_{\max})$. This window is mapped to a rectangular region in the device coordinate system (called the device or viewport window) with lower left corner $(U_{\min}, V_{\min})$ and upper right corner $(U_{\max}, V_{\max})$. The device coordinate transformation which maps the viewplane window with local coordinates (X, Y, Z) to homogeneous device coordinates (U, V, W) is obtained by a concatenation of planar transformations: a translation taking the lower left corner of the viewplane window to the origin, followed by a scaling about the origin (so that the translated viewplane window has the same size as the device window), followed by a translation mapping the origin to the lower left corner of the device window. The required device coordinate transformation is

$$\begin{aligned}\mathsf{DC} &= \mathsf{T}(-X_{\min}, -Y_{\min}) \cdot \mathsf{S}\left(\tfrac{U_{\max}-U_{\min}}{X_{\max}-X_{\min}}, \tfrac{V_{\max}-V_{\min}}{Y_{\max}-Y_{\min}}\right) \cdot \mathsf{T}(U_{\min}, V_{\min}) \\ &= \begin{pmatrix} 1 & 0 & 0 \\ 0 & 1 & 0 \\ -X_{\min} & -Y_{\min} & 1 \end{pmatrix} \begin{pmatrix} \frac{U_{\max}-U_{\min}}{X_{\max}-X_{\min}} & 0 & 0 \\ 0 & \frac{V_{\max}-V_{\min}}{Y_{\max}-Y_{\min}} & 0 \\ 0 & 0 & 1 \end{pmatrix} \\ &\quad \times \begin{pmatrix} 1 & 0 & 0 \\ 0 & 1 & 0 \\ U_{\min} & V_{\min} & 1 \end{pmatrix}.\end{aligned}$$

Thus

$$\mathsf{DC} = \begin{pmatrix} \frac{U_{\max}-U_{\min}}{X_{\max}-X_{\min}} & 0 & 0 \\ 0 & \frac{V_{\max}-V_{\min}}{Y_{\max}-Y_{\min}} & 0 \\ \frac{X_{\max}U_{\min}-X_{\min}U_{\max}}{X_{\max}-X_{\min}} & \frac{Y_{\max}V_{\min}-Y_{\min}V_{\max}}{Y_{\max}-Y_{\min}} & 1 \end{pmatrix}.$$

The matrix DC is most easily remembered by recalling the transformations which define it.

Example 4.9

Consider the projected prism of Example 4.8. Let a viewplane window be given by lower left corner $(X_{\min}, Y_{\min}) = (-3, -3)$ and upper right corner $(X_{\max}, Y_{\max}) = (3, 2)$. Suppose that the device is a computer screen with a resolution of 1280×1024 pixels, and that the origin of the device coordinate system is the lower left corner of the screen. Let the device window have lower left corner $(U_{\min}, V_{\min}) = (500, 400)$ and upper right corner $(U_{\max}, V_{\max}) = (980, 700)$. The device coordinate transformation matrix is

$$\begin{aligned}\mathsf{DC} &= \begin{pmatrix} 1 & 0 & 0 \\ 0 & 1 & 0 \\ 3 & 3 & 1 \end{pmatrix} \begin{pmatrix} \frac{980-500}{3-(-3)} & 0 & 0 \\ 0 & \frac{700-400}{2-(-3)} & 0 \\ 0 & 0 & 1 \end{pmatrix} \begin{pmatrix} 1 & 0 & 0 \\ 0 & 1 & 0 \\ 500 & 400 & 1 \end{pmatrix} \\ &= \begin{pmatrix} 80 & 0 & 0 \\ 0 & 60 & 0 \\ 740 & 580 & 1 \end{pmatrix}.\end{aligned}$$

To map the projected prism to the screen, the matrix DC is applied to the homogeneous viewplane coordinates of the vertices given in Example 4.8:

$$\begin{pmatrix} \mathbf{A}'' \\ \mathbf{B}'' \\ \mathbf{C}'' \\ \mathbf{D}'' \\ \mathbf{E}'' \\ \mathbf{F}'' \end{pmatrix} \mathsf{DC} = \begin{pmatrix} 33/5 & 6/5 & -3 \\ 3 & 6 & -3 \\ -21/5 & 3/5 & -3 \\ -3/5 & -21/5 & -3 \\ 12/5 & 9/5 & -2 \\ 24/5 & 18/5 & -2 \end{pmatrix} \begin{pmatrix} 80 & 0 & 0 \\ 0 & 60 & 0 \\ 740 & 580 & 1 \end{pmatrix}$$

$$= \begin{pmatrix} -1692 & -1668 & -3 \\ -1980 & -1380 & -3 \\ -2556 & -1704 & -3 \\ -2268 & -1992 & -3 \\ -1288 & -1052 & -2 \\ -1096 & -944 & -2 \end{pmatrix}.$$

The vertices in Cartesian device coordinates are $(564, 556)$, $(660, 460)$, $(852, 568)$, $(756, 664)$, $(644, 526)$, $(548, 472)$. The reader is left the exercise of sketching the screen and the device window containing the final image of the prism.

Remark 4.10

During the viewing pipeline, it is unnecessary to convert computed coordinates from homogeneous to Cartesian until after all three matrices $\mathsf{M}, \mathsf{VC}, \mathsf{DC}$ have been applied. It is generally most efficient to compute the viewing pipeline matrix VP and then to apply VP to the coordinates of points on the object. This is demonstrated in the following example.

Example 4.11

A viewing pipeline is specified (in world coordinates) by: viewpoint $(3, 3, 10)$, viewplane $3x - y + 2z + 2 = 0$, viewplane origin $\mathcal{O}(-1, 1, 1)$, X-axis direction $(1, 1, -1)$, Y-axis direction $(-1, 5, 4)$, viewplane window with lower left corner $(-4, -10)$, and upper right corner $(4, -2)$, device viewport with lower left corner $(0, 200)$, and upper right corner $(200, 400)$. The viewing pipeline matrix $\mathsf{VP} = \mathsf{M} \cdot \mathsf{VC} \cdot \mathsf{DC}$ is computed as follows. The homogeneous viewpoint is $\mathbf{V}(3, 3, 10, 1)$

and the viewplane vector is $\mathbf{n} = (3, -1, 2, 2)$. Hence the projection matrix is

$$\mathsf{M} = \begin{pmatrix} 3 \\ -1 \\ 2 \\ 2 \end{pmatrix} \begin{pmatrix} 3 & 3 & 10 & 1 \end{pmatrix} - (28) \begin{pmatrix} 1 & 0 & 0 & 0 \\ 0 & 1 & 0 & 0 \\ 0 & 0 & 1 & 0 \\ 0 & 0 & 0 & 1 \end{pmatrix}$$

$$= \begin{pmatrix} -19 & 9 & 30 & 3 \\ -3 & -31 & -10 & -1 \\ 6 & 6 & -8 & 2 \\ 6 & 6 & 20 & -26 \end{pmatrix}.$$

The unit vector in the direction of the X-axis is $\mathbf{r} = \left(1/\sqrt{3}, 1/\sqrt{3}, -1/\sqrt{3}\right)$, and the unit vector in the direction of the Y-axis is $\mathbf{s} = \left(-1/\sqrt{42}, 5/\sqrt{42}, 4/\sqrt{42}\right)$. Hence

$$\mathsf{K} = \begin{pmatrix} 1/\sqrt{3} & 1/\sqrt{3} & -1/\sqrt{3} & 0 \\ -1/\sqrt{42} & 5/\sqrt{42} & 4/\sqrt{42} & 0 \\ -1 & 1 & 1 & 1 \end{pmatrix}$$

$$= \begin{pmatrix} 0.577 & 0.577 & -0.577 & 0 \\ -0.154 & 0.772 & 0.617 & 0 \\ -1.0 & 1.0 & 1.0 & 1.0 \end{pmatrix},$$

$$\left(\mathsf{K}\mathsf{K}^T\right)^{-1} = \begin{pmatrix} 1.0 & 0.0 & -0.577 \\ 0.0 & 1.0 & 1.543 \\ -0.577 & 1.543 & 4.0 \end{pmatrix}^{-1}$$

$$= \begin{pmatrix} 1.259 & -0.693 & 0.449 \\ -0.693 & 2.852 & -1.200 \\ 0.449 & -1.200 & 0.778 \end{pmatrix}.$$

The formula $\mathsf{VC} = \mathsf{K}^T(\mathsf{K}\mathsf{K}^T)^{-1}$ gives the viewplane coordinate matrix

$$\mathsf{VC} = \begin{pmatrix} 0.385 & 0.360 & -0.333 \\ 0.642 & 0.600 & 0.111 \\ -0.706 & 0.960 & -0.222 \\ 0.449 & -1.200 & 0.778 \end{pmatrix}.$$

The device coordinate matrix is

$$\mathsf{DC} = \begin{pmatrix} 1 & 0 & 0 \\ 0 & 1 & 0 \\ 4 & 10 & 1 \end{pmatrix} \begin{pmatrix} 25 & 0 & 0 \\ 0 & 25 & 0 \\ 0 & 0 & 1 \end{pmatrix} \begin{pmatrix} 1 & 0 & 0 \\ 0 & 1 & 0 \\ 0 & 200 & 1 \end{pmatrix} = \begin{pmatrix} 25 & 0 & 0 \\ 0 & 25 & 0 \\ 100 & 450 & 1 \end{pmatrix}.$$

Hence

$$\mathsf{VP} = \mathsf{M} \cdot \mathsf{VC} \cdot \mathsf{DC} = \begin{pmatrix} -234.050 & 1944.067 & 3.000 \\ -460.844 & -1152.079 & -1.000 \\ 517.543 & 792.000 & 2.000 \\ -3090.752 & -10295.85 & -26.000 \end{pmatrix} .$$

Applying the viewing pipeline matrix VP to the vertices of the prism yields

$$\begin{pmatrix} 3 & 0 & 6 & 1 \\ 9 & 0 & 6 & 1 \\ 9 & 6 & 6 & 1 \\ 3 & 6 & 6 & 1 \\ 6 & 6 & 9 & 1 \\ 6 & 0 & 9 & 1 \end{pmatrix} \mathsf{VP} = \begin{pmatrix} -687.642 & 288.286 & -5.000 \\ -2091.941 & 11952.69 & 13.000 \\ -4857.002 & 5040.214 & 7.000 \\ -3452.703 & -6624.188 & -11.000 \\ -2602.223 & 1583.98 & 4.000 \\ 162.838 & 8496.454 & 10.000 \end{pmatrix} .$$

The Cartesian coordinates are

$$\mathbf{A}'''(137.528, -57.657), \mathbf{B}'''(-160.919, 919.438), \mathbf{C}'''(-693.857, 720.031),$$
$$\mathbf{D}'''(313.8821, 602.1989), \mathbf{E}'''(-650.556, 395.995), \mathbf{F}'''(16.284, 849.645) .$$

EXERCISES

4.15. A viewing pipeline is specified by: viewpoint $(2, 3, 8)$, viewplane $z+4 = 0$, viewplane origin $\mathcal{O}(-2, 1, -4)$, X-axis direction $(1, 1, 0)$, Y-axis direction $(-1, 1, 0)$, viewplane window corners $(-1, -7)$, $(4, -2)$, device viewport corners $(400, 300)$, $(800, 700)$. Determine the viewing pipeline matrix VP. A tetrahedron has vertices $\mathbf{A}(1, 0, 1)$, $\mathbf{B}(3, 0, 1)$, $\mathbf{C}(2, 2, 1)$, $\mathbf{D}(2, 1, 2)$. Apply VP to the tetrahedron, and sketch the projected image and the device window.

4.16. A viewing pipeline is specified by: viewpoint $(7, 0, 1)$, viewplane $x - y = 1$, viewplane origin $\mathcal{O}(2, 1, 1)$, X-axis direction $(1, 1, 0)$, Y-axis direction $(0, 0, 1)$, viewplane window corners lower left $(-2, -3)$, upper right $(6, 6)$, device corners lower left $(50, 50)$, upper right $(250, 150)$. Determine the viewing pipeline matrix VP. A tetrahedron has vertices $\mathbf{A}(2, 0, 1)$, $\mathbf{B}(2, -1, 4)$, $\mathbf{C}(4, 4, 3)$, $\mathbf{D}(1, 0, 4)$. Apply VP to the tetrahedron, and sketch the projected image and the device window.

4.17. A viewing pipeline is specified by: viewpoint $(6, 2, 0)$, viewplane $2x - 4y + 4 = 0$, viewplane origin $\mathcal{O}(-2, 0, 1)$, X-axis direction $(2, 1, 0)$, Y-axis direction $(0, 0, 1)$, viewplane window corners lower left $(-20, -15)$, upper right $(20, 15)$, device corners lower left $(50, 50)$, upper right $(150, 200)$. Determine the viewing pipeline matrix VP.

4.6 Classification of Projections

So far two types of projection, parallel and perspective, have been discussed. In the following sections further distinctions of these types are made according to how the viewpoint and viewplane are located with respect to the world coordinate axes. The directions of the world coordinate axes are called the *principal directions.*

Consider a line segment in space projected onto a viewplane. In general, it is expected that its image will be a line segment of a different length. The *foreshortening ratio* in the direction of that line is defined to be

$$\frac{\text{length of projected segment}}{\text{length of original segment}} .$$

Example 4.12

Consider the line segment $\mathbf{PQ}$ given by points $\mathbf{P}(0,1,1)$ and $\mathbf{Q}(2,1,3)$. The images of the points following a parallel projection onto the plane $z = 0$ in the direction of the negative z-axis are computed to be

$$\begin{aligned}
\begin{pmatrix} \mathbf{P}' \\ \mathbf{Q}' \end{pmatrix} &= \begin{pmatrix} \mathbf{P} \\ \mathbf{Q} \end{pmatrix} \mathsf{M} \\
&= \begin{pmatrix} 0 & 1 & 0 & 1 \\ 2 & 1 & 3 & 1 \end{pmatrix} \begin{pmatrix} 1 & 0 & 0 & 0 \\ 0 & 1 & 0 & 0 \\ 0 & 0 & 0 & 0 \\ 0 & 0 & 0 & 1 \end{pmatrix} \\
&= \begin{pmatrix} 0 & 1 & 0 & 1 \\ 2 & 1 & 0 & 1 \end{pmatrix} .
\end{aligned}$$

In Cartesian coordinates, the images are $\mathbf{P}'(0,1,0)$ and $\mathbf{Q}'(2,1,0)$. The distance from $\mathbf{P}$ to $\mathbf{Q}$ is $\sqrt{8} = 2\sqrt{2}$, and the distance from $\mathbf{P}'$ to $\mathbf{Q}'$ is 2. Thus the foreshortening ratio is $2 / (2\sqrt{2}) = 1 / \sqrt{2}$.

Exercise 4.18

Show that for any projection, line segments with the same direction are foreshortened by an equal amount.

4.6.1 Classification of Parallel Projections

Three types of parallel projection are distinguished, namely, orthographic, axonometric, and oblique.

Orthographic Projection

An *orthographic projection* is a parallel projection for which the direction of the projection is perpendicular to the viewplane. Thus if the viewplane vector is $\mathbf{n} = (n_1, n_2, n_3, n_4)$ then the centre of projection is $\mathbf{V}(-n_1, -n_2, -n_3, 0)$. Then

$$
\begin{aligned}
\mathsf{M} &= \begin{pmatrix} n_1 \\ n_2 \\ n_3 \\ n_4 \end{pmatrix} \begin{pmatrix} -n_1 & -n_2 & -n_3 & 0 \end{pmatrix} + \left(n_1^2 + n_2^2 + n_3^2\right) \begin{pmatrix} 1 & 0 & 0 & 0 \\ 0 & 1 & 0 & 0 \\ 0 & 0 & 1 & 0 \\ 0 & 0 & 0 & 1 \end{pmatrix} \\
&= \begin{pmatrix} n_2^2 + n_3^2 & -n_1 n_2 & -n_1 n_3 & 0 \\ -n_1 n_2 & n_1^2 + n_3^2 & -n_2 n_3 & 0 \\ -n_1 n_3 & -n_2 n_3 & n_1^2 + n_2^2 & 0 \\ -n_4 n_1 & -n_4 n_2 & -n_4 n_3 & n_1^2 + n_2^2 + n_3^2 \end{pmatrix}. \qquad (4.2)
\end{aligned}
$$

For instance, the projection matrix for an orthographic projection onto the plane $z = 0$ (for which $\mathbf{n} = (0, 0, 1, 0)$ and $\mathbf{V}(0, 0, 1, 0)$) is

$$
\mathsf{M} = \begin{pmatrix} 1 & 0 & 0 & 0 \\ 0 & 1 & 0 & 0 \\ 0 & 0 & 0 & 0 \\ 0 & 0 & 0 & 1 \end{pmatrix}.
$$

An orthographic projection can show the true dimensions and shape of a single planar face of an object. They are commonly used in engineering and architectural drawings and occur as “front, side, and planar elevations”.

Exercise 4.19

Suppose $\mathbf{n}$ is chosen so that $n_1^2 + n_2^2 + n_3^2 = 1$. Then the direction cosines of $\mathbf{n}$ with respect to the world coordinate system are n_1, n_2, and n_3. Show that an orthographic projection (4.2) yields foreshortening ratios in the principal directions of $\left(n_2^2 + n_3^2\right)^{1/2}$, $\left(n_1^2 + n_3^2\right)^{1/2}$, and $\left(n_1^2 + n_2^2\right)^{1/2}$. Deduce that two foreshortening ratios are equal if and only if the absolute values of two direction cosines of $\mathbf{n}$ are equal.

Axonometric Projection

Axonometric projections are orthographic projections which attempt to portray the general three-dimensional shape of an object. There are three types of axonometric projection, namely, trimetric, dimetric, and isometric. These are distinguished by whether none, two, or all three of the foreshortening ratios in the principal directions are equal. Exercise 4.19 implies that this distinction is equivalent to none, two, or all three of the direction cosines of the parallel projection direction vector being equal.

1. A *trimetric projection* is obtained when $|n_1|$, $|n_2|$, and $|n_3|$ are all different. Then the foreshortening ratios in the principal directions are all different. If measurements are taken from a trimetic projection of an object, then it is necessary to apply a scale factor in each of the principal directions in order to read off the correct dimensions of the object.

2. A *dimetric projection* is obtained when just one of $|n_1| = |n_2|$, $|n_2| = |n_3|$, or $|n_1| = |n_3|$ is true. Measurements along two of the three principal directions may be performed using a single scale factor, but a different scale factor is required in the third direction.

3. An *isometric projection* is obtained when $|n_1| = |n_2| = |n_3|$. Since all three foreshortening factors are equal, an isometric projection scales the object equally in all three principal directions.

Example 4.13

The effect of the various parallel projections on the unit cube with vertices $(0,0,0)$, $(1,0,0)$, $(1,1,0)$, $(0,1,0)$, $(0,0,1)$, $(1,0,1)$, $(1,1,1)$, $(0,1,1)$ are shown in Figures 4.11 and 4.12. The foreshortening factors in the x-, y-, and z-directions are denoted f_1, f_2, and f_3 respectively.

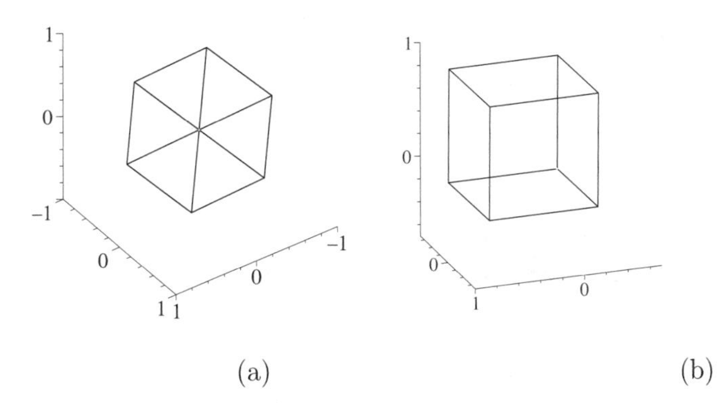

Figure 4.11 (a) Isometric projection with $\mathbf{n} = \left(1/\sqrt{3}, 1/\sqrt{3}, 1/\sqrt{3}, 0\right)$, $f_1 = f_2 = f_3 = \sqrt{6}/3$, and (b) dimetric projection with $\mathbf{n} = \left(1/3, \sqrt{7}/3, 1/3, 0\right)$, $f_1 = f_3 = \frac{2}{3}\sqrt{2}$, $f_2 = \sqrt{2}/3$

Oblique Projection

An *oblique projection* is a parallel projection for which the direction of the projection is not perpendicular to the viewplane. In general, oblique projections

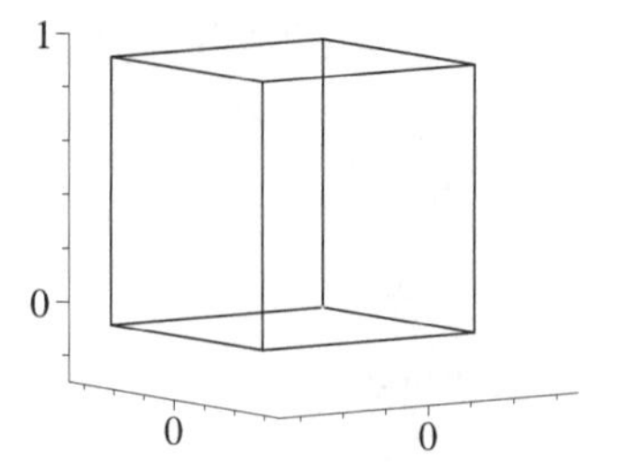

Figure 4.12 Trimetric projection with $\mathbf{n} = \left(\frac{1}{3}\sqrt{3}, \frac{7}{15}\sqrt{3}, \frac{1}{15}\sqrt{3}\right)$, $f_1 = 0.816$, $f_2 = 0.589$, $f_3 = 0.993$

give an impression of the depth of an object. The foreshortening ratio of line segments parallel to the viewplane is 1.

When the view direction (v_1, v_2, v_3) makes an angle of $\pi/4$ with the viewplane, a *cavalier projection* is obtained. This projection angle causes line segments perpendicular to the viewplane to have foreshortening ratio 1. The result is that an object with planar faces perpendicular to, or parallel to, the viewplane appears thicker than in reality. The angle θ between (v_1, v_2, v_3) and the viewplane normal (n_1, n_2, n_3) satisfies

$$(v_1, v_2, v_3) \cdot (n_1, n_2, n_3) = |(v_1, v_2, v_3)|\,|(n_1, n_2, n_3)| \cos\theta \,.$$

Thus a cavalier projection $(\theta = \pi/4)$ satisfies the identity

$$v_1 n_1 + v_2 n_2 + v_3 n_3 = \pm\frac{\sqrt{2}}{2}\sqrt{v_1^2 + v_2^2 + v_3^2}\sqrt{n_1^2 + n_2^2 + n_3^2}\,. \tag{4.3}$$

Example 4.14

Consider the cavalier projections onto the $z = 0$ plane. Then $\mathbf{n} = (0, 0, 1, 0)$ and identity (4.3) simplifies to $v_3^2 = v_1^2 + v_2^2$. A suitable view direction would be $(3, 4, 5)$ giving the projection matrix

$$\mathsf{M} = \begin{pmatrix} 0 \\ 0 \\ 1 \\ 0 \end{pmatrix} \begin{pmatrix} 3 & 4 & 5 & 0 \end{pmatrix} - 5 \begin{pmatrix} 1 & 0 & 0 & 0 \\ 0 & 1 & 0 & 0 \\ 0 & 0 & 1 & 0 \\ 0 & 0 & 0 & 1 \end{pmatrix} = \begin{pmatrix} -5 & 0 & 0 & 0 \\ 0 & -5 & 0 & 0 \\ 3 & 4 & 0 & 0 \\ 0 & 0 & 0 & -5 \end{pmatrix}.$$

Applying M to the line segment, perpendicular to the viewplane, joining the origin to the point $(0, 0, 10)$, gives the segment joining the origin to the point $(-6, -8, 0)$. The project line segment has length $\sqrt{(-6)^2 + (-8)^2} = 10$. Thus the foreshortening ratio is 1 as expected. The projection of the unit cube is illustrated in Figure 4.13(a).

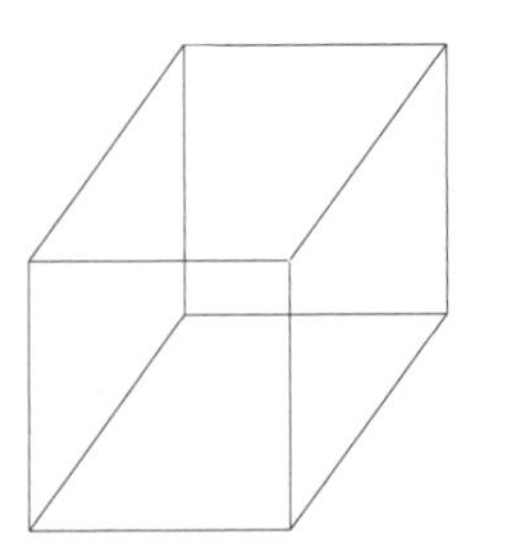

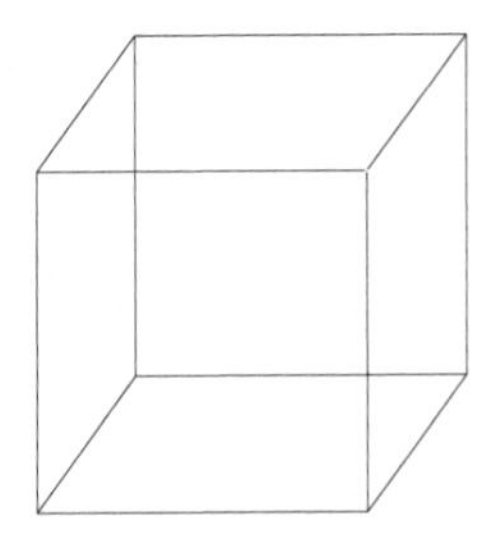

Figure 4.13 (a) Cavalier projection (b) Cabinet projection

A *cabinet projection* overcomes the "thickness" problem of a cavalier projection. The foreshortening factor for faces of the object perpendicular to the plane of projection is chosen to be $\frac{1}{2}$. This is achieved when the projection direction makes an angle of $\phi = \operatorname{arccot}(1/2)$ (approximately 1.107 radians) with the viewplane. If $\phi = \operatorname{arccot}(1/2)$, then $\sin(\phi) = 2\,/\sqrt{5}$ and $\cos(\phi) = 1\,/\sqrt{5}$. The angle between the viewplane normal and the projection direction is $\theta = \frac{\pi}{2} - \phi$ or $\theta = \frac{\pi}{2} + \phi$. Thus $\cos\theta = \cos(\frac{\pi}{2} \mp \phi) = \pm\sin\phi = \pm 2\,/\sqrt{5}$ and, therefore, a cabinet projection satisfies the identity

$$v_1 n_1 + v_2 n_2 + v_3 n_3 = \pm\frac{2}{\sqrt{5}}\sqrt{v_1^2 + v_2^2 + v_3^2}\sqrt{n_1^2 + n_2^2 + n_3^2}\,. \tag{4.4}$$

Example 4.15

Consider the cabinet projections onto the $z = 0$ plane. Then $\mathbf{n} = (0, 0, 1, 0)$ and condition (4.4) simplifies to $v_3^2 = 4\left(v_1^2 + v_2^2\right)$. A suitable view direction would be $(3, 4, 10)$. The projection matrix is

$$\begin{aligned} \mathsf{M} &= \begin{pmatrix} 0 \\ 0 \\ 1 \\ 0 \end{pmatrix} \begin{pmatrix} 3 & 4 & 10 & 0 \end{pmatrix} - 10 \begin{pmatrix} 1 & 0 & 0 & 0 \\ 0 & 1 & 0 & 0 \\ 0 & 0 & 1 & 0 \\ 0 & 0 & 0 & 1 \end{pmatrix} \\ &= \begin{pmatrix} -10 & 0 & 0 & 0 \\ 0 & -10 & 0 & 0 \\ 3 & 4 & 0 & 0 \\ 0 & 0 & 0 & -10 \end{pmatrix}. \end{aligned}$$

Applying M to the line segment, perpendicular to the viewplane, joining the origin to the point $(0, 0, 10)$, gives the segment joining the origin to the point $(-3, -4, 0)$. The project line segment has length $\sqrt{(-3)^2 + (-4)^2} = 5$. Thus the foreshortening ratio is $1/2$ as was expected. The projection of the unit cube is illustrated in Figure 4.13(b).

4.6.2 Classification of Perspective Projections

A perspective projection has the effect that the projected images of parallel lines in world coordinate space may be intersecting lines in the viewplane.

Theorem 4.16

1. Parallel projections map parallel lines in world coordinate space to parallel lines in the viewplane.
2. Perspective projections map parallel lines in world coordinate space to parallel lines in the viewplane if and only if the lines are parallel to the viewplane.
3. A projection which maps points at infinity in 3 or more linearly independent directions to points at infinity is a parallel projection.

Proof

Parallel lines in the direction (x, y, z) project to parallel lines in the viewplane if and only if the point at infinity $\mathbf{P}(x, y, z, 0)$ projects to a point at infinity in the viewplane. Let the projection matrix be the 4×4 matrix $\mathsf{M} = (m_{ij})$ given in Theorem 4.5. Then $\mathbf{P}' = (x\ y\ z\ 0)\mathsf{M} = (xm_{11}+ym_{21}+zm_{31}, xm_{12}+ym_{22}+zm_{32}, xm_{13}+ym_{23}+zm_{33}, xm_{14}+ym_{24}+zm_{34})$. Referring to Theorem 4.5, $m_{14} = v_4 n_1$, $m_{24} = v_4 n_2$, and $m_{34} = v_4 n_3$. Thus $\mathbf{P}'$ is infinite if and only if $v_4(n_1 x + n_2 y + n_3 z) = 0$.

1. If the projection is parallel, then $v_4 = 0$, and hence $\mathbf{P}'$ is infinite. Thus parallel projections map parallel lines in world coordinate space to parallel lines in the viewplane.
2. If the projection is perspective, then $v_4 \neq 0$. Hence $\mathbf{P}'$ is infinite if and only if $n_1 x + n_2 y + n_3 z = (n_1, n_2, n_3) \cdot (x, y, z) = 0$ which is true if and only if the lines with direction (x, y, z) are perpendicular to the viewplane vector, that is, parallel to the viewplane.
3. Consider three linearly independent directions (x_i, y_i, z_i), $i = 1, 2, 3$. Suppose the points at infinity in these directions map to points at infinity in the viewplane. Then $v_4(n_1 x + n_2 y + n_3 z) = 0$ for $i = 1, 2, 3$. If $v_4 \neq 0$ then $n_1 x_i + n_2 y_i + n_3 z_i = 0$ for $i = 1, 2, 3$, and since the vectors (x_i, y_i, z_i) are independent this implies $(n_1, n_2, n_3) = (0, 0, 0)$ which is impossible. So $v_4 = 0$ which implies the projection is parallel.

□

Corollary 4.17

Suppose (x_i, y_i, z_i), $i = 1, 2, 3$, are a set of mutually perpendicular vectors. Then the viewplane vector (n_1, n_2, n_3) of a perspective projection can be perpendicular to

1. **(three-point perspective)** none of the vectors. Then the family of parallel lines in each of the directions (x_i, y_i, z_i) maps to a family of non-parallel lines.
2. **(two-point perspective)** one of the vectors (x_1, y_1, z_1). Then the family of lines in the direction of (x_1, y_1, z_1) maps to a family of parallel lines, but the families of parallel lines with directions (x_2, y_2, z_2) and (x_3, y_3, z_3) map to families of non-parallel lines.
3. **(one-point perspective)** two of the vectors (x_1, y_1, z_1) and (x_2, y_2, z_2). Then the families of parallel lines with directions (x_1, y_1, z_1) and (x_2, y_2, z_2) map to families of parallel lines, but the family parallel to (x_3, y_3, z_3) maps to a family of non-parallel lines.

Proof

By Theorem 4.16, all parallel lines in the direction of a vector perpendicular to the viewplane normal (that is, parallel to the viewplane) map to parallel lines in the viewplane. Conversely, for a perspective projection, parallel lines which are not perpendicular to the viewplane normal do not map to parallel lines in the viewplane. Thus the number of independent vectors perpendicular to the viewplane normal determines the three cases.

□

If a perspective projection maps an infinite point $(x, y, z, 0)$ to a finite point $(x', y', z', 1)$ in the viewplane, then lines in the direction (x, y, z) in world coordinate space appear as lines converging to the point (x', y', z') in the (Cartesian) viewplane. The point (x', y', z') is called the *vanishing point* for the direction (x, y, z). Corollary 4.17 is generally applied to the principal directions to give:

1. *One-point perspective projection.* Parallel lines in one principal direction have a vanishing point.
2. *Two-point perspective projection.* Parallel lines in each of two principal directions have vanishing points.
3. *Three-point perspective projection.* Parallel lines in each of the three principal directions have vanishing points.

Example 4.18

Consider the one-point perspective projection onto the $z = 0$ plane from the viewpoint $(-1, 0, 2)$. Then $\mathbf{n} = (0, 0, 1, 0)$ and $\mathbf{V}(-1, 0, 2, 1)$. The projection matrix is

$$\mathsf{M} = \begin{pmatrix} 0 \\ 0 \\ 1 \\ 0 \end{pmatrix} \begin{pmatrix} -1 & 0 & 2 & 1 \end{pmatrix} - (2) \begin{pmatrix} 1 & 0 & 0 & 0 \\ 0 & 1 & 0 & 0 \\ 0 & 0 & 1 & 0 \\ 0 & 0 & 0 & 1 \end{pmatrix}$$

$$= \begin{pmatrix} -2 & 0 & 0 & 0 \\ 0 & -2 & 0 & 0 \\ -1 & 0 & 0 & 1 \\ 0 & 0 & 0 & -2 \end{pmatrix}.$$

The effect of the projection on the unit cube is shown in Figure 4.14. The position of the viewpoint is indicated by a small circle.

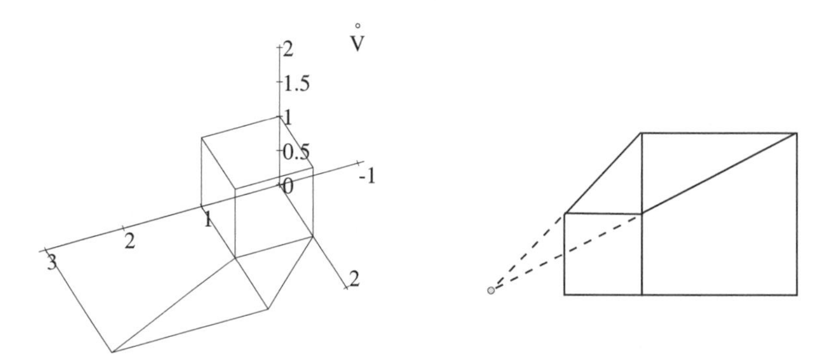

(a) One-point perspective projection (b) Vanishing point

Figure 4.14

Example 4.19

Consider the two-point perspective projection onto the $x - z = 0$ plane from the viewpoint $(-1, 0, 2)$. Then $\mathbf{n} = (1, 0, -1, 0)$ and $\mathbf{V}(-1, 0, 2, 1)$. The projection matrix is

$$\mathsf{M} = \begin{pmatrix} 1 \\ 0 \\ -3 \\ 0 \end{pmatrix} \begin{pmatrix} -1 & 0 & 2 & 1 \end{pmatrix} - (-7) \begin{pmatrix} 1 & 0 & 0 & 0 \\ 0 & 1 & 0 & 0 \\ 0 & 0 & 1 & 0 \\ 0 & 0 & 0 & 1 \end{pmatrix}$$

$$= \begin{pmatrix} 6 & 0 & 2 & 1 \\ 0 & 7 & 0 & 0 \\ 3 & 0 & 1 & -3 \\ 0 & 0 & 0 & 7 \end{pmatrix}.$$

The effect on the unit cube is shown in Figure 4.15.

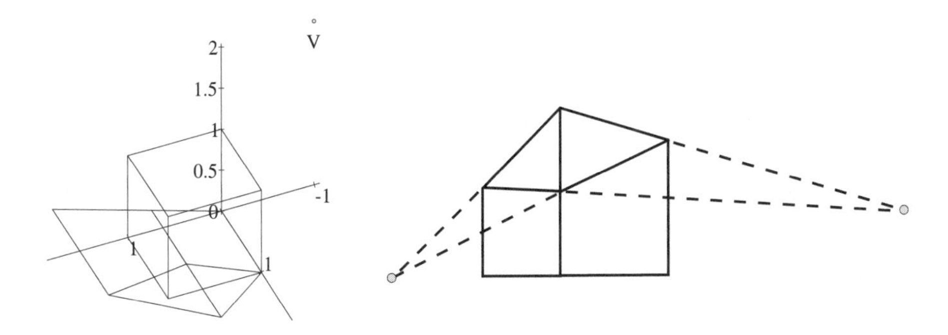

(a) Two-point perspective projection (b) Vanishing points

Figure 4.15

Example 4.20

Consider the three-point perspective projection onto the $x + 2y - z - 3 = 0$ plane from the viewpoint $(-1, 0, 2)$. Then $\mathbf{n} = (1, 2, -1, -3)$ and $\mathbf{V}(-1, 0, 2, 1)$. The projection matrix is

$$\mathsf{M} = \begin{pmatrix} 1 \\ 2 \\ -1 \\ -3 \end{pmatrix} \begin{pmatrix} -1 & 0 & 2 & 1 \end{pmatrix} - (-6) \begin{pmatrix} 1 & 0 & 0 & 0 \\ 0 & 1 & 0 & 0 \\ 0 & 0 & 1 & 0 \\ 0 & 0 & 0 & 1 \end{pmatrix}$$

$$= \begin{pmatrix} 5 & 0 & 2 & 1 \\ -2 & 6 & 4 & 2 \\ 1 & 0 & 4 & -1 \\ 3 & 0 & -6 & 3 \end{pmatrix}.$$

The effect on the unit cube is shown in Figure 4.16.

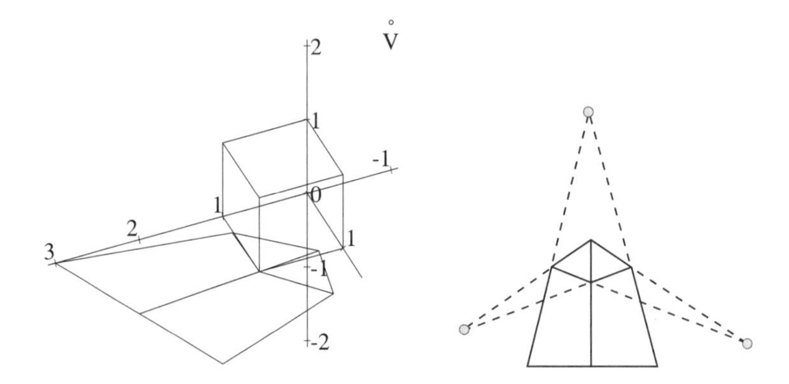

(a) Three-point perspective projection (b) The vanishing points

Figure 4.16

EXERCISES

4.20. Show that the viewplane coordinate mapping and the device coordinate transformation map points at infinity to points at infinity.

4.21. Show that the vanishing point of a perspective projection in the direction (x, y, z) is (p_1, p_2, p_3) where (using the notation of Theorem 4.16)

$$\begin{aligned} p_1 &= \frac{-xn_2v_2 - xn_3v_3 - xn_4v_4 + yn_2v_1 + zn_3v_1}{(xn_1 + yn_2 + zn_3)\,v_4}, \\ p_2 &= \frac{xn_1v_2 - yn_1v_1 - yn_3v_3 - yn_4v_4 + zn_3v_2}{(xn_1 + yn_2 + zn_3)\,v_4}, \\ p_3 &= \frac{xn_1v_3 + yn_2v_3 - zn_1v_1 - zn_2v_2 - zn_4v_4}{(xn_1 + yn_2 + zn_3)\,v_4}. \end{aligned}$$

4.22. Stereographic projection attempts to emulate human vision by combining two perspective projections of an object from two closely positioned viewpoints. One image is viewed by the left eye and the other by the right eye. The brain combines the two images to form a single image with a more realistic sense of depth. Investigate stereographic projection. See [21], [13], [12].

4.23. Image reconstruction is the process of obtaining a three-dimensional representation of an object from a number of projected images of the image. Investigate. See [12], [21].

5
Curves

5.1 Introduction

Curves arise in many applications such as art, industrial design, mathematics, architecture, and engineering, and numerous computer drawing packages and computer-aided design packages have been developed to facilitate the creation of curves. A particularly illustrative application is that of computer fonts which are defined by curves that specify the outline of each character in the font. Different font sizes are obtained by applying scaling transformations. Special font effects can be obtained by applying other transformations such as shears and rotations. Likewise, in other applications there is a need to perform various tasks such as modifying, analyzing, and visualizing the curves. In order to execute such operations a mathematical representation for curves is required. In this chapter, curve representations are introduced and the simplest types of curve, namely lines and conics, are described. Chapters 6–8 explore Bézier and B-spline curves, two important representations that are widely used in CAD and computer graphics. The representations of curves lead naturally to representations of surfaces in Chapter 9. Conics also emerge as the silhouettes of quadric surfaces in Section 11.5.

Definition 5.1

Three representations of curves are considered.

Parametric: The coordinates of points of a *parametric* curve are expressed as functions of a variable or *parameter* such as t. A curve in the plane has the form $\mathbf{C}(t) = (x(t), y(t))$, and a curve in space has the form $\mathbf{C}(t) = (x(t), y(t), z(t))$. The functions $x(t)$, $y(t)$, and $z(t)$ are called the *coordinate functions.* The image of $\mathbf{C}(t)$ is called the *trace* of $\mathbf{C}$, and $\mathbf{C}(t)$ is called a *parametrization* of $\mathbf{C}$. A subset of a curve $\mathbf{C}$ which is also a curve is called a *curve segment.* A parametric curve defined by polynomial coordinate functions is called a *polynomial curve.* The *degree* of a polynomial curve is the highest power of the variable occurring in any coordinate function. A function $p(t)/q(t)$ is said to be *rational* if $p(t)$ and $q(t)$ are polynomials. A parametric curve defined by rational coordinate functions is called a *rational curve.* The *degree* of a rational curve is the highest power of the variable occurring in the numerator or denominator of any coordinate function. Most of the curves considered in this book are parametric.

Non-parametric explicit: The coordinates (x, y) of points of a *non-parametric explicit* planar curve satisfy $y = f(x)$ or $x = g(y)$. Such curves have the parametric form $\mathbf{C}(t) = (t, f(t))$ or $\mathbf{C}(t) = (g(t), t)$. For non-parametric explicit spatial curves, two of the coordinates are expressed in terms of the third: for instance, $x = f(z)$, $y = g(z)$.

Implicit: The coordinates (x, y) of points of an *implicit* curve satisfy $F(x, y) = 0$, for some function F. When F is a polynomial in variables x and y the curve is called an *algebraic curve.* An implicitly defined spatial curve must satisfy (at least) two conditions $F(x, y, z) = 0$ and $G(x, y, z) = 0$ simultaneously. Implicit curves defined by polynomials of degree two are considered in Section 5.6.

Example 5.2 (Parametric Curves)

1. Parabola: (t, t^2), for $t \in \mathbb{R}$, is a polynomial curve of degree 2. See Figure 5.1(a).
2. Quarter circle: $\left(\frac{1-t^2}{1+t^2}, \frac{2t}{1+t^2}\right)$, for $t \in [0, 1]$, is a rational curve of degree 2.
3. Unit radius circle: $(\cos(t), \sin(t))$, for $t \in [0, 2\pi]$, see Figure 5.1(b).
4. Twisted space cubic: (t, t^2, t^3), for $t \in \mathbb{R}$ is a polynomial curve of degree 3. See Figure 5.1(c).
5. Helix: $(r\cos(t), r\sin(t), at)$, for $t \in \mathbb{R}$, $r > 0$, $a \neq 0$. See Figure 5.1(d).

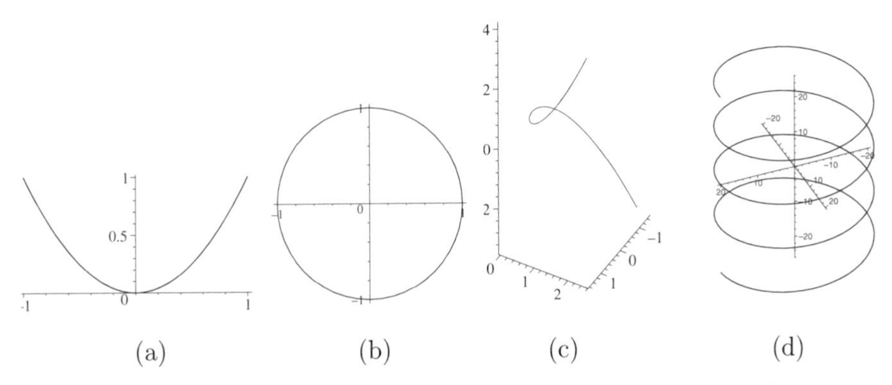

(a) (b) (c) (d)

Figure 5.1 Parametric curves: (a) parabola, (b) unit circle, (c) twisted cubic, and (d) helix

Example 5.3 (Non-parametric Implicit Curves)

1. Parabola: $y = x^2$, $x \in \mathbb{R}$.
2. Circular arc: $y = \sqrt{1 - x^2}$, $x \in [-1, 1]$.
3. Twisted space cubic: $y = x^2$, $z = x^3$, $x \in \mathbb{R}$.

Example 5.4 (Implicit Curves)

1. Unit radius circle: $x^2 + y^2 - 1 = 0$.
2. Cuspidal cubic: $y^2 - x^3 = 0$, see Figure 5.2.

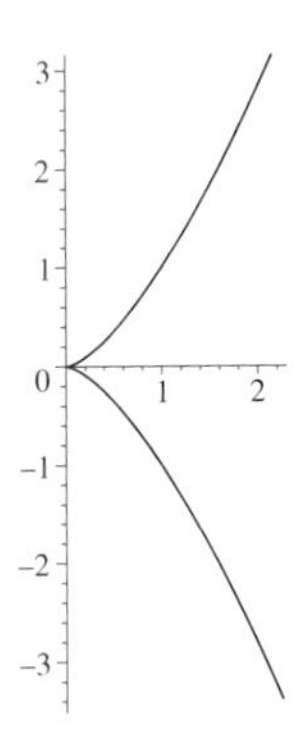

Figure 5.2 Cuspidal cubic $y^2 - x^3 = 0$

5.2 Curve Rendering

The process of drawing a curve is called *rendering*. Parametric curves are the most widely used in computer graphics and geometric modelling since points on the curve are easily computed. In contrast, the evaluation of points on an implicitly defined curve is substantially more difficult.

A curve of the form $\mathbf{C}(t) = (x(t), y(t))$ defined on the interval $[a, b]$ is rendered by evaluating $n+1$ points $(x(t_i), y(t_i))$, where $t_0 < t_1 < \cdots < t_n$ and $t_0 = a$, $t_n = b$. The points are joined in sequence by line segments to give a linear approximation to the curve as shown in Figure 5.3. If the resulting approximation is too jagged then a smoother curve can be obtained by increasing the number of evaluated points.

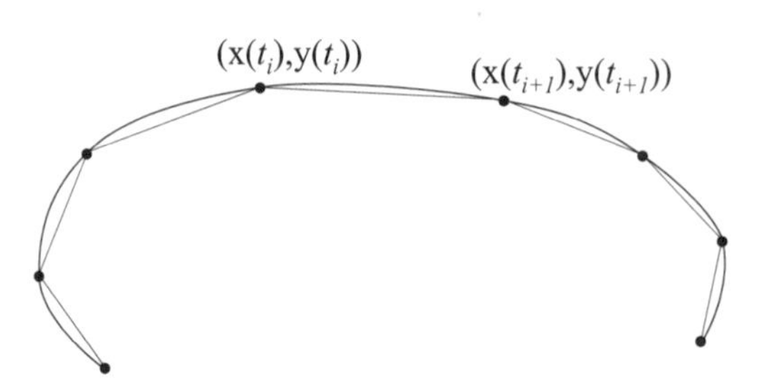

Figure 5.3 Linear approximation to a parametric curve

Points on polynomial and rational curves can be evaluated using a reasonable number of arithmetical operations. Points on curves defined by functions such as square roots, trigonometric functions, exponential and logarithmic functions are more computationally expensive to calculate.

The most economical way to evaluate a polynomial is to use *Horner's method.* Consider the polynomial $1 + 2t + 3t^2 + 4t^3$. If the polynomial is computed as $1+2\cdot t+3\cdot t\cdot t+4\cdot t\cdot t\cdot t$ then 3 additions and 6 multiplications are required. However, if the polynomial is computed as $((4\cdot t+3)\cdot t+2)\cdot t+1$, then only 3 additions and 3 multiplications are required yielding a saving of 3 multiplications. For polynomials of higher degree the saving is even greater.

In general, a polynomial of the form $a_0 + a_1t + a_2t^2 + \cdots + a_nt^n$ can be expressed in the form

$$(((a_nt + a_{n-1})\,t + a_{n-2})\,t + \cdots)\,t + a_0 \ .$$

A computer algorithm to evaluate a polynomial, based on Horner's method, is easily implemented.

EXERCISES

5.1. Express $3-5t+4t^2-2t^3+6t^4$ in Horner's form. Determine the difference in the number of $\pm$ and $\times$ required to evaluate the polynomial in its original and new form.

5.2. Write a computer program which takes as input the coefficients of a polynomial and a parameter value t, and which outputs the value of the polynomial at the given parameter value using Horner's method.

5.3. Determine the number of operations $\pm$ and $\times$ saved by evaluating a polynomial of degree n using Horner's method.

5.4. Write a computer program which renders a parametric curve. Alternatively, learn how to plot curves using a computer package. Plot some of the curves given in the examples.

5.3 Parametric Curves

Let $\mathbf{C}(t) = (x(t), y(t))$ be a curve defined on an open interval (a, b). Then $\mathbf{C}(t)$ is said to be C^k-continuous (or just C^k) if the first k derivatives of $x(t)$ and $y(t)$ exist and are continuous. If infinitely many derivatives exist then $\mathbf{C}(t)$ is said to be C^∞. A curve $\mathbf{C}(t) = (x(t), y(t))$ defined on a closed interval $[a, b]$ is said to be C^k-continuous if there exists an open interval (c, d) containing the interval $[a, b]$, and a C^k-continuous curve $\mathbf{D}(t)$ defined on (c, d), such that $\mathbf{C}(t) = \mathbf{D}(t)$ for all $t \in [a, b]$. Curves defined on a closed interval need to be "extendable" to a curve on an open interval in order to differentiate $x(t)$ and $y(t)$ at the ends of the interval. (Another type of continuity called G^k-continuity, which is important for CAD applications, is introduced in Definition 7.14.)

Suppose $\mathbf{C}(t)$ is a C^1 curve defined on an interval I, then the function $\nu(t) = \sqrt{(x'(t))^2 + (y'(t))^2}$ is called the *speed* of the curve $\mathbf{C}(t)$. If $\nu(t) \neq 0$, for all $t \in I$, then $\mathbf{C}(t)$ is said to be a *regular* curve. If $\nu(t) = 1$ for all $t \in I$, then $\mathbf{C}(t)$ is said to be a *unit speed curve.*

Example 5.5

1. Let $(x(t), y(t)) = (t, t^2)$. Then $(x'(t), y'(t)) = (1, 2t)$, and

$$\nu(t) = \sqrt{1^2 + (2t)^2} = \sqrt{1 + 4t^2} \, .$$

2. Let $(x(t), y(t), z(t)) = (\cos t, \sin t, t^2)$. Then
$$(x'(t), y'(t), z'(t)) = (-\sin t, \cos t, 2t) ,$$
and
$$\nu = \sqrt{(-\sin t)^2 + (\cos t)^2 + (2t)^2} = \sqrt{1 + 4t^2} .$$

Let $\mathbf{C}(t) = (x(t), y(t))$ be a regular parametric curve, and suppose $\mathbf{P}$ and $\mathbf{Q}$ are the points with coordinates $(x(t), y(t))$ and $(x(t+\delta t), y(t+\delta t))$ respectively. Let
$$\mathbf{t}_{\delta t} = \frac{\overrightarrow{\mathbf{PQ}}}{\delta t} = \frac{(x(t+\delta t), y(t+\delta t)) - (x(t), y(t))}{\delta t}$$
as shown in Figure 5.4. Then as $\delta t \to 0$, $\mathbf{Q} \to \mathbf{P}$ and $\mathbf{t}_{\delta t}$ tends to the limiting vector
$$\lim_{\delta t \to 0} \mathbf{t}_{\delta t} = \left(\lim_{\delta t \to 0} \frac{x(t+\delta t) - x(t)}{\delta t}, \lim_{\delta t \to 0} \frac{y(t+\delta t) - y(t)}{\delta t} \right) = (x'(t), y'(t)) .$$
$\mathbf{C}'(t) = (x'(t), y'(t))$ is called the *tangent vector*. The *unit tangent vector* is defined to be
$$\mathbf{t}(t) = \frac{1}{|(x'(t), y'(t))|}(x'(t), y'(t)) = \left(\frac{x'(t)}{\sqrt{x'(t)^2 + y'(t)^2}}, \frac{y'(t)}{\sqrt{x'(t)^2 + y'(t)^2}} \right) .$$
The line through the point $(x(t), y(t))$ in the direction of the tangent vector is called the *tangent line* and has the equation
$$y'(t)(x - x(t)) - x'(t)(y - y(t)) = 0 . \tag{5.1}$$

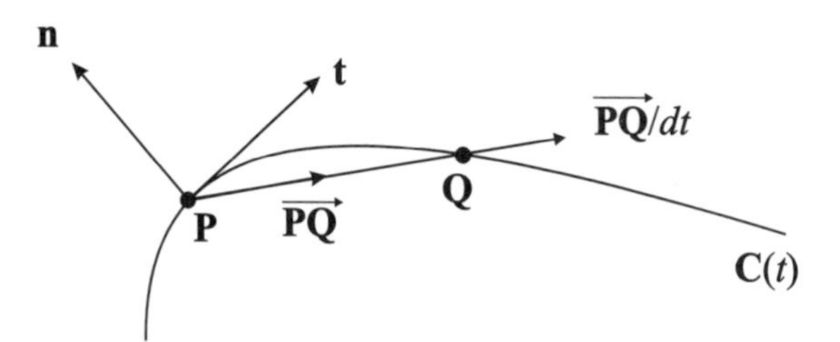

Figure 5.4 Tangent and normal to a curve

If the tangent vector $\mathbf{C}'(t) = (x'(t), y'(t))$ is rotated through an angle $\pi/2$ radians (in an anticlockwise direction), then the *normal vector* $(-y'(t), x'(t))$ is obtained. The *unit normal vector* of $\mathbf{C}(t)$ is defined to be
$$\mathbf{n}(t) = \frac{(-y'(t), x'(t))}{|(-y'(t), x'(t))|} = \left(\frac{-y'(t)}{\sqrt{x'(t)^2 + y'(t)^2}}, \frac{x'(t)}{\sqrt{x'(t)^2 + y'(t)^2}} \right) .$$

Example 5.6

Let $\mathbf{C}(t) = (\cos(t), \sin(t))$. Then the tangent vector is $\mathbf{C}'(t) = (-\sin(t), \cos(t))$ and the normal vector is $(-\cos(t), -\sin(t))$. Since $|\mathbf{C}'(t)| = 1$ these vectors are also the unit tangent and normal vectors. At the point $(\cos(\pi/4), \sin(\pi/4)) = \left(1/\sqrt{2}\,, 1/\sqrt{2}\right)$ the unit tangent vector is

$$(-\sin(\pi/4), \cos(\pi/4)) = \left(-\frac{1}{\sqrt{2}}, \frac{1}{\sqrt{2}}\right) ,$$

and the unit normal vector is

$$(-\cos(\pi/4), -\sin(\pi/4)) = \left(-\frac{1}{\sqrt{2}}, -\frac{1}{\sqrt{2}}\right) .$$

The tangent line to $\mathbf{C}(t)$ at $\left(1/\sqrt{2}\,, 1/\sqrt{2}\right)$ is

$$\cos(\pi/4)(x - \cos(\pi/4)) + \sin(\pi/4)(y - \sin(\pi/4)) = 0 ,$$

which simplifies to $x + y - \sqrt{2} = 0$.

The derivation of the tangent vector can be extended to space curves: for a curve $\mathbf{C}(t) = (x(t), y(t), z(t))$, the tangent vector is $\mathbf{C}'(t) = (x'(t), y'(t), z'(t))$, and the unit tangent vector is $\mathbf{t}(t) = \mathbf{C}'(t)/\sqrt{x'(t)^2 + y'(t)^2 + z'(t)^2}$.

EXERCISES

5.5. Let $\mathbf{C}(t) = (x(t), y(t))$ be a regular curve.

(a) Determine the parametric equation of the tangent line to $\mathbf{C}$ at the point $(x(t), y(t))$.

(b) Prove that the tangent line to $\mathbf{C}(t)$ at a point $(x(t), y(t))$ is given by Equation (5.1).

(c) The normal line to $\mathbf{C}$ at a point $\mathbf{p} = (x(t), y(t))$ is the line through $\mathbf{p}$ perpendicular to the tangent. Determine the implicit equation of the normal line.

5.6. For each of the curves below, determine (i) the unit tangent vector, (ii) the unit normal vector, and (iii) the implicit equation of the tangent line.

(a) (t, t^2) at the point $(1, 1)$.

(b) (t^2, t^3) at the point $(4, 8)$.

(c) *logarithmic spiral:* $\left(ae^{bt}\cos t, ae^{bt}\sin t\right)$.

(d) *cycloid:* $(t + \sin t, 1 - \cos t)$.

(e) $(\cos t + t\sin t, \sin t - t\cos t)$.

(f) *catenary:* $(c\cosh(t/c), t)$.

5.7. Show that a translation of a curve has no effect on the speed of a curve.

5.8. Show that a rotation has no effect on the speed of a curve.

5.4 Arclength and Reparametrization

Consider the following three parametrizations of the unit quarter circle (in the first quadrant) centred at the origin.

(1) $(\cos\frac{\pi}{2}t, \sin\frac{\pi}{2}t), t \in [0,1]$,

(2) $\left(\frac{1-t^2}{1+t^2}, \frac{2t}{1+t^2}\right), t \in [0,1]$, and

(3) $\left(\sqrt{1-t^2}, t\right), t \in [0,1]$.

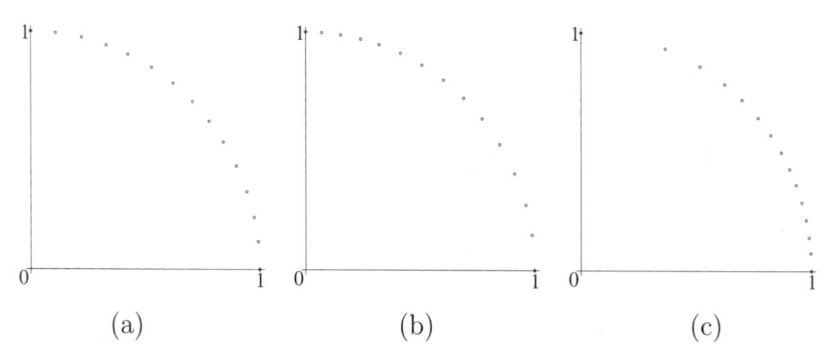

Figure 5.5 Different parametrizations of the quarter circle: (a) $(\cos\frac{\pi}{2}t, \sin\frac{\pi}{2}t)$, (b) $\left(\frac{1-t^2}{1+t^2}, \frac{2t}{1+t^2}\right)$, and (c) $\left(\sqrt{1-t^2}, t\right)$

Figure 5.5 shows the three parametrizations of the quarter circle evaluated at 15 equal parameter increments $t_i = i/14$, for $i = 0, \ldots, 14$. For parametrization (1) the points are equally spaced along the arc, for (2) the points are quite evenly spaced, and for (3) the points are unevenly spaced. The difference in the behaviour of the parametrizations corresponds to the fact that in (1) each parameter interval $[t_i, t_{i+1}]$ maps to a circular arc of equal length, whereas in (2) and (3) the parameter intervals map to circular arcs of varying lengths.

To explore this further a method to calculate the length of a curve is required. Consider a regular curve $\mathbf{C}(t) = (x(t), y(t))$, for $t \in [a, b]$, and a sequence of equally spaced parameter values $t_i = a + \frac{i}{n}(b - a)$ (for $i = 0, \ldots, n$) with $t_0 = a$ and $t_n = b$. The line segment from $(x(t_i), y(t_i))$ to $(x(t_{i+1}), y(t_{i+1}))$ approximates the curve on the interval $[t_i, t_{i+1}]$ and has length

$$\sqrt{\left(x(t_{i+1}) - x(t_i)\right)^2 + \left(y(t_{i+1}) - y(t_i)\right)^2} \ .$$

Thus the length $L(\mathbf{C})$ of the curve $\mathbf{C}$ on the interval $[a, b]$ is approximately

$$\sum_{i=0}^{n-1} \sqrt{\left(x(t_{i+1}) - x(t_i)\right)^2 + \left(y(t_{i+1}) - y(t_i)\right)^2} \ . \tag{5.2}$$

If the parameter increments $\delta t = t_{i+1} - t_i = (b - a)/n$ are sufficiently small, then $x'(t_i)$ is approximately equal to $(x(t_{i+1}) - x(t_i))/(t_{i+1} - t_i)$, $y'(t_i)$ is approximately equal to $(y(t_{i+1}) - y(t_i))/(t_{i+1} - t_i)$, and substitution into (5.2) yields that $L(\mathbf{C})$ is approximately

$$\sum_{i=0}^{n-1} \sqrt{\left(x'(t_i)\right)^2 + \left(y'(t_i)\right)^2}\, \delta t \ . \tag{5.3}$$

The true length of the curve over $[a, b]$ is realized by letting n tend to infinity. As n increases the line segments fit the curve more closely, and (5.3) becomes a better approximation to the length of the curve. The limit of (5.3) as n tends to infinity is

$$L(\mathbf{C}) = \int_a^b \sqrt{\left(x'(t)\right)^2 + \left(y'(t)\right)^2}\, dt = \int_a^b \nu(t)\, dt \ .$$

$L(\mathbf{C})$ is called the *arclength* of $\mathbf{C}(t)$ from $t = a$ to $t = b$. The *arclength function* $L_{\mathbf{C}}(t) = \int_{t_0}^t \sqrt{x'(u)^2 + y'(u)^2}\ du$, for $a \le t_0 \le b$, measures the length of the curve segment from an initial point $(x(t_0), y(t_0))$ to the point $(x(t), y(t))$. Then $L(\mathbf{C}) = L_{\mathbf{C}}(b) - L_{\mathbf{C}}(a)$.

Example 5.7

1. The speed of the quarter circle $\mathbf{C}(t) = (\cos t, \sin t)$, for $t \in \left[0, \frac{\pi}{2}\right]$, is $\nu(t) = \sqrt{(-\sin t)^2 + (\cos t)^2} = 1$. The parametrization is unit speed and the arclength function from $t_0 = 0$ is $L_{\mathbf{C}}(t) = \int_0^t 1\ du = t$. The curve has length $L_{\mathbf{C}}\left(\frac{\pi}{2}\right) - L_{\mathbf{C}}(0) = \frac{\pi}{2}$.

2. The speed of $\mathbf{C}(t) = (\cos \frac{\pi}{2}t, \sin \frac{\pi}{2}t)$, for $t \in [0, 1]$, is

$$\nu(t) = \sqrt{\left(-\frac{\pi}{2}\sin\frac{\pi}{2}t\right)^2 + \left(\frac{\pi}{2}\cos\frac{\pi}{2}t\right)^2} = \frac{\pi}{2} .$$

The parametrization has constant speed and the arclength function from $t_0 = 0$ is $L_{\mathbf{C}}(t) = \int_0^t \frac{\pi}{2}\, du = \frac{\pi}{2}t$. The curve has length $L_{\mathbf{C}}(1) - L_{\mathbf{C}}(0) = \frac{\pi}{2}$.

3. The speed of $\mathbf{C}(t) = \left(\frac{1-t^2}{1+t^2}, \frac{2t}{1+t^2}\right)$, for $t \in [0, 1]$, is

$$\nu(t) = \sqrt{\left(\frac{-4t}{(1+t^2)^2}\right)^2 + \left(\frac{2(1-t^2)}{(1+t^2)^2}\right)^2} = \frac{2}{1+t^2} ,$$

and the arclength function from $t_0 = 0$ is $L_{\mathbf{C}}(t) = \int_0^t \frac{2}{1+u^2}\, du = 2\arctan(t)$. Thus, with this parametrization, the unit quarter circle starts at point $(1, 0)$ with speed $\nu(0) = 2$. As t increases the speed decreases until the curve reaches the point $(0, 1)$ when the curve has speed $\nu(1) = 1$. The curve has length $L_{\mathbf{C}}(1) - L_{\mathbf{C}}(0) = \frac{\pi}{2}$.

4. Let the unit quarter circle be parametrized by $\mathbf{C}(t) = \left(\sqrt{1-t^2}, t\right)$, for $t \in [0, 1]$. Then $(x'(t), y'(t)) = (-t(1-t^2)^{-1/2}, 1)$, and

$$\nu(t) = \sqrt{\left(-t(1-t^2)^{-1/2}\right)^2 + 1} = (1-t^2)^{-1/2} .$$

Thus, with this parametrization, the unit quarter circle starts at point $(1, 0)$ with speed $\nu(0) = 1$. As t increases the speed increases until the curve reaches the point $(0, 1)$ when the curve has infinite speed. The arclength is computed in Exercise 5.9.

The arclength functions of the three parametrizations of the quarter circle are illustrated in Figure 5.6. In each plot the vertical axis shows the length of the curve traced from $(x(0), y(0))$ to $(x(t), y(t))$. Naturally, the total curve length in each case is $\pi/2$. Parametrization (1) traces the curve uniformly. The speed of parametrization (2) is decreasing, so the curve is traced more quickly in the beginning than at the end. The speed of parametrization (3) is increasing so the curve is traced more quickly at the end than at the beginning.

Definition 5.8

Let $\mathbf{C}(t)$ and $\mathbf{D}(t)$ be curves defined on intervals I and J respectively. Then $\mathbf{D}$ is said to be a *reparametrization* of $\mathbf{C}$ if there exists a differentiable function $h : J \to I$ such that $h'(t) \neq 0$ and $\mathbf{D}(t) = \mathbf{C}(h(t))$ for all $t \in J$. The function $h(t)$ is referred to as a *reparametrization*.

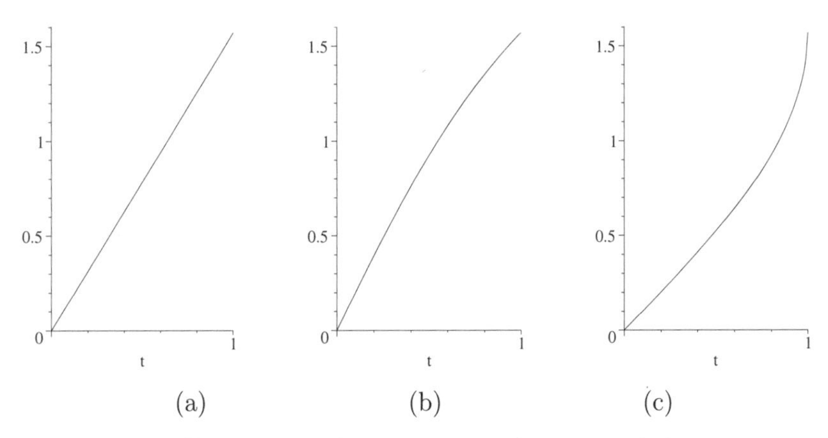

Figure 5.6 Comparison of the arclength functions of three parametrizations of the quarter circle

Example 5.7(1) is a unit-speed parametrization of the unit quarter circle. Parametrization (1) can be obtained from the unit speed parametrization by a reparametrization with $h(t) = \frac{\pi}{2}t$ and $J = [0, 1]$. Parametrizations (2) and (3) are also reparametrizations of the unit speed quarter circle. The next theorem shows that the arclength function can be used to reparametrize a curve to give a unit speed curve with the same trace.

Theorem 5.9

Let $\mathbf{C}(t) = (x(t), y(t))$ be a regular curve defined on an interval I with arclength function $s(t) = L_{\hat{\mathbf{C}}}(t)$. Then $\mathbf{C}(s) = \left(x\left(L_{\mathbf{C}}^{-1}(s)\right), y\left(L_{\mathbf{C}}^{-1}(s)\right)\right)$ is a unit speed curve.

Proof

Let $s = L_{\mathbf{C}}(t)$. Differentiating with respect to t gives $L'_{\mathbf{C}}(t) = |\mathbf{C}'(t)| = \nu(t)$. Since $\mathbf{C}(t)$ is regular, $L'_{\mathbf{C}}(t) \neq 0$ for all $u \in I$, and the inverse function theorem implies that the inverse $t = L_{\mathbf{C}}^{-1}(s)$ exists. Let $h(s) = L_{\mathbf{C}}^{-1}(s)$. Then $\frac{dh}{ds} = 1/L'_{\mathbf{C}} \neq 0$, and $t = h(s)$ may be used to reparametrize $\mathbf{C}(t)$ to give the curve $\hat{\mathbf{C}}(s) = \mathbf{C}(L_{\mathbf{C}}^{-1}(s)) = \left(x\left(L_{\mathbf{C}}^{-1}(s)\right), y\left(L_{\mathbf{C}}^{-1}(s)\right)\right)$. Then the chain rule gives

$$\frac{d\hat{\mathbf{C}}}{ds}(s) = \frac{dL_{\mathbf{C}}^{-1}(s)}{ds}\left(x'\left(L_{\mathbf{C}}^{-1}(s)\right), y'\left(L_{\mathbf{C}}^{-1}(s)\right)\right),$$

and

$$\begin{aligned}\left|\frac{d\hat{\mathbf{C}}}{ds}(s)\right| &= \left|\frac{dL_{\mathbf{C}}^{-1}(s)}{ds}\right| \left|\left(x'\left(L_{\mathbf{C}}^{-1}(s)\right), y'\left(L_{\mathbf{C}}^{-1}(s)\right)\right)\right| \\ &= \left|\frac{1}{L_{\mathbf{C}}'(t)}\right| |\mathbf{C}'(t)| = \frac{1}{|\mathbf{C}'(t)|} |\mathbf{C}'(t)| = 1 \,.\end{aligned}$$

Hence $\hat{\mathbf{C}}(s)$ is unit speed, and the proof is complete.

□

Example 5.10

Consider parametrization (3) of the unit quarter circle. Exercise 5.9 will determine the arclength function to be $s = \arcsin(u)$. Substituting $t = \sin(s)$ into $\left(\sqrt{1-t^2}, t\right)$ gives the unit speed parametrization of the circle $(\cos(s), \sin(s))$.

EXERCISES

5.9. Show that the arclength function of $\mathbf{C}(t) = \left(\sqrt{1-t^2}, t\right), t \in [0,1]$ is $L_{\mathbf{C}}(t) = \arcsin t$ (assume arcsin has range $\left[-\frac{1}{2}\pi, \frac{1}{2}\pi\right]$).

5.10. Determine the speed and arclength function for each of the following curves

(a) *cycloid:* $(t + \sin t, 1 - \cos t)$, $t \in [-\pi, \pi]$.

(b) $(\cos t + t \sin t, \sin t - t \cos t)$, $t > 0$.

(c) *catenary:* $(c \cosh(t/c), t)$, $t \in \mathbb{R}$.

(d) *astroid:* $\left(\cos^3 t, \sin^3 t\right)$, $t \in [0, \pi/2]$.

(e) *logarithmic spiral:* $\left(ae^{bt} \cos t, ae^{bt} \sin t\right)$.

5.11. Determine the length of the cycloid and the astroid of the previous exercise.

5.12. Write a program to determine the arclength of a curve using the summation formula (5.2) to within a user specified accuracy $\epsilon > 0$. This is achieved by increasing the number of increments n until the difference between successive computed approximate lengths is less than ϵ.

5.13. Determine the unit speed parametrization of the unit quarter circle from parametrization (2) by reparametrizing with the arclength function.

5.14. Obtain a unit speed reparametrization of

(a) *cycloid:* $(t + \sin t, 1 - \cos t)$.

(b) $(\cos t + t\sin t, \sin t - t\cos t)$.

(c) *catenary:* $(c\cosh(t/c), t)$.

(d) *logarithmic spiral:* $\left(ae^{bt}\cos t, ae^{bt}\sin t\right)$.

(e) $y = \cosh(x - 1)$.

5.5 Application: Numerical Controlled Machining and Offsets

Numerically controlled (NC) milling machines are used to make products and parts, or the moulds and dies from which the parts are manufactured. A CAD definition of a curve, describing the shape of a part, can be converted into a sequence of commands which are used to drive the milling machine cutting tool. NC machines can be programmed to move the tool in various ways. For instance, a five-axis machine can perform both translations and orientations of the tool, whereas a two-axis machine can translate the tool freely in the x- and y-directions, but a fixed orientation of the tool is maintained. The NC machine is programmed to move the cutter along a path so that the unwanted portion of the material is removed, and the remaining material has the desired shape.

In many applications the tool is a *ball-end* or *ball-nose cutter*. For a two-axis machine, cutting in a specified plane, the ball-end cutter can be considered a circular disk of fixed radius d. Suppose the shape to be cut is given by a regular curve $\mathbf{C}(t) = (x(t), y(t))$, with unit normal $\mathbf{n}(t)$. Referring to Figure 5.7, the cutter disk is required to be perpendicular to the curve, which implies that the disk centre is a distance d along the curve normal. Therefore, as the shape is cut, the disk centre follows the path of the curve $\mathbf{O}_d(t) = \mathbf{C}(t) + d \cdot \mathbf{n}(t)$ called the *offset* or *parallel* of $\mathbf{C}(t)$ at a distance d. The sign of d determines which side of the curve the cutter lies. Offset curves generalise to offset surfaces which are discussed in Section 9.2.1.

Example 5.11

Consider the curve $\mathbf{C}(t) = (x(t), y(t)) = (t, t^2)$. Then $(x'(t), y'(t)) = (1, 2t)$.

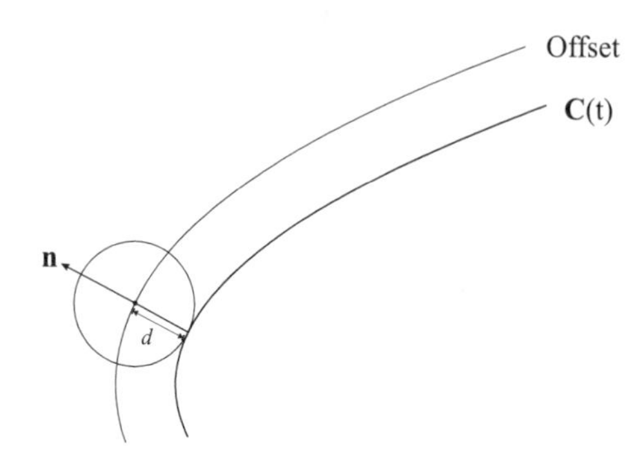

Figure 5.7 Path of centre of ball-end cutter along offset

Hence, $\mathbf{n}(t) = \left(-\frac{2t}{\sqrt{1+4t^2}}, \frac{1}{\sqrt{1+4t^2}}\right)$, and the offset at a distance d is

$$\mathbf{O}_d(t) = \left(t - d\frac{2t}{\sqrt{1+4t^2}}, t^2 + d\frac{1}{\sqrt{1+4t^2}}\right) .$$

Figure 5.8 shows the offsets at distances $d = -2, -1, 0.5, 1$. Note that the offsets are not parabolas. The $d = 1$ offset exhibits *cusp singularities.* If the cutting is to be executed on the same side as the normal to the parabola, then the cutting disk must have a radius less than 1 in order to avoid singularities of the offset. Such singularities indicate that the cutting tool is too big to cut the desired shape. (See Exercise 10.7.)

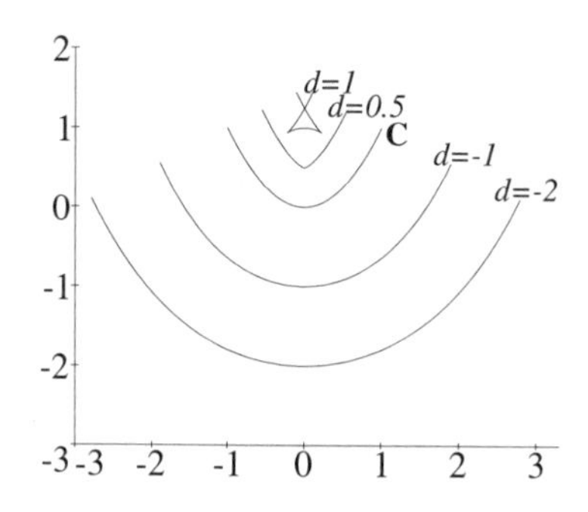

Figure 5.8 Offsets of the parabola (t, t^2)

EXERCISES

5.15. Determine the offset of $\mathbf{C}(t) = \left(1 - 3t + 3t^2, 3t^2 - 2t^3\right)$ at a distance d. Plot the curve and its offset at a distance $d = 1$.

5.16. Determine the offsets at a distance d for the following curves:

(a) $(c\cosh(t/c), t)$.

(b) $(e^{bt}\cos t, e^{bt}\sin t)$.

(c) $(\cos t + t\sin t, \sin t - t\cos t)$.

5.17. Show that the offset at a distance d of a circle of radius r is a circle of radius $r + d$.

5.6 Conics

The simplest implicitly defined planar curve is a straight line given by a linear equation $ax + by + c = 0$. Curves defined implicitly by a quadratic polynomial equation

$$ax^2 + 2bxy + cy^2 + 2dx + 2ey + f = 0 \qquad (5.4)$$

are called *conics*. Circles, ellipses, hyperbolas, and parabolas are all types of conic. "Conics" or "conic sections" receive their name from a classical geometrical method of construction, namely, as the curve of intersection of a plane with a cone.

A *cone* is the surface formed by rotating a line $\mathbf{L}$ through a fixed point $\mathbf{O}$ about a fixed axis $\mathbf{OA}$ so that $\mathbf{L}$ maintains a constant angle $\alpha < \pi/2$ with the axis. The point $\mathbf{O}$ is called the *vertex* of the cone. The cone consists of two parts called *sheets* which meet at the vertex. Consider a plane, not passing through $\mathbf{O}$, making an angle β with the axis. When $\beta > \alpha$, the intersection curve of the plane and the cone is an *ellipse* lying entirely in one sheet. When $\beta = \pi/2$ (so the axis is perpendicular to the plane) the intersection is a *circle*, a special case of the ellipse. When $\beta < \alpha$, the plane intersects both sheets of the cone resulting in a curve of two separate branches called a *hyperbola*. When $\beta = \alpha$, the plane intersects the cone in one sheet to give a curve called a *parabola*. The ellipse, parabola, and hyperbola are illustrated in Figure 5.9. There are also degenerate conics which arise when the plane passes through the vertex. The degenerate cases are a union of two lines when $\beta > \alpha$, two coincident lines when $\beta = \alpha$, and the point $\mathbf{O}$ when $\beta < \alpha$.

If $\mathbf{L}$ is a line parallel to the axis, then the resulting surface is a cylinder which may be considered a cone with its vertex at infinity. A plane intersects

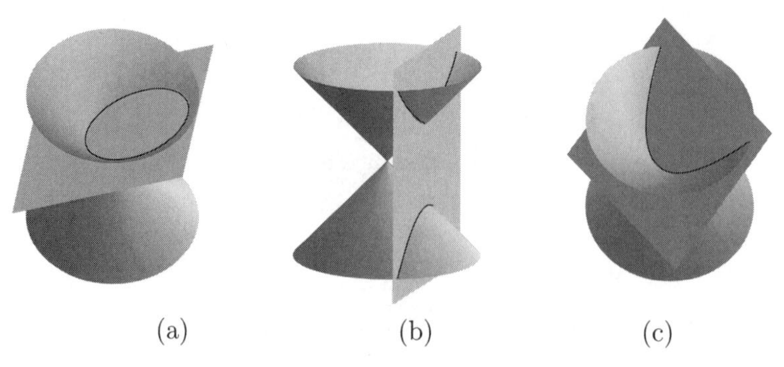

Figure 5.9 Conic sections

the cylinder in an ellipse, or in the degenerate cases of two distinct parallel lines, two coincident lines, or no intersection. The reader is referred to [26] and [5] for a historical account of conics and a proof that the sections of a cone are expressible by quadratic equations.

There is a second geometric construction for conics called the focus–directrix construction [26]. Given a fixed line $\mathbf{D}$ in the plane, called the *directrix*, and a fixed point $\mathbf{F}$, called the *focus*, the locus of all points $\mathbf{P}$ such that the distance $\mathbf{PF}$ from $\mathbf{P}$ to $\mathbf{F}$ is proportional to the distance $\mathbf{PD}$ from $\mathbf{P}$ to the directrix, is a conic. Thus there exists a constant ϵ, called the *eccentricity*, such that $\mathbf{PF} = \epsilon\mathbf{PD}$.

Example 5.12

Let a conic have directrix the x-axis, focus $\mathbf{F}(2,3)$, and eccentricity $\epsilon = 4$. Let $\mathbf{P}(x,y)$ be a general point on the conic. Then

$$\mathbf{PD} = y, \ \mathbf{PF} = \sqrt{(x-2)^2 + (y-3)^2} \ .$$

Hence $\mathbf{PF} = \epsilon\mathbf{PD}$ implies

$$\sqrt{(x-2)^2 + (y-3)^2} = 4y \ ,$$

giving the conic with the equation $x^2 - 15y^2 - 4x - 6y + 13 = 0$ shown in Figure 5.10.

To prove that the focus–directrix construction gives a conic it is sufficient to show that the curve satisfies the general equation (5.4). Suppose the directrix is the line $lx + my + n = 0$, and the focus is $(x_\mathbf{F}, y_\mathbf{F})$. Then

$$\mathbf{PD} = (lx_\mathbf{F} + my_\mathbf{F} + n) \Big/ \sqrt{l^2 + m^2}$$

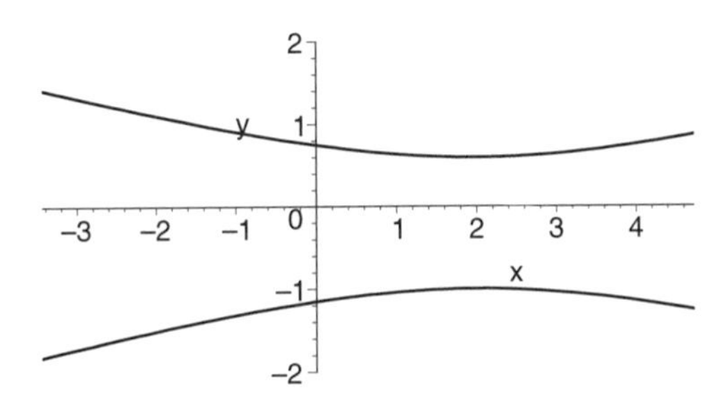

Figure 5.10 Conic $x^2 - 15y^2 - 4x - 6y + 13 = 0$

and

$$\mathbf{PF} = \sqrt{(x - x_{\mathbf{F}})^2 + (y - y_{\mathbf{F}})^2}\,.$$

Hence,

$$\epsilon(lx + my + n) \Big/ \sqrt{l^2 + m^2} = \sqrt{(x - x_{\mathbf{F}})^2 + (y - y_{\mathbf{F}})^2}\,.$$

Squaring both sides and multiplying through by $l^2 + m^2$ yields

$$\epsilon^2(lx + my + n)^2 = (l^2 + m^2)((x - x_{\mathbf{F}})^2 + (y - y_{\mathbf{F}})^2)$$

which is a quadratic equation in x and y of the form (5.4) where $a = \epsilon^2 l^2 - l^2 - m^2$, $b = \epsilon^2 lm$, $c = \epsilon^2 m^2 - \left(l^2 + m^2\right)$, $d = \epsilon^2 nl + x_{\mathbf{F}}(l^2 + m^2)$, $e = \epsilon^2 mn + y_{\mathbf{F}}\left(l^2 + m^2\right)$, $f = \epsilon^2 n^2 - \left(l^2 + m^2\right)\left(x_{\mathbf{F}}^2 + y_{\mathbf{F}}^2\right)$.

The converse, that any non-degenerate conic can be obtained by a focus–directrix construction, can be proved in two steps: (i) computation of the eccentricity of a conic expressed in implicit form, and (ii) computation of the focus and directrix. The first step is proved in Exercise 5.18, while the remainder of the proof can be found in [26].

Exercise 5.18

Let a conic have focus $(x_{\mathbf{F}}, y_{\mathbf{F}})$, eccentricity ϵ, and directrix with equation $x\cos\theta + y\sin\theta - p = 0$. Expand the expression

$$\left((x - x_{\mathbf{F}})^2 + (y - y_{\mathbf{F}})^2\right) = \epsilon^2(x\cos\theta + y\sin\theta - p)^2$$

and compare the coefficients with a scalar multiple of the coefficients of (5.4). Show that

$$\frac{\left(2 - \epsilon^2\right)^2}{1 - \epsilon^2} = \frac{(a + c)^2}{ac - b^2}\,.$$

5.6.1 Classification of Conics

Consider a conic defined by Equation (5.4). If (5.4) is a product of two linear factors, then the conic is a union of two lines and it is said to be a *reducible* conic. Otherwise, the conic is said to be *irreducible*. A condition on the coefficients of (5.4) for the conic to be reducible is determined as follows. Suppose that $a \neq 0$. Then multiply (5.4) through by a and complete the square to give

$$(ax + by + d)^2 - ((b^2 - ac)y^2 + 2(bd - ae)y + (d^2 - af)) = 0 \,. \tag{5.5}$$

Let $A = b^2 - ac$, $B = 2(bd - ae)$, and $C = (d^2 - af)$. Then (5.5) can be written

$$(ax + by + d)^2 - \left(Ay^2 + By + C\right) = 0 \,. \tag{5.6}$$

The expression (5.6) has two linear factors if and only if it can be written as the difference of two squares. Thus $Ay^2 + By + C$ must be a perfect square, which is possible if and only if $B^2 - 4AC = 0$. Hence the condition for the conic to be reducible is

$$B^2 - 4AC = 4(bd - ae)^2 - 4(b^2 - ac)(d^2 - af) = 0 \,.$$

Dividing through by $-4a$, the condition for reducibility can be expressed as the following determinant Δ which is called the *discriminant* of the conic.

$$\Delta = \begin{vmatrix} a & b & d \\ b & c & e \\ d & e & f \end{vmatrix} = 0 \,.$$

When $\Delta = 0$, $Ay^2 + By + C = A(y + B/2A)^2$ and two cases can be distinguished. (1) When $A = b^2 - ac \geq 0$, (5.6) has two real linear factors and the conic is a pair of lines. (2) When $A = b^2 - ac < 0$, (5.6) has two imaginary linear factors and the conic is an isolated point. The reader is left the exercise of showing that $\Delta = 0$ is also the condition for reducibility in the case when $a = 0$.

Next, suppose that (5.4) has two real linear factors ($a \neq 0$, $b^2 - ac \geq 0$)

$$a(x - \alpha_1 y + \beta_1)(x - \alpha_2 y + \beta_2) = 0 \,.$$

Expanding the brackets and comparing the coefficients of the resulting expression with (5.4) gives $\alpha_1\alpha_2 = c/a, \alpha_1 + \alpha_2 = -2b/a$. A simple computation yields that the angle θ between the two lines is given by $\tan\theta = 2\sqrt{b^2 - ac}/\,(a + c)$. It follows that the conic is a pair of perpendicular lines when $\Delta = 0$ and $a + c = 0$, and a pair of parallel lines whenever $\Delta = 0$ and $b^2 - ac = 0$. This concludes the study of the reducible conics.

The irreducible conics are as follows: (1) hyperbolas when $b^2 - ac > 0$, (2) ellipses when $b^2 - ac < 0$, and (3) parabolas when $b^2 - ac = 0$. The distinction can be explained by the conic's behaviour at infinity. Let (X, Y, W) be homogeneous coordinates of a point (x, y), so that $x = X/W$ and $y = Y/W$. Substituting into (5.4) and multiplying through by W^2 yields that the homogeneous coordinates of any point on the conic satisfies the homogeneous equation

$$\mathbf{C}(X, Y, W) = aX^2 + 2bXY + cY^2 + 2dXW + 2eYW + fW^2 = 0\,. \tag{5.7}$$

The points at infinity of the conic are obtained by setting $W = 0$ in (5.7) to give

$$aX^2 + 2bXY + cY^2 = 0\,. \tag{5.8}$$

When $b^2 - ac > 0$, (5.8) can be expressed as two distinct real linear factors $a\,(X + \mu_1 Y)\,(X + \mu_2 Y) = 0$, and it follows that the conic has two distinct real points at infinity, $(\mu_1, -1, 0)$ and $(\mu_2, -1, 0)$. Likewise, when $b^2 - ac < 0$, (5.8) can be expressed as two complex conjugate linear factors which give rise to two complex conjugate points at infinity. When $b^2 - ac = 0$, (5.8) can be expressed as a perfect square $a\,(X + \mu_1 Y)^2 = 0$, and hence the conic has a repeated real point at infinity, $(\mu_1, -1, 0)$.

The tangent lines to a curve at points at infinity are the *asymptotes* of the curve. Therefore, when $b^2 - ac > 0$ the irreducible conic has two real asymptotes and the curve is a hyperbola, and when $b^2 - ac < 0$, there are no real asymptotes and the conic is an ellipse. When $b^2 - ac = 0$, the asymptote is the line at infinity $W = 0$ and the conic is a parabola.

Definition 5.13

The *centre* of a conic $\mathbf{C}(x, y) = 0$ is the point (x, y) satisfying

$$\frac{\partial \mathbf{C}}{\partial x}(x, y) = 2ax + 2by + 2d = 0, \quad \frac{\partial \mathbf{C}}{\partial y}(x, y) = 2bx + 2cy + 2e = 0\,.$$

If $b^2 - ac \neq 0$ then the conic has centre

$$(x, y) = \left((be - cd) \,/\, \left(ac - b^2\right), (bd - ae) \,/\, \left(ac - b^2\right)\right)\,,$$

otherwise there is no centre. Conics with a centre are called *central* conics. Of the irreducible conics, the ellipse and hyperbola are central conics but the parabola is not.

In addition to the implicit form (5.4), conics have a parametric form

$$(x(t), y(t)) = \left(\frac{a_0 + a_1 t + a_2 t^2}{c_0 + c_1 t + c_2 t^2}, \frac{b_0 + b_1 t + b_2 t^2}{c_0 + c_1 t + c_2 t^2}\right)\,, \tag{5.9}$$

where the coefficients c_0, c_1, c_2 are not all zero. Conics can also be defined parametrically by other functions such as trigonometric and hyperbolic functions.

Sections 5.6.4 and 5.6.5 show how to convert a non-degenerate conic from implicit to parametric form and vice versa. In particular, Theorem 5.26 shows that a parametric curve of the form (5.9) can be expressed in the implicit form (5.4), and is therefore a conic.

Recall that the irreducible types hyperbola/parabola/ellipse are distinguished by the fact that they have two real/one real/two complex conjugate points at infinity. For a parametric conic (5.9), the points at infinity occur at parameter values for which the denominator of the coordinate functions vanishes, that is, when $c_2t^2 + c_1t + c_0 = 0$. When $c_1^2 - 4c_0c_2 > 0$ the denominator vanishes at two real values of t which give rise to two real points at infinity. The conic is therefore a hyperbola. Similarly, it can be shown that the conic is an ellipse when $c_1^2 - 4c_0c_2 < 0$, and a parabola when $c_1^2 - 4c_0c_2 = 0$. In particular, if a conic is parametrized by quadratic polynomials, then $c_1 = c_2 = 0$, $c_0 = 1$, and the conic is a parabola.

Summary. A conic is either *irreducible* when $\Delta \neq 0$, or *reducible* when $\Delta = 0$. The irreducible conics have three distinct types, namely, ellipse, parabola, and hyperbola. Reducible conics are a union of two lines (real or imaginary) which may be distinct and non-parallel, distinct and parallel, or coincident. The types of conic are summarized in Table 5.1.

Table 5.1 Summary of conic types

$b^2 - ac$	Δ	**Central**	**Conic type**	$c_1^2 - 4c_0c_2$
> 0	$\neq 0$	Yes	hyperbola	> 0
< 0	$\neq 0$	Yes	ellipse	< 0
$= 0$	$\neq 0$	No	parabola	$= 0$
> 0	$= 0$	Yes	two real distinct intersecting lines	
< 0	$= 0$	Yes	two complex conjugate lines intersecting in a real point	
$= 0$	$= 0$	No	two real distinct parallel lines	
$= 0$	$= 0$	No	two real coincident lines	

Example 5.14

The following examples show how to determine whether a conic is irreducible or reducible, and whether an irreducible conic is an ellipse, a hyperbola, or a parabola.

1. Consider the conic given by $x^2+2xy-3y^2+4x-5=0$. Then $a=1$, $b=1$, $c=-3$, $d=2$, $e=0$, $f=-5$,

$$\Delta = \begin{vmatrix} 1 & 1 & 2 \\ 1 & -3 & 0 \\ 2 & 0 & -5 \end{vmatrix} = 32 \,.$$

   Since $\Delta \neq 0$ the conic is irreducible. Further, $b^2-ac=4>0$, hence the conic is a hyperbola.

2. Consider the conic given by $-2x^2+xy-x-y+3=0$. Then $a=-2$, $b=1/2$, $c=0$, $d=-1/2$, $e=-1/2$, $f=3$,

$$\Delta = \begin{vmatrix} -2 & 1/2 & -1/2 \\ 1/2 & 0 & -1/2 \\ -1/2 & -1/2 & 3 \end{vmatrix} = 0 \,.$$

   Since $\Delta = 0$ the conic is reducible. Completing the square of the conic yields

$$-\frac{1}{2}\left(-2x+\frac{1}{2}y-\frac{1}{2}\right)^2+\left(-\frac{5}{4}y+\frac{25}{8}+\frac{1}{8}y^2\right)=0 \,.$$

   Factorize the quadratic in y to give the difference of two squares

$$-\frac{1}{2}\left(-2x+\frac{1}{2}y-\frac{1}{2}\right)^2+\frac{1}{8}(y-5)^2=0 \,.$$

   Factorizing,

$$(x-1)(-2x+y-3)=0 \,.$$

   Hence, the conic is the union of the two lines $x-1=0$ and $2x-y+3=0$.

EXERCISES

5.19. For each of the conics below determine whether the conic is irreducible or reducible. If it is irreducible then determine whether it is an ellipse, a hyperbola, or a parabola. If the conic is reducible then determine the linear factors.

(a) $4x^2-3xy+y^2-x+2y+7=0$.

(b) $-2x^2 + y^2 + 3x - 4y + 1 = 0$.

(c) $3x^2 - 5xy - x - 2y^2 + 9y - 4 = 0$.

(d) $x^2 - 3xy + 5y^2 - 2x + 6 = 0$.

(e) $2x^2 + 2xy - 5x - 3y + 3 = 0$.

(f) $2x^2 - 2y^2 + 3x + 4y + 7 = 0$.

(g) $2xy + 3y = 5$.

(h) $3x^2 - xy - 2y^2 + 6x + 4y = 0$.

(i) $2x^2 + 2xy - 3x - 3y + 1 = 0$.

(j) $2x^2 - 4xy + 2y^2 - 9 = 0$.

5.20. Let $\mathbf{C}(x, y) = 0$ be a central conic. Any line through the centre is called a *diameter*. The centre is the midpoint of the two points of intersection of $\mathbf{C}$ with any diameter. Verify this fact for

(a) the hyperbola $4x^2 - 9y^2 = 16$, and

(b) the ellipse $4x^2 + 9y^2 + x - 6y = 0$.

5.21. There are conics which have no real points. For example, $x^2 + y^2 = -1$. Determine others.

5.6.2 Conics in Standard Form

It can be shown that hyperbolas and ellipses have two lines of reflectional symmetry, and that a parabola has one. A conic for which the lines of symmetry coincide with the coordinate axes is said to be in *standard form*. It will be shown that any irreducible conic can be obtained by applying a composite transformation consisting of a rotation and a translation to a conic in standard form. Conversely, any conic can be obtained by applying a transformation to a conic in standard form. The implicit and parametric standard forms of the irreducible conics are given in Table 5.2.

Recall that in homogeneous coordinates, the conic (5.4) is given by (5.7). Let $\mathbf{x} = (X, Y, W)$ then (5.7) can be expressed in the matrix form

$$\mathbf{C}(X, Y, W) = \mathbf{x}\mathsf{M}\mathbf{x}^T = \begin{pmatrix} X & Y & W \end{pmatrix} \begin{pmatrix} a & b & d \\ b & c & e \\ d & e & f \end{pmatrix} \begin{pmatrix} X \\ Y \\ W \end{pmatrix}.$$

The discriminant of $\mathbf{C}$ is denoted $\Delta_{\mathbf{C}} = \det(\mathsf{M})$.

Table 5.2 Standard forms for the irreducible conics

Conic	**Implicit forms**	**Parametric forms**
Hyperbola	$x^2/a^2 - y^2/b^2 = 1$	$\left(\frac{a(b^2+a^2t^2)}{a^2t^2-b^2}, \frac{2ab^2t}{a^2t^2-b^2}\right)$, $t \in \mathbb{R}$, $t \neq \frac{b}{a}$; $\pm(a\cosh\theta, b\sinh\theta)$, $\theta \in \mathbb{R}$; $(a\sec\theta, b\tan\theta)$, $\theta \in \left(-\frac{\pi}{2}, \frac{\pi}{2}\right)$, and $\theta \in \left(\frac{\pi}{2}, \frac{3\pi}{2}\right)$.
Parabola	$y = mx^2$ $x = my^2$	(t, mt^2) , $t \in \mathbb{R}$; (mt^2, t) , $t \in \mathbb{R}$.
Ellipse	$x^2/a^2 + y^2/b^2 = 1$	$\left(\frac{a(1-t^2)}{1+t^2}, \frac{2bt}{1+t^2}\right)$, $t \in \mathbb{R}$; $(a\cos\theta, b\sin\theta)$, $\theta \in [0, 2\pi]$.

Theorem 5.15

Let $\hat{\mathbf{C}}$ be the image of the conic $\mathbf{C} = \mathbf{x}\mathsf{M}\mathbf{x}^T$ following the application of a non-singular planar transformation with transformation matrix A. Then

$$\Delta_{\hat{\mathbf{C}}} = \det(\mathsf{A})^2 \Delta_{\mathbf{C}} \, .$$

So a non-singular transformation does not affect the irreducibility of a conic.

Proof

Let the conic be $\mathbf{C} = \mathbf{x}\mathsf{M}\mathbf{x}^T$. Then the transformation $\mathbf{x} = \mathbf{y}\mathsf{A}$ yields a conic $\hat{\mathbf{C}} = (\mathbf{y}\mathsf{A})\mathsf{M}(\mathbf{y}\mathsf{A})^T = \mathbf{y}\mathsf{A}\mathsf{M}\mathsf{A}^T\mathbf{y}^T = \mathbf{y}\hat{\mathsf{M}}\mathbf{y}^T$, where $\hat{\mathsf{M}} = \mathsf{A}\mathsf{M}\mathsf{A}^T$. Hence

$$\Delta_{\hat{\mathbf{C}}} = \det(\mathsf{A}\mathsf{M}\mathsf{A}^T) = \det(\mathsf{A})\det(\mathsf{M})\det(\mathsf{A}^T) = \det(\mathsf{A})^2\det(\mathsf{M}) = \det(\mathsf{A})^2\Delta_{\mathbf{C}}.$$

Since A is non-singular, $\det(\mathsf{A}) \neq 0$. Therefore $\Delta_{\hat{\mathbf{C}}} = 0$ if and only if $\Delta_{\mathbf{C}} = 0$. Hence $\mathbf{C}$ is irreducible if and only if $\hat{\mathbf{C}}$ is irreducible.

□

The distinctions of hyperbola, parabola, and ellipse apply with respect to a particular Cartesian coordinate system. The effect on a conic of an application of an orthogonal change of coordinates (see Exercise 2.21) is expressed by the following theorem.

Theorem 5.16

Let $\hat{\mathbf{C}}$ be the image of the conic $\mathbf{C}$ (given by (5.7)) following the application of an orthogonal change of coordinates. Then

$$\begin{aligned} b^2 - ac &= \hat{b}^2 - \hat{a}\hat{c}\,, \\ a + c &= \hat{a} + \hat{c}\,, \end{aligned}$$

where $\hat{a}$, $\hat{b}$ and $\hat{c}$ denote the corresponding coefficients of $\hat{\mathbf{C}}$.

Proof

Let the orthogonal change of coordinates be $X = \hat{X}\cos\theta - \hat{Y}\sin\theta + g\hat{W}$, $Y = \hat{X}\sin\theta + \hat{Y}\cos\theta + h\hat{W}$, $W = \hat{W}$ (expressed in homogeneous coordinates). Then substituting for X and Y in (5.7) yields

$$\hat{a}\hat{X}^2 + 2\hat{b}\hat{X}\hat{Y} + \hat{c}\hat{Y}^2 + \hat{d}\hat{X}\hat{W} + \hat{e}\hat{Y}\hat{W} + \hat{f}\hat{W}^2 = 0$$

where

$$\begin{aligned} \hat{a} &= a\cos^2\theta + 2b\cos\theta\sin\theta + c\sin^2\theta\,, \\ \hat{b} &= b\left(\cos^2\theta - \sin^2\theta\right) + (c-a)\sin\theta\cos\theta\,, \\ \hat{c} &= a\sin^2\theta - 2b\cos\theta\sin\theta + c\cos^2\theta\,, \\ \hat{d} &= bg\sin\theta + e\sin\theta + bh\cos\theta + d\cos\theta + ag\cos\theta + ch\sin\theta\,, \\ \hat{e} &= ch\cos\theta - ag\sin\theta + bg\cos\theta - bh\sin\theta - d\sin\theta + e\cos\theta\,, \text{ and} \\ \hat{f} &= ag^2 + ch^2 + 2dg + 2bgh + f + 2eh\,. \end{aligned}$$

Then

$$\begin{aligned} \hat{b}^2 - \hat{a}\hat{c} &= \left(b\left(\cos^2\theta - \sin^2\theta\right) + (c-a)\sin\theta\cos\theta\right)^2 \\ &\quad - \left(a\cos^2\theta + 2b\cos\theta\sin\theta + c\sin^2\theta\right) \\ &\quad \times \left(a\sin^2\theta - 2b\cos\theta\sin\theta + c\cos^2\theta\right) \\ &= b^2 - ac\,, \end{aligned}$$

and

$$\begin{aligned} \hat{a} + \hat{c} &= \left(a\cos^2\theta + 2b\cos\theta\sin\theta + c\sin^2\theta\right) \\ &\quad + \left(a\sin^2\theta - 2b\cos\theta\sin\theta + c\cos^2\theta\right) \\ &= a + c\,. \end{aligned}$$

□

Let A be the transformation matrix of an orthogonal change of coordinates. Then, by Exercise 2.21, $\det(\mathsf{A}) = 1$. Since $\hat{b}^2 - \hat{a}\hat{c} = b^2 - ac$ and $\Delta_{\hat{\mathsf{C}}} = \det(\mathsf{A})^2 \Delta_{\mathsf{C}} = \Delta_{\mathsf{C}}$ the type of conic is unaffected by a change of coordinates. If (5.4) is multiplied through by a constant μ, then the quantities $a+c$, $b^2 - ac$, and Δ become $\mu\,(a+c)$, $\mu^2\left(b^2 - ac\right)$, and $\mu^3 \Delta$. Thus the ratios $(a+c) : \left(b^2 - ac\right)^{1/2} : \Delta^{1/3}$ are *absolute invariants*. A conic expressed in any rectangular Cartesian coordinate system has the same absolute invariants.

Theorem 5.17

An irreducible conic can be mapped to a conic in standard form by applying an orthogonal change of coordinates.

Proof

Let the conic be given by (5.7). First, apply a rotation $X = \hat{X}\cos\theta - \hat{Y}\sin\theta$, and $Y = \hat{X}\sin\theta + \hat{Y}\cos\theta$. Then (5.7) has the form

$$\hat{a}\hat{X}^2 + 2\hat{b}\hat{X}\hat{Y} + \hat{c}\hat{Y}^2 + 2\hat{d}\hat{X}\hat{W} + 2\hat{e}\hat{Y}\hat{W} + \hat{f}\hat{W}^2 = 0\,, \tag{5.10}$$

where the coefficients are given by the expressions in Theorem 5.16, but with g and h set equal to zero. The $\hat{X}\hat{Y}$ term can be eliminated by choosing the angle θ so that the coefficient $\hat{b}$ vanishes: the required angle satisfies $\tan 2\theta = 2b/(a-c)$ if $a \neq c$, and $\theta = \pi/4$ or $3\pi/4$ if $a = c$. Provided $\hat{a} \neq 0$ and $\hat{c} \neq 0$, then (5.10) has the form

$$\hat{a}\left(\hat{X} + \frac{\hat{d}}{\hat{a}}\hat{W}\right)^2 + \hat{c}\left(\hat{Y} + \frac{\hat{e}}{\hat{c}}\hat{W}\right)^2 + \left(\hat{f} - \frac{\hat{d}^2}{\hat{a}} - \frac{\hat{e}^2}{\hat{c}}\right)\hat{W}^2 = 0\,.$$

Applying the translation $\mathsf{T}\left(-\frac{\hat{d}}{\hat{a}}, -\frac{\hat{e}}{\hat{c}}\right)$, yields a standard form for the hyperbola or ellipse given by

$$\hat{a}\hat{X}^2 + \hat{c}\hat{Y}^2 + \left(\hat{f} - \frac{\hat{d}^2}{\hat{a}} - \frac{\hat{e}^2}{\hat{c}}\right)\hat{W}^2 = 0\,.$$

If $\hat{a} = 0$, $\hat{c} \neq 0$, then (5.10) has the form

$$\hat{c}\left(\hat{Y} + \frac{\hat{e}}{\hat{c}}\hat{W}\right)^2 + 2\hat{d}\hat{X}\hat{W} + \hat{f}\hat{W}^2 = 0\,.$$

Applying the translation $\mathsf{T}\left(-\frac{\hat{f}}{2\hat{d}}, -\frac{\hat{e}}{\hat{c}}\right)$ gives

$$\hat{c}\hat{Y}^2 + 2\hat{d}\hat{X}\hat{W}\,,$$

a standard form for the parabola. Similarly, a standard form for the parabola is obtained when $\hat{a} \neq 0$, $\hat{c} = 0$. The case $\hat{a} = 0$, $\hat{c} = 0$ is not considered since the conic is reducible.

□

Example 5.18

To determine the standard form for the ellipse

$$2x^2 - 2\sqrt{3}xy + 4y^2 + 5x + 6y - 1 = 0 \,.$$

Then $a = 2$, $b = -\sqrt{3}$, $c = 4$, and the required rotation angle is given by $\tan 2\theta = \sqrt{3}$, yielding $\theta = \pi/6$. Then $\cos(\pi/6) = \sqrt{3}/2$, $\sin(\pi/6) = \frac{1}{2}$, and the required rotation is $X = \frac{\sqrt{3}}{2}\hat{X} - \frac{1}{2}\hat{Y}$, and $Y = \frac{1}{2}\hat{X} + \frac{\sqrt{3}}{2}\hat{Y}$, giving the conic

$$\begin{aligned} &2\left(\frac{\sqrt{3}}{2}\hat{X} - \frac{1}{2}\hat{Y}\right)^2 - 2\sqrt{3}\left(\frac{\sqrt{3}}{2}\hat{X} - \frac{1}{2}\hat{Y}\right)\left(\frac{1}{2}\hat{X} + \frac{\sqrt{3}}{2}\hat{Y}\right) \\ &+ 4\left(\frac{1}{2}\hat{X} + \frac{\sqrt{3}}{2}\hat{Y}\right)^2 + 5\left(\frac{\sqrt{3}}{2}\hat{X} - \frac{1}{2}\hat{Y}\right)\hat{W} \\ &+ 6\left(\frac{1}{2}\hat{X} + \frac{\sqrt{3}}{2}\hat{Y}\right)\hat{W} - \hat{W}^2 = 0 \,. \end{aligned}$$

Simplifying yields

$$\hat{X}^2 + 5\hat{Y}^2 + \left(\frac{5}{2}\sqrt{3} + 3\right)\hat{X}\hat{W} + \left(3\sqrt{3} - \frac{5}{2}\right)\hat{Y}\hat{W} - \hat{W}^2 = 0 \,.$$

Then, completing the squares in $\hat{X}$ and $\hat{Y}$,

$$\begin{aligned} &\left(\hat{X} + \left(\frac{5}{4}\sqrt{3} + \frac{3}{2}\right)\hat{W}\right)^2 + 5\left(\hat{Y} + \left(\frac{3}{10}\sqrt{3} - \frac{1}{4}\right)\right)^2 \\ &- \left(1 + \left(\frac{5}{4}\sqrt{3} + \frac{3}{2}\right)^2 + \left(\frac{3}{10}\sqrt{3} - \frac{1}{4}\right)^2\right)\hat{W}^2 = 0 \,, \end{aligned}$$

and making the translation $\bar{X} = \hat{X} + \left(\frac{5}{4}\sqrt{3} + \frac{3}{2}\right)\hat{W}$, $\bar{Y} = \hat{Y} + \left(\frac{3}{10}\sqrt{3} - \frac{1}{4}\right)\hat{W}$, $\bar{W} = \hat{W}$ gives

$$\bar{X}^2 + 5\bar{Y}^2 - \left(\frac{827}{100} + \frac{18}{5}\sqrt{3}\right)\bar{W}^2 = 0 \,.$$

In Cartesian coordinates, the standard form of the conic is $x^2 + 5y^2 = \left(\frac{827}{100} + \frac{18}{5}\sqrt{3}\right)$. Figure 5.11 shows the original conic and the computed standard form conic.

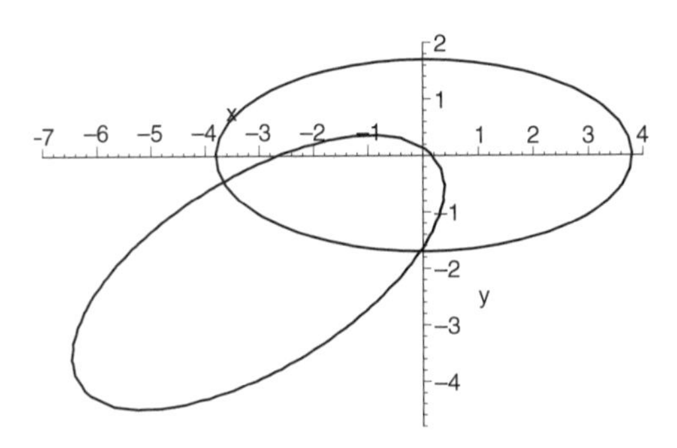

Figure 5.11

EXERCISES

5.22. Determine the standard form of the following conics:

(a) $13x^2 - 10xy + 13y^2 - 12\sqrt{2}x + 60\sqrt{2}y + 72 = 0$.

(b) $6x^2 + 12xy + 6y^2 - 35\sqrt{2}x - 37\sqrt{2}y + 118 = 0$.

(c) $11x^2 - 6x\sqrt{3}y - 6x\sqrt{3} + y^2 + 2y - 63 = 0$.

5.23. Determine the absolute invariants of the standard forms. Compute the absolute invariants for each of the conics of the previous exercise, and verify that the computed standard form has the same invariants.

5.24. Use Exercise 5.18 to show that a conic with eccentricity $\epsilon > 0$ is a hyperbola if $\epsilon > 1$, an ellipse if $\epsilon < 1$, or a parabola if $\epsilon = 1$.

5.25. Show that a translation leaves the values of a, b, c unaltered. These quantities are called *translational invariants*.

5.6.3 Intersections of a Conic with a Line

The points of intersection of a conic and a line are found by a process of elimination of variables to give a quadratic polynomial equation in one of the variables. The equation is solved, and backward substitution of the solutions is used to determine the points of intersection. The derivation of the quadratic depends on whether the conic and the line are given in implicit or in parametric form. The procedure is explained by means of examples.

Example 5.19

To find the intersections of the conic $x^2 - 3xy + y^2 + 4x - 2 = 0$ and the line parametrized by $(x(t), y(t)) = (2t+1, t-1)$, shown in Figure 5.12, substitute $x = 2t+1$ and $y = t-1$ into the conic equation to give

$$(2t+1)^2 + (t-1)^2 - 3(2t+1)(t-1) + 4(2t+1) - 2 = -t^2 + 13t + 7 = 0 .$$

The solutions are approximately $t = -0.518$ and $t = 13.518$. Substituting the values of t into the parametric equation $(x(t), y(t)) = (2t+1, t-1)$ gives the two points of intersection $(-0.036, -1.518)$ and $(28.036, 12.518)$.

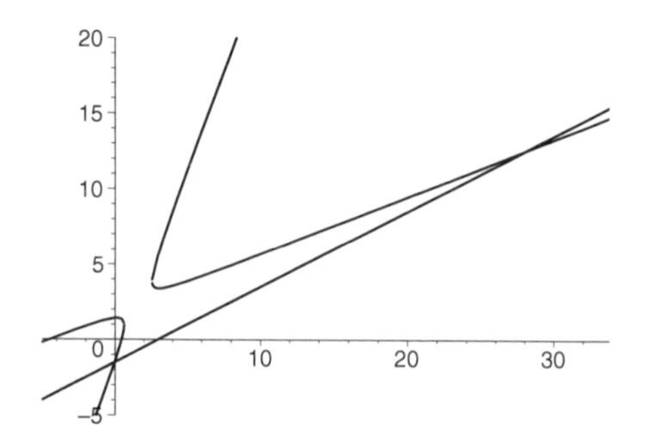

Figure 5.12 Intersection of conic $x^2 - 3xy + y^2 + 4x - 2 = 0$ and line $(2t+1, t-1)$

Example 5.20

To find the intersection of the conic $(x(t), y(t)) = \left(\frac{3t+t^2}{1+2t^2}, \frac{3t}{1+2t^2}\right)$ with the line $2x + y + 1 = 0$, shown in Figure 5.13(a), substitute $x = \frac{3t+t^2}{1+2t^2}$ and $y = \frac{3t}{1+2t^2}$ into the equation of the line to give

$$2\left(\frac{3t+t^2}{1+2t^2}\right) + \left(\frac{3t}{1+2t^2}\right) + 1 = \frac{4t^2+9t+1}{1+2t^2} = 0 .$$

The solutions of $4t^2 + 9t + 1 = 0$ are approximately $t = -2.133$ and $t = -0.117$. Substituting for t in $\left(\frac{3t+t^2}{1+2t^2}, \frac{3t}{1+2t^2}\right)$ yields the two points of intersection $(-0.183, -0.634)$ and $(-0.329, -0.342)$.

Example 5.21

To determine the intersection of the line $x + y - 1 = 0$ and the conic

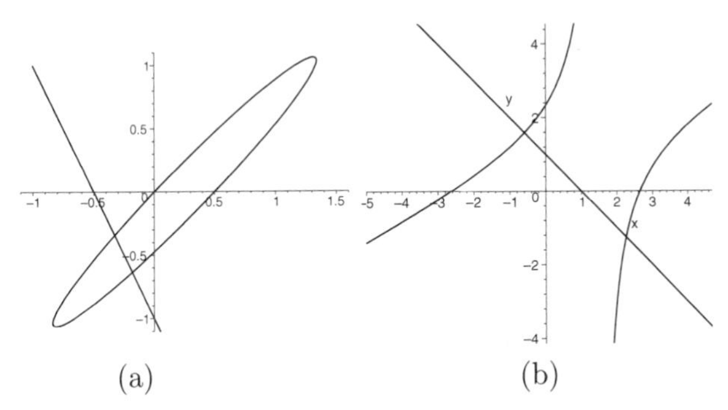

Figure 5.13 (a) Intersection of conic $\left(\frac{3t+t^2}{1+2t^2}, \frac{3t}{1+2t^2}\right)$ and line $2x+y+1=0$, and (b) intersection of conic $x^2-2xy+3y-7=0$ and line $x+y-1=0$

$x^2-2xy+3y-7=0$, shown in Figure 5.13(b), substitute $y=1-x$ into the conic equation to give

$$x^2-2x(1-x)+3(1-x)-7=3x^2-5x-4=0\ .$$

Solving yields $x=-0.591$ and $x=2.257$. Substituting the solutions into $y=1-x$ yields the points $(-0.591, 1.591)$ and $(2.257, -1.257)$.

If both the line and the conic are expressed in parametric form, then the line is converted to implicit form, and the method of Example 5.20 is applied.

EXERCISES

5.26. Find the points of intersection of the following conics and lines:

(a) conic $9x^2-xy+y^2-4x+2y+1=0$, line $(x(t), y(t))=(2t-3, -3t+4)$.

(b) line $x+3y-6=0$, conic $3x^2-2xy+y^2-5x+6y-16=0$.

(c) line $-2x+5y+7=0$, conic $(x(t), y(t))=(3t^2-4t+1, 2t^2-9t)$.

(d) line $(t+1, t-1)$, conic $x^2+2xy+x-y-1=0$.

(e) line $-3x-2y+4=0$, conic $(t^2+1, t-1)$.

5.27. The conic segment $(x(t), y(t))=(3t^2-4t-1, 2t^2-9t+10)$, $t\in[-1,4]$ is to be clipped by the rectangle with bottom left corner at $(0,0)$ and upper right corner $(20,20)$ as shown in Figure 5.14. The clipping operation removes the parts of the conic contained outside the rectangle. Determine the parameter values where the conic intersects each side of the rectangle. For each conic segment inside the rectangle, determine the parameter interval on which it is defined. For any parametrized conic, what is the maximum number of segments that arise following a rectangular clipping operation?

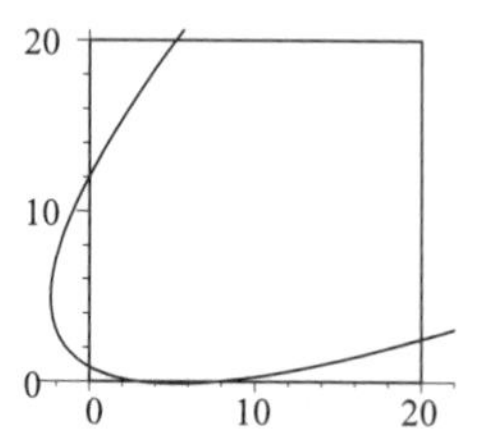

Figure 5.14

5.6.4 Parametrization of an Irreducible Conic

An irreducible conic $\mathbf{C}(x, y) = 0$ can be parametrized by performing the following steps.

1. Determine a point $\mathbf{P}(p_1, p_2)$ on the conic.
2. Consider the family of lines $y = (x - p_1)t + p_2$, parametrized by t, consisting of all lines in the plane through $\mathbf{P}$. Each parameter value t corresponds to a line in the family.
3. Determine the points of intersection of the line $y = (x - p_1)t + p_2$ and the conic as follows. Substitute $y = (x - p_1)t + p_2$ in $\mathbf{C}(x, y) = 0$ to give a quadratic polynomial (dependent on t) $q_t(x) = 0$. The roots of $q_t(x) = 0$ are the x–coordinates of the intersection points. Since $\mathbf{P}(p_1, p_2)$ is known to be an intersection point of the conic and every line in the family, it follows that $x = p_1$ is a root of $q_t(x)$. Hence $x - p_1$ is a factor of $q_t(x)$.
4. Factorise $q_t(x)$ as $(x - p_1)(\beta(t)x - \alpha(t))$ for some quadratic polynomials $\alpha(t)$ and $\beta(t)$. Then the second root of $q_t(x) = 0$ is $x = \alpha(t)/\beta(t)$, giving the x-coordinate of the other point of intersection as a function of t: $x = x(t)$.
5. Substitute $x = \alpha(t)/\beta(t)$ in $y = (x - p_1)t + p_2$ to give $y = y(t)$.
6. It follows that $(x(t), y(t))$ parametrizes the conic, since every point $\mathbf{Q}$ on the conic, distinct from $\mathbf{P}$, is the intersection of the conic and the line $\overline{\mathbf{PQ}}$ through $\mathbf{P}$.

Different choices of the point $\mathbf{P}$ will give rise to alternative parametrizations of the conic.

Example 5.22

Find a parametrization of the hyperbola $-2x^2 - 5xy + 4y^2 + x - 5y + 15 = 0$ by considering lines through the point $(2, 3)$.

It is easily checked that $(2, 3)$ is a point on the conic. The family of lines through $(2, 3)$ is given by $y = t(x - 2) + 3$. Substituting for y in the equation of the conic gives

$$-2x^2 - 5x(t(x-2)+3) + 4(t(x-2)+3)^2 + x - 5(t(x-2)+3) + 15 = 0\ ,$$

which factorizes as

$$(x-2)\left(-2x - 5tx + 4t^2x - 8t^2 + 19t - 18\right) = 0\ .$$

Solving for x yields $x = 2$ and

$$x = \frac{18 - 19t + 8t^2}{-2 - 5t + 4t^2}\ . \tag{5.11}$$

The solution $x = 2$ corresponds to the known intersection $(2, 3)$. Using (5.11) to substitute for x in $y = t(x - 2) + 3$ gives

$$y = t\left(\frac{18 - 19t + 8t^2}{-2 - 5t + 4t^2} - 2\right) + 3 = \frac{-6 + 7t + 3t^2}{-2 - 5t + 4t^2}\ .$$

It follows that

$$(x(t), y(t)) = \left(\frac{18 - 19t + 8t^2}{-2 - 5t + 4t^2}, \frac{-6 + 7t + 3t^2}{-2 - 5t + 4t^2}\right)$$

is a parametrization of the conic. The values of t for which the denominator vanishes are the solutions of $4t^2 - 5t - 2 = 0$, that is, $t = -0.319$ and $t = 1.569$. Therefore the curve is defined on the parameter intervals $(-\infty, -0.319)$, $(-0.319, 1.569)$, and $(1.569, \infty)$. Each interval corresponds to a branch or a part of a branch of the conic.

Example 5.23

Find a parametrization of the conic $\mathbf{C}(x, y) = x^2 - 2xy + 4y^2 + 2x + y + 1 = 0$.

First determine a point on the conic. One way to do this is to find an intersection of the conic with one of the axes. If this fails one can try intersecting with other lines parallel to one of the axes. Set $y = 0$ in $\mathbf{C}(x, y) = 0$. Then $x^2 + 2x + 1 = 0$. Thus $x = -1$. Hence $(-1, 0)$ is a point on the conic. Next consider the family of lines $y = (x+1)t$ through $(-1, 0)$. Substituting $y = (x+1)t$ in $\mathbf{C}(x, y) = 0$, gives

$$x^2 - 2x(x+1)t + 4((x+1)t)^2 + 2x + (x+1)t + 1 = 0 ,$$

and factorizing gives

$$(x+1)\left(-2xt + 4t^2x + x + 1 + 4t^2 + t\right) = 0 .$$

Each line in the family intersects the conic in two points: $(-1, 0)$ and one other. Setting the second factor equal to zero, and solving for x, gives the x-coordinate of the unknown intersection point

$$x = -\frac{4t^2 + t + 1}{4t^2 - 2t + 1} .$$

The y-coordinate is obtained by substituting for x in $y = (x+1)t$ giving

$$y = \left(\left(-\frac{4t^2 + t + 1}{4t^2 - 2t + 1}\right) + 1\right) t = -\frac{3t^2}{4t^2 - 2t + 1} .$$

Thus the conic is defined parametrically by

$$(x(t), y(t)) = \left(-\frac{4t^2 + t + 1}{4t^2 - 2t + 1}, -\frac{3t^2}{4t^2 - 2t + 1}\right) .$$

The denominator $4t^2 - 2t + 1$ does not vanish for real values of t, and hence the parametrization is defined for all t.

EXERCISES

5.28. Determine another parametrization of the conic $x^2 - 2xy + 4y^2 + 2x + y + 1 = 0$ of Example 5.23 by considering lines through the point $(-2, -1)$.

5.29. Convert the following conics from implicit to parametric form:

(a) $x^2 + 2y^2 - 2xy + 2y = 0$; consider lines through $(0, 0)$.

(b) $x^2 - 2xy + 5y^2 - 2x + 3y + 1 = 0$.

(c) $x^2 + 2xy - y^2 - 1 = 0$.

(d) $2x^2 - y^2 + 4x - 2y = 0$.

5.6.5 Converting from Parametric Form to Implicit Form

Conics defined by polynomial coordinate functions are easily converted to implicit form.

Example 5.24

Consider the conic defined by $x = 2t^2 - 3t$ and $y = t^2 + t - 2$. Add scalar multiples of the equations to eliminate the quadratic terms

$$\begin{aligned} x - 2y &= \left(2t^2 - 3t\right) - 2\left(t^2 + t - 2\right) \\ &= -5t + 4 \,. \end{aligned}$$

Solving for t in terms of x and y gives $t = (-x + 2y + 4)/\,5$. Substituting for t in $x = 2t^2 - 3t$ (or alternatively, in $y = t^2 + t - 2$) gives

$$x = 2\left(\frac{-x + 2y + 4}{5}\right)^2 - 3\left(\frac{-x + 2y + 4}{5}\right) .$$

Expanding and simplifying gives an implicit equation for the conic

$$x^2 - 4xy - 13x + 4y^2 + y - 14 = 0 \,.$$

In general, a conic is parametrized by rational functions and the approach indicated in Example 5.24 is tedious. A more general method follows from the following result.

Theorem 5.25

A necessary and sufficient condition that two quadratics

$$a_0 + a_1 t + a_2 t^2 = 0, \text{ and} \tag{5.12}$$

$$b_0 + b_1 t + b_2 t^2 = 0, \tag{5.13}$$

have a common solution is

$$\begin{vmatrix} a_0 b_2 - a_2 b_0 & a_1 b_2 - a_2 b_1 \\ a_0 b_1 - a_1 b_0 & a_0 b_2 - a_2 b_0 \end{vmatrix} = 0 \,. \tag{5.14}$$

Proof

Suppose (5.12) and (5.13) have a common solution. Then $b_2 \times (5.12) - a_2 \times (5.13)$ yields

$$(a_0 b_2 - a_2 b_0) + (a_1 b_2 - a_2 b_1)\, t = 0 \,. \tag{5.15}$$

Similarly, $b_0 \times (5.12) - a_0 \times (5.13)$ yields

$$((a_1 b_0 - a_0 b_1) + (a_2 b_0 - a_0 b_2)\, t)\, t = 0 \,. \tag{5.16}$$

Eliminating t from (5.15) and (5.16) gives

$$(a_0 b_2 - a_2 b_0)\,(a_2 b_0 - a_0 b_2) - (a_1 b_0 - a_0 b_1)\,(a_1 b_2 - a_2 b_1) = 0 \,,$$

and hence (5.14). The proof of the converse is left as an exercise for the reader.

□

Theorem 5.26

The conic with parametrization

$$x = \frac{a_0 + a_1 t + a_2 t^2}{c_0 + c_1 t + c_2 t^2}\,, \quad y = \frac{b_0 + b_1 t + b_2 t^2}{c_0 + c_1 t + c_2 t^2} \tag{5.17}$$

has an implicit equation of the form

$$(A_1 x + B_1 y + C_1)^2 - (A_0 x + B_0 y + C_0)(A_2 x + B_2 y + C_2) = 0$$

where the coefficients A_i, B_i, C_i are the signed 2×2 minors of the matrix

$$\mathsf{Q} = \begin{pmatrix} a_0 & a_1 & a_2 \\ b_0 & b_1 & b_2 \\ c_0 & c_1 & c_2 \end{pmatrix} .$$

Proof

For all t at which the conic is defined, (5.17) can be multiplied through by the denominator to give

$$a_0 - c_0 x + (-c_1 x + a_1)\, t + (a_2 - c_2 x)\, t^2 = 0 \,, \text{ and} \tag{5.18}$$

$$b_0 - c_0 y + (-c_1 y + b_1)\, t + (b_2 - c_2 y)\, t^2 = 0 \,. \tag{5.19}$$

Applying Theorem 5.25 to (5.18) and (5.19) gives the necessary and sufficient condition $D_1^2 - D_2 D_3 = 0$ where

$$\begin{aligned} D_1 &= (b_0 c_2 - b_2 c_0)\, x + (a_2 c_0 - a_0 c_2)\, y + a_0 b_2 - a_2 b_0 \,, \\ D_2 &= (b_1 c_2 - b_2 c_1)\, x + (a_2 c_1 - a_1 c_2)\, y + a_1 b_2 - a_2 b_1 \,, \text{ and} \\ D_3 &= (b_0 c_1 - b_1 c_0)\, x + (a_1 c_0 - a_0 c_1)\, y + a_0 b_1 - a_1 b_0 \,. \end{aligned}$$

The proof is now complete since every point (x, y) of the conic satisfies $D_1^2 - D_2 D_3 = 0$, a quadratic polynomial in x and y of the form

$$(A_1 x + B_1 y + C_1)^2 - (A_0 x + B_0 y + C_0)(A_2 x + B_2 y + C_2) = 0$$

where the coefficients A_i, B_i, C_i are the signed minors of the matrix Q.

□

Example 5.27

To determine an implicit equation for the conic

$$(x(t), y(t)) = \left(\frac{18 - 19t + 8t^2}{-2 - 5t + 4t^2}, \frac{-6 + 7t + 3t^2}{-2 - 5t + 4t^2}\right) .$$

Apply Theorem 5.26 to

$$\mathsf{Q} = \begin{pmatrix} 18 & -19 & 8 \\ -6 & 7 & 3 \\ -2 & -5 & 4 \end{pmatrix} .$$

The required minors are

$$\begin{aligned} A_0 &= (7)(4) - (-5)(3) = 43 , \\ A_1 &= -((-6)(4) - (-2)(3)) = 18 \quad \text{etc.} \end{aligned}$$

Alternatively, compute the transpose of the adjugate matrix of Q, to give

$$\begin{pmatrix} A_0 & A_1 & A_2 \\ B_0 & B_1 & B_2 \\ C_0 & C_1 & C_2 \end{pmatrix} = \begin{pmatrix} 43 & 18 & 44 \\ 36 & 88 & 128 \\ -113 & -102 & 12 \end{pmatrix} .$$

Therefore

$$\begin{aligned} &(A_1x + B_1y + C_1)^2 - (A_0x + B_0y + C_0)(A_2x + B_2y + C_2) \\ &= (18x + 88y - 102)^2 - (43x + 36y - 113)(44x + 128y + 12) . \end{aligned}$$

Expanding and simplifying gives $784\left(-2x^2 - 5xy + x + 4y^2 - 5y + 15\right) = 0$. The solution reverses the computation of Example 5.22.

Exercise 5.30

Convert the following conics from parametric to implicit form:

(a) $\left(t^2 - 1, t + 2\right)$,

(b) $\left(2t^2 - 1, t + 3\right)$,

(c) $\left(2t^2 + t - 1, t^2 - 3t + 3\right)$,

(d) $\left(\frac{t^2+1}{t}, 2t\right)$,

(e) $\left(-\frac{4t^2+t+1}{4t^2-2t+1}, -\frac{3t^2}{4t^2-2t+1}\right)$.

Theorem 5.26 can be generalized to planar rational curves of any degree (see [23], [13]):

Theorem 5.28

Let

$$(x(t), y(t)) = \left(\frac{\sum_{i=0}^{n} a_i t^i}{\sum_{i=0}^{n} c_i t^i}, \frac{\sum_{i=0}^{n} b_i t^i}{\sum_{i=0}^{n} c_i t^i} \right)$$

be a rational curve of degree of degree n. Then an implicit form is obtained from the *Bezout resultant*

$$\begin{vmatrix} D_{0,0} & \cdots & D_{0,n} \\ \vdots & & \vdots \\ D_{n,0} & \cdots & D_{n,n} \end{vmatrix} = 0 \, ,$$

where $D_{i,j} = \sum_{\substack{k \le \min(i,j) \\ m=i+j-k+1}} (b_m c_k - c_m b_k)\, x + (a_k c_m - a_m c_k)\, y + (a_m b_k - a_k b_m)$.

□

5.7 Conics in Space

A conic in three-dimensional space is given parametrically by

$$\left(\frac{a_0 + a_1 t + a_2 t^2}{d_0 + d_1 t + d_2 t^2}, \frac{b_0 + b_1 t + b_2 t^2}{d_0 + d_1 t + d_2 t^2}, \frac{c_0 + c_1 t + c_2 t^2}{d_0 + d_1 t + d_2 t^2} \right) .$$

Any conic in space is contained in a plane. To verify this, suppose every point of the conic lies in the plane $Ax + By + Cz + D = 0$. Then

$$\begin{aligned} A\left(a_0 + a_1 t + a_2 t^2\right) &+ B\left(b_0 + b_1 t + b_2 t^2\right) \\ &+ C\left(c_0 + c_1 t + c_2 t^2\right) + D\left(d_0 + d_1 t + d_2 t^2\right) = 0 \, , \end{aligned}$$

that is,

$$\begin{aligned} (Aa_2 + Bb_2 + Cc_2 + Dd_2)\, t^2 &+ (Aa_1 + Bb_1 + Cc_1 + Dd_1)\, t \\ &+ Aa_0 + Bb_0 + Cc_0 + Dd_0 = 0 \, . \quad (5.20) \end{aligned}$$

Since this holds for all t (in an interval) the coefficients of (5.20) must be identically zero, implying

$$\begin{aligned} Aa_0 + Bb_0 + Cc_0 + Dd_0 &= 0 \, , \\ Aa_1 + Bb_1 + Cc_1 + Dd_1 &= 0 \, , \text{ and} \\ Aa_2 + Bb_2 + Cc_2 + Dd_2 &= 0 \, . \end{aligned}$$

The equations can be interpreted as defining three planes in the three-dimensional projective space with homogeneous coordinates (A, B, C, D). Thus the coefficients A, B, C, D can be determined using the method for computing the intersection of three planes given in Section 3.4. This yields

$$(A, B, C, D) = \begin{vmatrix} \mathbf{e}_1 & \mathbf{e}_2 & \mathbf{e}_3 & \mathbf{e}_4 \\ a_0 & b_0 & c_0 & d_0 \\ a_1 & b_1 & c_1 & d_1 \\ a_2 & b_2 & c_2 & d_2 \end{vmatrix}$$

or any multiple of the determinant.

A planar representation of the conic can be obtained by applying a viewplane coordinate mapping. The origin and the X- and Y-axis directions in the derived plane are specified, and the viewplane coordinate matrix VC is computed. Then VC is applied to $(x(t), y(t), z(t))$ to give a conic in the specified Cartesian coordinate system.

Example 5.29

Consider the conic $(x(t), y(t), z(t)) = \left(\frac{3+6t-4t^2}{-1-6t+2t^2}, \frac{9-6t^2}{-1-6t+2t^2}, \frac{1-3t+t^2}{-1-6t+2t^2}\right)$. Then

$$\begin{aligned} (A, B, C, D) &= \begin{vmatrix} \mathbf{e}_1 & \mathbf{e}_2 & \mathbf{e}_3 & \mathbf{e}_4 \\ 3 & 9 & 1 & -1 \\ 6 & 0 & -3 & -6 \\ -4 & -6 & 1 & 2 \end{vmatrix} \\ &= 54\mathbf{e}_1 - 18\mathbf{e}_2 + 36\mathbf{e}_3 + 36\mathbf{e}_4 \\ &= (54, -18, 36, 36) \ . \end{aligned}$$

After dividing (A, B, C, D) through by 18, the conic is found to lie in the plane $3x - y + 2z + 2 = 0$. This can be verified by substituting $x = x(t), y = y(t), z = z(t)$ into the equation of the plane,

$$3\left(\frac{3+6t-4t^2}{-1-6t+2t^2}\right) - \left(\frac{9-6t^2}{-1-6t+2t^2}\right) + 2\left(\frac{1-3t+t^2}{-1-6t+2t^2}\right) + 2 = 0 \ .$$

Multiplying through by the denominator gives

$$3\left(3+6t-4t^2\right) - \left(9-6t^2\right) + 2\left(1-3t+t^2\right) + 2\left(-1-6t+2t^2\right) = 0 \ .$$

Expanding the brackets yields that the left-hand side of the equation is identically zero, and hence the conic lies in the derived plane.

Let the plane have origin and axes as specified in Example 4.11. Applying the viewing coordinate mapping matrix VC to the conic gives

$$\begin{pmatrix} 3+6t-4t^2 & 9-6t^2 & 1-3t+t^2 & -1-6t+2t^2 \end{pmatrix}$$
$$\times \begin{pmatrix} 0.385 & 0.360 & -0.333 \\ 0.642 & 0.600 & 0.111 \\ -0.706 & 0.960 & -0.222 \\ 0.449 & -1.200 & 0.778 \end{pmatrix}$$
$$= \begin{pmatrix} 5.778+1.734t-5.2t^2 & 8.64+6.48t-6.48t^2 & -1-6t+2t^2 \end{pmatrix} .$$

The planar representation of the conic is $\left(\frac{5.778+1.734t-5.2t^2}{-1-6t+2t^2}, \frac{8.64+6.48t-6.48t^2}{-1-6t+2t^2}\right)$. The vector form for the conic in space is

$$\mathcal{O} + \left(\frac{5.778+1.734t-5.2t^2}{-1-6t+2t^2}\right)\mathbf{X} + \left(\frac{8.64+6.48t-6.48t^2}{-1-6t+2t^2}\right)\mathbf{Y} ,$$

where $\mathcal{O}$, $\mathbf{X}$ and $\mathbf{Y}$ are as given in Example 4.11.

Exercise 5.31

By applying Theorem 5.6, show that the conic

$$(x(t), y(t), z(t)) = \left(\frac{9+3t-4t^2}{3+4t+4t^2}, \frac{4t-3t^2}{3+4t+4t^2}, \frac{1+t^2}{3+4t+4t^2}\right)$$

lies in the plane $-16x + 55y + 273z - 43 = 0$.

5.8 Applications of Conics

Example 5.30 (Headlights and Radar)

Parabolas have the special property that rays of light, emanating from a light source positioned at the focus, are reflected in the parabola along parallel lines, as illustrated in Figure 5.15. This property is used in the design of car headlight reflectors. A reflector has the shape of a paraboloid, that is, a surface obtained by rotating a parabola about its axis of symmetry. If a headlight bulb is positioned at the focus of the parabola then it produces a beam of light consisting of the reflected parallel rays of light.

The same property is used in the design of radar or satellite dishes. Signals from a distant point travel along (nearly) parallel rays. Signals which reach a paraboloid shaped dish are reflected along linear paths which pass through the focus. The satellite receiver is positioned at the focus in order to obtain the best reception of the signals.

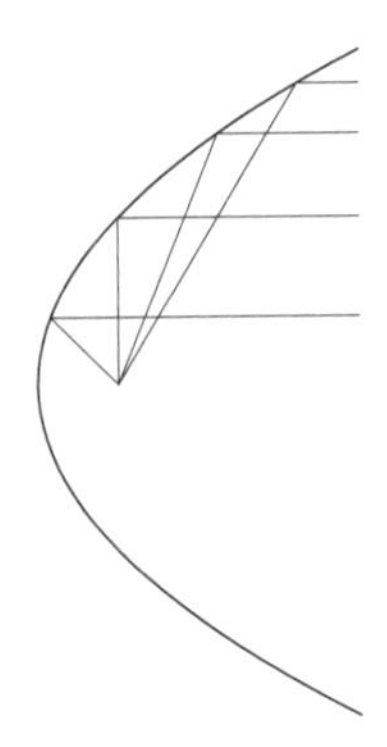

Figure 5.15

EXERCISES

5.32. Show that a ray of light emanating from the focus $\mathbf{F}(0, m)$ of the parabola $y = 4mx^2$ is reflected parallel to the axis of symmetry, that is the y-axis, as follows.

(a) Determine the tangent vector/line at a point $\mathbf{P}(x, y)$ on the parabola.

(b) Determine the angle α between the line $\mathbf{FP}$ and the tangent.

(c) Show that the angle between the tangent and y-axis is also α and deduce that the reflection of the ray is parallel to the y-axis.

5.33. Show that all rays of light emanating from one focus of an ellipse are reflected in lines containing the other focus. The reader may wish to contemplate an elliptical snooker table with the cue ball positioned at one focus and a single pocket positioned at the other focus. If the ball is struck (without spin) with sufficient strength then the ball will hit the elliptical cushion and rebound along a line containing the pocket.

Example 5.31 (Suspension Bridges)

Suspension bridges are designed so that a cable hanging from two pillars or towers carries the weight of the bridge uniformly along the cable. The resulting shape of the cable is a parabola. Suppose the pillars are $1,410$ metres apart (the span of the Humber Bridge, Hull, UK) and the height of the pillars is h metres. Let the origin be the lowest point of the parabola, and let the horizontal plane of the bridge be the x-axis. Then the parabola is symmetrical about the y-axis and passes through the points $(-705, h)$, $(705, h)$, $(0, 0)$. Let the parabola be $y = ax^2 + bx + c$. Then, clearly $c = 0$, and

$$\begin{aligned} h &= a\,(-705)^2 + b\,(-705)\,, \text{ and} \\ h &= a\,(705)^2 + b\,(705)\,. \end{aligned}$$

Thus $a = h/\,(705)^2$ and $b = 0$ giving the parabola $y = \frac{h}{705^2}x^2$.

In reality, one must account for the earth's curvature when modelling large structures. So in the example of the bridge the distance between the base of the pillars is $1,410$ metres, but the distance between the tops of the pillars is greater.

Example 5.32 (Radar)

Discovered as recently as the 1940s, the method known as hyperbolic navigation has had a considerable influence on sea and air navigation. A receiver records the radio signals transmitted from two fixed stations. Assuming that the velocity v of radio energy is constant, the distances travelled by radio energy are proportional to the time taken. Suppose the times taken to receive the signals, sent at the same time, from each station are t_1 and t_2. Then the distance from each station is vt_1 and vt_2. Hence the difference in distance of the receiver from the stations is $v(t_1 - t_2)$. The locus of all possible positions of the receiver relative to the fixed stations is a branch of a hyperbola with the stations positioned at the foci (see Exercise 5.34). There are two points on the hyperbola a given distance vt_1 from the first station and it remains to decide which is the correct location. Commercial hyperbolic navigation systems include the Decca Navigation System, LORAN, Omega, and Global Positioning Systems (GPS).

Exercise 5.34

Let $\mathbf{F}_1$ and $\mathbf{F}_2$ be fixed points, and let d_1 and d_2 be the distances of a point $\mathbf{P}$ from $\mathbf{F}_1$ and $\mathbf{F}_2$ respectively. Show that the locus of all points $\mathbf{P}$ such that $d_1 - d_2$ is constant is a hyperbola.

6
Bézier Curves I

6.1 Introduction

Two of the most important mathematical representations of curves and surfaces used in computer graphics and computer-aided design are the Bézier and B-spline forms. The original development of Bézier curves took place in the automobile industry during the period 1958–60 by two Frenchmen, Pierre Bézier at Renault and Paul de Casteljau at Citröen. The development of B-splines followed the publication in 1946 of a landmark paper [22] on splines. B-splines will be discussed in detail in Chapter 8. Further discussion of the historical development of Bézier curves may be found in [2], [6] and [10].

Bézier curves are polynomial curves (see Definition 5.1) which have a particular mathematical representation. Their popularity is due to the fact that they possess a number of mathematical properties which facilitate their manipulation and analysis, and yet no mathematical knowledge is required in order to use the curves. A Bézier curve of degree n is specified by a sequence of $n+1$ points which are called the *control points*. The polygon obtained by joining the control points with line segments in the prescribed order is called the *control polygon.*

Control of the shape of a Bézier curve is facilitated by the fact that the control polygon reflects the basic shape of the curve. In many drawing and CAD packages the control points of a Bézier curve may be specified by clicking with a mouse at the desired locations within a document window. The control points are visible on the computer screen, and modification of a control point is executed with a simple click and drag operation of the mouse. As an aid to design, many packages display the changing curve as a control point is modified.

6.2 Bézier Curves of Low Degree

6.2.1 Linear Bézier Curves

A *linear Bézier curve* is a line segment joining two control points $\mathbf{b}_0(p_0, q_0)$ and $\mathbf{b}_1(p_1, q_1)$, and parametrized by

$$(x(t), y(t)) = (1-t)(p_0, q_0) + t(p_1, q_1), \quad \text{for } t \in [0, 1] ,$$

so that $x(t) = (1-t)p_0 + tp_1$, and $y(t) = (1-t)q_0 + tq_1$. Letting $\mathbf{B}(t) = (x(t), y(t))$, the curve can be written in the vector form

$$\mathbf{B}(t) = (1-t)\mathbf{b}_0 + t\mathbf{b}_1 . \tag{6.1}$$

The curve is defined on the interval $[0, 1]$, so the starting point of the curve is $\mathbf{B}(0) = \mathbf{b}_0$ and the finishing point is $\mathbf{B}(1) = \mathbf{b}_1$, that is, the Bézier curve *interpolates* the first and last control points.

Example 6.1

The Bézier form for the linear segment passing through points $\mathbf{b}_0(1, 2)$ and $\mathbf{b}_1(3, 4)$ is $\mathbf{B}(t) = (1-t)\mathbf{b}_0 + t\mathbf{b}_1 = (1-t)(1, 2) + t(3, 4)$. Hence $x(t) = (1-t) + 3t = 1 + 2t$ and $y(t) = 2(1-t) + 4t = 2 + 2t$.

6.2.2 Quadratic Bézier Curves

Suppose three control points $\mathbf{b}_0(p_0, q_0)$, $\mathbf{b}_1(p_1, q_1)$, and $\mathbf{b}_2(p_2, q_2)$ are specified. Then the *quadratic Bézier curve* is defined to be

$$\mathbf{B}(t) = (1-t)^2(p_0, q_0) + 2(1-t)t(p_1, q_1) + t^2(p_2, q_2), \quad \text{for } t \in [0, 1] .$$

The starting point of the curve is $\mathbf{B}(0) = \mathbf{b}_0$ and the finishing point is $\mathbf{B}(1) = \mathbf{b}_2$. The curve can be expressed in the parametric form $(x(t), y(t))$ where

$$\begin{aligned} x(t) &= (1-t)^2 p_0 + 2(1-t)tp_1 + t^2 p_2 , \text{ and} \\ y(t) &= (1-t)^2 q_0 + 2(1-t)tq_1 + t^2 q_2 . \end{aligned}$$

The triangle $\mathbf{b}_0\mathbf{b}_1\mathbf{b}_2$ obtained by joining the control points with line segments, in their prescribed order, is called the *control polygon*.

Example 6.2

The parametric form of the quadratic Bézier curve $\mathbf{B}(t)$ with control points $\mathbf{b}_0(1,2)$, $\mathbf{b}_1(4,-1)$, and $\mathbf{b}_2(8,6)$ is $(x(t), y(t))$ where

$$\begin{aligned} x(t) &= (1-t)^2(1) + 2(1-t)t(4) + t^2(8) = 1 + 6t + t^2 \ , \text{ and} \\ y(t) &= (1-t)^2(2) + 2(1-t)t(-1) + t^2(6) = 2 - 6t + 10t^2 \ . \end{aligned}$$

The point $\mathbf{B}(0.5)$ is obtained by substituting $t = 0.5$ into the equations to give $x(0.5) = 4.25$ and $y(0.5) = 1.5$, that is, $\mathbf{B}(0.5) = (4.25, 1.5)$. Alternatively, the coordinates of the point $\mathbf{B}(0.5)$ can be evaluated using the vector form of the curve

$$\begin{aligned} \mathbf{B}(t) &= (1-0.5)^2(1,2) + 2(1-0.5)(0.5)(4,-1) + (0.5)^2(8,6) \\ &= 0.25(1,2) + 0.5(4,-1) + 0.25(8,6) = (4.25, 1.5) \ . \end{aligned}$$

A plot of the curve is obtained by evaluating $\mathbf{B}(t)$ at a sequence of parameter values in the interval $[0,1]$. The curve and its control polygon are illustrated in Figure 6.1.

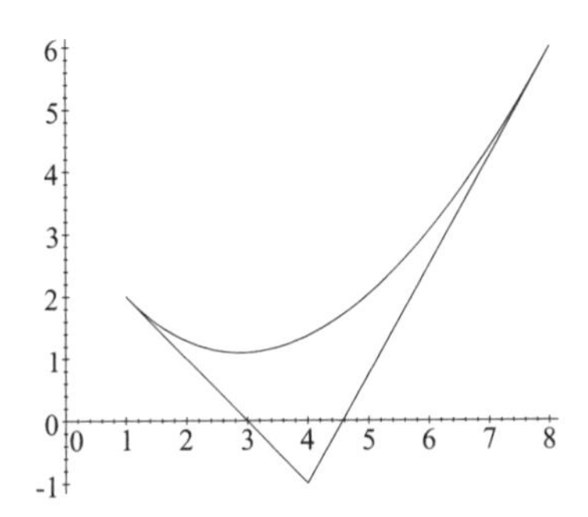

Figure 6.1 Quadratic Bézier curve with control points $\mathbf{b}_0(1,2)$, $\mathbf{b}_1(4,-1)$, and $\mathbf{b}_2(8,6)$

6.2.3 Cubic Bézier Curves

Suppose *four* control points $\mathbf{b}_0$, $\mathbf{b}_1$, $\mathbf{b}_2$, and $\mathbf{b}_3$ are specified, then the *cubic Bézier curve* is defined to be

$$\mathbf{B}(t) = (1-t)^3\mathbf{b}_0 + 3(1-t)^2t\mathbf{b}_1 + 3(1-t)t^2\mathbf{b}_2 + t^3\mathbf{b}_3, \quad t \in [0,1] \ . \tag{6.2}$$

As in the quadratic case, the polygon obtained by joining the control points in the specified order is called the *control polygon*.

Cubic Bézier curves provide a greater range of shapes than quadratic Bézier curves, since they can exhibit loops as shown in Figure 6.2(a), sharp corners (called *cusps*) as shown in Figure 6.2(b), and inflections (see Exercise 6.12).

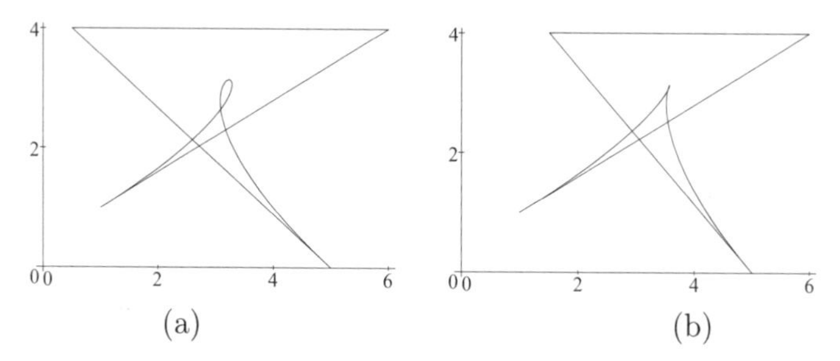

Figure 6.2 (a) Cubic Bézier curve containing a loop, and (b) cubic Bézier curve containing a cusp

A further geometric property is obtained by determining the tangent vector to the cubic Bézier curve at each of its endpoints. The derivative of Equation (6.2) is

$$\mathbf{B}'(t) = -3(1-t)^2\mathbf{b}_0 + 3(1-4t+3t^2)\mathbf{b}_1 + 3t(2-3t)\mathbf{b}_2 + 3t^2\mathbf{b}_3, \quad t \in [0,1] .$$

Thus $\mathbf{B}'(0) = 3(\mathbf{b}_1 - \mathbf{b}_0)$. This implies that the tangent vector of $\mathbf{B}(t)$ at $\mathbf{b}_0$ has the same direction as the vector $\overrightarrow{\mathbf{b}_0\,\mathbf{b}_1}$ joining control points $\mathbf{b}_0$ and $\mathbf{b}_1$. Further, the magnitude of the tangent vector is 3 times the length of $\overrightarrow{\mathbf{b}_0\,\mathbf{b}_1}$. Likewise, $\mathbf{B}'(1) = 3(\mathbf{b}_3 - \mathbf{b}_2)$. Hence the tangent vector of $\mathbf{B}(t)$ at $\mathbf{b}_3$ has direction equal to the direction of the line segment $\overrightarrow{\mathbf{b}_2\,\mathbf{b}_3}$ joining the last pair of control points. Therefore, the choice of the first two control points determines the starting point and the starting direction of the Bézier curve, and the choice of the last two control points determines the finishing point and direction. The shape of the curve is controlled by the user's choice of the *control* points. The geometric property of the starting and finishing directions of Bézier curves is referred to as the *endpoint tangent property.*

Two further Bézier cubics, together with their associated control polygons, are shown in Figure 6.3. Note in particular the endpoint interpolation and tangent properties.

EXERCISES

6.1. Write down the parametric form of the quadratic Bézier curve $\mathbf{B}(t)$ with control points $\mathbf{b}_0(-1,5)$, $\mathbf{b}_1(2,0)$, and $\mathbf{b}_2(4,6)$. Evaluate $\mathbf{B}(0.75)$ and $\mathbf{B}(1.25)$.

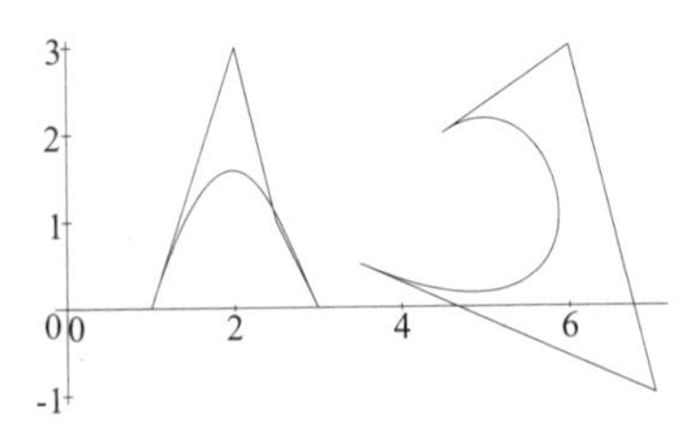

Figure 6.3 Cubic Bézier curves with their associated control polygons

6.2. Show that a quadratic Bézier curve is a parabola.

6.3. Let $\mathbf{b}_0(1,0)$, $\mathbf{b}_1(2,3)$, $\mathbf{b}_2(5,4)$, and $\mathbf{b}_3(2,1)$ be the control points of a cubic Bézier curve $\mathbf{B}(t)$. Determine $\mathbf{B}(t)$, $\mathbf{B}(0)$, $\mathbf{B}(0.5)$, and $\mathbf{B}(1)$.

6.4. Show that a cubic Bézier curve satisfies the *endpoint interpolation property*: $\mathbf{B}(0) = \mathbf{b}_0$ and $\mathbf{B}(1) = \mathbf{b}_3$.

6.5. Determine the tangent vectors at the endpoints of (a) a linear Bézier curve, and (b) a quadratic Bézier curve.

6.6. Let $\mathbf{b}_0(1,3)$, $\mathbf{b}_1(4,6)$, $\mathbf{b}_2(5,1)$, and $\mathbf{b}_3(2,1)$ be the control points of a cubic Bézier curve $\mathbf{B}(t)$. Determine the end tangent vectors. Make a sketch of the curve together with its control polygon (without the assistance of a computer or graphic calculator).

6.7. Whenever the control points $\mathbf{b}_0$, $\mathbf{b}_1$, $\mathbf{b}_2$, and $\mathbf{b}_3$ of a cubic Bézier curve are collinear, the curve is a straight line. In particular, let $\mathbf{b}_1 = (2\mathbf{b}_0 + \mathbf{b}_3)/3$, and $\mathbf{b}_2 = (\mathbf{b}_0 + 2\mathbf{b}_3)/3$. Show that the cubic Bézier curve simplifies to the linear Bézier curve $(1-t)\mathbf{b}_0 + t\mathbf{b}_3$.

6.8. Let the control points of a cubic Bézier curve satisfy $\mathbf{b}_1 = (\mathbf{b}_0 + 2\mathbf{b}^*)/3$ and $\mathbf{b}_2 = (\mathbf{b}_3 + 2\mathbf{b}^*)/3$ for some point $\mathbf{b}^*$. Show that the cubic Bézier curve simplifies to the quadratic Bézier curve $(1-t)^2\mathbf{b}_0 + 2t(1-t)\mathbf{b}^* + t^2\mathbf{b}_3$.

6.9. Suggest plausible control points for the cubic Bézier curve illustrated in Figure 6.4(a).

6.10. Figure 6.4(b) shows two cubic Bézier curves. Write down the control points of a third cubic curve such that (a) the curves join at the points indicated in the figure to form a single continuous curve, and (b) at each point where two curves join the tangent vectors to the curves have the same direction.

6.11. Write a program or use a computer package to draw quadratic and cubic Bézier curves. Construct examples of Bézier curves and adjust

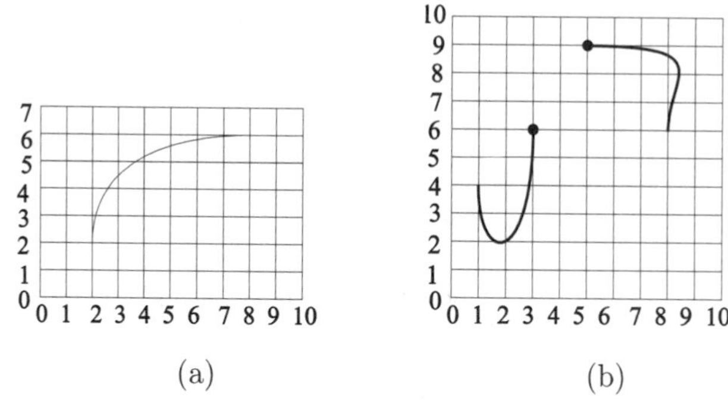

Figure 6.4

the control points to develop a "feel" for the relationship between the control polygon and the curve.

6.12. Plot a cubic Bézier curve which has a point of inflection. (A point of inflection is a point at which the direction of the curve changes from being convex to concave, or vice versa, to give a curve with an "S"-shape.) Hint: Choose an "S"-shaped control polygon.

6.3 The Effect of Adjusting a Control Point

Consider a cubic Bézier curve with control points $\mathbf{b}_0$, $\mathbf{b}_1$, $\mathbf{b}_2$, and $\mathbf{b}_3$. The shape of the curve can be changed by adjusting the position of one or more control points. If $\mathbf{b}_0$ or $\mathbf{b}_3$ are adjusted, then the endpoint interpolation property implies that the starting or finishing point of the curve *will* change. If $\mathbf{b}_1$ or $\mathbf{b}_2$ are adjusted, then the start and finishing points remain unchanged, but the start or finishing directions *may* change. It is possible to change control points $\mathbf{b}_1$ or $\mathbf{b}_2$ without affecting the end directions. If $\mathbf{b}_1$ is adjusted to a new point on the line through $\mathbf{b}_0$ and the original position of $\mathbf{b}_1$, then the magnitude of the tangent vector will change but its direction will not. Hence the initial direction of the curve remains unchanged. The adjustment of a control point is illustrated in Figure 6.5. Likewise, if $\mathbf{b}_2$ is adjusted to a new point on the line through $\mathbf{b}_3$ and the original position of $\mathbf{b}_2$, then the final direction of the curve will not change. However, *the adjustment of any control point always changes the shape of the entire curve.* The effect of adjusting a control point of a general Bézier curve, which will be introduced in Section 6.4, is similar.

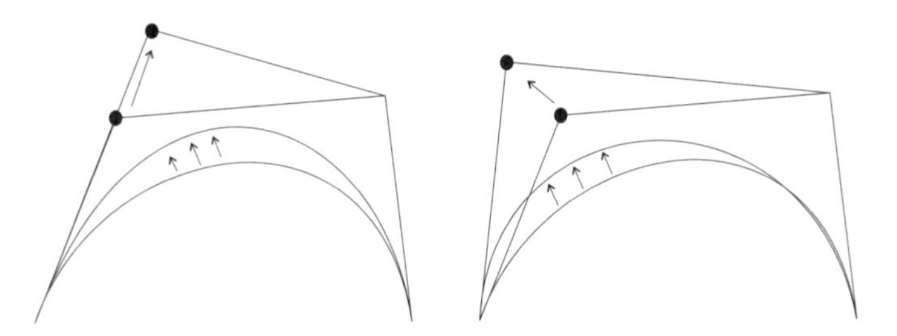

Figure 6.5 Adjustment of a control point so that the starting direction of the curve is (a) unchanged, and (b) changed

6.4 The General Bézier Curve

Given $n+1$ control points $\mathbf{b}_0, \mathbf{b}_1, \ldots, \mathbf{b}_n$ the Bézier curve of degree n is defined to be

$$\mathbf{B}(t) = \sum_{i=0}^{n} \mathbf{b}_i B_{i,n}(t) , \tag{6.3}$$

where

$$B_{i,n}(t) = \begin{cases} \frac{n!}{(n-i)!i!}(1-t)^{n-i}t^i, & \text{if } 0 \le i \le n \\ 0, & \text{otherwise} \end{cases} \tag{6.4}$$

are called the *Bernstein polynomials* or *Bernstein basis functions* of degree n. To distinguish Bézier curves from "rational" Bézier curves which will be introduced in Section 7.5, they are often referred to as *integral* Bézier curves. The original application of Bernstein polynomials is explored in Exercise 6.17. The polygon formed by joining the control points $\mathbf{b}_0, \ldots, \mathbf{b}_n$ in the specified order is called the Bézier *control polygon*. It is a straightforward exercise to show that the cases $n = 1$, $n = 2$, and $n = 3$ correspond to the linear, quadratic, and cubic Bézier curves encountered in the previous sections. The Bernstein polynomials of degrees 2 and 3 are illustrated in Figure 6.6.

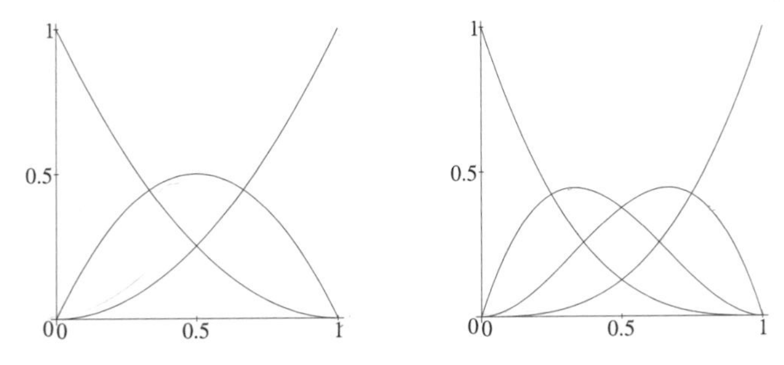

Figure 6.6 Bernstein polynomials of (a) degree 2, and (b) degree 3

The quantities $\frac{n!}{(n-i)!i!}$ are called *binomial coefficients* and are denoted by $\binom{n}{i}$ or ${}^nC_i$. Recall the convention that $0! = 1$, and therefore $\binom{n}{0} = \frac{n!}{n!0!} = 1$ and $\binom{n}{n} = \frac{n!}{0!n!} = 1$.

Example 6.3

For a Bézier cubic $n = 3$, and $B_{0,3}(t) = (1-t)^3$, $B_{1,3}(t) = 3(1-t)^2 t$, $B_{2,3}(t) = 3(1-t)t^2$, and $B_{3,3}(t) = t^3$.

Example 6.4

The Bernstein polynomials of degree 4 are $B_{0,4}(t) = (1-t)^4$, $B_{1,4}(t) = 4(1-t)^3 t$, $B_{2,4}(t) = 6(1-t)^2 t^2$, $B_{3,4}(t) = 4(1-t)t^3$, and $B_{4,4}(t) = t^4$, as illustrated in Figure 6.7.

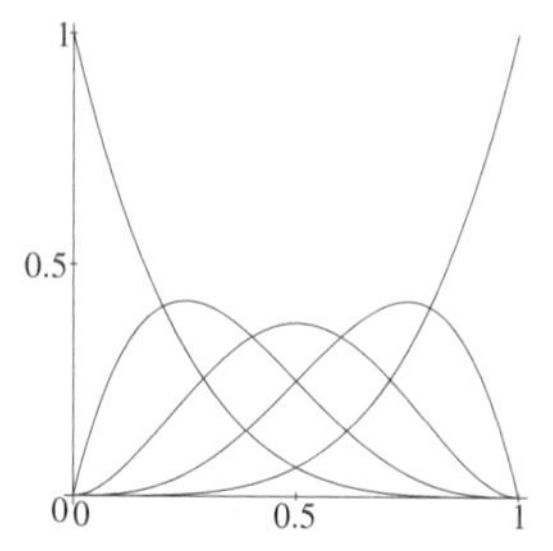

Figure 6.7 Bernstein polynomials of degree 4

The binomial coefficients arise in the result known as the *binomial theorem.*

Theorem 6.5 (Binomial)

For any natural number n, and any real numbers x and y

$$(x+y)^n = \sum_{i=0}^{n} \binom{n}{i} x^{n-i} y^i \, .$$

Example 6.6

Expand $(x+y)^3$ using the binomial theorem. Then

$$\begin{aligned}(x+y)^3 &= \binom{3}{0}x^3 + \binom{3}{1}x^2y + \binom{3}{2}xy^2 + \binom{3}{3}y^3 \\ &= x^3 + 3x^2y + 3xy^2 + y^3 \, .\end{aligned}$$

Example 6.7

To expand $((1-t)+t)^3$ using the binomial theorem, let $x = 1-t$ and $y = t$, and apply the result obtained in Example 6.6

$$((1-t)+t)^3 = (1-t)^3 + 3(1-t)^2 t + 3(1-t)t^2 + t^3 \ .$$

It follows that $B_{0,3}(t) + B_{1,3}(t) + B_{2,3}(t) + B_{3,3}(t) = 1$.

EXERCISES

6.13. By expanding $((1-t)+t)^4$, show that the Bernstein basis functions of degree 4 sum to 1.

6.14. Determine the Bernstein polynomials of degree 5.

6.15. Show that $\binom{n}{i} + \binom{n}{i+1} = \binom{n+1}{i+1}$.

6.16. Show that $\int_0^1 B_{i,3}(t)\, dt = \frac{1}{4}$, for $i = 0, \ldots, 3$.

6.17. Bernstein polynomials first appeared in a proof of the Weierstrass theorem which states that any continuous function can be approximated by a polynomial function to within any specified tolerance. The Bernstein approximation $B(t)$ of degree n of a function $f(t)$ over an interval $[0, 1]$ is defined to be

$$B(t) = \sum_{i=0}^{n} f(t_i) B_{i,n}(t) \ ,$$

where $t_i = \frac{i}{n}$. The proof of the theorem states that for any tolerance ε there is a choice of n for which

$$|f(t) - B(t)| < \varepsilon \ ,$$

that is, the approximation deviates from the actual function by less than the tolerance ε. The main limitation of the approximation is that for a given ε, the choice of n is not easily determined.

(a) Plot the Bernstein approximations of degree 5, 9, and 13 for the function $f(t) = \sin(\pi t)$ over the interval $[0, 1]$.

(b) For each approximation, plot the error function

$$\mathrm{err}(t) = |\sin(\pi t) - B(t)| \ ,$$

and hence determine the maximum absolute error of the approximations.

(c) Make a guess at the value of n for which the Bernstein approximation has error less than 0.01 over the interval $[0, 1]$.

6.5 Properties of the Bernstein Polynomials

The Bernstein polynomials have a number of important properties which give rise to properties of Bézier curves.

Partition of Unity: The Bernstein polynomials of degree n sum to one

$$\sum_{i=0}^{n} B_{i,n}(t) = 1, \quad t \in [0, 1] .$$

Positivity: The Bernstein polynomials are non-negative on the interval $[0, 1]$,

$$B_{i,n}(t) \geq 0, \quad t \in [0, 1] .$$

Symmetry:

$$B_{n-i,n}(t) = B_{i,n}(1 - t) \ , \ \text{for } i = 0, \ldots, n .$$

Therefore, the graph of $B_{n-i,n}(t)$ is a reflection of the graph of $B_{i,n}(1 - t)$. This can be observed in Figures 6.6 and 6.7 which show plots of the quadratic, cubic, and quartic Bernstein polynomials.

Recursion: The Bernstein polynomials of degree n can be expressed in terms of the polynomials of degree $n - 1$

$$B_{i,n}(t) = (1 - t)B_{i,n-1}(t) + tB_{i-1,n-1}(t) ,$$

for $i = 0, \ldots, n$, where $B_{-1,n-1}(t) = 0$ and $B_{n,n-1}(t) = 0$.

The partition of unity and positivity properties give rise to two important properties of Bézier curves, namely, invariance under transformations and the convex hull property. These properties are derived in Section 6.7. As a consequence of the symmetry property, a symmetrical control polygon gives rise to a symmetrical curve. The recursion property gives rise to the de Casteljau algorithm described in Section 6.8.

Proof

(Partition of unity) Applying the binomial theorem to $((1 - t) + t)^n = 1$ gives

$$((1 - t) + t)^n = \sum_{i=0}^{n} \binom{n}{i} (1 - t)^{n-i} t^i = \sum_{i=0}^{n} B_{i,n}(t) = 1 .$$

(Recursion) The recursion property is proved as follows. By definition,

$$B_{i,n-1}(t) = \binom{n-1}{i}(1-t)^{n-1-i}t^i \text{ , and}$$

$$B_{i-1,n-1}(t) = \binom{n-1}{i-1}(1-t)^{n-i}t^{i-1} \text{ .}$$

For $i = 0$,

$$B_{0,n}(t) = (1-t)^n = (1-t)B_{0,n-1}(t) + tB_{-1,n-1}(t)$$

since $B_{-1,n-1}(t) = 0$. Similarly, for $i = n$,

$$B_{n,n}(t) = t^n = (1-t)B_{n,n-1}(t) + tB_{n-1,n-1}(t)$$

since $B_{n,n-1}(t) = 0$. For $1 \leq i \leq n-1$,

$$\begin{aligned}(1-t)B_{i,n-1}(t) + tB_{i-1,n-1}(t) &= \binom{n-1}{i}(1-t)^{n-i}t^i + \binom{n-1}{i-1}(1-t)^{n-i}t^i \\ &= \left(\binom{n-1}{i} + \binom{n-1}{i-1}\right)(1-t)^{n-i}t^i \text{ .}\end{aligned}$$

Applying Exercise 6.15,

$$(1-t)B_{i,n-1}(t) + tB_{i-1,n-1}(t) = \binom{n}{i}(1-t)^{n-i}t^i = B_{i,n}(t) \text{ .}$$

□

The proofs of the properties of positivity and symmetry are left as exercises.

EXERCISES

6.18. Prove the positivity property.

6.19. Prove the symmetry property.

6.20. Show that $\sum_{i=0}^{n} \frac{i}{n} B_{i,n}(t) = t$. Deduce the *linear precision property* that if $\mathbf{b}_i = \left(1 - \frac{i}{n}\right)\mathbf{a} + \frac{i}{n}\mathbf{b}$ for some fixed points $\mathbf{a}$ and $\mathbf{b}$ (so the control points are evenly distributed along the line segment $\overline{\mathbf{ab}}$), then the resulting Bézier curve $\mathbf{B}(t) = \sum_{i=0}^{n} \mathbf{b}_i B_{i,n}(t)$ is the straight line segment $\overline{\mathbf{ab}}$.

6.21. Let $\mathbf{B}(t)$ be a Bézier curve of degree n with control points $\mathbf{b}_0, \ldots, \mathbf{b}_n$. Let $\mathbf{C}(t)$ be the Bézier curve of degree $n+1$ with control points $\mathbf{c}_0 = \mathbf{b}_0$, $\mathbf{c}_{n+1} = \mathbf{b}_n$, and $\mathbf{c}_i = (1 - \alpha_i)\mathbf{b}_i + \alpha_i \mathbf{b}_{i-1}$ where $\alpha_i = i/(n+1)$, for $i = 1, \ldots, n$. Show that $\mathbf{C}(t) = \mathbf{B}(t)$ for all $t \in [0, 1]$. The process of representing a Bézier curve of degree n by a Bézier curve of higher degree is called *degree raising*. Degree-raising algorithms are used to increase the number of control points to give greater freedom for designing curve shapes.

6.6 Convex Hulls

An important and useful property of Bézier curves is that of the convex hull property (CHP) which will be derived in Section 6.7. The CHP and the de Casteljau algorithm, derived in Section 6.8, lead naturally to geometric algorithms for rendering a Bézier curve, and for finding the points of intersection of two Bézier curves. In order to describe the CHP it is necessary to define the convex hull of a set of points. Given a set of points $X = \{\mathbf{x}_0, \mathbf{x}_1, \ldots, \mathbf{x}_n\}$ the *convex hull* of X, denoted by $\mathrm{CH}\{X\}$, is defined to be the set of points

$$\mathrm{CH}\{X\} = \left\{ a_0\mathbf{x}_0 + \cdots + a_n\mathbf{x}_n \;\middle|\; \sum_{i=0}^{n} a_i = 1 \,, a_i \geq 0 \right\} . \tag{6.5}$$

For points in a plane, the convex hull $\mathrm{CH}\{X\}$ may be visualized as follows. Imagine an "elastic band" placed around the entire set of points. The band is permitted to shrink around the points to form a polygon, the vertices of which are a subset of the original set of points. The region bounded by the polygon is the convex hull of the set of points.

The definition of the convex hull is valid for points in space. The intuitive elastic band is replaced by an "elastic balloon" which is permitted to shrink around the points to form a polyhedron. The convex hull is the region bounded by the polyhedron. Several examples of convex hulls are illustrated in Figure 6.8.

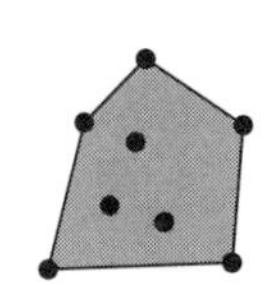
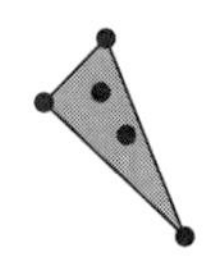

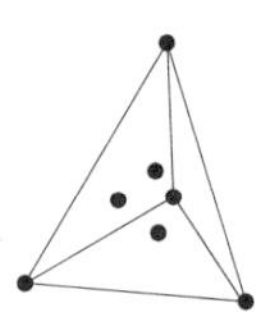

Figure 6.8

6.7 Properties of Bézier Curves

Theorem 6.8

A Bézier curve $\mathbf{B}(t)$ of degree n with control points $\mathbf{b}_0, ..., \mathbf{b}_n$ satisfies the following properties.

Endpoint Interpolation Property: $\mathbf{B}(0) = \mathbf{b}_0$ and $\mathbf{B}(1) = \mathbf{b}_n$.

Endpoint Tangent Property:

$$\mathbf{B}'(0) = n(\mathbf{b}_1 - \mathbf{b}_0) \quad \text{and} \quad \mathbf{B}'(1) = n(\mathbf{b}_n - \mathbf{b}_{n-1}).$$

Convex Hull Property (CHP): For all $t \in [0,1]$, $\mathbf{B}(t) \in \mathrm{CH}\{\mathbf{b}_0, ..., \mathbf{b}_n\}$. Thus *every point of a Bézier curve lies inside the convex hull of its defining control points.* The convex hull of the control points is often referred to as the convex hull of the Bézier curve.

Invariance under Affine Transformations: Let T be an (affine) transformation (for example, a rotation, reflection, translation, or scaling). Then

$$\mathsf{T}\left(\sum_{i=0}^{n} \mathbf{b}_i B_{i,n}(t)\right) = \sum_{i=0}^{n} \mathsf{T}(\mathbf{b}_i) B_{i,n}(t).$$

Variation Diminishing Property (VDP): For a planar Bézier curve $\mathbf{B}(t)$, the VDP states that the number of intersections of a given line with $\mathbf{B}(t)$ is less than or equal to the number of intersections of that line with the control polygon.

Proof

The proof of the endpoint interpolation property is Exercise 6.23. The endpoint tangent property follows from Theorem 7.3 which will be proved later.

(Convex Hull Property) From the definition of the convex hull expressed in Equation (6.5) it is sufficient to show that every point $\mathbf{B}(t)$ on a Bézier curve has the form $a_0\mathbf{b}_0 + \cdots + a_n\mathbf{b}_n$ for some constants a_i satisfying $\sum_{i=0}^{n} a_i = 1$. Let $a_i = B_{i,n}(t)$, then positivity implies $a_i \geq 0$, the partition of unity implies that $\sum_{i=0}^{n} a_i = 1$, and the proof is complete. Figure 6.9 illustrates the CHP for a cubic Bézier curve.

(Affine Invariance) Let an affine transformation T be given by

$$(x', y') = (ax + by + c, dx + ey + f),$$

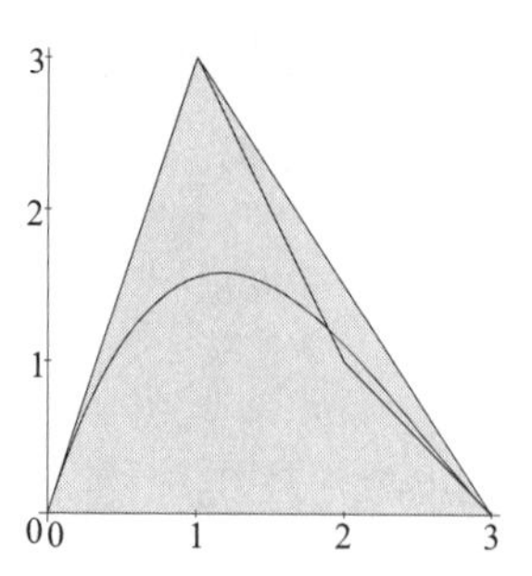

Figure 6.9 Convex hull property for a cubic Bézier curve

and let a Bézier curve of degree n have control points $\mathbf{b}_i(p_i, q_i)$ for $i = 0, \ldots, n$. Then

$$\mathbf{B}(t) = (x(t), y(t)) = \left(\sum_{i=0}^{n} p_i B_{i,n}(t), \sum_{i=0}^{n} q_i B_{i,n}(t) \right) .$$

Applying the transformation yields

$$\begin{aligned} \mathsf{T}(\mathbf{B}(t)) \;\; = \;\; & \left(a \sum_{i=0}^{n} p_i B_{i,n}(t) + b \sum_{i=0}^{n} q_i B_{i,n}(t) + c, \right. \\ & \left. d \sum_{i=0}^{n} p_i B_{i,n}(t) + e \sum_{i=0}^{n} q_i B_{i,n}(t) + f \right) . \end{aligned}$$

Then, by partition of unity, $\sum_{i=0}^{n} B_{i,n}(t) = 1$, and

$$\begin{aligned} \mathsf{T}(\mathbf{B}(t)) = & \left(a \sum_{i=0}^{n} p_i B_{i,n}(t) + b \sum_{i=0}^{n} q_i B_{i,n}(t) + c \sum_{i=0}^{n} B_{i,n}(t), \right. \\ & \left. d \sum_{i=0}^{n} p_i B_{i,n}(t) + e \sum_{i=0}^{n} q_i B_{i,n}(t) + f \sum_{i=0}^{n} B_{i,n}(t) \right) \\ = & \left(\sum_{i=0}^{n} (ap_i + bq_i + c) B_{i,n}(t), \sum_{i=0}^{n} (dp_i + eq_i + f) B_{i,n}(t) \right) \\ = & \sum_{i=0}^{n} (ap_i + bq_i + c, dp_i + eq_i + f) B_{i,n}(t) \\ = & \sum_{i=0}^{n} \mathsf{T}(\mathbf{b}_i) B_{i,n}(t) . \end{aligned}$$

□

Example 6.9

Consider a cubic Bézier curve with vertices $\mathbf{b}_0(1,0)$, $\mathbf{b}_1(2,3)$, $\mathbf{b}_2(5,4)$, and $\mathbf{b}_3(2,1)$. To apply a rotation through an angle $\pi/4$ about the origin in the anticlockwise direction to the curve, it is sufficient to apply the rotation matrix $\mathsf{Rot}(\pi/4)$ to the homogeneous coordinates of the control points:

$$\begin{pmatrix} 1 & 0 & 1 \\ 2 & 3 & 1 \\ 5 & 4 & 1 \\ 2 & 1 & 1 \end{pmatrix} \begin{pmatrix} \cos\pi/4 & \sin\pi/4 & 0 \\ -\sin\pi/4 & \cos\pi/4 & 0 \\ 0 & 0 & 1 \end{pmatrix} = \begin{pmatrix} 0.707 & 0.707 & 1.0 \\ -0.707 & 3.536 & 1.0 \\ 0.707 & 6.364 & 1.0 \\ 0.707 & 2.121 & 1.0 \end{pmatrix}.$$

The control points of the rotated curve are $\mathbf{b}_0(0.707, 0.707)$, $\mathbf{b}_1(-0.707, 3.536)$, $\mathbf{b}_2(0.707, 6.364)$, and $\mathbf{b}_3(0.707, 2.121)$. The curve and its rotated image are illustrated in Figure 6.10.

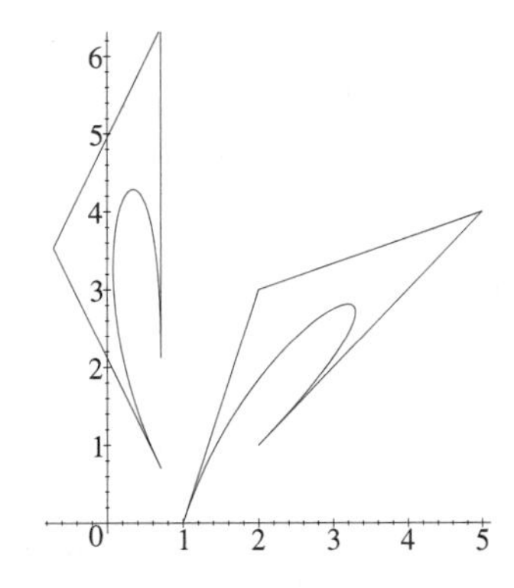

Figure 6.10 Application of a rotation to a cubic Bézier curve

Figure 6.11 illustrates two lines intersecting a Bézier curve and its control polygon. The upper line intersects the polygon in two points but does not intersect the curve. The lower line intersects both the polygon and the curve in two points. In both cases, the number of intersections with the given line is equal to or greater than the number of intersections of the line with the curve. Thus the variation diminishing property is satisfied. The proof of the variation diminishing property is beyond the scope of this book, and the reader is referred to [15].

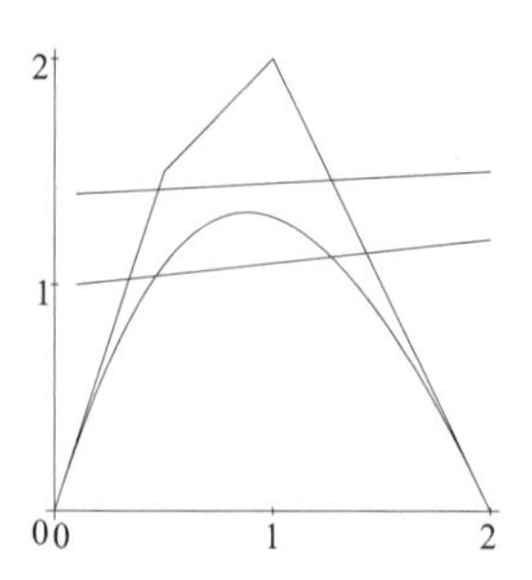

Figure 6.11 Variation diminishing property

EXERCISES

6.22. Plot the cubic Bézier curve defined by control points $\mathbf{b}_0(0,1)$, $\mathbf{b}_1(2,5)$, $\mathbf{b}_2(4,6)$, and $\mathbf{b}_3(8,1)$. On the same plot, draw the control polygon. Observe that the resulting curve satisfies the convex hull property. Next plot the Bézier cubic with control points $\mathbf{b}_0(1,1)$, $\mathbf{b}_1(3.4,1.8)$, $\mathbf{b}_2(6,6.5)$, and $\mathbf{b}_3(9,1)$. Does the newly displayed curve violate the convex hull property? Explain.

6.23. Prove the endpoint interpolation property for the general Bézier curve: $\mathbf{B}(0) = \mathbf{b}_0$ and $\mathbf{B}(1) = \mathbf{b}_n$.

6.24. Prove that when the control points are collinear, the resulting Bézier curve is a straight line segment.

6.25. Determine the control points of the image of the Bézier curve with control points $\mathbf{b}_0(0,0)$, $\mathbf{b}_1(2,1)$, $\mathbf{b}_2(3,-1)$, and $\mathbf{b}_3(1,-2)$ when the following transformations have been applied

(a) a translation of 3 units in the x-direction and 4 units in the y-direction,

(b) a rotation about the origin through an angle of $\pi/2$ radians in an anti-clockwise direction,

(c) a reflection in the line $y = x$.

For each transformation plot the image curve and its control polygon.

6.26. The basis functions $B_{0,3}(t) = (1-t)^2$, $B_{1,3}(t) = 2t(1-t)^2$, $B_{2,3}(t) = 2t^2(1-t)$, $B_{0,3}(t) = t^2$ give rise to a representation for cubic curves $\mathbf{B}(t) = \sum_{i=0}^{3} \mathbf{b}_i B_{i,3}(t)$.

(a) Show that if $\mathbf{b}_1 = \mathbf{b}_2$ then $\mathbf{B}(t)$ is a quadratic curve with control polygon $\mathbf{b}_0$, $\mathbf{b}_1$ and $\mathbf{b}_3$.

(b) Show that the representation satisfies end interpolation and tangent properties similar to Bézier curves.

6.8 The de Casteljau Algorithm

The de Casteljau algorithm provides a method for evaluating the point on a Bézier curve corresponding to the parameter value $t \in [0,1]$. In Section 6.9 it will be shown that the same algorithm can be used to divide a curve into two curve segments. For the case of a cubic Bézier curve with control points $\mathbf{b}_0$, $\mathbf{b}_1$, $\mathbf{b}_2$, and $\mathbf{b}_3$, and for a specified parameter value $t \in [0,1]$, the de Casteljau algorithm is expressed by the recursive formula

$$\begin{cases} \mathbf{b}_i^0 = \mathbf{b}_i, \\ \mathbf{b}_i^j = (1-t)\mathbf{b}_i^{j-1} + t\mathbf{b}_{i+1}^{j-1}, \end{cases}$$

for $j = 1, 2, 3$ and $i = 0, \ldots, 3-j$. The formula generates a triangular set of values (6.6) for which $\mathbf{b}_0^3 = \mathbf{B}(t)$ for the specified value of t:

$$\begin{array}{cccc} \mathbf{b}_0^0 & \mathbf{b}_1^0 & \mathbf{b}_2^0 & \mathbf{b}_3^0 \\ \mathbf{b}_0^1 & \mathbf{b}_1^1 & \mathbf{b}_2^1 & \\ \mathbf{b}_0^2 & \mathbf{b}_1^2 & & \\ \mathbf{b}_0^3 & & & \end{array} \tag{6.6}$$

Example 6.10

A cubic Bézier curve has control points $\mathbf{b}_0(1.0, 1.0)$, $\mathbf{b}_1(2.0, 7.0)$, $\mathbf{b}_2(8.0, 6.0)$, and $\mathbf{b}_3(12.0, 2.0)$. The point $\mathbf{B}(0.25)$ is determined by applying the de Casteljau algorithm with $t = 0.25$. Then

$$\begin{aligned} \mathbf{b}_0^1 &= \tfrac{3}{4}(1.0, 1.0) + \tfrac{1}{4}(2.0, 7.0) = (1.25, 2.5), \\ \mathbf{b}_1^1 &= \tfrac{3}{4}(2.0, 7.0) + \tfrac{1}{4}(8.0, 6.0) = (3.5, 6.75), \\ \mathbf{b}_2^1 &= \tfrac{3}{4}(8.0, 6.0) + \tfrac{1}{4}(12.0, 2.0) = (9.0, 5.0), \\ \mathbf{b}_0^2 &= \tfrac{3}{4}(1.25, 2.5) + \tfrac{1}{4}(3.5, 6.75) = (1.8125, 3.5625), \text{ etc.} \end{aligned}$$

The algorithm gives the following table of points:

$$\begin{array}{ccccccc}
(1.0, 1.0) & & (2.0, 7.0) & & (8.0, 6.0) & & (12.0, 2.0) \\
& \frac{3}{4} \downarrow \swarrow \frac{1}{4} & & \frac{3}{4} \downarrow \swarrow \frac{1}{4} & & \frac{3}{4} \downarrow \swarrow \frac{1}{4} & \\
(1.25, 2.5) & & (3.5, 6.75) & & (9.0, 5.0) & & \\
& \frac{3}{4} \downarrow \swarrow \frac{1}{4} & & \frac{3}{4} \downarrow \swarrow \frac{1}{4} & & & \\
(1.8125, 3.5625) & & (4.875, 6.3125) & & & & \\
& \frac{3}{4} \downarrow \quad \swarrow \frac{1}{4} & & & & & \\
(2.578, 4.25) & & & & & &
\end{array}$$

The algorithm yields $\mathbf{B}(0.25) = (2.578, 4.25)$. Geometrically, each step of the algorithm is a linear interpolation of the control polygon as illustrated in Figure 6.12.

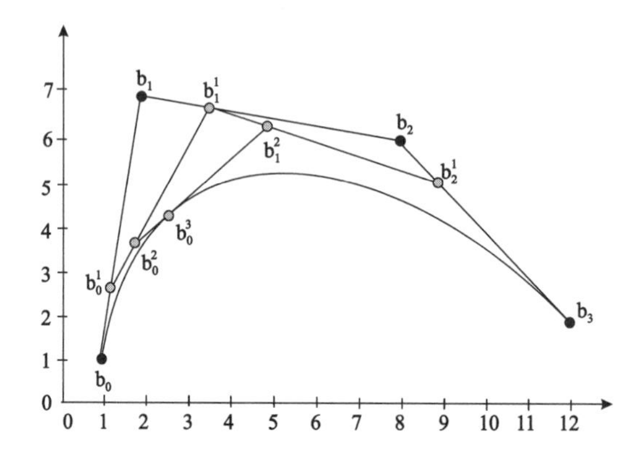

Figure 6.12 The de Casteljau algorithm with $t = 0.25$

Theorem 6.11

Let a Bézier curve of degree n be given by control points $\mathbf{b}_0, \ldots, \mathbf{b}_n$, and let $t \in [0, 1]$ be any parameter value. Then $\mathbf{B}(t) = \mathbf{b}_0^n$, where $\mathbf{b}_i^0 = \mathbf{b}_i$, and

$$\mathbf{b}_i^j = \mathbf{b}_i^{j-1}(1-t) + \mathbf{b}_{i+1}^{j-1} t \,,$$

for $j = 1, \ldots, n$, and $i = 0, \ldots, n-j$.

Proof

The de Casteljau algorithm follows from the recursion property of the Bernstein polynomials

$$B_{i,n}(t) = (1-t)B_{i,n-1}(t) + tB_{i-1,n-1}(t) \,. \tag{6.7}$$

Then

$$\begin{aligned}\mathbf{B}(t) &= \sum_{i=0}^{n} \mathbf{b}_i B_{i,n}(t) = \sum_{i=0}^{n} \mathbf{b}_i \left((1-t)B_{i,n-1}(t) + tB_{i-1,n-1}(t)\right) \\ &= \sum_{i=0}^{n} \mathbf{b}_i (1-t) B_{i,n-1}(t) + \sum_{i=0}^{n} \mathbf{b}_i t B_{i-1,n-1}(t) \,.\end{aligned}$$

Since $B_{n,n-1}(t) = 0$, and $B_{-1,n-1}(t) = 0$ it follows that

$$\mathbf{B}(t) = \sum_{i=0}^{n-1} \mathbf{b}_i (1-t) B_{i,n-1}(t) + \sum_{i=1}^{n} \mathbf{b}_i t B_{i-1,n-1}(t) \,.$$

Next renumber the second summation by replacing i by $i+1$,

$$\begin{aligned}\mathbf{B}(t) &= \sum_{i=0}^{n-1} \mathbf{b}_i (1-t) B_{i,n-1}(t) + \sum_{i=0}^{n-1} \mathbf{b}_{i+1} t B_{i,n-1}(t) \\ &= \sum_{i=0}^{n-1} \left(\mathbf{b}_i (1-t) + \mathbf{b}_{i+1} t\right) B_{i,n-1}(t) \,.\end{aligned}$$

Set $\mathbf{b}_i^1 = \mathbf{b}_i(1-t) + \mathbf{b}_{i+1}t = \mathbf{b}_i^0(1-t) + \mathbf{b}_{i+1}^0 t$ for $i = 0, \ldots, n-1$, then

$$\mathbf{B}(t) = \sum_{i=0}^{n-1} \mathbf{b}_i^1 B_{i,n-1}(t) \,. \tag{6.8}$$

Equation (6.8) expresses $\mathbf{B}(t)$ as a Bézier curve of degree $n-1$ with control points $\mathbf{b}_0^1, \ldots, \mathbf{b}_{n-1}^1$. Applying a similar argument yields

$$\mathbf{B}(t) = \sum_{i=0}^{n-2} \mathbf{b}_i^2 B_{i,n-2}(t) \,,$$

where $\mathbf{b}_{i+1}^2 = \mathbf{b}_i^1(1-t) + \mathbf{b}_{i+1}^1 t$ for $i = 0, \ldots, n-2$. In general,

$$\mathbf{B}(t) = \sum_{i=0}^{n-j} \mathbf{b}_i^j B_{i,n-j}(t) \,,$$

where $\mathbf{b}_i^j = \mathbf{b}_i^{j-1}(1-t) + \mathbf{b}_{i+1}^{j-1} t$ for $i = 0, \ldots, n-j$. In particular, $j = n$ gives

$$\mathbf{B}(t) = \sum_{i=0}^{0} \mathbf{b}_i^n B_{i,n-n}(t) = \mathbf{b}_0^n \,.$$

□

EXERCISES

6.27. A cubic Bézier curve has control points $\mathbf{b}_0(1, 0)$, $\mathbf{b}_1(3, 3)$, $\mathbf{b}_2(5, 5)$, and $\mathbf{b}_3(7, 2)$. Evaluate the point $\mathbf{B}(0.25)$ by (a) applying the de Casteljau algorithm, and (b) substituting $t = 0.25$ into the defining equation of the Bézier curve. Make a sketch illustrating the points derived in applying de Casteljau algorithm.

6.28. Apply the de Casteljau algorithm to the quartic Bézier curve with control points $\mathbf{b}_0(3.0, 3.0)$, $\mathbf{b}_1(4.0, 2.0)$, $\mathbf{b}_2(-1.0, 0.0)$, $\mathbf{b}_3(6.0, 1.0)$, and $\mathbf{b}_4(8.0, 5.0)$, and evaluate the point $\mathbf{B}(0.6)$.

6.29. (Used in Theorem 6.13) Prove that the intermediate control points defined in the de Casteljau algorithm satisfy

$$\mathbf{b}_k^j = \sum_{i=0}^{j} B_{i,j}(t)\mathbf{b}_{i+k}.$$

6.30. Show that

$$\begin{aligned}(1-t)B_i^n(t) &= \left(\frac{n+1-i}{n+1}\right) B_i^{n+1}(t)\,, \\ tB_i^n(t) &= \left(\frac{i+1}{n+1}\right) B_{i+1}^{n+1}(t)\,.\end{aligned}$$

6.31. (Used in Theorem 6.13) Use Exercise 6.29 (or otherwise) to show that $B_{i,n}(\alpha t) = \sum_{j=0}^{n} B_{i,j}(\alpha)B_{j,n}(t)$.

6.9 Subdivision of a Bézier Curve

A Bézier curve is generally defined over the interval $[0, 1]$ and given by $\mathbf{B}(t) = \sum_{i=0}^{n} \mathbf{b}_i B_{i,n}(t)$. On occasions, only a part of a curve is of interest. For instance, suppose that a Bézier curve is "cut" at the parameter value $t = \alpha$ to give two curve segments, denoted by $\mathbf{B}_{\text{left}}(t)$ and $\mathbf{B}_{\text{right}}(t)$, defined over the intervals $[0, \alpha]$, and $[\alpha, 1]$ as shown in Figure 6.13. Since $\mathbf{B}_{\text{left}}(t)$ and $\mathbf{B}_{\text{right}}(t)$ are polynomial curves they can be represented in Bézier form over the interval $[0, 1]$. Theorem 6.13 will show that to determine the control points of $\mathbf{B}_{\text{left}}(t)$ and $\mathbf{B}_{\text{right}}(t)$ it is sufficient to apply the de Casteljau algorithm to $\mathbf{B}(t)$ with $t = \alpha$. For a cubic Bézier curve, the theorem implies that the control points of $\mathbf{B}_{\text{left}}(t)$ are $\mathbf{b}_0^0, \mathbf{b}_0^1, \mathbf{b}_0^2, \mathbf{b}_0^3$, and the control points of $\mathbf{B}_{\text{right}}(t)$ are $\mathbf{b}_0^3, \mathbf{b}_1^2, \mathbf{b}_2^1, \mathbf{b}_3^0$. The two sets of points are observed to be two edges of the triangle of control points (6.6). Subdivision is one way of creating extra control points in order to give additional freedom for curve design. For instance, one segment of the curve can be left untouched while the other part of the curve is changed.

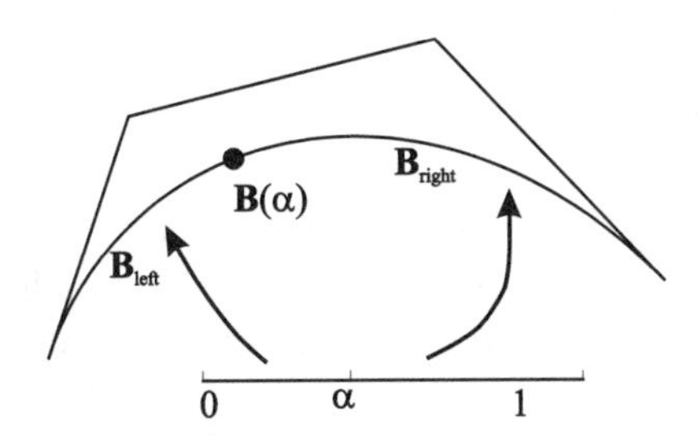

Figure 6.13

Example 6.12

A Bézier cubic $\mathbf{B}(t)$ has control points $\mathbf{b}_0(1.0, 1.0)$, $\mathbf{b}_1(2.0, 7.0)$, $\mathbf{b}_2(8.0, 6.0)$, and $\mathbf{b}_3(12.0, 2.0)$. The control points of the two curve segments $\mathbf{B}_{\text{left}}(t)$ and $\mathbf{B}_{\text{right}}(t)$, obtained by cutting $\mathbf{B}(t)$ at the parameter value $t = 0.25$, are determined from the triangle of points computed in Example 6.10. $\mathbf{B}_{\text{left}}$ has control points $\mathbf{b}_0(1.0, 1.0)$, $\mathbf{b}_1(1.25, 2.5)$, $\mathbf{b}_2(1.8125, 3.5625)$, $\mathbf{b}_3(2.578, 4.25)$, and $\mathbf{B}_{\text{right}}$ has control points $\mathbf{b}_0(2.578, 4.25)$, $\mathbf{b}_1(4.875, 6.3125)$, $\mathbf{b}_2(9.0, 5.0)$, $\mathbf{b}_3(12.0, 2.0)$.

Theorem 6.13 (Subdivision)

For a general Bézier curve $\mathbf{B}(t) = \sum_{i=0}^{n} \mathbf{b}_i B_{i,n}(t)$, the control points of the two curve segments obtained by subdivision at parameter value t are $\mathbf{b}_0^0, \mathbf{b}_0^1, \ldots, \mathbf{b}_0^{n-1}, \mathbf{b}_0^n$ for $\mathbf{B}_{\text{left}}$ and $\mathbf{b}_0^n, \mathbf{b}_1^{n-1}, \ldots, \mathbf{b}_{n-1}^1, \mathbf{b}_n^0$ for $\mathbf{B}_{\text{right}}$, where the $\mathbf{b}_i^j$ are the points computed in the de Casteljau algorithm (Theorem 6.11).

Proof

Suppose $\mathbf{B}(t)$ is subdivided at $t = \alpha$. The segment $\mathbf{B}_{\text{left}}$ is defined by $\mathbf{B}_{\text{left}}(t) = \sum_{i=0}^{n} \mathbf{b}_i B_{i,n}(t)$ over the interval $[0, \alpha]$. Thus the curve can be reparametrized as $\mathbf{B}_{\text{left}}(t) = \sum_{i=0}^{n} \mathbf{b}_i B_{i,n}(\alpha t)$, over the interval $[0, 1]$. Hence Exercise 6.30 gives

$$\mathbf{B}_{\text{left}}(t) = \sum_{i=0}^{n} \mathbf{b}_i \left(\sum_{j=0}^{n} B_{i,j}(\alpha) B_{j,n}(t) \right) = \sum_{j=0}^{n} \left(\sum_{i=0}^{n} \mathbf{b}_i B_{i,j}(\alpha) \right) B_{j,n}(t) \, .$$

Finally, Exercise 6.28 (with $k = 0$) and the fact that $B_{i,j}(\alpha) = 0$ whenever $i > j$, gives

$$\mathbf{B}_{\text{left}}(t) = \sum_{j=0}^{n} \left(\sum_{i=0}^{j} \mathbf{b}_i B_{i,j}(\alpha) \right) B_{j,n}(t) = \sum_{j=0}^{n} \mathbf{b}_0^j B_{j,n}(t) .$$

Therefore the segment is defined by control points $\mathbf{b}_0^j$ ($j = 0, \ldots, n$) over the interval $[0, 1]$ as required.

The result for $\mathbf{B}_{\text{right}}$ follows from an application of the symmetry property as follows. Substitute t for $1-t$ which maps the interval $[\alpha, 1]$ onto the interval $[0, 1-\alpha]$. Apply the result for $\mathbf{B}_{\text{left}}$ with the control points in the reverse order and with $1 - \alpha$ in place of α.

□

EXERCISES

6.32. A cubic Bézier curve $\mathbf{B}(t)$ is given by the four control points $\mathbf{b}_0(0.2, 0.0)$, $\mathbf{b}_1(1.0, 0.4)$, $\mathbf{b}_2(1.8, 1.2)$, and $\mathbf{b}_3(3.4, 0.0)$.

(a) Use the de Casteljau algorithm to evaluate the point $\mathbf{B}(0.25)$.

(b) Use the triangular array of points evaluated in part (a) to write down the sets of control points, defining the segments $\mathbf{B}_{\text{left}}$ and $\mathbf{B}_{\text{right}}$, that are obtained when $\mathbf{B}(t)$ is subdivided at $t = 0.25$.

6.33. Plot the curves $\mathbf{B}(t)$, $\mathbf{B}_{\text{left}}(t)$, and $\mathbf{B}_{\text{right}}(t)$ obtained in Example 6.12 and verify that the union of the two segments is equal to the original curve.

6.34. A Bézier curve $\mathbf{B}(t)$ is given by the four control points $\mathbf{b}_0(0.3, 0.1)$, $\mathbf{b}_1(0.9, 0.6)$, $\mathbf{b}_2(1.3, -0.1)$, $\mathbf{b}_3(0.7, -0.4)$.

(a) Use the de Casteljau algorithm to evaluate the point $\mathbf{B}(1/3)$.

(b) Write down the control points defining $\mathbf{B}_{\text{left}}$ and $\mathbf{B}_{\text{right}}$ obtained by subdividing $\mathbf{B}(t)$ at $t = 1/3$.

6.35. Determine the number of additions and multiplications that are required to compute the coordinates of one point of a cubic Bézier curve by (a) using the de Casteljau algorithm, (b) evaluating the equation of $\mathbf{B}(t)$ (assume that the value of $1 - t$ is computed just once). Repeat the calculation for a quartic Bézier curve. Deduce the number of additions and multiplications that are required for a general Bézier curve. Is the de Casteljau algorithm the most efficient method of computing a point?

6.36. By writing a program, or using a computer package, implement the de Casteljau algorithm for a general Bézier curve to (a) obtain the coordinate of any point on the curve, and (b) determine the control points of the two segments obtained by subdivision.

6.10 Applications

In this section the de Casteljau algorithm is applied to three problems: (i) rendering a Bézier curve, (ii) finding the points of intersection of a Bézier curve and a line, and (iii) finding the points of intersection of two Bézier curves. The reader should note that there are alternative methods which solve these problems. The algorithms discussed in this section indicate how the properties of the Bézier representation can be applied to these problems.

6.10.1 Rendering

To *render* a curve means to obtain a plot of it. The main step of the rendering algorithm is an application of the de Casteljau algorithm to subdivide the curve.

Step 1: Apply the de Casteljau algorithm with $t = 1/2$ to subdivide the Bézier curve into two curve segments denoted $\mathbf{B}_{\text{left}}$ and $\mathbf{B}_{\text{right}}$.

Step 2: If $\mathbf{B}_{\text{left}}$ is "near linear" (using the criterion described below) then go to step 3; else, go to step 1 and apply the algorithm to $\mathbf{B}_{\text{left}}$. Similarly, if $\mathbf{B}_{\text{right}}$ is near linear go to step 3; else go to step 1 and apply the algorithm to $\mathbf{B}_{\text{right}}$.

The algorithm continues to subdivide the newly obtained curve segments that are not near linear. Eventually, the subdivision produces curve segments that are near linear and no further subdivisions take place.

Step 3: The segment is near linear and can be approximated by its control polygon. Draw the control polygon.

Each time step 3 is executed, the control polygon of a segment of the curve is drawn. The union of all these control polygons gives a linear approximation of the original Bézier curve.

Test to Determine Whether a Bézier Curve is Near Linear

There are a number of ways of deciding whether or not a curve is close to being linear. One method requires the user to specify a tolerance $\epsilon > 0$. For a plane curve, the control points are enclosed in a rectangle (see Figure 6.14(a)), called a *minmax bounding box,* with lower left corner $(x_{\min}, y_{\min})$ and upper right corner $(x_{\max}, y_{\max})$, where $x_{\min}/x_{\max}$ is the minimum/maximum x-coordinate of any control point, and $y_{\min}/y_{\max}$ is the minimum/maximum y-coordinate of any control point. The curve is considered linear if the horizontal or vertical dimensions of the box are less than ϵ. The smaller the tolerance the greater the number of subdivisions computed, and the smoother the resulting approximation to the curve.

A more sophisticated alternative is to determine the largest distance of any interior control point (i.e. a control point which is not an endpoint) from the line through the endpoints of the curve as shown in Figure 6.14(b). This is a more computationally expensive method than the minmax box method, but generally it will result in fewer subdivisions which will give a saving in computations. A further improvement of the algorithm can be obtained by

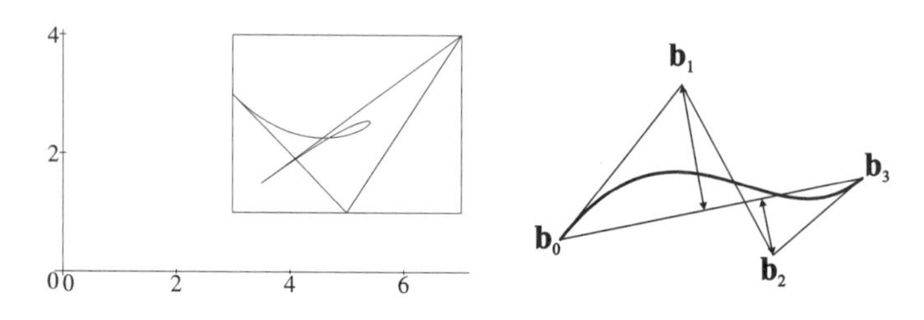

Figure 6.14 (a) Minmax bounding box, and (b) test for near linearity

subdividing at values other than $t = 1/2$. In general, the point $\mathbf{B}(1/2)$ is not exactly half-way along the curve. So the algorithm could be improved if the subdivision takes place nearer to half-way as this would reduce the number of subdivisions. To determine the value of t which corresponds to half-way requires considerable additional computation, and it is not obvious whether this results in a more efficient algorithm.

6.10.2 Intersection of a Planar Bézier Curve and a Line

A planar Bézier curve $\mathbf{B}(t)$ of degree n can intersect a line ℓ in the plane in up to n points. (It is assumed that $\mathbf{B}(t)$ is not a line segment and intersecting ℓ in an infinite set.) A simple algorithm to compute the points of intersection is as follows.

Step 1: Test whether the convex hull of the control points intersects the line (see below for details). If so, go to step 2 as there may be an intersection; else, the curve does not intersect the line and the curve may be disregarded.

Step 2: Test to see if the curve is near linear. If so, go to step 3; else, apply the de Casteljau algorithm to subdivide the curve into two Bézier curve segments and repeat step 1 with each segment.

Step 3: The curve (or curve segment) is linear and may be approximated by a line segment (for example, by the line joining the first and last control points), and intersected with ℓ using the algorithm implemented in Exercise 6.36. The intersection is a point of intersection of the Bézier curve and ℓ.

Test to Determine Whether a Convex Hull Intersects a Line

Suppose the line ℓ has the equation $ax+by+c=0$. A line partitions the plane into two regions, one on either side of the line. The regions are distinguished by the fact that points (x, y) in one region satisfy $ax + by + c > 0$, and points in the other satisfy $ax + by + c < 0$. All control points of a curve lie on one side of the line if and only if the convex hull does not intersect the line. Thus a simple check for intersection would be to determine whether $ax_i + by_i + c$ has the same sign for every control point $\mathbf{b}_i(x_i, y_i)$ $(i = 0, \ldots, n)$ as illustrated in Figure 6.15.

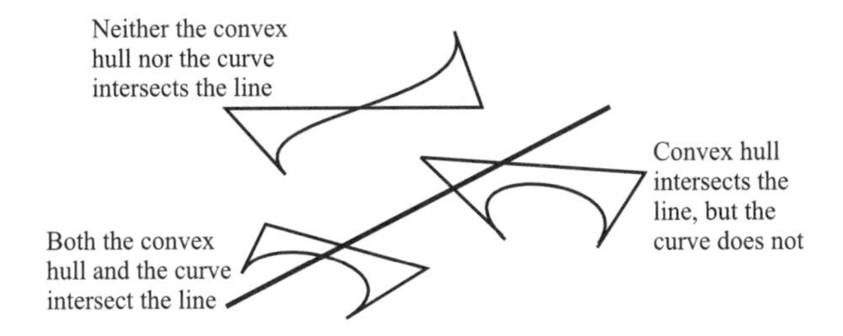

Figure 6.15 Convex hull test

6.10.3 Intersection of Two Bézier Curves

Determining the points of intersection of two curves is a complex problem. A curve of degree m intersects another of degree n in up to $m \times n$ points. (This result follows from a more general result known as Bezout's theorem [11].)

For example, two cubic curves can meet in as many as *nine* points. A simple algorithm to compute the intersection points of two Bézier curves is as follows.

Step 1: Test to see if the convex hulls of the control polygons intersect (see below for details). If so, go to step 2 as the curves may intersect; else the curves cannot intersect and may be disregarded.

Step 2: Test whether the curves are linear. If so, go to step 3; else, apply the de Casteljau algorithm to subdivide each curve into two segments, and to give a total of four pairs of curve segments. Go to step 1 and apply the algorithm to each pair.

Thus, each pair of subdivided curve segments is treated in a similar manner to the original pair. Each subdivision produces curves with control points that have successively smaller convex hulls. The subdivision process will stop when the curve segments are near linear. Alternatively, the subdivision process could be coded to stop after a fixed number of iterations (obviously with some loss of precision in the computation of the intersection points).

Step 3: Since both curves are near linear, the curves may be approximated by line segments (for example, by the line joining the first and last control points). The two linear segments are intersected (using the algorithm of Exercise 6.36) to determine the point of intersection of the two segments.

Test to Determine Whether Two Convex Hulls are Intersecting

The simplest method is to enclose each convex hull in a minmax bounding box. The two boxes are easily checked for overlap.

EXERCISES

6.37. Implement the Bézier rendering algorithm using (a) the simple rectangular bounding box criterion, and (b) the distance to line criterion. Compare the performance of the two algorithms for curves of small and large degrees, and for nearly linear curves which are parallel or at an angle to the x- or y-axes.

6.38. Determine algorithms which (a) approximate a nearly linear Bézier curve by a line, and (b) determine the intersection of two linear segments. Use the algorithms to implement the line/Bézier intersection, and Bézier/Bézier intersection algorithms.

7
Bézier Curves II

7.1 Spatial Bézier Curves

A spatial Bézier curve $\mathbf{B}(t) = \sum_{i=0}^{n} \mathbf{b}_i B_{i,n}(t)$ is obtained when the control points $\mathbf{b}_i$ are three-dimensional. Spatial Bézier curves satisfy the properties of planar Bézier curves given in Section 6.7, namely, the endpoint interpolation and endpoint tangent conditions, invariance under affine transformations, the convex hull property (CHP), and the variation diminishing property (VDP). In general, the convex hull of the set of control points is a volume. In the special case when the control points are coplanar the convex hull is a planar region and the CHP implies that the Bézier curve is contained in a plane. The VDP in the spatial case states that a plane intersects a Bézier curve in less than or equal to the number of intersections of that plane with the control polygon.

The de Casteljau algorithm is executed in a similar manner to the two-dimensional case, except that the linear interpolation is applied to three coordinates rather than two (as illustrated in the next example).

Example 7.1

Let a spatial cubic Bézier curve be specified by control points $\mathbf{b}_0(1,2,1)$, $\mathbf{b}_1(3,0,4)$, $\mathbf{b}_2(6,-3,2)$, and $\mathbf{b}_3(4,2,3)$. The endpoints of the curve are $\mathbf{b}_0(1,2,1)$ and $\mathbf{b}_3(4,2,3)$. The endpoint tangent vectors are $3\left((3,0,4)-(1,2,1)\right) = (6,-6,9)$ and $3\left((4,2,3)-(6,-3,2)\right) = (-6,15,3)$. The point $\mathbf{B}(0.3)$, for instance, is obtained by applying the de Casteljau algorithm with $t = 0.3$ to the three-dimensional control points

$$\begin{array}{llll} (1,2,1) & (3,0,4) & (6,-3,2) & (4,2,3) \\ (1.6,1.4,1.9) & (3.9,-0.9,3.4) & (5.4,-1.5,2.3) & \\ (2.29,0.71,2.35) & (4.35,-1.08,3.07) & & \\ (2.908,0.173,2.566) & & & \end{array}$$

Hence $\mathbf{B}(0.3) = (2.908, 0.173, 2.566)$. The de Casteljau algorithm also subdivides the curve into two segments (as described in Section 6.9): $\mathbf{B}_{\text{left}}$ defined by control points

$$(1,2,1),\ (1.6,1.4,1.9)\,,\ (2.29,0.71,2.35)\,,\ (2.908,0.173,2.566)$$

and $\mathbf{B}_{\text{right}}$ defined by control points

$$(2.908,0.173,2.566)\,,\ (4.35,-1.08,3.07)\,,\ (5.4,-1.5,2.3)\,,\ (4,2,3)\,.$$

7.2 Derivatives of Bézier Curves

Many computations involving curves, such as determining tangents and normals, require the calculation of derivatives. Derivatives of Bézier curves are obtained from the derivatives of the Bernstein basis functions. For instance, the derivatives of the cubic Bernstein basis functions $B_{0,3}(t) = (1-t)^3$, $B_{1,3}(t) = 3(1-t)^2 t$, $B_{2,3}(t) = 3(1-t)t^2$, and $B_{3,3}(t) = t^3$ are

$$\begin{aligned} B'_{0,3}(t) &= -3(1-t)^2 = -3B_{0,2}(t)\,, \\ B'_{1,3}(t) &= 3(1-t)^2 - 6t(1-t) = 3(1-4t+3t^2) = 3B_{0,2}(t) - 3B_{1,2}(t)\,, \\ B'_{2,3}(t) &= 3t(2-3t) = 3B_{1,2}(t) - 3B_{2,2}(t)\,, \text{ and} \\ B'_{3,3}(t) &= 3t^2 = 3B_{2,2}(t)\,. \end{aligned}$$

Hence, the derivative of a cubic Bézier curve $\mathbf{B}(t) = \sum_{i=0}^{3} \mathbf{b}_i B_{i,3}(t)$ is

$$\begin{aligned} \mathbf{B}'(t) &= -3\mathbf{b}_0 B_{0,2}(t) + 3\mathbf{b}_1(B_{0,2}(t) - B_{1,2}(t)) \\ &\qquad + 3\mathbf{b}_2(B_{1,2}(t) - B_{2,2}(t)) + 3\mathbf{b}_3 B_{2,2}(t)\,, \\ &= 3(\mathbf{b}_1 - \mathbf{b}_0)B_{0,2}(t) + 3(\mathbf{b}_2 - \mathbf{b}_1)B_{1,2}(t) + 3(\mathbf{b}_3 - \mathbf{b}_2)B_{2,2}(t)\,. \end{aligned}$$

The generalizations of the above formulae for Bézier curves of degree n are expressed in Theorems 7.2 and 7.3.

Theorem 7.2

The first and second derivatives of the Bernstein basis functions $B_{i,n}(t)$ of degree n satisfy

$$
\begin{aligned}
B'_{i,n}(t) &= \frac{(i-nt)}{t(1-t)} B_{i,n}(t) , \\
B''_{i,n}(t) &= \left(\frac{i(i-1) - 2i(n-1)t + n(n-1)t^2}{t^2(1-t)^2} \right) B_{i,n}(t) , \\
B'_{i,n}(t) &= n\left(B_{i-1,n-1}(t) - B_{i,n-1}(t)\right) .
\end{aligned}
$$

Proof

Differentiating $B_{i,n}(t) = \binom{n}{i}(1-t)^{n-i}t^i$ by the product rule gives

$$
\begin{aligned}
B'_{i,n}(t) &= \binom{n}{i}\left(-(n-i)(1-t)^{n-i-1}t^i + i(1-t)^{n-i}t^{i-1}\right) \\
&= \left(\frac{i-nt}{t(1-t)}\right)\binom{n}{i}(1-t)^{n-i}t^i ,
\end{aligned}
$$

which establishes the first formula. The second formula is obtained by differentiating the first formula,

$$
\begin{aligned}
B''_{i,n}(t) &= \left(\frac{i-nt}{t(1-t)}\right)' B_{i,n}(t) + \frac{(i-nt)}{t(1-t)} B'_{i,n}(t) , \\
&= \frac{(2it - nt^2 - i)}{t^2(1-t)^2} B_{i,n}(t) + \frac{(i-nt)^2}{t^2(1-t)^2} B_{i,n}(t) \\
&= \left(\frac{i(i-1) - 2i(n-1)t + n(n-1)t^2}{t^2(1-t)^2} \right) B_{i,n}(t) .
\end{aligned}
$$

The third formula is Exercise 7.5. □

Theorem 7.3

The first derivative of a Bézier curve of degree n is

$$
\mathbf{B}'(t) = \sum_{i=0}^{n-1} \mathbf{b}_i^{(1)} B_{i,n-1}(t) , \tag{7.1}
$$

where $\mathbf{b}_i^{(1)} = n\left(\mathbf{b}_{i+1} - \mathbf{b}_i\right)$.

Proof

Applying the third formula $B'_{i,n}(t) = n\left(B_{i-1,n-1}(t) - B_{i,n-1}(t)\right)$ of Theorem 7.2 and using the fact that $B_{-1,n-1}(t) = B_{n,n-1}(t) = 0$, gives

$$\begin{aligned}
\mathbf{B}'(t) &= \sum_{i=0}^{n} \mathbf{b}_i B'_{i,n}(t) = \sum_{i=0}^{n} \mathbf{b}_i n\left(B_{i-1,n-1}(t) - B_{i,n-1}(t)\right) \\
&= \sum_{i=0}^{n} n\mathbf{b}_i B_{i-1,n-1}(t) - \sum_{i=0}^{n} n\mathbf{b}_i B_{i,n-1}(t) \\
&= \sum_{i=1}^{n} n\mathbf{b}_i B_{i-1,n-1}(t) - \sum_{i=0}^{n-1} n\mathbf{b}_i B_{i,n-1}(t) .
\end{aligned}$$

Renumbering the first summation of the previous line gives

$$\mathbf{B}'(t) = \sum_{i=0}^{n-1} n\mathbf{b}_{i+1} B_{i,n-1}(t) - \sum_{i=0}^{n-1} n\mathbf{b}_i B_{i,n-1}(t) = \sum_{i=0}^{n-1} n\left(\mathbf{b}_{i+1} - \mathbf{b}_i\right) B_{i,n-1}(t) .$$

□

The second and higher order derivatives of $\mathbf{B}(t)$ are obtained by repeated applications of the first derivative formula. Note that the formulae apply to spatial as well as planar Bézier curves.

Corollary 7.4

The second derivative of a Bézier curve of degree n is

$$\mathbf{B}''(t) = \sum_{i=0}^{n-2} \mathbf{b}_i^{(2)} B_{i,n-2}(t) ,$$

where $\mathbf{b}_i^{(2)} = (n-1)\left(\mathbf{b}_{i+1}^{(1)} - \mathbf{b}_i^{(1)}\right) = n(n-1)\left(\mathbf{b}_{i+2} - 2\mathbf{b}_{i+1} + \mathbf{b}_i\right)$.

Corollary 7.5

The rth derivative of a Bézier curve of degree n is

$$\mathbf{B}^{(r)}(t) = \sum_{i=0}^{n-r} \mathbf{b}_i^{(r)} B_{i,n-r}(t) ,$$

where

$$\mathbf{b}_i^{(r)} = n(n-1)\ldots(n-r+1)\sum_{j=0}^{r}(-1)^{r-j}\binom{r}{j}\mathbf{b}_{i+j} .$$

Example 7.6

Consider the cubic Bézier curve defined by control points $\mathbf{b}_0(2,1)$, $\mathbf{b}_1(5,6)$, $\mathbf{b}_2(6,2)$, and $\mathbf{b}_3(9,3)$. The differences of the control points are

$$(5,6)-(2,1)=(3,5),\ (6,2)-(5,6)=(1,-4),\ (9,3)-(6,2)=(3,1)\ .$$

Multiply each difference by 3 to give the control points of the first derivative

$$\mathbf{b}_0^{(1)}(9,15),\ \mathbf{b}_1^{(1)}(3,-12),\ \mathbf{b}_2^{(1)}(9,3)\ .$$

Therefore, the derivative of the cubic is the quadratic Bézier curve

$$(1-t)^2(9,15)+2(1-t)t(3,-12)+t^2(9,3)\ .$$

To determine the second derivative, take the differences of the control points of the first derivative

$$(3,-12)-(9,15)=(-6,-27)\,,\quad (9,3)-(3,-12)=(6,15)\ ,$$

and multiply by $(n-1)=2$ to give $\mathbf{b}_0^{(2)}(-12,-54)$ and $\mathbf{b}_1^{(2)}(12,30)$. Hence the second derivative of the cubic is the linear Bézier curve

$$(1-t)(-12,-54)+t(12,30)\ . \tag{7.2}$$

Then, for instance, the tangent vector of the curve at the point corresponding to parameter $t=0.5$ is obtained by substituting $t=0.5$ in the first derivative

$$(1-0.5)^2(9,15)+2(1-0.5)0.5(3,-12)+0.5^2(9,3)=(6.0,-1.5)\ .$$

EXERCISES

7.1. Apply the de Casteljau algorithm with $t=0.3$ to the spatial cubic Bézier curve $\mathbf{B}(t)$ with control points $\mathbf{b}_0(2,7,4)$, $\mathbf{b}_1(4,6,5)$, $\mathbf{b}_2(5,8,4)$, and $\mathbf{b}_3(3,5,3)$. Determine $\mathbf{B}(0.3)$ and the control points of the two curve segments obtained by subdividing at $t=0.3$.

7.2. Determine the first and second derivatives of the cubic Bézier curve with control points $\mathbf{b}_0(6,3)$, $\mathbf{b}_1(4,3)$, $\mathbf{b}_2(1,2)$, and $\mathbf{b}_3(-1,2)$.

7.3. Determine the first and second derivatives of the quartic Bézier curve with control points $\mathbf{b}_0(1,1)$, $\mathbf{b}_1(1,3)$, $\mathbf{b}_2(5,6)$, $\mathbf{b}_3(6,2)$, and $\mathbf{b}_4(4,-1)$.

7.4. Determine an expression, in terms of the control points, for the acceleration vectors (second derivatives) at the endpoints of a Bézier curve of degree n.

7.5. Prove the final result of Theorem 7.2 that the Bernstein basis functions satisfy $B'_{i,n}(t) = n\,(B_{i-1,n-1}(t) - B_{i,n-1}(t))$.

7.6. Extend your computer implementation of the de Casteljau algorithm to apply to spatial Bézier curves.

7.3 Conversions Between Representations

All polynomial curves can be represented in Bézier form. Suppose a polynomial curve of degree n is expressed in the *monomial* form

$$\mathbf{a}_0 + \mathbf{a}_1 t + \cdots + \mathbf{a}_n t^n = (p_0 + p_1 t + \cdots + p_n t^n, q_0 + q_1 t + \cdots + q_n t^n) \quad (7.3)$$

over the interval $[0, 1]$. The points $\mathbf{a}_i$ are called *monomial control points.* The curve can be converted into Bézier form by multiplying the monomial control points by a conversion matrix. For instance, expanding the expression for a quadratic Bézier curve gives

$$\mathbf{b}_0(1-t)^2 + \mathbf{b}_1 2(1-t)t + \mathbf{b}_2 t^2 = (\mathbf{b}_0 - 2\mathbf{b}_1 + \mathbf{b}_2)t^2 + (-2\mathbf{b}_0 + 2\mathbf{b}_1)t + \mathbf{b}_0 \,. \quad (7.4)$$

A comparison of the coefficients with those of $\mathbf{a}_2 t^2 + \mathbf{a}_1 t + \mathbf{a}_0$ gives

$$\mathbf{a}_0 = \mathbf{b}_0, \quad \mathbf{a}_1 = 2(\mathbf{b}_1 - \mathbf{b}_0), \quad \text{and} \quad \mathbf{a}_2 = \mathbf{b}_0 - 2\mathbf{b}_1 + \mathbf{b}_2 \,.$$

The relationship between the control points of the two representations can be expressed in the matrix form

$$\begin{pmatrix} \mathbf{a}_0 \\ \mathbf{a}_1 \\ \mathbf{a}_2 \end{pmatrix} = \begin{pmatrix} 1 & 0 & 0 \\ -2 & 2 & 0 \\ 1 & -2 & 1 \end{pmatrix} \begin{pmatrix} \mathbf{b}_0 \\ \mathbf{b}_1 \\ \mathbf{b}_2 \end{pmatrix} . \quad (7.5)$$

The inverse matrix can be used to express the Bézier control points in terms of the monomial control points

$$\begin{pmatrix} \mathbf{b}_0 \\ \mathbf{b}_1 \\ \mathbf{b}_2 \end{pmatrix} = \begin{pmatrix} 1 & 0 & 0 \\ 1 & \frac{1}{2} & 0 \\ 1 & 1 & 1 \end{pmatrix} \begin{pmatrix} \mathbf{a}_0 \\ \mathbf{a}_1 \\ \mathbf{a}_2 \end{pmatrix} . \quad (7.6)$$

The two conversion matrices are denoted

$$\mathsf{Bez} = \begin{pmatrix} 1 & 0 & 0 \\ -2 & 2 & 0 \\ 1 & -2 & 1 \end{pmatrix}, \text{ and } \mathsf{Bez}^{-1} = \begin{pmatrix} 1 & 0 & 0 \\ 1 & \frac{1}{2} & 0 \\ 1 & 1 & 1 \end{pmatrix} .$$

Let $\mathsf{a} = \begin{pmatrix} \mathbf{a}_0 & \mathbf{a}_1 & \mathbf{a}_2 \end{pmatrix}^T$ and $\mathsf{b} = \begin{pmatrix} \mathbf{b}_0 & \mathbf{b}_1 & \mathbf{b}_2 \end{pmatrix}^T$, then the conversions (7.5) and (7.6) may be written

$$\mathsf{a} = \mathsf{Bez}\cdot\mathsf{b} \,, \text{ and } \mathsf{b} = \mathsf{Bez}^{-1}\cdot\mathsf{a} \,. \quad (7.7)$$

Example 7.7

The curve $(2+5t-3t^2, 4-t+6t^2)$ is converted into quadratic Bézier form as follows

$$\begin{pmatrix} 1 & 0 & 0 \\ 1 & \frac{1}{2} & 0 \\ 1 & 1 & 1 \end{pmatrix} \begin{pmatrix} 2 & 4 \\ 5 & -1 \\ -3 & 6 \end{pmatrix} = \begin{pmatrix} 2.0 & 4.0 \\ 4.5 & 3.5 \\ 4.0 & 9.0 \end{pmatrix} .$$

Hence the control points of the quadratic Bézier curve are

$$\mathbf{b}_0(2.0, 4.0), \quad \mathbf{b}_1(4.5, 3.5), \quad \text{and} \quad \mathbf{b}_2(4.0, 9.0) .$$

EXERCISES

7.7. Convert the curve $(1-t^2, 4-2t+3t^2)$ into Bézier form.

7.8. Convert the quadratic Bézier curve with control points $\mathbf{b}_0(-3.0, -3.0)$, $\mathbf{b}_1(1.0, -1.0)$, and $\mathbf{b}_2(-1.0, 2.0)$ into monomial form.

7.9. Let $\mathsf{B} = \begin{pmatrix} B_{0,2}(t) & B_{1,2}(t) & B_{2,2}(t) \end{pmatrix}$ and $\mathsf{T} = \begin{pmatrix} 1 & t & t^2 \end{pmatrix}$. Show that

$$\mathsf{B} = \mathsf{T} \cdot \mathsf{Bez}, \text{ and } \mathbf{B}(t) = \mathsf{T} \cdot \mathsf{Bez} \cdot \mathsf{b} . \tag{7.8}$$

The procedure for conversion between the general Bézier curve and a curve in monomial form is similar. Let

$$\mathsf{B} = \begin{pmatrix} B_{0,n}(t) & B_{1,n}(t) & \dots & B_{n,n}(t) \end{pmatrix}, \quad \mathsf{T} = \begin{pmatrix} 1 & t & \dots & t^n \end{pmatrix},$$
$$\mathsf{a} = \begin{pmatrix} \mathbf{a}_0 & \mathbf{a}_1 & \dots & \mathbf{a}_n \end{pmatrix}^T, \qquad \mathsf{b} = \begin{pmatrix} \mathbf{b}_0 & \mathbf{b}_1 & \dots & \mathbf{b}_n \end{pmatrix}^T,$$

and let the matrices $\mathsf{Bez} = (\mathsf{Bez}_{i,j})$ and $\mathsf{Bez}^{-1} = \left(\mathsf{Bez}^{-1}_{i,j}\right)$, $0 \le i, j \le n$, have entries respectively defined by

$$\mathsf{Bez}_{i,j} = \begin{cases} (-1)^{i-j}\binom{n}{i}\binom{i}{j}, & \text{if } i \ge j \\ 0, & \text{otherwise} \end{cases}, \text{ and}$$

$$\mathsf{Bez}^{-1}_{i,j} = \begin{cases} \binom{i}{j} \Big/ \binom{n}{j}, & \text{if } j \le i \\ 0, & \text{otherwise} \end{cases} .$$

Then identities (7.7) and (7.8) hold. The most efficient computer implementation for conversion between $\mathbf{a}_i$ and $\mathbf{b}_i$ is yielded by the equivalent formulae

$$\mathbf{a}_i = \sum_{j=0}^{i} (-1)^{i-j} \binom{n}{i} \binom{i}{j} \mathbf{b}_j ,$$

$$\mathbf{b}_i = \sum_{j=0}^{i} \binom{i}{j} \Big/ \binom{n}{j} \mathbf{a}_j .$$

EXERCISES

7.10. Following the approach taken for conversions of quadratic polynomial curves, show that for cubic curves

$$\mathsf{Bez} = \begin{pmatrix} 1 & 0 & 0 & 0 \\ -3 & 3 & 0 & 0 \\ 3 & -6 & 3 & 0 \\ -1 & 3 & -3 & 1 \end{pmatrix}.$$

7.11. Show that the matrix

$$\mathsf{Bez}^{-1} = 1/3 \begin{pmatrix} 3 & 0 & 0 & 0 \\ 3 & 1 & 0 & 0 \\ 3 & 2 & 1 & 0 \\ 3 & 3 & 3 & 3 \end{pmatrix}$$

is the inverse of the matrix Bez given in Exercise 7.10.

7.12. Convert the curve $(2-3t-4t^2+7t^3, -4+8t-5t^3)$ into Bézier form.

7.13. Convert the Bézier curve with control points $\mathbf{b}_0(2,-1)$, $\mathbf{b}_1(5,2)$, $\mathbf{b}_2(7,3)$, and $\mathbf{b}_3(6,-1)$ into monomial form.

7.4 Piecewise Bézier Curves

A Bézier curve of degree n has $n+1$ control points. Curves of high degree are not often used since there is only a weak relationship between the shape of the curve and the shape of the control polygon. Further, operations such as the evaluation of points require a large number of arithmetical operations, and so there is an increased risk of computational errors. In contrast, curves of low degree have few control points, and therefore yield a limited range of curve shapes. To widen the range of shapes without increasing the degree of the curve, a number of Bézier curves can be joined end to end to form a single continuous curve called a piecewise Bézier curve. In practice, the joins of the curves are required to be smooth.

Definition 7.8

An *arbitrary interval* Bézier curve $\mathbf{B}(t)$ of degree n with control points

$\mathbf{b}_0, \ldots, \mathbf{b}_n$ defined on an interval $[t_{\min}, t_{\max}]$ is given by

$$\mathbf{B}(t) = \sum_{i=0}^{n} \mathbf{b}_i B_{i,n}\left(\frac{t - t_{\min}}{t_{\max} - t_{\min}}\right)$$

where $B_{i,n}$ denote the Bernstein basis functions of degree n. The arbitrary interval Bézier curve $\mathbf{B}(t)$ is a reparametrization of the ordinary Bézier curve $\hat{\mathbf{B}}(t) = \sum_{i=0}^{n} \mathbf{b}_i B_{i,n}(t)$, $t \in [0,1]$. $\hat{\mathbf{B}}(t)$ is said to be the *normalization* of $\mathbf{B}(t)$.

Definition 7.9

Let $I = [a, b]$. $\mathbf{P}(t)$ is said to be a *piecewise Bézier curve* if there exist $t_0 < t_1 < \cdots < t_{r-1} < t_r$ such that $a = t_0$ and $b = t_r$, and arbitrary interval Bézier curves $\mathbf{B}_j(t)$ defined on $[t_j, t_{j+1}]$ $(j = 0, \ldots, r-1)$, such that (i) $\mathbf{P}(t) = \mathbf{B}_j(t)$ for $t \in (t_j, t_{j+1})$, (ii) $\mathbf{P}(t_j) = \mathbf{B}_{j-1}(t_j)$ or $\mathbf{P}(t_j) = \mathbf{B}_j(t_j)$ (possibly both) for $j = 1, \ldots, r-1$, and (iii) $\mathbf{P}(t_0) = \mathbf{B}_0(t_0)$ and $\mathbf{P}(t_r) = \mathbf{B}_{r-1}(t_r)$. The parameter values t_j are called *breakpoints*. If the largest degree of the curves $\mathbf{B}_j(t)$ is n, then the piecewise Bézier curve is said to have *degree n*. The terms quadratic, cubic, etc. are used to describe piecewise curves of degrees two, three, etc.

Remark 7.10

The definition ensures that $\mathbf{P}(t)$ is single valued on the interval $[a, b]$. In practice, a piecewise Bézier curve is considered simply as a union of Bézier curves and therefore $\mathbf{P}(t_j)$ has *two* values $(j = 1, \ldots, r-1)$, namely $\mathbf{B}_{j-1}(t_j)$ and $\mathbf{B}_j(t_j)$. In many applications only continuous curves are considered, in which case $\mathbf{P}(t_j) = \mathbf{B}_{j-1}(t_j) = \mathbf{B}_j(t_j)$ and so $\mathbf{P}(t)$ is single valued everywhere.

Example 7.11

Consider three arbitrary interval Bézier curves $\mathbf{B}_0(t)$, $t \in [-2, 0]$; $\mathbf{B}_1(t)$, $t \in [0, 3]$; $\mathbf{B}_2(t)$, $t \in [3, 4]$ with the following control points

$$\begin{aligned}
\mathbf{B}_0(t) &: \mathbf{b}_0(2,-1), \mathbf{b}_1(5,2), \mathbf{b}_2(7,3), \mathbf{b}_3(8,-1)\\
\mathbf{B}_1(t) &: \mathbf{b}_0(8,-1), \mathbf{b}_1(8,-3), \mathbf{b}_2(7,-4), \mathbf{b}_3(5,-4)\\
\mathbf{B}_2(t) &: \mathbf{b}_0(5,-4), \mathbf{b}_1(3,-4), \mathbf{b}_2(4,-2), \mathbf{b}_3(6,-2)\,.
\end{aligned}$$

Let $\mathbf{P}(t)$ be the piecewise Bézier curve defined on the interval $[-2, 4]$ given by

$$\mathbf{P}(t) = \begin{cases} \mathbf{B}_0(t), & -2 \le t < 0 \\ \mathbf{B}_1(t), & 0 \le t < 3 \\ \mathbf{B}_2(t), & 3 \le t \le 4 \end{cases}.$$

The breakpoints are $t_0 = -2$, $t_1 = 0$, $t_3 = 3$, $t_4 = 4$. Since the parametrization does not affect the trace of a curve, $\mathbf{P}(t)$ can be plotted by drawing the normalizations of $\mathbf{B}_j(t)$ which are the Bézier curves defined by the same sets of control points. The curve is shown in Figure 7.1.

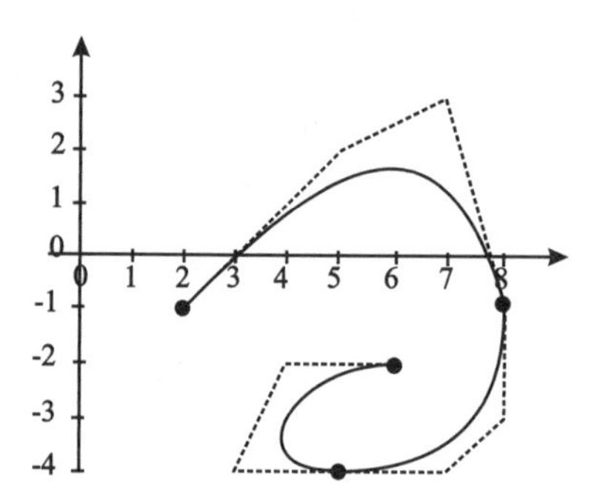

Figure 7.1

Recall that a parametric curve $\mathbf{C}(t)$ is said to be C^k-continuous whenever all of the coordinate functions are C^k-continuous. Since a polynomial function is C^∞, a piecewise polynomial function is C^∞ everywhere except at the parameter values corresponding to the joins of the individual functions. It can be shown that, at the join of two polynomial functions of degrees p and q, the piecewise function is at most C^k-continuous where $k = \min(p, q)$. It follows that at the join of two polynomial curves of degrees p and q, the piecewise curve is at most C^k-continuous where $k = \min(p, q)$.

Example 7.12

Consider the curve $(x(t), y(t)) = (t, f(t))$ where

$$f(t) = \begin{cases} 3t^2 + 2t + 1, & \text{if } t \leq 0 \\ t + 1, & \text{if } t > 0 \end{cases} .$$

The curve is shown in Figure 7.2(a). Since any polynomial function is C^∞, $x(t) = t$ is C^∞ everywhere, and $y(t) = f(t)$ is C^∞ everywhere except possibly at $t = 0$, the parameter value at which the second polynomial function takes over from the first in the definition of f. It is easily checked that $f(t)$ is C^0. To determine whether $f(t)$ is C^1 it is necessary to consider its first derivative

$$f'(t) = \begin{cases} 6t + 2, & \text{if } t \leq 0 \\ 1, & \text{if } t > 0 \end{cases} .$$

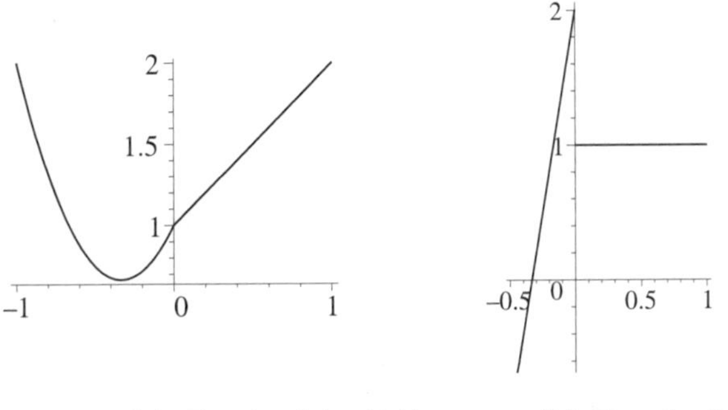

Figure 7.2 (a) Graph of $(t, f(t))$ (b) Graph of $f'(t)$

The graph of $f'(t)$ is shown in Figure 7.2(b). It is clear that f is not C^1: geometrically, the graph of $f'(t)$ has a break at $t = 0$. Hence $(t, f(t))$ is a C^0-continuous curve, but not C^1.

The piecewise curve in this example is the join of the two polynomial curves $(t, 3t^2+2t+1)$ and $(t, t+1)$. Since $k = \min(2,1) = 1$ for curves of degrees $p = 2$ and $q = 1$, the maximum continuity possible is C^1; the example only achieves C^0-continuity.

Suppose $\mathbf{P}(t)$ is a piecewise Bézier curve defined on $I = [a, b]$, consisting of Bézier curves $\mathbf{B}_j(t)$ defined on intervals $I_j = [t_j, t_{j+1}]$ $(j = 0, \ldots, r-1)$ as expressed in Definition 7.9. Since the coordinate functions of $\mathbf{P}(t)$ are piecewise polynomial functions, $\mathbf{P}(t)$ is C^∞ at all parameter values which are not breakpoints. Suppose $\mathbf{B}_j(t)$ has degree n_j and control points $\mathbf{b}_0^j, \ldots, \mathbf{b}_{n_j}^j$. Then $\mathbf{P}(t)$ is continuous at $t = t_j$ if and only if $\lim_{t\to t_j^+} \mathbf{P}(t) = \lim_{t\to t_j^-} \mathbf{P}(t) = \mathbf{P}(t_j)$. (This is a standard criterion for continuous functions, see [25].) But

$$\begin{aligned} \lim_{t\to t_j^+} \mathbf{P}(t) &= \lim_{t\to t_j^+} \mathbf{B}_j(t) = \mathbf{b}_0^j, \quad \text{and} \\ \lim_{t\to t_j^-} \mathbf{P}(t) &= \lim_{t\to t_j^-} \mathbf{B}_{j-1}(t) = \mathbf{b}_{n_{j-1}}^{j-1} . \end{aligned}$$

Thus $\mathbf{P}(t)$ is continuous at $t = t_j$ if and only if $\mathbf{P}(t_j) = \mathbf{b}_{n_{j-1}}^{j-1} = \mathbf{b}_0^j$. Hence the last control point of $\mathbf{B}_{j-1}(t)$ should equal the first control point of $\mathbf{B}_j(t)$. Therefore $\mathbf{P}(t)$ is continuous if and only if $\mathbf{b}_{n_{j-1}}^{j-1} = \mathbf{b}_0^j$ for all $j = 1, \ldots, r-1$.

Example 7.13

Consider the piecewise cubic Bézier curve consisting of two cubic Bézier curves $\mathbf{B}(t)$ and $\mathbf{C}(t)$ with control points $\mathbf{b}_0, \mathbf{b}_1, \mathbf{b}_2, \mathbf{b}_3$ and $\mathbf{c}_0, \mathbf{c}_1, \mathbf{c}_2, \mathbf{c}_3$ respectively.

Assuming that $\mathbf{B}(t)$ is the first curve and $\mathbf{C}(t)$ is the second, the piecewise cubic Bézier curve is C^0 if and only if $\mathbf{b}_3 = \mathbf{c}_0$.

Let $\mathbf{P}^{(p)}$ denote the pth derivative of $\mathbf{P}(t)$. Then

$$\begin{aligned} \lim_{t \to t_j^+} \mathbf{P}^{(p)}(t) &= \lim_{t \to t_j^+} \mathbf{B}_j^{(p)}(t) \,, \\ \lim_{t \to t_j^-} \mathbf{P}^{(p)}(t) &= \lim_{t \to t_j^-} \mathbf{B}_{j-1}^{(p)}(t) \,. \end{aligned}$$

Thus $\mathbf{B}(t)$ is C^k if and only if for all $p \le k$ and for all $j = 1, \ldots, r-1$,

$$\mathbf{B}_{j-1}^{(p)}(t_j) = \mathbf{B}_j^{(p)}(t_j) \,. \tag{7.9}$$

The condition (7.9) for C^k-continuity can be expressed in terms of the normalized Bézier curves. Since $\mathbf{B}_j(t) = \hat{\mathbf{B}}_j(\frac{t - t_{\min}}{t_{\max} - t_{\min}})$ the chain rule yields

$$\frac{1}{(t_j - t_{j-1})^p} \hat{\mathbf{B}}_{j-1}^{(p)}(1) = \frac{1}{(t_{j+1} - t_j)^p} \hat{\mathbf{B}}_j^{(p)}(0) \,. \tag{7.10}$$

This is a more useful formula since the derivatives of $\hat{\mathbf{B}}_j^{(p)}$ are easily obtained in terms of the control points using Theorem 7.3 and its corollaries.

Definition 7.14

Suppose two regular curves $\mathbf{B}(s)$, $s \in [s_0, s_1]$, and $\mathbf{C}(t)$, $t \in [t_0, t_1]$, meet at a point $\mathbf{P} = \mathbf{B}(s_1) = \mathbf{C}(t_0)$. Then the two curves are said to meet with G^k-*continuity* whenever there is a reparametrization $\alpha : [u_0, u_1] \to [s_0, s_1]$ such that $s_1 = \alpha(u_1)$ and

$$\left. \frac{d^i \mathbf{B}}{du^i}(\alpha(u)) \right|_{u=u_1} = \left. \frac{d^i \mathbf{C}}{dt^i}(t) \right|_{t=t_0}$$

for all $i = 0, \ldots, k$. This type of continuity is called *geometric continuity.* (The notation $|_{t=t_0}$ indicates that the function is evaluated at t_0.)

If two curves meet with G^1-continuity at a point $\mathbf{P}$, then their tangent vectors at $\mathbf{P}$ have the same direction, but they may have different magnitudes. To the eye, the curves meet smoothly and the curves are said to be *visually tangent continuous* at $\mathbf{P}$. It is left as an exercise to the reader to show that $\mathbf{B}(s)$ and $\mathbf{C}(t)$ are G^1-continuous at $\mathbf{P}$ if and only if

$$\mu \frac{d\mathbf{B}}{ds}(s_1) = \frac{d\mathbf{C}}{dt}(t_0) \tag{7.11}$$

for some $\mu \neq 0$ (using the notation of Definition 7.14), and that C^1-continuity is obtained whenever (7.11) is satisfied for $\mu = 1$.

Suppose $\mathbf{P}(t)$ is a continuous piecewise Bézier curve. Then $\mathbf{P}(t)$ is visually tangent continuous at $t = t_j$ whenever there exist constants $\mu_j > 0$, such that $\mu_j \mathbf{B}'_{j-1}(t_j) = \mathbf{B}'_j(t_j)$. When the piecewise Bézier curve is given by Bézier curves $\mathbf{B}_j(t)$ defined on intervals of unit length (that is $t_j - t_{j-1} = 1$ for all j) condition (7.10) simplifies to $\hat{\mathbf{B}}^{(p)}_{j-1}(1) = \hat{\mathbf{B}}^{(p)}_j(0)$. Then Theorem 7.3 yields the conditions for C^k-continuity and visual tangent continuity in terms of the control points alone. So, in effect the intervals can be ignored and it is common practice to work as if ordinary Bézier curves are being used to construct the piecewise curve. Examples are given below.

Example 7.15

Consider a piecewise Bézier curve consisting of two cubic Bézier curves $\mathbf{B}(t)$ and $\mathbf{C}(t)$ (defined on intervals of unit length) with control points $\mathbf{b}_0, \mathbf{b}_1, \mathbf{b}_2, \mathbf{b}_3$, and $\mathbf{c}_0, \mathbf{c}_1, \mathbf{c}_2, \mathbf{c}_3$, respectively. Suppose that $\mathbf{b}_3 = \mathbf{c}_0$ so that the piecewise curve is continuous. Visual tangent continuity is obtained when $\mu \mathbf{B}'(1) = \mathbf{C}'(0)$. Then $\mathbf{B}'(1) = 3(\mathbf{b}_3 - \mathbf{b}_2)$ and $\mathbf{C}'(0) = 3(\mathbf{c}_1 - \mathbf{c}_0)$, giving $\mu 3(\mathbf{b}_3 - \mathbf{b}_2) = 3(\mathbf{c}_1 - \mathbf{c}_0)$. Substituting $\mathbf{c}_0 = \mathbf{b}_3$ and simplifying gives $\mathbf{c}_1 = (1+\mu)\,\mathbf{b}_3 - \mu\mathbf{b}_2$. Hence $\mathbf{b}_2$, $\mathbf{b}_3$, and $\mathbf{c}_1$ are collinear, and $\mathbf{b}_3$ lies between $\mathbf{b}_2$ and $\mathbf{c}_1$. Geometrically, the visual tangent continuity implies that the tangent direction at the end of the first segment equals the tangent direction at the beginning of the second segment. The resulting join of the two curves appears smooth, but the underlying parametrization is not C^1-continuous. C^1-continuity is achieved when $\mu = 1$, so that the magnitudes and directions of the tangents are equal. Hence $\mathbf{c}_0 = \mathbf{b}_3$ and $\mathbf{c}_1 = 2\mathbf{b}_3 - \mathbf{b}_2$. The arguments above are easily generalized to give continuity conditions for Bézier curves of degree n expressed in Theorem 7.17.

Example 7.16

Consider the piecewise curve consisting of two cubic Bézier curves (defined on intervals of unit length) with control points $\mathbf{b}_0(2,5), \mathbf{b}_1(3,1), \mathbf{b}_2(5,1), \mathbf{b}_3(6,3)$, and $\mathbf{c}_0(6,3), \mathbf{c}_1(8,7), \mathbf{c}_2(5,8), \mathbf{c}_3(3,6)$. Since $\mathbf{b}_3 = \mathbf{c}_0 = (6,3)$, the two curves join to form a continuous curve. Further, the condition $\mathbf{c}_1 = (1+\mu)\,\mathbf{b}_3 - \mu\mathbf{b}_2$ gives

$$(8,7) = (1+\mu)\,(6,3) - \mu(5,1)$$

which is satisfied for $\mu = 2 > 0$, and hence the curve is visually tangent continuous. C^1-continuity can be obtained by adjusting a control point. For instance, if $\mathbf{c}_1$ is changed to $(7,5)$, then $\mathbf{c}_1 = 2\mathbf{b}_3 - \mathbf{b}_2$ is satisfied.

Theorem 7.17

Two Bézier curves of degree n,

$$\mathbf{B}(t) = \sum_{i=0}^{n} \mathbf{b}_i B_{i,n}(t), \quad \text{and} \quad \mathbf{C}(t) = \sum_{i=0}^{n} \mathbf{c}_i B_{i,n}(t),$$

defined on intervals of unit length, join to form a piecewise curve with

1. C^0-continuity if and only if $\mathbf{b}_n = \mathbf{c}_0$ (or $\mathbf{c}_n = \mathbf{b}_0$);
2. C^1-continuity if and only if $\mathbf{b}_n = \mathbf{c}_0$ and $\mathbf{c}_1 = 2\mathbf{b}_n - \mathbf{b}_{n-1}$ (or $\mathbf{c}_n = \mathbf{b}_0$ and $\mathbf{b}_1 = 2\mathbf{c}_n - \mathbf{c}_{n-1}$);
3. visual tangent continuity if and only if $\mathbf{b}_n = \mathbf{c}_0$, and

$$\mathbf{c}_1 = (1+\mu)\,\mathbf{b}_n - \mu\mathbf{b}_{n-1},$$

for some μ (or $\mathbf{c}_n = \mathbf{b}_0$ and $\mathbf{b}_1 = (1+\mu)\,\mathbf{c}_n - \mu\mathbf{c}_{n-1}$).

Corollary 7.18

Using the earlier notation, a piecewise Bézier curve $\mathbf{B}(t)$ for which the Bézier curves $\mathbf{B}_j(t)$ are defined on unit length intervals has

1. C^0-continuity if and only if $\mathbf{b}_n^i = \mathbf{b}_0^{i+1}$ for $i = 1, \ldots, r-1$;
2. C^1-continuity if and only if $\mathbf{b}_n^i = \mathbf{b}_0^{i+1}$ and $\mathbf{b}_1^{i+1} = 2\mathbf{b}_n^i - \mathbf{b}_{n-1}^i$ for $i = 1, \ldots, r-1$;
3. visual tangent continuity if and only if $\mathbf{b}_n^i = \mathbf{b}_0^{i+1}$ and $\mathbf{b}_1^{i+1} = (1+\mu_i)\,\mathbf{b}_n^i - \mu_i \mathbf{b}_{n-1}^i$, for some μ_i, for $i = 1, \ldots, r-1$.

EXERCISES

7.14. Consider the piecewise curve consisting of two cubic Bézier curves with control points $\mathbf{b}_0(2,1)$, $\mathbf{b}_1(4,2)$, $\mathbf{b}_2(5,4)$, $\mathbf{b}_3(3,6)$, and $\mathbf{c}_0(3,6)$, $\mathbf{c}_1(2,7)$, $\mathbf{c}_2(0,5)$, $\mathbf{c}_3(0,3)$. Show that the curves have visual tangent continuity. Plot the curves and their control polygons. Alter the control point $\mathbf{b}_2$ so that the curves join with C^1-continuity.

7.15. Determine the conditions on the control points for C^0-, C^1-, and C^2-continuity of two quadratic Bézier curves (defined on intervals of unit length).

7.16. Determine the conditions on the control points for C^2-continuity of two cubic Bézier curves (defined on unit intervals). Determine the conditions for C^2-continuity of two Bézier curves of degree n.

7.5 Rational Bézier Curves

In Section 5.6.1 it was shown that there are three types of irreducible conics, namely, hyperbolas, parabolas, and ellipses. Parabolas can be parametrized by polynomial functions, whereas hyperbolas and ellipses are parametrized by rational functions. Thus quadratic Bézier curves, which have polynomial parametrizations, exclude hyperbolas and ellipses. In order to represent such curves it is necessary to introduce rational Bézier curves.

Definition 7.19

A *rational Bézier curve* of degree n with control points $\mathbf{b}_0, \ldots, \mathbf{b}_n$ and corresponding scalar *weights* w_i is defined to be

$$\mathbf{B}(t) = \frac{\sum_{i=0}^{n} w_i \mathbf{b}_i B_{i,n}(t)}{\sum_{i=0}^{n} w_i B_{i,n}(t)}, \quad t \in [0,1],$$

with the understanding that if $w_i = 0$, then $w_i \mathbf{b}_i$ is to be replaced by $\mathbf{b}_i$. It is assumed that not all the weights are zero. When $\mathbf{b}_i \in \mathbb{R}^2$ $(i = 0, \ldots, n)$ then the curve is planar, and when $\mathbf{b}_i \in \mathbb{R}^3$ the curve is spatial. Note that the term *integral* Bézier curve is used to describe non-rational Bézier curves.

Let $\mathbf{b}_i = (x_i, y_i, z_i)$. Define *homogeneous control points* $\hat{\mathbf{b}}_i$ by

$$\hat{\mathbf{b}}_i = \begin{cases} (w_i x_i, w_i y_i, w_i z_i, w_i), & \text{if } w_i \neq 0 \\ (x_i, y_i, z_i, 0), & \text{if } w_i = 0 \end{cases}.$$

In homogeneous coordinates the rational Bézier curve is given by

$$\mathbf{B}(t) = \sum_{i=0}^{n} \hat{\mathbf{b}}_i B_{i,n}(t)$$

which takes the form of an integral Bézier curve, but with homogeneous control points. Since $\sum_{i=0}^{n} B_{i,n}(t) = 1$, by the partition of unity property, integral Bézier curves are obtained whenever $w_0 = \cdots = w_n$.

A rational Bézier curve can also be written in the basis form

$$\mathbf{B}(t) = \sum_{i=0}^{n} \mathbf{b}_i R_{i,n}(t)$$

where

$$R_{i,n}(t) = \begin{cases} \frac{w_i B_{i,n}(t)}{\sum_{j=0}^{n} w_j B_{j,n}(t)}, & \text{if } w_i \neq 0 \\ \frac{B_{i,n}(t)}{\sum_{j=0}^{n} w_j B_{j,n}(t)}, & \text{if } w_i = 0 \end{cases}.$$

Example 7.20

The rational quadratic with control points $\mathbf{b}_0(-1,0)$, $\mathbf{b}_1(2,1)$, and $\mathbf{b}_2(4,-1)$ and corresponding weights $1,1,2$ is shown in Figure 7.3(a), and with weights $1,0.6,2$ is shown in Figure 7.3(b).

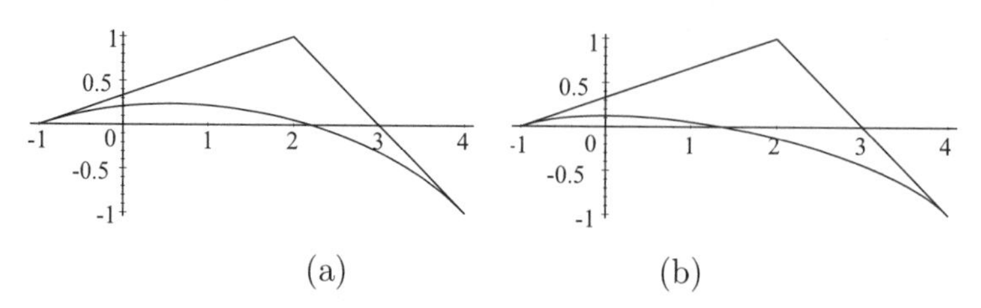

Figure 7.3 Rational Bézier curve with control points $(-1,0)$, $(2,1)$, $(4,-1)$, and (a) weights $1,1,2$, and (b) weights $1,0.6,2$

Example 7.21

The unit quarter circle in the first quadrant can be represented as a quadratic rational Bézier curve with control points $\mathbf{b}_0(1,0)$, $\mathbf{b}_1(1,1)$, and $\mathbf{b}_2(0,1)$ and weights $w_0 = 1$, $w_1 = 1$, and $w_2 = 2$. Then

$$\begin{aligned}
&(1-t)^2 w_0 \mathbf{b}_0 + 2t(1-t) w_1 \mathbf{b}_1 + t^2 w_2 \mathbf{b}_2 \\
&\qquad = (1-t)^2(1,0) + 2t(1-t)(1,1) + 2t^2(0,1) \\
&\qquad = \left(1-t^2, 2t\right) , \\
&(1-t)^2 w_0 + 2t(1-t) w_1 + t^2 w_2 \\
&\qquad = (1-t)^2 + 2t(1-t) + 2t^2 = 1 + t^2 .
\end{aligned}$$

Hence $\mathbf{B}(t) = \left(\frac{1-t^2}{1+t^2}, \frac{2t}{1+t^2}\right)$ which is a familiar parametrization of the unit quarter circle.

Example 7.22

The spatial rational cubic Bézier curve with control points $\mathbf{b}_0(1,0,1)$, $\mathbf{b}_1(2,1,-1)$, $\mathbf{b}_2(5,4,2)$, and $\mathbf{b}_3(2,-3,1)$ and weights $1,2,2,1$ is shown in Figure 7.4.

A quadratic rational Bézier curve has the form

$$\mathbf{B}(t) = \frac{w_0 \mathbf{b}_0 B_{0,2}(t) + w_1 \mathbf{b}_1 B_{1,2}(t) + w_2 \mathbf{b}_2 B_{2,2}(t)}{w_0 B_{0,2}(t) + w_1 B_{1,2}(t) + w_2 B_{2,2}(t)} . \tag{7.12}$$

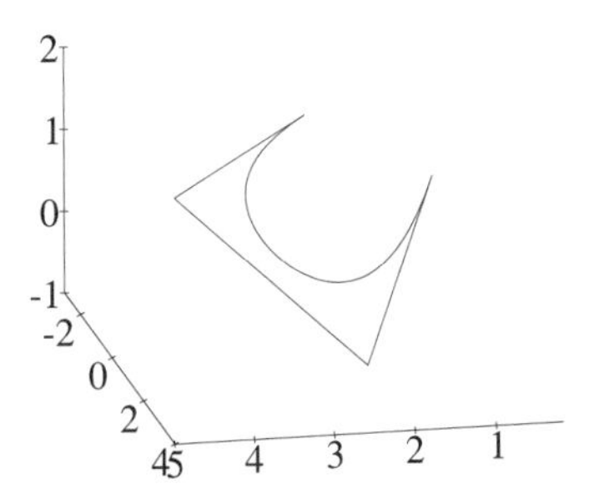

Figure 7.4 Spatial rational Bézier curve

It was noted in Sections 5.6.4 and 5.6.5 that any conic can be parametrized by quadratic rational functions, and conversely, that any curve parametrized by quadratic rational functions is a conic. Thus quadratic rational Bézier curves are conics. The classification of conics, given in Table 5.1, determines the type of a quadratic Bézier curve (assumed not to be a line) to be as follows:

ellipse when $w_1^2 - w_0 w_2 < 0$;

parabola when $w_1^2 - w_0 w_2 = 0$;

hyperbola when $w_1^2 - w_0 w_2 > 0$.

7.5.1 Properties of Rational Bézier Curves

Rational Bézier curves inherit a number of the properties of integral Bézier curves.

Convex Hull Property: Suppose $w_i > 0$ for all $i = 0, \ldots, n$. Then every point on the curve lies in the convex hull of the control polygon.

Invariance under Affine Transformations: If T is an affine transformation, then

$$\mathsf{T}\left(\frac{\sum_{i=0}^{n} w_i \mathbf{b}_i B_{i,n}(t)}{\sum_{i=0}^{n} w_i B_{i,n}(t)}\right) = \frac{\sum_{i=0}^{n} w_i \mathsf{T}\left(\mathbf{b}_i\right) B_{i,n}(t)}{\sum_{i=0}^{n} w_i B_{i,n}(t)}.$$

Variation Diminishing Property: Suppose $w_i > 0$ for all i. The VDP holds as for integral Bézier curves. See Section 6.7.

Endpoint Interpolation: $\mathbf{B}(0) = \mathbf{b}_0$, $\mathbf{B}(1) = \mathbf{b}_n$.

Endpoint Tangent: $\mathbf{B}'(0) = n\frac{w_1}{w_0}\left(\mathbf{b}_1 - \mathbf{b}_0\right)$ and $\mathbf{B}'(1) = n\frac{w_{n-1}}{w_n}\left(\mathbf{b}_n - \mathbf{b}_{n-1}\right)$. See Section 7.5.4.

Invariance under Projective Transformations: If T is a projective transformation, then

$$\mathsf{T}\left(\sum_{i=0}^{n} \hat{\mathbf{b}}_i B_{i,n}(t)\right) = \sum_{i=0}^{n} \mathsf{T}\left(\hat{\mathbf{b}}_i\right) B_{i,n}(t) .$$

See Section 7.5.3 for details.

Lemma 7.23

The circular arc, radius r, centred at the origin with endpoints $(r, 0)$ and $(r\cos\theta, r\sin\theta)$, $\theta \in [-\pi, \pi]$, has a rational quadratic Bézier representation given by control points

$$\mathbf{b}_0(r, 0),\ \mathbf{b}_1\left(r, r\tan\frac{\theta}{2}\right),\ \mathbf{b}_2\left(r\cos\theta, r\sin\theta\right) ,$$

and weights $w_0 = w_2 = 1$ and $w_1 = \cos\frac{\theta}{2}$.

Proof

Then

$$\begin{aligned} x(t) &= \frac{r(1-t)^2 + 2rt(1-t)\cos\frac{\theta}{2} + rt^2\cos\theta}{(1-t)^2 + 2t(1-t)\cos\frac{\theta}{2} + t^2}, \quad \text{and} \\ y(t) &= \frac{2rt(1-t)\sin\frac{\theta}{2} + rt^2\sin\theta}{(1-t)^2 + 2t(1-t)\cos\frac{\theta}{2} + t^2} . \end{aligned}$$

The tedious task of showing that the curve is circular, that is $x(t)^2 + y(t)^2 = r^2$, is left to the reader. The CHP implies that the curve lies within the triangular region defined by the control points, and thus the curve is a circular arc.

□

Example 7.24

The unit quarter circle in the first quadrant obtained by taking $r = 1$, $\theta = \frac{\pi}{2}$ in Lemma 7.1 does not give the parametrization of Example 7.1. The curve is given by $\mathbf{b}_0(1, 0)$, $\mathbf{b}_1(1, 1)$, $\mathbf{b}_2(0, 1)$, $w_0 = w_2 = 1$, $w_1 = 1/\sqrt{2}$, which yields

$$\begin{aligned} x(t) &= \frac{(1-t)^2 + \sqrt{2}t(1-t)}{(1-t)^2 + t(1-t)\sqrt{2} + t^2} = \frac{\sqrt{2}\,(1-t)\left(t + 1 + \sqrt{2}\right)}{\left(2t^2 - 2t + 2 + \sqrt{2}\right)} , \\ y(t) &= \frac{\sqrt{2}t(1-t) + t^2}{(1-t)^2 + \sqrt{2}t(1-t) + t^2} = \frac{\sqrt{2}\left(\sqrt{2} + 2 - t\right)t}{\left(2t^2 - 2t + 2 + \sqrt{2}\right)} . \end{aligned}$$

The arclength function for this parametrization is $s(t) = 2\arctan\left(\frac{2t-1}{\sqrt{2}+1}\right) + \pi/4$ which, surprisingly, differs from the unit speed arclength function $s(t) = \frac{\pi}{2}t$ by less than 0.0167 over the interval $[0, 1]$. The parametrization of Example 7.21 has arclength function $s(t) = 2\arctan t$ which differs from $s(t) = \frac{\pi}{2}t$ by as much as 0.1451.

Rational Bézier curves offer greater flexibility for curve design since for a given set of control points there are an infinite number of curves depending on the choice of weights. When a weight is adjusted the whole curve changes, but in a predictable manner, as described in the next theorem.

Theorem 7.25

Suppose a weight w_k is changed to $w_k + \delta w_k$, then every point $\mathbf{b} = \mathbf{B}(t)$ moves to the point $\mathbf{b}_w = (1-\alpha)\mathbf{b} + \alpha \mathbf{b}_k$ where

$$\alpha = \frac{\delta w_k B_{k,n}(t)}{\sum_{i=0}^n w_i B_{i,n}(t) + \delta w_k B_{k,n}(t)} .$$

Proof

For α defined as above

$$1 - \alpha = \frac{\sum_{i=0}^n w_i B_{i,n}(t)}{\sum_{i=0}^n w_i B_{i,n}(t) + \delta w_k B_{k,n}(t)}$$

Then

$$\begin{aligned}
\mathbf{b}_w &= \frac{\sum_{i=0}^n w_i \mathbf{b}_i B_{i,n}(t) + \delta w_k \mathbf{b}_k B_{k,n}(t)}{\sum_{i=0}^n w_i B_{i,n}(t) + \delta w_k B_{k,n}(t)} \\
&= \frac{\sum_{i=0}^n w_i \mathbf{b}_i B_{i,n}(t)}{\sum_{i=0}^n w_i B_{i,n}(t) + \delta w_k B_{k,n}(t)} + \frac{\delta w_k B_{k,n}(t)}{\sum_{i=0}^n w_i B_{i,n}(t) + \delta w_k B_{k,n}(t)} \mathbf{b}_k \\
&= \frac{\sum_{i=0}^n w_i B_{i,n}(t)}{\sum_{i=0}^n w_i B_{i,n}(t) + \delta w_k B_{k,n}(t)} \left(\frac{\sum_{i=0}^n w_i \mathbf{b}_i B_{i,n}(t)}{\sum_{i=0}^n w_i B_{i,n}(t)} \right) \\
&\quad + \frac{\delta w_k B_{k,n}(t)}{\sum_{i=0}^n w_i B_{i,n}(t) + \delta w_k B_{k,n}(t)} \mathbf{b}_k \\
&= (1-\alpha)\mathbf{b} + \alpha \mathbf{b}_k .
\end{aligned}$$

□

EXERCISES

7.17. Prove the convex hull property for rational Bézier curves of degree n with positive weights.

7.18. Prove the endpoint interpolation property for rational Bézier curves of degree n.

7.19. Show that when $w_1^2 - w_0 w_2 = 0$, a rational quadratic Bézier curve can be reparametrized to give an integral quadratic Bézier curve.

7.20. Consider a rational quadratic Bézier curve given by (7.12). Suppose $w_0 = w_2 = 1$, and define the *midpoint* $\mathbf{M} = \frac{1}{2}(\mathbf{b}_0 + \mathbf{b}_2)$ and the *shoulder point* $\mathbf{S} = (1-s)\mathbf{M} + s\mathbf{b}_1$ where $s = w_1/(1+w_1)$. Show that the curve is (a) an ellipse when $-1 < w_1 < 1$ ($s < 1/2$), (b) a parabola when $w_1 = 1$ or $w_1 = -1$ ($s = 1/2$ or ∞), or (c) a hyperbola when $w_1 > 1$ or $w_1 < -1$ ($s > 1/2$). Further show that $\mathbf{S}$ is the point on the curve corresponding to $t = 1/2$.

7.5.2 de Casteljau Algorithm for Rational Curves

The de Casteljau algorithm for integral Bézier curves extends to the rational case. There are two ways of performing the algorithm.

Method 1. Suppose

$$\mathbf{B}(t) = \frac{\sum_{i=0}^{n} w_i \mathbf{b}_i B_{i,n}(t)}{\sum_{i=0}^{n} w_i B_{i,n}(t)}$$

where $\mathbf{b}_i = (x_i, y_i)$ for planar curves and $\mathbf{b}_i = (x_i, y_i, z_i)$ for spatial curves. Let $\hat{\mathbf{b}}_i = (w_i x_i, w_i y_i, w_i)$ for planar curves and $\hat{\mathbf{b}}_i = (w_i x_i, w_i y_i, w_i z_i, w_i)$ for spatial curves. Apply the de Casteljau algorithm for integral Bézier curves (described in Section 6.8), treating the weight w_i as an additional coordinate.

Method 2. The first method, though straightforward to implement and computationally efficient, is prone to computational errors under certain conditions. The problem is avoided if the homogeneous control points are converted to Cartesian coordinates at the end of each iteration. The new algorithm is

$$\begin{cases} \mathbf{b}_i^j = (1-t)\dfrac{w_i^{j-1}}{w_i^j}\mathbf{b}_i^{j-1} + t\dfrac{w_{i+1}^{j-1}}{w_i^j}\mathbf{b}_{i+1}^{j-1} \\ w_i^j = (1-t)w_i^{j-1} + t w_{i+1}^{j-1} \end{cases}, \qquad (7.13)$$

for $j = 1, \ldots, n$ and $i = 0, \ldots, n-j$.

The rational de Casteljau algorithm evaluates the point $\mathbf{B}(t)$ and subdivides the curve at the point corresponding to the parameter value t.

Example 7.26

Let a rational cubic Bézier curve have control points $\mathbf{b}_0(1,1)$, $\mathbf{b}_1\,(2,7)$, $\mathbf{b}_2\,(8,6)$, $\mathbf{b}_3\,(12,1)$ and weights $w_0 = 1$, $w_1 = 2$, $w_2 = 2$, $w_3 = 1$. Then the rational de Casteljau algorithm with $t = 0.25$ yields the triangles of weights

$$\begin{array}{llll} 1.0 & 2.0 & 2.0 & 1.0 \\ 1.25 & 2.0 & 1.75 & \\ 1.4375 & 1.9375 & & \\ 1.5625 & & & \end{array}$$

and control points

$$\begin{array}{llll} (1.0, 1.0) & (2.0, 7.0) & (8.0, 6.0) & (12.0, 1.0) \\ (1.4, 3.4) & (3.5, 6.75) & (8.5714, 5.2857) & \\ (2.1304, 4.5652) & (4.6452, 6.4194) & & \\ (2.91, 5.14) & & & \end{array}$$

Thus $\mathbf{B}(0.25) = (2.91, 5.14)$.

7.5.3 Projections of Rational Bézier Curves

The property of invariance under projective transformations is a useful feature for the computer display of rational Bézier curves. To apply a projective transformation to a rational Bézier curve, it is sufficient to apply the transformation to the control points and weights in the manner described below. The transformed images of the control points and weights define the rational Bézier curve which is the image of the original curve. The property is proved as follows. Let M be a 4×4 projective transformation matrix. In homogeneous coordinates the curve

$$\mathbf{B}(t) = \frac{\sum_{i=0}^{n} w_i \mathbf{b}_i B_{i,n}(t)}{\sum_{i=0}^{n} w_i B_{i,n}(t)}$$

with $\mathbf{b}_i = (x_i, y_i, z_i)$ is expressed as

$$\mathbf{B}(t) = \sum_{i=0}^{n} \hat{\mathbf{b}}_i B_{i,n}(t)\ ,$$

where $\hat{\mathbf{b}}_i = (x_i w_i, y_i w_i, z_i w_i, w_i)$ if $w_i \neq 0$, and $\hat{\mathbf{b}}_i = (x_i w_i, y_i w_i, z_i w_i, 0)$ if $w_i = 0$. Applying M yields

$$\mathbf{B}(t)\mathsf{M} = \left(\sum_{i=0}^{n} \hat{\mathbf{b}}_i B_{i,n}(t) \right) \mathsf{M} = \sum_{i=0}^{n} \left(\hat{\mathbf{b}}_i \mathsf{M} \right) B_{i,n}(t) = \sum_{i=0}^{n} \hat{\mathbf{c}}_i B_{i,n}(t)\ , \qquad (7.14)$$

defining a rational Bézier curve with control points $\mathbf{c}_i$ and weights defined by the homogeneous control points $\hat{\mathbf{c}}_i = \hat{\mathbf{b}}_i\mathsf{M}$. In particular, if M is the 4×4 projection matrix of a perspective or parallel projection, then the image of a rational Bézier curve is determined by projecting the homogeneous control points. The image curve has control points which are the projected images of the original control points but the weights have changed. Thus the projected image of an integral Bézier curve is, in general, a rational Bézier curve.

It is easily shown that the argument expressed in (7.14) also applies to the viewplane coordinate mapping matrix VC. Furthermore, the invariance of planar rational Bézier curves under affine transformations applies to the device coordinate mapping DC . Thus the whole process of viewing a rational Bézier curve can be executed by applying the complete viewing pipeline matrix $\mathsf{VP} = \mathsf{M} \cdot \mathsf{VC} \cdot \mathsf{DC}$ to the control points.

Example 7.27

Consider the perspective projection of Example 4.7 onto the xy-plane with viewpoint $\mathbf{V}(1, 5, 3)$. Further, suppose that the viewplane origin is $(1, 2, 0)$, the X-axis has direction $(3, 4, 0)$, and the Y-axis has direction $(8, -6, 0)$, as in Example 4.8. The product $\mathsf{M} \cdot \mathsf{VC}$ of the projection matrix M and viewplane coordinate matrix VC is

$$\begin{pmatrix} -3 & 0 & 0 & 0 \\ 0 & -3 & 0 & 0 \\ 1 & 5 & 0 & 1 \\ 0 & 0 & 0 & -3 \end{pmatrix} \begin{pmatrix} 0.6 & 0.8 & 0.0 \\ 0.8 & -0.6 & 0.0 \\ 0.0 & 0.0 & 0.0 \\ -2.2 & 0.4 & 1.0 \end{pmatrix} = \begin{pmatrix} -1.8 & -2.4 & 0.0 \\ -2.4 & 1.8 & 0.0 \\ 2.4 & -1.8 & 1.0 \\ 6.6 & -1.2 & -3.0 \end{pmatrix}.$$

(The viewing pipeline can be completed by applying a device coordinate mapping D.) The projection and conversion to viewplane coordinates of the cubic rational Bézier curve $\mathbf{B}(t)$ with control points $\mathbf{b}_0(0, 0, 0)$, $\mathbf{b}_1(1, 0, 0)$, $\mathbf{b}_2(1, 0, 1)$, $\mathbf{b}_3(1, 1, 1)$, and weights $1, 2, 2, 1$ is obtained as follows. The homogeneous control points are $\hat{\mathbf{b}}_0(0, 0, 0, 1)$, $\hat{\mathbf{b}}_1(2, 0, 0, 2)$, $\hat{\mathbf{b}}_2(2, 0, 2, 2)$, $\hat{\mathbf{b}}_3(1, 1, 1, 1)$. Thus,

$$\begin{pmatrix} \hat{\mathbf{b}}_0 \\ \hat{\mathbf{b}}_1 \\ \hat{\mathbf{b}}_2 \\ \hat{\mathbf{b}}_3 \end{pmatrix} \mathsf{M} \cdot \mathsf{VC} = \begin{pmatrix} 0 & 0 & 0 & 1 \\ 2 & 0 & 0 & 2 \\ 2 & 0 & 2 & 2 \\ 1 & 1 & 1 & 1 \end{pmatrix} \mathsf{M} \cdot \mathsf{VC} = \begin{pmatrix} 6.6 & -1.2 & -3.0 \\ 9.6 & -7.2 & -6.0 \\ 14.4 & -10.8 & -4.0 \\ 4.8 & -3.6 & -2.0 \end{pmatrix}.$$

The projected curve is given by homogeneous control points $\hat{\mathbf{c}}_0(6.6, -1.2, -3.0)$, $\hat{\mathbf{c}}_1(9.6, -7.2, -6.0)$, $\hat{\mathbf{c}}_2(14.4, -10.8, -4.0)$, $\hat{\mathbf{c}}_3(4.8, -3.6, -2.0)$. Since all the weights are negative, the $\hat{\mathbf{c}}_i$ can be multiplied by -1 to give points with corresponding positive weights. After dividing through by the weights the projected curve is found to be the planar quadratic Bézier curve with control points $\mathbf{c}_0\,(-2.2, 0.4)$, $\mathbf{c}_1\,(-1.6, 1.2)$, $\mathbf{c}_2\,(-3.6, 2.7)$, and $\mathbf{c}_3\,(-2.4, 1.8)$, and weights $3, 6, 4$, and 2.

For instance, the point $\mathbf{B}(0.75) = (0.99, 0.27, 0.81)$ projects to the point with homogeneous coordinates

$$\begin{pmatrix} 0.99 & 0.27 & 0.81 & 1.0 \end{pmatrix} \mathsf{M} \cdot \mathsf{VC} = \begin{pmatrix} 6.114 & -4.548 & -2.19 \end{pmatrix} ,$$

and Cartesian coordinates $(-2.791781, 2.076712)$. The point can also be computed using the control points and weights of the projected curve

$$\begin{aligned} \sum_{i=0}^{3} w_i \mathbf{b}_i B_{i,n}(t) &= (-6.6, 1.2)(0.25)^3 + (-9.6, 7.2)3(0.25)^2(0.75) \\ &\quad + (-14.4, 10.8)3(0.25)(0.75)^2 + (-4.8, 3.6)(0.75)^3 \\ &= (-9.553125, 7.10625) , \end{aligned}$$

$$\begin{aligned} \sum_{i=0}^{3} w_i B_{i,n}(t) &= 3(0.25)^3 + 18(0.25)^2(0.75) + 12(0.25)(0.75)^2 + 2(0.75)^3 \\ &= 3.421875 . \end{aligned}$$

Hence $\mathbf{B}(0.75) = (-9.553125, 7.10625)/3.421875 = (-2.791781, 2.076712)$. The projection of the curve is shown in Figure 7.5.

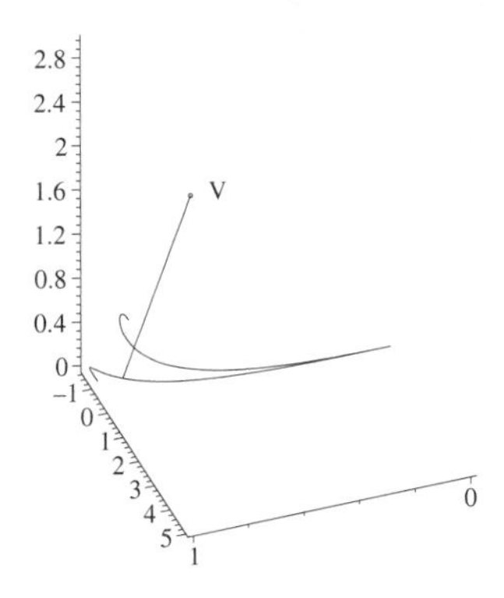

Figure 7.5

EXERCISES

7.21. Apply the rational de Casteljau algorithm to the cubic curve with control points $\mathbf{b}_0(3,2)$, $\mathbf{b}_1(7,6)$, $\mathbf{b}_2(5,3)$, $\mathbf{b}_3(3,0)$ and weights $w_0 = 2$, $w_1 = 3$, $w_2 = 5$, $w_3 = 1$ to determine (a) the point $\mathbf{B}(0.6)$, and (b) the control points of the two Bézier curve segments obtained following a subdivision at the point $\mathbf{B}(0.6)$.

7.22. A rational Bézier curve $\mathbf{B}(t)$ with $w_0 \neq 0$ and $w_n \neq 0$ can be reparametrized to give a rational Bézier curve for which $w_0 = w_n = 1$. Prove this by (a) dividing the denominator and numerator of $\mathbf{B}(t)$ by w_0 to give a rational curve with $w_0 = 1$, and (b) verifying that the transformation

$$t = t_1 / (a + (1-a)t_1) , \quad (1-t) = a(1-t_1) / (a + (1-a)t_1) ,$$

where $a = \sqrt[n]{w_n}$, yields a new rational curve in the variable t_1 with $w_0 = w_n = 1$.

7.23. Implement the rational de Casteljau algorithm and the operation of projecting a rational Bézier curve.

7.24. Compute the control points and weights of the image of the rational Bézier curve with control points $(1,2,-1)$, $(3,5,4)$, $(-1,3,3)$, $(0,1,2)$, and weights $2, 1/2, 4, 3$, when projected from the point $(9,7,5)$ onto the plane $3x + 3y + 12 = 0$. Assume that the viewplane coordinate system has origin $(-4,0,0)$, and that the X- and Y-axes have directions $(-1,1,0)$ and $(0,0,1)$ respectively.

7.25. Show that the application of a projective transformation $\mathsf{M} = (m_{ij})$ of an integral Bézier curve with control points $\mathbf{b}_i = (x_i, y_i, z_i)$ yields a rational Bézier curve with control points

$$\mathbf{c}_i = (c_{i,1}/w_i, c_{i,2}/w_i, c_{i,3}/w_i) ,$$

and weights w_i, where $c_{i,j} = x_i m_{1j} + y_i m_{2j} + z_i m_{3j} + m_{4j}$ and $w_i = x_i m_{14} + y_i m_{24} + z_i m_{34} + m_{44}$. Using the notation of Theorem 4.5, deduce that a projection M yields weights $w_i = (x_i n_1 + y_i n_2 + z_i n_3)\, v_4 + (-n_1 v_1 - n_2 v_2 - n_3 v_3)$. Hence show that a projection of an integral curve is an integral curve if and only if (a) the projection is parallel, or (b) the projection is perspective and the curve lies in a plane parallel to the viewplane (not containing the viewpoint).

7.5.4 Derivatives of Rational Bézier Curves

A recursive formula to determine the derivative of a rational Bézier curve is obtained from the following method for differentiating rational functions. Let $F(t) = f(t)/g(t)$. Then the quotient rule gives

$$F'(t) = \frac{g(t)f'(t) - g'(t)f(t)}{g(t)^2} = \frac{f'(t) - g'(t)F(t)}{g(t)} . \tag{7.15}$$

The Leibnitz rule [27] for obtaining the derivatives of a product of two functions yields that the rth derivative of $f(t) = g(t)F(t)$ is

$$\begin{aligned} f^{(r)}(t) &= \sum_{i=0}^{r} \binom{r}{i} g^{(i)}(t) F^{(r-i)}(t) \\ &= g(t)F^{(r)}(t) + \sum_{i=1}^{r} \binom{r}{i} g^{(i)}(t) F^{(r-i)}(t) . \end{aligned}$$

Hence

$$F^{(r)}(t) = \frac{f^{(r)}(t) - \sum\limits_{i=1}^{r} \binom{r}{i} g^{(i)}(t) F^{(r-i)}(t)}{g(t)} . \tag{7.16}$$

Thus the rth derivative of $F(t)$ can be obtained in terms of the first $r-1$ derivatives of $F(t)$, and the first r derivatives of $f(t)$ and $g(t)$.

Consider a rational Bézier curve of degree n

$$\mathbf{B}(t) = \frac{\sum_{i=0}^{n} w_i \mathbf{b}_i B_{i,n}(t)}{\sum_{i=0}^{n} w_i B_{i,n}(t)} .$$

Let $f(t) = \sum_{i=0}^{n} w_i \mathbf{b}_i B_{i,n}(t)$ and $g(t) = \sum_{i=0}^{n} w_i B_{i,n}(t)$. The derivatives of $f(t)$ and $g(t)$ are obtained by applying the algorithm to determine the derivatives of integral Bézier curves given in Section 7.2, where $w_i \mathbf{b}_i$ are considered to be the control points of $f(t)$, and the weights w_i are considered to be the control points of $g(t)$. In particular, for $n = 1$ and $n = 2$

$$\begin{aligned} \mathbf{B}'(t) &= \frac{\left(\sum_{i=0}^{n} w_i \mathbf{b}_i B_{i,n}(t)\right)' - \left(\sum_{i=0}^{n} w_i B_{i,n}(t)\right)' \mathbf{B}(t)}{\sum_{i=0}^{n} w_i B_{i,n}(t)} , \\ \mathbf{B}''(t) &= \frac{\left(\sum_{i=0}^{n} w_i \mathbf{b}_i B_{i,n}(t)\right)'' - 2\left(\sum_{i=0}^{n} w_i B_{i,n}(t)\right)' \mathbf{B}'(t) - \left(\sum_{i=0}^{n} w_i B_{i,n}(t)\right)'' \mathbf{B}(t)}{\sum_{i=0}^{n} w_i B_{i,n}(t)} . \end{aligned}$$

Therefore,

$$\mathbf{B}'(0) = \frac{n\left(w_1 \mathbf{b}_1 - w_0 \mathbf{b}_0\right) - n\left(w_1 - w_0\right) \mathbf{b}_0}{w_0} = n \frac{w_1}{w_0} \left(\mathbf{b}_1 - \mathbf{b}_0\right) .$$

Similarly, $\mathbf{B}'(1) = n \frac{w_{n-1}}{w_n} \left(\mathbf{b}_n - \mathbf{b}_{n-1}\right)$. The endpoint tangent condition of Section 7.5.1 is now proved.

Example 7.28

Consider the rational cubic Bézier curve with control points $\mathbf{b}_0(2,1)$, $\mathbf{b}_1(5,6)$, $\mathbf{b}_2(6,2)$, $\mathbf{b}_3(9,3)$ and weights $w_0 = 3$, $w_1 = 2$, $w_2 = 1$, $w_3 = 4$. Multiply $\mathbf{b}_i$ by w_i to give the control points of $f(t) = \sum_{i=0}^{3} w_i \mathbf{b}_i B_{i,n}(t)$: $(6,3)$, $(10,12)$, $(6,2)$, $(36,12)$. Hence $f'(t)$ has control points $3((10,12) - (6,3)) = (12,27)$, $3((6,2) - (10,12)) = (-12,-30)$, and $3((36,12) - (6,2)) = (90,30)$. The control values of $g(t) = \sum_{i=0}^{3} w_i B_{i,n}(t)$ are $3, 2, 1, 4$, and hence the control values of $g'(t)$ are $3(2-3) = -3$, $3(1-2) = -3$, $3(4-1) = 9$. Then $\mathbf{B}'(0.25)$ is computed as follows:

$$\begin{aligned}
f(0.25) &= (1-0.25)^3(6,3) + 3(1-0.25)^2(0.25)(10,12) \\
&\quad + 3(1-0.25)(0.25)^2(6,2) + (0.25)^3(36,12) \\
&= (8.156, 6.797) \ , \\
g(0.25) &= (1-0.25)^3 3 + 3(1-0.25)^2(0.25)2 \\
&\quad + 3(1-0.25)(0.25)^2 1 + (0.25)^3 4 = 2.313 \ , \\
f'(0.25) &= (1-0.25)^2 \,(12,27) + 2(1-0.25)\,(0.25)\,(-12,-30) \\
&\quad + (0.25)^2 \,(90,30) = (7.875, 5.8125) \ , \\
g'(0.25) &= (1-0.25)^2\,(-3) + 2(1-0.25)\,(0.25)\,(-3) + (0.25)^2\,(9) = -2.25 \ .
\end{aligned}$$

Hence

$$\mathbf{B}(0.25) = f(0.25)/g(0.25) = (8.156, 6.797)/\,2.313 = (3.526, 2.939) \ ,$$

and

$$\begin{aligned}
\mathbf{B}'(0.25) &= \frac{f'(0.25) - g'(0.25)\mathbf{B}(0.25)}{g(0.25)} \\
&= \frac{(7.875, 5.813) - (-2.25)\,(3.526, 2.939)}{2.313} \\
&\approx (6.835, 5.372) \ .
\end{aligned}$$

EXERCISES

7.26. Evaluate, at $t = 0.5$, the first derivative of the rational cubic Bézier curve with control points $\mathbf{b}_0(0,0)$, $\mathbf{b}_1(2,1)$, $\mathbf{b}_2(3,3)$, $\mathbf{b}_3(2,0)$ and weights $w_0 = 1$, $w_1 = 2$, $w_2 = 2$, $w_3 = 1$.

7.27. Evaluate, at $t = 0.25$, the second derivative of the rational Bézier curve given in Example 7.28.

7.28. Implement an algorithm to determine the first derivative of a rational Bézier curve. If time is available then implement an algorithm to determine higher order derivatives.

8
B-splines

8.1 Integral B-spline Curves

A piecewise polynomial curve has a *B-spline* basis representation with properties similar to those of a Bézier curve. A B-spline curve defined on the interval $[a, b]$ is specified by the following information:

1. The *degree* d (or *order* $d+1$), so that each segment of the piecewise polynomial curve has degree d or less.
2. A sequence of $m+1$ real numbers $t_0, t_1, \ldots, t_m$, called the *knot vector*, such that $t_i \leq t_{i+1}$ $(i = 0, \ldots, m-1)$, $t_d = a$ and $t_{m-d} = b$. The knots $t_0, t_1, ..., t_d$ and $t_{m-d}, t_{m-d+1}, \ldots, t_m$ are called *end knots*, and the knots $t_{d+1}, t_{d+2}, ..., t_{m-d-1}$ are called *interior knots.*
3. *Control points* $\mathbf{b}_0, \ldots, \mathbf{b}_n$.

A B-spline curve is defined in terms of B-spline basis functions.

Definition 8.1

The *B-spline basis functions* of degree d, denoted $N_{i,d}(t)$, defined by the knot vector $t_0, t_1, \ldots, t_m$ are defined recursively as follows:

$$N_{i,0}(t) = \begin{cases} 1, & \text{if } t \in [t_i, t_{i+1}) \\ 0, & \text{otherwise} \end{cases}, \tag{8.1}$$

$$N_{i,d}(t) = \frac{t - t_i}{t_{i+d} - t_i} N_{i,d-1}(t) + \frac{t_{i+d+1} - t}{t_{i+d+1} - t_{i+1}} N_{i+1,d-1}(t), \tag{8.2}$$

for $i = 0, \ldots, n$ and $d \geq 1$. If the knot vector contains a sufficient number of repeated knot values, then a division of the form $N_{i,d-1}(t)/(t_{i+d} - t_i) = 0/0$ (for some i) may be encountered during the execution of the recursion. Whenever this occurs, it is assumed that $0/0 = 0$.

Definition 8.2

The *B-spline curve* of degree d (or order $d + 1$) with control points $\mathbf{b}_0, \ldots, \mathbf{b}_n$ and knots $t_0, \ldots, t_m$ is defined on the interval $[a, b] = [t_d, t_{m-d}]$ by

$$\mathbf{B}(t) = \sum_{i=0}^{n} \mathbf{b}_i N_{i,d}(t),$$

where $N_{i,d}(t)$ are the B-spline basis functions of degree d. To distinguish B-spline curves from their rational form (which will be introduced in Section 8.2) they are often referred to as *integral* B-splines.

Example 8.3

Let $d = 2$, and $t_0 = 2$, $t_1 = 4$, $t_2 = 5$, $t_3 = 7$, $t_4 = 8$, $t_5 = 10$, $t_6 = 11$, with control points $\mathbf{b}_0(1, 2)$, $\mathbf{b}_1(3, 5)$, $\mathbf{b}_2(6, 2)$, $\mathbf{b}_3(9, 4)$. Then the $k = 0$ basis functions are

$$N_{0,0}(t) = \begin{cases} 1, & \text{if } t \in [2, 4) \\ 0, & \text{otherwise} \end{cases}, \quad N_{1,0}(t) = \begin{cases} 1, & \text{if } t \in [4, 5) \\ 0, & \text{otherwise} \end{cases},$$

$$N_{2,0}(t) = \begin{cases} 1, & \text{if } t \in [5, 7) \\ 0, & \text{otherwise} \end{cases}, \quad N_{3,0}(t) = \begin{cases} 1, & \text{if } t \in [7, 8) \\ 0, & \text{otherwise} \end{cases},$$

$$N_{4,0}(t) = \begin{cases} 1, & \text{if } t \in [8, 10) \\ 0, & \text{otherwise} \end{cases}, \quad N_{5,0}(t) = \begin{cases} 1, & \text{if } t \in [10, 11) \\ 0, & \text{otherwise} \end{cases}.$$

The $k = 1$ basis functions are determined in terms of these

$$\begin{aligned}
N_{0,1}(t) &= \tfrac{t-t_0}{t_1-t_0} N_{0,0}(t) + \tfrac{t_2-t}{t_2-t_1} N_{1,0}(t) = \tfrac{t-2}{4-2} N_{0,0}(t) + \tfrac{5-t}{5-4} N_{1,0}(t) \\
&= \tfrac{t-2}{2} N_{0,0}(t) + (5-t)\, N_{1,0}(t)\,, \\
N_{1,1}(t) &= \tfrac{t-t_1}{t_2-t_1} N_{1,0}(t) + \tfrac{t_3-t}{t_3-t_2} N_{2,0}(t) = \tfrac{t-4}{5-4} N_{1,0}(t) + \tfrac{7-t}{7-5} N_{2,0}(t) \\
&= (t-4)\, N_{1,0}(t) + \tfrac{1}{2}\,(7-t)\, N_{2,0}(t)\,, \\
N_{2,1}(t) &= \tfrac{t-t_2}{t_3-t_2} N_{i,0}(t) + \tfrac{t_4-t}{t_4-t_3} N_{3,0}(t) = \tfrac{1}{2}\,(t-5)\, N_{2,0}(t) + (8-t)\, N_{3,0}(t)\,, \\
N_{3,1}(t) &= \tfrac{t-t_3}{t_4-t_3} N_{3,0}(t) + \tfrac{t_5-t}{t_5-t_4} N_{4,0}(t) = (t-7)\, N_{3,0}(t) + \tfrac{1}{2}\,(10-t)\, N_{4,0}(t)\,, \\
N_{4,1}(t) &= \tfrac{t-t_4}{t_5-t_4} N_{4,0}(t) + \tfrac{t_6-t}{t_6-t_5} N_{5,0}(t) = \tfrac{1}{2}\,(t-8)\, N_{4,0}(t) + (11-t)\, N_{5,0}(t)\,.
\end{aligned}$$

Finally, the $k = 2$ basis functions can be computed

$$\begin{aligned}
N_{0,2}(t) &= \tfrac{t-t_0}{t_2-t_0} N_{0,1}(t) + \tfrac{t_3-t}{t_3-t_1} N_{1,1}(t) = \tfrac{t-2}{3} N_{0,1}(t) + \tfrac{7-t}{3} N_{1,1}(t) \\
&= \tfrac{t-2}{3} \left(\tfrac{t-2}{2} N_{0,0}(t) + (5-t)\, N_{1,0}(t)\right) \\
&\quad + \tfrac{7-t}{3} \left((t-4)\, N_{1,0}(t) + \tfrac{7-t}{2} N_{2,0}(t)\right) \\
&= \tfrac{1}{6}\,(t-2)^2\, N_{0,0}(t) + \tfrac{1}{3}\left(-2t^2 + 18t - 38\right) N_{1,0}(t) \\
&\quad + \tfrac{1}{6}\,(7-t)^2\, N_{2,0}(t)\,, \\
N_{1,2}(t) &= \tfrac{t-t_1}{t_3-t_1} N_{1,1}(t) + \tfrac{t_4-t}{t_4-t_2} N_{2,1}(t) = \tfrac{t-4}{3} N_{1,1}(t) + \tfrac{8-t}{3} N_{2,1}(t) \\
&= \tfrac{t-4}{3} \left((t-4)\, N_{1,0}(t) + \tfrac{7-t}{2} N_{2,0}(t)\right) \\
&\quad + \tfrac{8-t}{3} \left(\tfrac{t-5}{2} N_{2,0}(t) + (8-t)\, N_{3,0}(t)\right) \\
&= \tfrac{1}{3}\,(t-4)^2\, N_{1,0}(t) + \tfrac{1}{3}\left(-t^2 + 12t - 34\right) N_{2,0}(t) \\
&\quad + \tfrac{1}{3}\,(8-t)^2\, N_{3,0}(t)\,, \\
N_{2,2}(t) &= \tfrac{t-t_2}{t_4-t_2} N_{2,1}(t) + \tfrac{t_5-t}{t_5-t_3} N_{3,1}(t) = \tfrac{t-5}{3} N_{2,1}(t) + \tfrac{10-t}{3} N_{3,1}(t) \\
&= \tfrac{t-5}{3} \left(\tfrac{t-5}{2} N_{2,0}(t) + (8-t)\, N_{3,0}(t)\right) \\
&\quad + \tfrac{10-t}{3} \left((t-7)\, N_{3,0}(t) + \tfrac{10-t}{2} N_{4,0}(t)\right) \\
&= \tfrac{1}{6}\,(t-5)^2\, N_{2,0}(t) + \tfrac{1}{3}\left(-2t^2 + 30t - 110\right) N_{3,0}(t) \\
&\quad + \tfrac{1}{6}\,(10-t)^2\, N_{4,0}(t)\,, \\
N_{3,2}(t) &= \tfrac{t-t_3}{t_5-t_3} N_{3,1}(t) + \tfrac{t_6-t}{t_6-t_4} N_{4,1}(t) = \tfrac{t-7}{3} N_{3,1}(t) + \tfrac{11-t}{3} N_{4,1}(t) \\
&= \tfrac{t-7}{3} \left((t-7)\, N_{3,0}(t) + \tfrac{10-t}{2} N_{4,0}(t)\right) \\
&\quad + \tfrac{11-t}{3} \left(\tfrac{t-8}{2} N_{4,0}(t) + (11-t)\, N_{5,0}(t)\right) \\
&= \tfrac{1}{3}\,(t-7)^2\, N_{3,0}(t) + \tfrac{1}{3}\left(-t^2 + 18t - 79\right) N_{4,0}(t) \\
&\quad + \tfrac{1}{3}\,(11-t)^2\, N_{5,0}(t)\,.
\end{aligned}$$

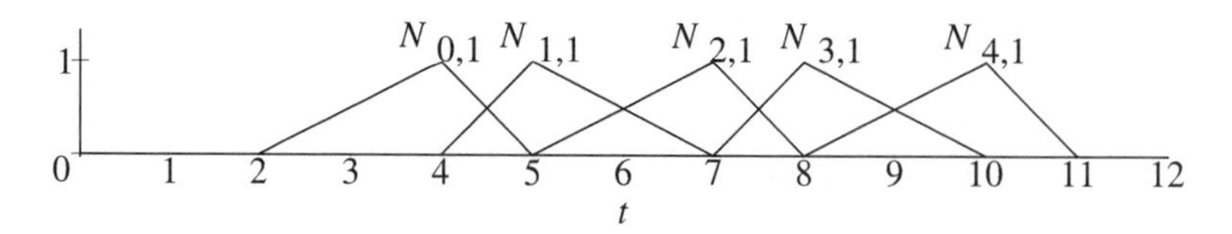

Figure 8.1 Basis functions of degree 1 for the knot vector $t_0 = 2$, $t_1 = 4$, $t_2 = 5$, $t_3 = 7$, $t_4 = 8$, $t_5 = 10$, $t_6 = 11$

The $k = 2$ basis functions may be expressed in the form

$$N_{0,2}(t) = \begin{cases} 0, & t < 2 \\ \frac{1}{6}(t-2)^2, & 2 \le t < 4 \\ \frac{1}{3}\left(-2t^2 + 18t - 38\right), & 4 \le t < 5 \\ \frac{1}{6}(7-t)^2, & 5 \le t < 7 \\ 0, & 7 \le t \end{cases},$$

$$N_{1,2}(t) = \begin{cases} 0, & t < 4 \\ \frac{1}{3}(t-4)^2, & 4 \le t < 5 \\ \frac{1}{3}\left(-t^2 + 12t - 34\right), & 5 \le t < 7 \\ \frac{1}{3}(8-t)^2, & 7 \le t < 8 \\ 0, & 8 \le t \end{cases},$$

$$N_{2,2}(t) = \begin{cases} 0, & t < 5 \\ \frac{1}{6}(t-5)^2, & 5 \le t < 7 \\ \frac{1}{3}\left(-2t^2 + 30t - 110\right), & 7 \le t < 8 \\ \frac{1}{6}(10-t)^2, & 8 \le t < 10 \\ 0, & 10 \le t \end{cases}, \quad \text{and}$$

$$N_{3,2}(t) = \begin{cases} 0, & t < 7 \\ \frac{1}{3}(t-7)^2, & 7 \le t < 8 \\ \frac{1}{3}\left(-t^2 + 18t - 79\right), & 8 \le t < 10 \\ \frac{1}{3}(11-t)^2, & 10 \le t < 11 \\ 0, & 11 \le t \end{cases}.$$

Plots of the degree 1 and 2 basis functions are shown in Figures 8.1 and 8.2. Observe that the basis functions satisfy $N_{i,2}(t) > 0$ for $t \in (t_i, t_{i+3})$ and $N_{i,2}(t) = 0$ elsewhere. General B-spline basis functions satisfy similar "positivity" and "local support" properties (see Theorem 8.5). The B-spline curve

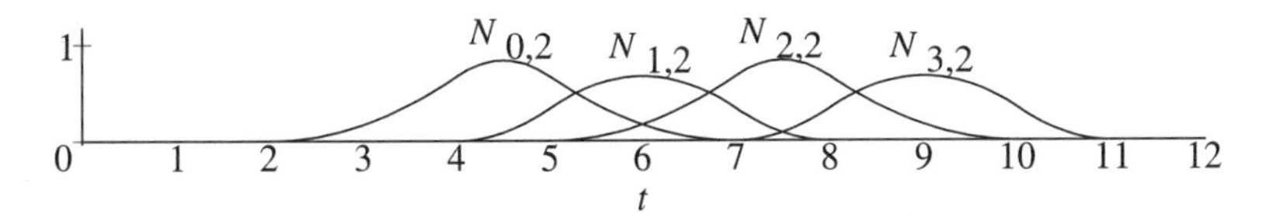

Figure 8.2 Basis functions of degree 2 for the knot vector $t_0 = 2$, $t_1 = 4$, $t_2 = 5$, $t_3 = 7$, $t_4 = 8$, $t_5 = 10$, $t_6 = 11$

is defined on the interval $[5, 8]$ by

$$\begin{aligned}
\mathbf{B}(t) &= (1,2)N_{0,2}(t) + (3,5)N_{1,2}(t) + (6,2)N_{2,2}(t) + (9,4)N_{3,2}(t) \\
&= \begin{cases}
\begin{array}{r} \frac{1}{6}(7-t)^2(1,2) + \frac{1}{3}\left(-t^2+12t-34\right)(3,5) \\ +\frac{1}{6}(t-5)^2(6,2), \end{array} & \text{if } 5 \le t < 7, \\
\begin{array}{r} \frac{1}{3}(8-t)^2(3,5) + \frac{1}{3}\left(-2t^2+30t-110\right)(6,2) \\ +\frac{1}{3}(t-7)^2(9,4), \end{array} & \text{if } 7 \le t \le 8.
\end{cases}
\end{aligned} \tag{8.3}$$

The B-spline curve, shown in Figure 8.3, is the union of two polynomial curve segments.

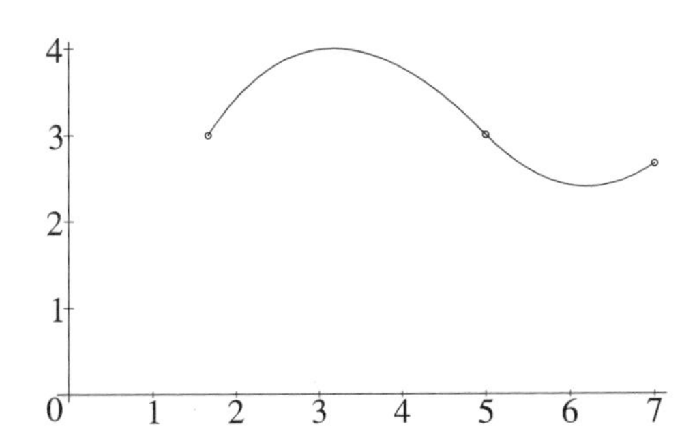

Figure 8.3 B-spline of Example 8.3

Whereas a Bézier curve of degree d has exactly $d + 1$ control points, a B-spline of degree d can have any number of control points provided a sufficient number of knots are specified. Therefore, in order to define complex curve shapes, B-splines can be given additional freedom by increasing the number of control points, yet without increasing the degree of the curve.

Each basis function $N_{i,d}(t)$ is defined by $d + 2$ knots $t_i, \ldots, t_{i+d+1}$. So if $n+1$ control points are required, then it is necessary to specify $n+d+2$ knots $t_0, \ldots, t_{n+d+1}$. Therefore, *the number of knots equals the number of control points plus the degree plus one,* giving the identity

$$m = n + d + 1 .$$

A knot vector can have repeated knot values. The number of times a knot value occurs is called the *multiplicity* of the knot. Define a new sequence $u_0, \dots, u_r$ ($u_0 < \cdots < u_r$), called the *breakpoints*, consisting of the distinct values of the interior knots. Then the B-spline is the union of the polynomial curve segments $\mathbf{B}_i(t)$ of degree d, $t \in [u_i, u_{i+1})$.

Example 8.4

The breakpoints of Example 8.3 are $u_0 = 5$, $u_1 = 7$, and $u_2 = 8$, and the B-spline consists of two segments. In Figure 8.3, the start and end points of each segment are indicated by a • on the curve. The parametric equations of the two polynomial curve segments are obtained from Equation (8.3)

$$\begin{aligned}\mathbf{B}_0(t) &= \tfrac{1}{6}(7-t)^2(1,2) + \tfrac{1}{3}\left(-t^2+12t-34\right)(3,5) + \tfrac{1}{6}(t-5)^2(6,2) \\ &= \left(-\tfrac{65}{3} + 8t - \tfrac{2}{3}t^2, -\tfrac{701}{18} + \tfrac{133}{9}t - \tfrac{23}{18}t^2\right) ,\end{aligned}$$

defined on $[u_0, u_1) = [5, 7)$, and

$$\begin{aligned}\mathbf{B}_1(t) &= \tfrac{1}{3}(8-t)^2(3,5) + \tfrac{1}{3}\left(-2t^2+30t-110\right)(6,2) + \tfrac{1}{3}(t-7)^2(9,4) \\ &= \left(-\tfrac{155}{3} + \tfrac{38}{3}t - \tfrac{2}{3}t^2, -\tfrac{340}{9} + \tfrac{100}{9}t - \tfrac{7}{9}t^2\right) ,\end{aligned}$$

defined on $[u_1, u_2) = [7, 8]$.

Theorem 8.5

The B-spline basis functions $N_{i,k}(t)$ satisfy the following properties.

Positivity: $N_{i,k}(t) > 0$ for $t \in (t_i, t_{i+k+1})$.

Local Support: $N_{i,k}(t) = 0$ for $t \notin (t_i, t_{i+k+1})$.

Piecewise Polynomial: $N_{i,k}(t)$ are piecewise polynomial functions of degree k.

Partition of Unity: $\sum_{j=r-k}^{r} N_{j,k}(t) = 1$, for $t \in [t_r, t_{r+1})$.

Continuity: If the interior knot t_i has *multiplicity* p_i, then $N_{i,k}(t)$ is C^{k-p_i} at $t = t_i$. $N_{i,k}(t)$ is C^∞ elsewhere.

Proof

The first three properties are proved by induction on k. The initial induction step $k = 0$ is satisfied since it is clear from (8.1) that the basis functions $N_{i,0}(t)$ satisfy the positivity and local support properties, and that they are piecewise polynomial functions.

The induction hypothesis is that all basis functions of degree k, $N_{i,0}(t), \ldots,$ $N_{i,k}(t)$, satisfy the three properties. Then

$$N_{i,k+1}(t) = \frac{t - t_i}{t_{i+k+1} - t_i} N_{i,k}(t) + \frac{t_{i+k+2} - t}{t_{i+k+2} - t_{i+1}} N_{i+1,k}(t) \ ,$$

where $N_{i,k}(t) > 0$ for $t \in (t_i, t_{i+k+1})$, $N_{i,k}(t) = 0$ for $t \notin (t_i, t_{i+k+1})$, $N_{i+1,k}(t) > 0$ for $t \in (t_{i+1}, t_{i+k+2})$, $N_{i+1,k}(t) = 0$ for $t \notin (t_{i+1}, t_{i+k+2})$. Suppose $t \notin (t_i, t_{i+k+2})$, then $t \notin (t_i, t_{i+k+1})$, and $t \notin (t_{i+1}, t_{i+k+2})$. Thus $N_{i,k}(t) = 0$ and $N_{i+1,k}(t) = 0$, and hence $N_{i,k+1}(t) = 0$ as required. Next, suppose $t \in (t_i, t_{i+k+2})$. If $t \in (t_i, t_{i+k+1})$, then

$$\frac{t - t_i}{t_{i+k+1} - t_i} > 0, \ \frac{t_{i+k+2} - t}{t_{i+k+2} - t_{i+1}} > 0, \ N_{i,k}(t) > 0, \ N_{i+1,k}(t) \geq 0 \ ,$$

which imply $N_{i,k+1}(t) > 0$. Otherwise, $t \in (t_{i+1}, t_{i+k+2})$ and

$$\frac{t - t_i}{t_{i+k+1} - t_i} > 0, \ \frac{t_{i+k+2} - t}{t_{i+k+2} - t_{i+1}} > 0, \ N_{i,k}(t) \geq 0, \ N_{i+1,k}(t) > 0 \ ,$$

which imply $N_{i,k+1}(t) > 0$. In either case $N_{i,k+1}(t) > 0$ as required.

Since the product of a polynomial and a piecewise polynomial is piecewise polynomial, and the sum of two piecewise polynomials is piecewise polynomial, it follows from (8.2), and the fact that $N_{i,k}(t)$ and $N_{i,k+1}(t)$ are piecewise polynomial, that $N_{i,k+1}(t)$ is piecewise polynomial. Hence, by induction, the first three properties are proved.

The partition of unity property is also proved by induction. The initial step $k = 0$ is trivial. The induction hypothesis is that the partition of unity property holds for the basis functions of degree $k - 1$. Then

$$\sum_{j=r-k}^{r} N_{j,k}(t) = N_{r-k,k}(t) + \cdots + N_{r-1,k}(t) + N_{r,k}(t)$$

$$
\begin{aligned}
&= \left(\tfrac{t-t_{r-k}}{t_r-t_{r-k}}N_{r-k,k-1}(t) + \tfrac{t_{r+1}-t}{t_{r+1}-t_{r-k+1}}N_{r-k+1,k-1}(t)\right) + \\
&\quad \cdots + \left(\tfrac{t-t_{r-1}}{t_{r+k-1}-t_{r-1}}N_{r-1,k-1}(t) + \tfrac{t_{r+k}-t}{t_{r+k}-t_r}N_{r,k-1}(t)\right) \\
&\quad + \left(\tfrac{t-t_r}{t_{r+k}-t_r}N_{r,k-1}(t) + \tfrac{t_{r+k+1}-t}{t_{r+k+1}-t_{r+1}}N_{r+1,k-1}(t)\right) \\
&= \tfrac{t-t_{r-k}}{t_r-t_{r-k}}N_{r-k,k-1}(t) + \left(\tfrac{t_{r+1}-t}{t_{r+1}-t_{r-k+1}} + \tfrac{t-t_{r-k+1}}{t_{r+1}-t_{r-k+1}}\right)N_{r-k+1,k-1}(t) + \\
&\quad \cdots + \left(\tfrac{t_{r+k}-t}{t_{r+k}-t_r} + \tfrac{t-t_r}{t_{r+k}-t_r}\right)N_{r,k-1}(t) + \tfrac{t_{r+k+1}-t}{t_{r+k+1}-t_{r+1}}N_{r+1,k-1}(t) \\
&= \tfrac{t-t_{r-k}}{t_r-t_{r-k}}N_{r-k,k-1}(t) + N_{r-k+1,k-1}(t) + \\
&\quad \cdots + N_{r,k-1}(t) + \tfrac{t_{r+k+1}-t}{t_{r+k+1}-t_{r+1}}N_{r+1,k-1}(t) \; .
\end{aligned}
$$

Since $N_{r-k,k-1}(t) = 0$ for $t \notin (t_{r-k}, t_r)$, and $N_{r+1,k-1}(t) = 0$ for $t \notin (t_{r+1}, t_{r+k+1})$, it follows that

$$
\sum_{j=r-k}^{r} N_{j,k}(t) = N_{r-k+1,k-1}(t) + \cdots + N_{r,k-1}(t) = \sum_{j=r-k-1}^{r} N_{j,k-1}(t) \; .
$$

The induction hypothesis implies

$$
\sum_{j=r-k-1}^{r} N_{j,k-1}(t) = 1 \; .
$$

Hence

$$
\sum_{j=r-k}^{r} N_{j,k}(t) = 1 \; ,
$$

and the partition of unit property is proved.

The property of continuity is proved in Lemma 8.15.

□

8.1.1 Properties of the B-spline Curve

A number of properties of B-spline curves are expressed in the following theorem.

Theorem 8.6

A B-spline curve $\mathbf{B}(t) = \sum_{i=0}^{n} \mathbf{b}_i N_{i,d}(t)$ of degree d defined on the knot vector $t_0, \ldots, t_m$ satisfies the following properties.

Local Control: Each segment is determined by $d+1$ control points. If $t \in [t_r, t_{r+1})$ $(d \leq r \leq m-d-1)$, then

$$\mathbf{B}(t) = \sum_{i=r-d}^{r} \mathbf{b}_i N_{i,d}(t) \ .$$

Thus to evaluate $\mathbf{B}(t)$ it is sufficient to evaluate $N_{r-d,d}(t), \ldots, N_{r,d}(t)$.

Convex Hull: If $t \in [t_r, t_{r+1})$ $(d \leq r \leq m-d-1)$, then

$$\mathbf{B}(t) \in \mathrm{CH}\{\mathbf{b}_{r-d}, ..., \mathbf{b}_r\} \ .$$

Continuity: If p_i is the multiplicity of the breakpoint $t = u_i$, then $\mathbf{B}(t)$ is C^{d-p_i} (or greater) at $t = u_i$, and C^∞ elsewhere.

Invariance under Affine Transformations: Let T be an affine transformation. Then $\mathsf{T}\left(\sum_{i=0}^{n} \mathbf{b}_i N_{i,d}(t)\right) = \sum_{i=0}^{n} \mathsf{T}\left(\mathbf{b}_i\right) N_{i,d}(t)$.

Proof

Suppose $t \in [t_r, t_{r+1})$. Then the positivity property implies that $N_{i,d}(t) = 0$ for all $i \leq r-d-1$ and for all $i \geq r+1$. Hence $\mathbf{B}(t) = \sum_{i=0}^{n} \mathbf{b}_i N_{i,d}(t) = \sum_{j=r-d}^{r} \mathbf{b}_j N_{j,d}(t)$, and the local control property is proved.

Further, the partition of unity property gives $\sum_{j=r-d}^{r} N_{j,d}(t) = 1$. It follows from the definition of the convex hull (Section 6.6) and the local control property that $\mathbf{B}(t) \in \mathrm{CH}\{\mathbf{b}_{r-d}, ..., \mathbf{b}_r\}$ for all t, thus establishing the convex hull property.

Since $\mathbf{B}(t)$ is piecewise polynomial, it is C^∞ everywhere except at the breakpoints $t = u_i$ where the individual polynomial segments join. If u_i is a breakpoint of multiplicity p_i, then $N_{i,d}(t)$ is C^{d-p_i} at $t = u_i$ and C^∞ elsewhere. Hence, at $t = u_i$, $\mathbf{B}(t)$ is a sum of functions which are either C^{d-p_i} or C^∞. Hence $\mathbf{B}(t)$ has continuity C^{d-p_i}.

Invariance under affine transformations is proved in a similar manner to the corresponding result for Bézier curves. □

8.1.2 B-spline Types

Open B-splines

In general, B-spline curves do not interpolate the first and last control points $\mathbf{b}_0$ and $\mathbf{b}_n$. For curves of degree d, endpoint interpolation and an endpoint tangent condition are obtained by *open B-splines* for which the end knots satisfy $t_0 = t_1 = ... = t_d$ and $t_{m-d} = t_{m-d+1} = ... = t_m$. A minor modification of the definition of the basis functions (8.1) is required in order to accommodate the multiplicity of the knots: N_{m-d-1} should take the value 1 at $t = m - d$ (and 0 elsewhere). Since $t_d \in [t_d, t_{d+1})$, the local control property with $r = d$ gives

$$\mathbf{B}(t_d) = \sum_{j=0}^{d} \mathbf{b}_j N_{j,d}(t_d) \, .$$

For $0 \leq j \leq d$,

$$N_{j,d}(t_d) = \tfrac{t_d - t_j}{t_{j+d} - t_j} N_{j,d-1}(t_d) + \tfrac{t_{j+1+d} - t_d}{t_{j+1+d} - t_{j+1}} N_{j+1,d-1}(t_d) \, .$$

Since $t_0 = t_1 = ... = t_d$, then $t_j = t_d$, and therefore (applying the convention $0/0 = 0$ for the case $j = 0$)

$$N_{j,d}(t_d) = (0)\, N_{j,d-1}(t_d) + \tfrac{t_{j+1+d} - t_d}{t_{j+1+d} - t_{j+1}} N_{j+1,d-1}(t_d) \, .$$

Therefore

$$N_{j,d}(t_d) = \left(\tfrac{t_{j+1+d} - t_d}{t_{j+1+d} - t_{j+1}}\right) \left(\tfrac{t_d - t_{j+1}}{t_{j+d} - t_{j+1}} N_{j+1,d-2}(t_d) + \tfrac{t_{j+2+d} - t_d}{t_{j+2+d} - t_{j+2}} N_{j+2,d-2}(t_d)\right) \, .$$

Repeated similar simplifications and replacements of basis functions by ones of lower order yields

$$N_{j,d}(t_d) = \left(\tfrac{t_{j+1+d} - t_d}{t_{j+1+d} - t_{j+1}}\right) \left(\tfrac{t_{j+2+d} - t_d}{t_{j+2+d} - t_{j+2}}\right) \cdots \left(\tfrac{t_{d+j+d} - t_d}{t_{d+j+d} - t_{j+d}}\right) N_{j+d,0}(t_d) \, . \tag{8.4}$$

Since $N_{j+d,0}(t_d) = 0$ for $j > 0$, it follows from (8.4) that $N_{j,d}(t_d) = 0$ for $j > 0$. When $j = 0$, identities (8.4) and $N_{d,0}(t_d) = 1$ give

$$N_{0,d}(t_d) = \left(\tfrac{t_{1+d} - t_d}{t_{1+d} - t_1}\right) \left(\tfrac{t_{2+d} - t_d}{t_{2+d} - t_2}\right) \cdots \left(\tfrac{t_{d+d} - t_d}{t_{d+d} - t_d}\right) N_{d,0}(t_d) = 1 \, .$$

Hence

$$\mathbf{B}(t_d) = \sum_{j=0}^{d} \mathbf{b}_j N_{j,d}(t_d) = \mathbf{b}_0 \, .$$

Similarly,

$$\mathbf{B}(t_{m-d}) = \mathbf{b}_n \, .$$

In Example 8.19 it is shown that open B-splines also satisfy

$$\mathbf{B}'(t_d) = \tfrac{d}{t_{d+1}-t_1}(\mathbf{b}_1 - \mathbf{b}_0) \text{ and } \mathbf{B}'(t_{m-d}) = \tfrac{d}{t_{m-1}-t_{m-d-1}}(\mathbf{b}_n - \mathbf{b}_{n-1}) . \quad (8.5)$$

Thus $\mathbf{b}_0, \mathbf{b}_1$ define the initial tangent direction, and $\mathbf{b}_{n-1}, \mathbf{b}_n$ define the final tangent direction of an open B-spline curve. The endpoint interpolation and endpoint tangent properties imply that open B-splines behave in a similar manner to Bézier curves.

Example 8.7

Let $t_0 = 0$, $t_1 = 0$, $t_2 = 0$, $t_3 = 1$, $t_4 = 2$, $t_5 = 3$, $t_6 = 4$, $t_7 = 4$, and $t_8 = 4$. The $k = 0$ basis functions are

$$N_{i,0} = \begin{cases} 1, & \text{if } t \in [t_i, t_{i+1}) \\ 0, & \text{otherwise} \end{cases} .$$

The $k = 1$ basis functions are

$$\begin{aligned}
N_{0,1}(t) &= \tfrac{(t-0)}{(0-0)}N_{0,0} + \tfrac{(0-t)}{(0-0)}N_{1,0} = 0 \quad \text{(using the convention } 0/0 = 0) , \\
N_{1,1}(t) &= \tfrac{(t-0)}{(0-0)}N_{1,0} + \tfrac{(1-t)}{(1-0)}N_{2,0} = (1-t)N_{2,0}(t) \quad \text{(using } 0/0 = 0) , \\
N_{2,1}(t) &= \tfrac{(t-0)}{(1-0)}N_{2,0} + \tfrac{(2-t)}{(2-0)}N_{3,0} = tN_{2,0}(t) + (2-t)N_{3,2}(t) , \\
N_{3,1}(t) &= \tfrac{(t-1)}{(2-1)}N_{3,0} + \tfrac{(3-t)}{(3-1)}N_{4,0} = (t-1)N_{3,0}(t) + (3-t)N_{4,0}(t) , \\
N_{4,1}(t) &= \tfrac{(t-2)}{(3-2)}N_{4,0} + \tfrac{(4-t)}{(4-2)}N_{5,0} = (t-2)N_{4,0}(t) + (4-t)N_{5,0}(t) , \\
N_{5,1}(t) &= \tfrac{(t-3)}{(4-3)}N_{5,0} + \tfrac{(4-t)}{(4-4)}N_{6,0} = (t-3)N_{5,0} \quad \text{(using } 0/0 = 0) , \\
N_{6,1}(t) &= \tfrac{(t-4)}{(4-4)}N_{6,0} + \tfrac{(4-t)}{(4-4)}N_{7,0} = 0 \quad \text{(using } 0/0 = 0) .
\end{aligned}$$

The $k = 2$ basis functions are

$$\begin{aligned}
N_{0,2}(t) &= (1-t)^2 N_{2,0}(t) , \\
N_{1,2}(t) &= \tfrac{1}{2}(4-3t)tN_{2,0}(t) + \tfrac{1}{2}(2-t)^2 N_{3,0}(t) , \\
N_{2,2}(t) &= \tfrac{1}{2}t^2 N_{2,0}(t) + (-t^2 + 3t - \tfrac{3}{2})N_{3,0} + (\tfrac{9}{2} - 3t + \tfrac{1}{2}t^2)N_{4,0} , \\
N_{3,2}(t) &= \tfrac{1}{2}(t-1)^2 N_{3,0}(t) + (-\tfrac{11}{2} + 5t - t^2)N_{4,0} + (8 - 4t + \tfrac{1}{2}t^2)N_{5,0} , \\
N_{4,2}(t) &= \tfrac{1}{2}(t-2)^2 N_{4,0}(t) + (-16 + 10t - \tfrac{3}{2}t^2)N_{5,0} , \\
N_{5,2}(t) &= (t-3)^2 N_{5,0}(t) .
\end{aligned}$$

An open B-spline of degree $d = 2$, defined on the given knot vector, consists of the four segments

$$\begin{aligned}
\mathbf{B}_1(t) &= (1-t)^2\mathbf{b}_0 + \tfrac{1}{2}t(4-3t)\mathbf{b}_1 + \tfrac{1}{2}t^2\mathbf{b}_2, \quad t \in [0,1] , \\
\mathbf{B}_2(t) &= \tfrac{1}{2}(2-t)^2\mathbf{b}_1 + \tfrac{1}{2}(-2t^2 + 6t - 3)\mathbf{b}_2 + \tfrac{1}{2}(t-1)^2\mathbf{b}_3, \quad t \in [1,2] , \\
\mathbf{B}_3(t) &= \tfrac{1}{2}(3-t)^2\mathbf{b}_2 + \tfrac{1}{2}(-2t^2 + 10t - 11)\mathbf{b}_3 + \tfrac{1}{2}(t-2)^2\mathbf{b}_4, \quad t \in [2,3] , \\
\mathbf{B}_4(t) &= \tfrac{1}{2}(4-t)^2\mathbf{b}_3 + \tfrac{1}{2}(-3t^2 + 20t - 32)\mathbf{b}_4 + (t-3)^2\mathbf{b}_5, \quad t \in [3,4] .
\end{aligned}$$

Note that $\mathbf{B}(0) = \mathbf{b}_0$ and $\mathbf{B}(4) = \mathbf{b}_5$, and hence the B-spline interpolates the first and last control points.

EXERCISES

8.1. Let $t_0 = 0$, $t_1 = 1$, $t_2 = 2$, $t_3 = 3$, $t_4 = 4$, $t_5 = 5$, and $t_6 = 6$. Determine the basis functions for a B-spline of degree 2. Use the method of Examples 8.3 and 8.4 to obtain the equations of the segments of a B-spline of degree 2 defined on this knot vector.

8.2. Let $t_0 = 0$, $t_1 = 1$, $t_2 = 2$, $t_3 = 3$, $t_4 = 4$, $t_5 = 5$, $t_6 = 6$, $t_7 = 7$. Determine the basis functions required for a B-spline of degree 3. (The results of the previous exercise are useful!) Obtain the segments of a B-spline of degree 3 defined on this knot vector.

8.3. Let $t_0 = 0$, $t_1 = 0$, $t_2 = 0$, $t_3 = 1$, $t_4 = 2$, $t_5 = 3$, $t_6 = 3$, and $t_7 = 3$. Determine the segments of an open B-spline of degree 2 defined on this knot vector.

8.4. The open cubic B-spline with knot vector $t_0 = t_1 = t_2 = t_3 = 0$ and $t_4 = t_5 = t_6 = t_7 = 1$ has just one segment which satisfies the endpoint conditions. Determine the basis functions, and show that the open B-spline is a cubic Bézier curve.

Uniform B-splines

A B-spline is said to be *uniform* whenever its knots are equally spaced, and *non-uniform* otherwise. Let the knot vector be $t_0 = 0, t_1 = 1, t_2 = 2, \ldots, t_m = m$. The basis functions for the uniform B-spline of degree 2 on this knot vector are obtained as follows:

$$N_{i,0}(t) = \begin{cases} 1, & \text{if} \quad t \in [t_i, t_{i+1}) \\ 0, & \text{otherwise} \end{cases} .$$

$$\begin{aligned} N_{0,1}(t) &= \tfrac{(t-0)}{(1-0)} N_{0,0} + \tfrac{(2-t)}{(2-1)} N_{1,0} = tN_{0,0} + (2-t)N_{1,0} \,, \\ N_{1,1}(t) &= \tfrac{(t-1)}{(2-1)} N_{1,0} + \tfrac{(3-t)}{(3-2)} N_{2,0} = (t-1)\, N_{1,0} + (3-t)N_{2,0} \,, \\ N_{2,1}(t) &= \tfrac{(t-2)}{(3-2)} N_{2,0} + \tfrac{(4-t)}{(4-3)} N_{3,0} = (t-2)\, N_{2,0} + (4-t)N_{3,0} \,. \end{aligned}$$

$$
\begin{aligned}
N_{0,2}(t) &= \tfrac{(t-0)}{(2-0)} N_{0,1}(t) + \tfrac{(3-t)}{(3-1)} N_{1,1}(t) = \tfrac{1}{2} t N_{0,1}(t) + \tfrac{1}{2}(3-t) N_{1,1}(t) \\
&= \tfrac{1}{2} t \left(t N_{0,0} + (2-t) N_{1,0} \right) + \tfrac{1}{2}(3-t) \left((t-1) N_{1,0} + (3-t) N_{2,0} \right) \\
&= \tfrac{1}{2} t^2 N_{0,0} + \left(\tfrac{1}{2} t(2-t) + \tfrac{1}{2}(3-t)(t-1) \right) N_{1,0} + \tfrac{1}{2}(3-t)^2 N_{2,0} \\
&= \begin{cases} \tfrac{1}{2} t^2, & t \in [0,1] \\ -\tfrac{1}{2}\left(3 - 6t + 2t^2\right), & t \in [1,2] \\ \tfrac{1}{2}(3-t)^2, & t \in [2,3] \end{cases} .
\end{aligned}
$$

The ith basis function is

$$
\begin{aligned}
N_{i,2}(t) &= \tfrac{t-i}{(i+2)-i} N_{i,1}(t) + \tfrac{i+3-t}{(i+3)-(i+1)} N_{i+1,1}(t) \\
&= \tfrac{t-i}{2} N_{i,1}(t) + \tfrac{i+3-t}{2} N_{i+1,1}(t) \\
&= \tfrac{t-i}{2} \left(\tfrac{t-i}{(i+1)-i} N_{i,0}(t) + \tfrac{(i+2)-t}{(i+2)-(i+1)} N_{i+1,0}(t) \right) \\
&\quad + \tfrac{i+3-t}{2} \left(\tfrac{t-(i+1)}{(i+2)-(i+1)} N_{i+1,0}(t) + \tfrac{i+3-t}{(i+3)-(i+2)} N_{i+2,0}(t) \right) \\
&= \tfrac{1}{2}(t-i)^2 N_{i,0}(t) \\
&\quad + \tfrac{1}{2} \left((t-i)(i+2-t) + (i+3-t)(t-i-1) \right) N_{i+1,0}(t) \\
&\quad + \tfrac{1}{2}(i+3-t)^2 N_{i+2,0}(t) .
\end{aligned}
$$

Thus the ith segment $\mathbf{B}_i(t)$, defined on $[i+2, i+3)$, is given by

$$
\begin{aligned}
\mathbf{B}_i(t) &= \sum_{j=i}^{i+2} \mathbf{b}_j N_{j,2}(t) \\
&= \tfrac{1}{2}(i+3-t)^2 \mathbf{b}_i \\
&\quad + \tfrac{1}{2} \left((t-i-1)(i+3-t) + (i+4-t)(t-i-2) \right) \mathbf{b}_{i+1} \\
&\quad + \tfrac{1}{2}(t-i-2)^2 \mathbf{b}_{i+2} .
\end{aligned}
$$

Finally, the reparametrization $t \mapsto t+i+2$ defines the segment on the interval $[0,1]$ by

$$
\begin{aligned}
\mathbf{B}_i(t) &= \tfrac{1}{2}(1-t)^2 \mathbf{b}_i + \tfrac{1}{2}(1+2t-2t^2) \mathbf{b}_{i+1} + \tfrac{1}{2} t^2 \mathbf{b}_{i+2} \\
&= \tfrac{1}{2} \begin{pmatrix} t^2 & t & 1 \end{pmatrix} \begin{pmatrix} 1 & -2 & 1 \\ -2 & 2 & 0 \\ 1 & 1 & 0 \end{pmatrix} \begin{pmatrix} \mathbf{b}_i \\ \mathbf{b}_{i+1} \\ \mathbf{b}_{i+2} \end{pmatrix} .
\end{aligned} \tag{8.6}
$$

Points on each segment are efficiently computed since the basis functions $\frac{1}{2}(1-t)^2$, $\frac{1}{2}(1+2t-2t^2)$, $\frac{1}{2}t^2$ have only to be evaluated once for each t.

For uniform B-splines of degree $d = 3$, a similar method gives

$$
\mathbf{B}_i(t) = \tfrac{1}{6} \begin{pmatrix} t^3 & t^2 & t & 1 \end{pmatrix} \begin{pmatrix} -1 & 3 & -3 & 1 \\ 3 & -6 & 3 & 0 \\ -3 & 0 & 3 & 0 \\ 1 & 4 & 1 & 0 \end{pmatrix} \begin{pmatrix} \mathbf{b}_i \\ \mathbf{b}_{i+1} \\ \mathbf{b}_{i+2} \\ \mathbf{b}_{i+3} \end{pmatrix} .
$$

Example 8.8

Consider the uniform B-spline $\mathbf{B}(t)$ of degree $d = 2$ defined on the knot vector $t_0 = 0$, $t_1 = 1$, $t_2 = 2$, $t_3 = 3$, $t_4 = 4$, $t_5 = 5$, $t_6 = 6$, $t_7 = 7$, and with control points $\mathbf{b}_0(3,2)$, $\mathbf{b}_1(7,-1)$, $\mathbf{b}_2(5,2)$, $\mathbf{b}_3(4,5)$, $\mathbf{b}_4(2,3)$. The curve is defined on the interval $[t_d, t_{m-d}] = [2,5]$. There are three curve segments defined on the sub-intervals $[2,3]$, $[3,4]$, and $[4,5]$. For instance, to determine the point $\mathbf{B}(3.6)$ which lies on the $i = 1$ segment, $t = 3.6$ is translated into the interval $[0,1]$ using $t \mapsto t - i - 2$. The required parameter is $t = 3.6 - 1 - 2 = 0.6$, and (8.6) gives

$$\begin{aligned} \mathbf{B}_1(0.6) &= \tfrac{1}{2}(1-0.6)^2(7,-1) + \tfrac{1}{2}(1+2(0.6)-2(0.6)^2)(5,2) + \tfrac{1}{2}(0.6)^2(4,5) \\ &= (4.98, 2.3) \ . \end{aligned}$$

Periodic B-splines and Closed Periodic B-splines

In a number of applications, it is desirable to represent *closed* curves for which the starting point equals the finishing point. A closed Bézier curve can be obtained by choosing control points which form a closed control polygon. But, in general, B-splines do not interpolate the first and last control points, and therefore a closed control polygon does not yield a closed curve. Closure of the curve is obtained by imposing conditions on the control points and knots. For instance, an open B-spline could be used for this purpose. An alternative is to use a closed periodic B-spline.

A *periodic* B-spline of degree d and with $n+1$ control points is obtained by choosing knots $t_0 \le \ldots \le t_n$ arbitrarily, and then setting

$$t_{n+i} = t_{n+i-1} + (t_i - t_{i-1}) \ ,$$

for $i = 1, \ldots, d+1$. A knot vector of this form is called a *periodic knot vector*. In particular, a uniform B-spline is a special case of a periodic B-spline.

A *closed periodic* B-spline of degree d and control points $\mathbf{b}_0, \ldots, \mathbf{b}_n, \mathbf{b}_{n+1} = \mathbf{b}_0, \mathbf{b}_{n+2} = \mathbf{b}_1, \ldots, \mathbf{b}_{n+d} = \mathbf{b}_{d-1}$ is obtained by choosing knots $t_0 \le \ldots \le t_{n+1}$ arbitrarily, and forming a periodic knot vector with $n + 2d + 2$ knots.

Example 8.9

Let $d = 3$ and $n = 4$. Let the first five control points be $\mathbf{b}_0(1,2)$, $\mathbf{b}_1(3,7)$, $\mathbf{b}_2(6,6)$, $\mathbf{b}_3(6,-2)$, $\mathbf{b}_4(1,-1)$, and let the remaining control points be $\mathbf{b}_5(1,2)$, $\mathbf{b}_6(3,7)$, $\mathbf{b}_7(6,6)$. Suppose the first $n+2 = 6$ knots are $t_0 = 0.0$, $t_1 = 0.5$, $t_2 = 2.0$, $t_3 = 3.0$, $t_4 = 3.1$, $t_5 = 3.4$. The periodic knot vector is obtained by taking

$$\begin{aligned} t_6 &= 3.4 + (0.5 - 0.0) = 3.9\,, \quad t_7 = 3.9 + (2.0 - 0.5) = 5.4\,, \\ t_8 &= 5.4 + (3.0 - 2.0) = 6.4\,, \quad t_9 = 6.4 + (3.1 - 3.0) = 6.5\,, \\ t_{10} &= 6.5 + (3.4 - 3.1) = 6.8\,, \quad t_{11} = 6.8 + (5.4 - 3.9) = 8.3\,. \end{aligned}$$

The B-spline and its control polygon are illustrated in Figure 8.4.

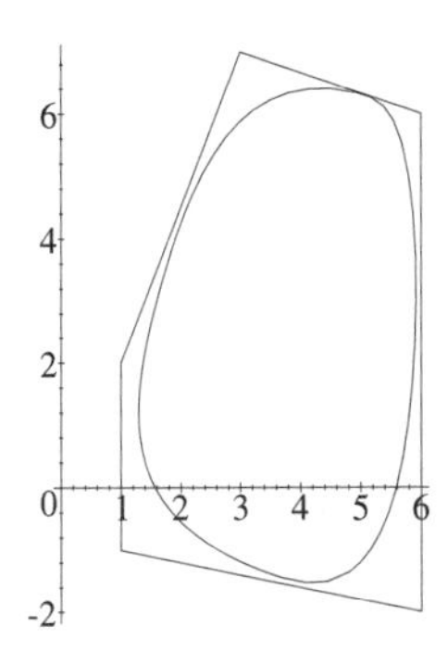

Figure 8.4 Closed periodic B-spline of degree 3

EXERCISES

8.5. Evaluate the uniform B-spline of Example 8.8 at $t = 2.5$ and $t = 4.2$.

8.6. Let $d = 3$, and $\mathbf{b}_0(0,0)$, $\mathbf{b}_1(2,0)$, $\mathbf{b}_2(4,2)$, $\mathbf{b}_3(2,4)$, $\mathbf{b}_4(0,2)$. Suppose the first knots are $t_0 = 0$, $t_1 = 1$, $t_2 = 2$, $t_3 = 3$, $t_4 = 4$, $t_5 = 5$. Compute the knots and control points required to form a closed B-spline curve.

8.7. Let $d = 3$, and $\mathbf{b}_0(3,5)$, $\mathbf{b}_1(4,8)$, $\mathbf{b}_2(7,2)$, $\mathbf{b}_3(6,-3)$, $\mathbf{b}_4(3,-1)$. Suppose the first knots are $t_0 = 0.3$, $t_1 = 0.4$, $t_2 = 1.3$, $t_3 = 2.5$, $t_4 = 2.9$, $t_5 = 3.7$. Compute the knots and control points required to form a closed B-spline curve.

Open Uniform B-splines

Open B-splines for which the interior knots are uniform are referred to as *open uniform B-splines*. Example 8.7 is an open uniform B-spline.

Example 8.10

Let $t_0 = 0$, $t_1 = 0$, $t_2 = 0$, $t_3 = 1$, $t_4 = 2$, $t_5 = 3$, $t_6 = 3$, and $t_7 = 3$. The basis functions for open uniform B-splines of degree 2 defined on this knot vector are

$$N_{0,2} = \begin{cases} 0, & t < 0 \\ (t-1)^2, & 0 \le t < 1 \\ 0, & 1 \le t \end{cases},$$

$$N_{1,2} = \begin{cases} 0, & t < 0 \\ 2t - \frac{3}{2}t^2, & 0 \le t < 1 \\ \frac{3}{2} - t + \frac{1}{2}(t-1)^2, & 1 \le t < 2 \\ 0, & 2 \le t \end{cases},$$

$$N_{2,2} = \begin{cases} 0, & t < 0 \\ \frac{1}{2}t^2, & 0 \le t < 1 \\ -\frac{1}{2} + t - (t-1)^2, & 1 \le t < 2 \\ \frac{5}{2} - t + \frac{1}{2}(t-2)^2, & 2 \le t < 3 \\ 0, & 3 \le t \end{cases},$$

$$N_{3,2} = \begin{cases} 0, & t < 1 \\ \frac{1}{2}(t-1)^2, & 1 \le t < 2 \\ -\frac{3}{2} + t - \frac{3}{2}(t-2)^2, & 2 \le t < 3 \\ 0, & 3 \le t \end{cases},$$

$$N_{4,2} = \begin{cases} 0, & t < 2 \\ (t-2)^2, & 2 \le t < 3 \\ 0, & 3 \le t \end{cases}.$$

EXERCISES

8.8. Let $t_0 = 0$, $t_1 = 0$, $t_2 = 0$, $t_3 = 2$, $t_4 = 4$, $t_5 = 6$, $t_6 = 8$, $t_7 = 8$, $t_8 = 8$. Obtain the basis functions for open uniform B-splines of degree 2 defined on this knot vector.

8.9. Let $t_0 = 0$, $t_1 = 0$, $t_2 = 0$, $t_3 = 0$, $t_4 = 1$, $t_5 = 2$, $t_6 = 3$, $t_7 = 4$, $t_8 = 4$, $t_9 = 4$, $t_{10} = 4$. Obtain the basis functions for open uniform B-splines of degree 3 defined on this knot vector.

8.10. Use a computer package to plot the cubic open uniform B-spline defined on the knot vector of Exercise 8.8 with control points $(2,4)$, $(-2,4)$, $(-2,0)$, $(0,0)$, $(2,0)$, $(2,-4)$, $(-2,-4)$.

8.1.3 Application: Font Design

An interesting application of B-splines is to font design. The boundary of each character in a font is specified by B-spline (or Bézier) curves. Figures 8.5(a) and (b) show a letter 'G' together with its defining control polygons. Different font sizes are obtained by applying scaling transformations to the control points, and making use of the property of invariance under affine transformations. Italic fonts may be obtained by applying a shear transformation (Figure 8.5(c)) or a projection (Figure 8.5(d)). Projections of B-splines are discussed in Section 8.2.1. The B-spline data required to store the font definition is considerably less than, for instance, storing a bitmap representation for each character.

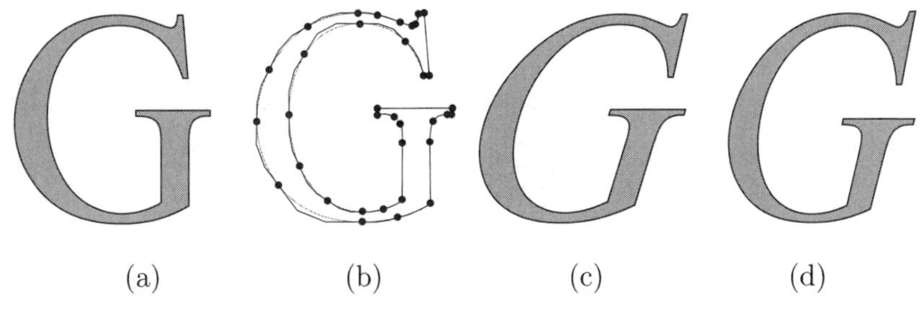

(a) (b) (c) (d)

Figure 8.5 B-spline font definition

8.1.4 Application: Morphing or Soft Object Animation

Morphing is a technique used in computer graphics in which a shape is gradually deformed over an period of time. Morphing has been used in animation sequences of feature films. In practice, morphing can involve a number of advanced computational methods including surface deformation, surface rendering, and texture mapping. In this section the process of deformation is exemplified by a simple version of the technique where the initial and final shapes are B-spline curves (including Bézier curves as a special case). It is assumed that the B-splines have the same degree and knot vector, though these restrictions can be removed by applying knot insertion (Section 8.3) and degree raising algorithms [13].

Let $\mathbf{B}(t)$ and $\mathbf{C}(t)$ be B-splines of degree d with control points $\mathbf{b}_0, \ldots, \mathbf{b}_n$ and $\mathbf{c}_0, \ldots, \mathbf{c}_n$ respectively, and defined on the knot vector $t_0, \ldots, t_m$.

Definition 8.11

An N-step *deformation* of $\mathbf{B}(t)$ into $\mathbf{C}(t)$ is a sequence of B-splines $\mathbf{D}_0(t), \mathbf{D}_1(t), \ldots, \mathbf{D}_N(t)$ such that $\mathbf{D}_0(t) = \mathbf{B}(t)$ and $\mathbf{D}_N(t) = \mathbf{C}(t)$.

The $N+1$ curves $\mathbf{D}_k(t)$ $(k = 0, \ldots, N+1)$ are the "in-between" curves which define the gradual change from curve $\mathbf{B}$ into curve $\mathbf{C}$. In order to describe a deformation it is sufficient to prescribe the control points $\mathbf{d}_0^k, \ldots, \mathbf{d}_n^k$ of $\mathbf{D}_k(t)$. A *linear* N-step deformation $\mathbf{D}_k(t)$ is given by

$$\mathbf{d}_i^k = \mathbf{b}_i + \tfrac{k}{N}(\mathbf{c}_i - \mathbf{b}_i) \text{ for } k = 0, \ldots, N \; . \tag{8.7}$$

Thus $\mathbf{d}_i^0 = \mathbf{b}_i$ and $\mathbf{d}_i^N = \mathbf{c}_i$.

Example 8.12

Let the control points of a curve $\mathbf{B}(t)$ be $\mathbf{b}_0(1,0)$, $\mathbf{b}_1(0,0)$, $\mathbf{b}_2(-1,5)$, $\mathbf{b}_3(1,3)$, $\mathbf{b}_4(3,5)$, $\mathbf{b}_5(2,0)$, $\mathbf{b}_6(1,0)$, and the control points of a curve $\mathbf{C}(t)$ be $\mathbf{c}_0(1,0)$, $\mathbf{c}_1(-3,1)$, $\mathbf{c}_2(-4,5)$, $\mathbf{c}_3(1,4)$, $\mathbf{c}_4(6,5)$, $\mathbf{c}_5(5,1)$, $\mathbf{c}_6(1,0)$. Let $t_0 = \cdots = t_6 = 0$ and $t_7 = \cdots = t_{13} = 1$. Then both B-splines are Bézier curves of degree 6 and

$$\begin{aligned}
\mathbf{d}_0^k &= (1,0) + ((1,0) - (1,0))\,\tfrac{k}{N} = (1,0) \; , \\
\mathbf{d}_1^k &= (0,0) + ((-3,1) - (0,0))\,\tfrac{k}{N} = \left(\tfrac{-3k}{N}, \tfrac{k}{N}\right) \; , \\
\mathbf{d}_2^k &= (-1,5) + ((-4,5) - (-1,5))\,\tfrac{k}{N} = \left(-1 - \tfrac{3k}{N}, 5\right) \; , \\
\mathbf{d}_3^k &= (1,3) + ((1,4) - (1,3))\,\tfrac{k}{N} = \left(1, 3 + \tfrac{k}{N}\right) \; , \\
\mathbf{d}_4^k &= (3,5) + ((6,5) - (3,5))\,\tfrac{k}{N} = \left(3 + \tfrac{3k}{N}, 5\right) \; , \\
\mathbf{d}_5^k &= (2,0) + ((5,1) - (2,0))\,\tfrac{k}{N} = \left(2 + \tfrac{3k}{N}, \tfrac{k}{N}\right) \; , \\
\mathbf{d}_6^k &= (1,0) + ((1,0) - (1,0))\,\tfrac{k}{N} = (1,0) \; .
\end{aligned}$$

The in-between curves for $N = 4$ are shown in Figure 8.6. More general deformations can be obtained by replacing $\frac{k}{N}$ in (8.7) by more general functions of k. For example, let $\lambda_{i,k}(s)$ be continuous functions such that $\lambda_{i,0}(0) = 0$, $\lambda_{i,N}(1) = 1$. Then a deformation $\mathbf{D}_i(t)$ is given by

$$\mathbf{d}_i^k = \mathbf{b}_i + \lambda_{i,k}\left(\tfrac{k}{N}\right)(\mathbf{c}_i - \mathbf{b}_i) \text{ for } k = 0, \ldots, N \; . \tag{8.8}$$

Exercise 8.11

Implement the B-spline deformation given by (8.8). Experiment with different choices of functions $\lambda_{i,k}(s)$.

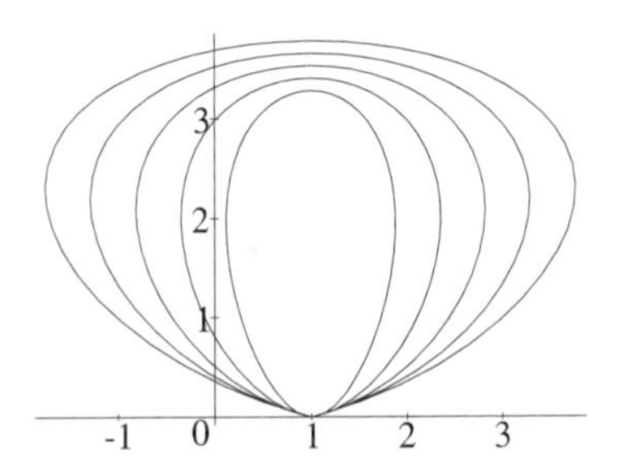

Figure 8.6 Linear deformation of a B-spline curve

8.1.5 The de Boor Algorithm

Evaluations of points on a B-spline curve can be performed using a method known as the de Boor algorithm. Just as the de Casteljau algorithm for Bézier curves is a consequence of the recursive property of the Bernstein basis functions, the de Boor algorithm follows from the recursion property of the B-spline basis functions

$$N_{i,k}(t) = \frac{t - t_i}{t_{i+k} - t_i} N_{i,k-1}(t) + \frac{t_{i+k+1} - t}{t_{i+k+1} - t_{i+1}} N_{i+1,k-1}(t) \,. \tag{8.9}$$

Suppose $t \in [t_r, t_{r+1})$. Then (8.9) implies

$$\begin{aligned} \mathbf{B}(t) &= \sum_{i=r-k}^{r} \mathbf{b}_i N_{i,k}(t) \\ &= \sum_{i=r-k}^{r} \mathbf{b}_i \tfrac{t-t_i}{t_{i+k}-t_i} N_{i,k-1}(t) + \sum_{i=r-k}^{r} \mathbf{b}_i \tfrac{t_{i+k+1}-t}{t_{i+k+1}-t_{i+1}} N_{i+1,k-1}(t) \,. \end{aligned}$$

Replacing i by $i-1$ in the second sum gives

$$\mathbf{B}(t) = \sum_{i=r-k}^{r} \mathbf{b}_i \frac{t - t_i}{t_{i+k} - t_i} N_{i,k-1}(t) + \sum_{i=r-k+1}^{r+1} \mathbf{b}_{i-1} \frac{t_{i+k} - t}{t_{i+k} - t_i} N_{i,k-1}(t) \,,$$

and since $N_{r+1,k-1}(t) = N_{r-k,k-1}(t) = 0$ on $[t_r, t_{r+1})$,

$$\mathbf{B}(t) = \sum_{i=r-k+1}^{r} \left(\mathbf{b}_i \frac{t - t_i}{t_{i+k} - t_i} + \mathbf{b}_{i-1} \frac{t_{i+k} - t}{t_{i+k} - t_i} \right) N_{i,k-1}(t) \,.$$

Let

$$\mathbf{b}_i^1(t) = \mathbf{b}_{i-1} \frac{t_{i+k} - t}{t_{i+k} - t_i} + \mathbf{b}_i \frac{t - t_i}{t_{i+k} - t_i} \,,$$

for $i = r - k + 1, \ldots, r$. Note that $\mathbf{b}_i^1$ is dependent on the parameter value t. In a similar manner, $N_{i,k-1}(t)$ can be expressed in terms of the basis functions of degree $k-2$ and so on. For a curve of degree d, the result is a recursive procedure

$$\left.\begin{aligned} \mathbf{b}_i^j(t) &= (1-\alpha_i^j(t))\mathbf{b}_{i-1}^{j-1}(t) + \alpha_i^j(t)\mathbf{b}_i^{j-1}(t) \,, \\ \alpha_i^j(t) &= \frac{t-t_i}{t_{i+d-j+1}-t_i} \,, \end{aligned}\right\} \tag{8.10}$$

for $j=1,\ldots,d$, $i=r-d+j,\ldots,r$, where $\mathbf{b}_j^0(t)=\mathbf{b}_j$, $\mathbf{b}_{-1}=\mathbf{0}$, and $\mathbf{b}_{m-d+1}=\mathbf{0}$ (where $\mathbf{0}$ denotes $(0,0)$ for a plane curve and $(0,0,0)$ for a spatial curve). The jth step yields $\mathbf{B}(t)$ in terms of the basis functions of degree $d-j$ (note $\mathbf{b}_i^j$ is a function of t)

$$\mathbf{B}(t) = \sum_{i=r-d+j}^{r} \mathbf{b}_i^j N_{i,d-j}(t) \,.$$

Thus when $j=d$, the algorithm yields the point $\mathbf{B}(t)=\sum_{i=r}^{r}\mathbf{b}_i^d N_{i,0}(t)=\mathbf{b}_r^d$ on the curve.

To summarize, for a given parameter value t, the de Boor algorithm (8.10) yields a triangular array of points such that $\mathbf{b}_r^d=\mathbf{B}(t)$.

$$\begin{array}{lllll} \mathbf{b}_{r-d}^0 & \mathbf{b}_{r-d+1}^0 & \cdots & \cdots & \mathbf{b}_r^0 \\ \mathbf{b}_{r-d+1}^1 & \cdots & & \cdots\ \mathbf{b}_r^1 & \\ \vdots & \cdots & & & \\ \mathbf{b}_{r-1}^{d-1} & \mathbf{b}_r^{d-1} & & & \\ \mathbf{b}_r^d = \mathbf{B}(t) \,. & & & & \end{array}$$

Example 8.13

The de Boor algorithm can be applied to evaluate the uniform B-spline of Example 8.8 at $t=3.6$. Then $d=2$, and since $3.6\in[3,4)=[t_3,t_4)$, it follows that $r=3$. The first row of points is $\mathbf{b}_1^0(7,-1)$, $\mathbf{b}_2^0(5,2)$, $\mathbf{b}_3^0(4,5)$. The algorithm with $j=1\ldots 2$, $i=(1+j)\ldots 3$, yields

$$\begin{aligned} \alpha_2^1 &= \frac{t-t_2}{t_4-t_2} = \frac{3.6-2}{4-2} = 0.8, \quad \alpha_3^1 = \frac{t-t_3}{t_5-t_3} = \frac{3.6-3}{5-3} = 0.3 \,, \\ \alpha_3^2 &= \frac{t-t_3}{t_4-t_3} = \frac{3.6-3}{4-3} = 0.6 \,, \end{aligned}$$

$$\begin{aligned} \mathbf{b}_2^1 &= (1-0.8)\,(7,-1)+0.8\,(5,2) = (5.4,1.4) \,, \\ \mathbf{b}_3^1 &= (1-0.3)\,(5,2)+0.3\,(4,5) = (4.7,2.9) \,, \\ \mathbf{b}_3^2 &= (1-0.6)\,(5.4,1.4)+0.6\,(4.7,2.9) = (4.98,2.3) \,. \end{aligned}$$

Hence $\mathbf{B}(3.6)=(4.98,2.3)$ verifying the result of Example 8.8.

EXERCISES

8.12. Apply the de Boor algorithm to evaluate the uniform B-spline of Example 8.8 at $t = 2.5$ and $t = 4.2$.

8.13. An open B-spline $\mathbf{B}(t)$ of degree $d = 2$ is defined on the knot vector $t_0 = 0,\ t_1 = 0,\ t_2 = 0,\ t_3 = 1,\ t_4 = 2,\ t_5 = 3,\ t_6 = 3,\ t_7 = 3,$ and with control points $\mathbf{b}_0(1,1)$, $\mathbf{b}_1(3,4)$, $\mathbf{b}_2(6,2)$, $\mathbf{b}_3(4,2)$, $\mathbf{b}_4(2,5)$. Apply the de Boor algorithm to evaluate the point $\mathbf{B}(2.4)$.

8.14. Let $\mathbf{B}(t)$ be a B-spline of degree d defined on the knot vector $t_i = 0$ for $i = 0, \ldots, d$, and $t_i = 1$ for $i = d+1, \ldots, 2d+1$. Show that the de Boor algorithm specializes to the de Casteljau algorithm (that is, $\alpha_i^j = t$). Deduce that any B-spline curve defined on this knot vector is a Bézier curve of degree d.

8.15. Let $N_{i,d}(t)$ be the B-spline basis functions of degree d defined on the knot vector $t_i = 0$ for $i = 0, \ldots, d$, and $t_i = 1$ for $i = d+1, \ldots, 2d+1$. Show that $N_{i,d}(t)$ are the Bernstein basis functions and deduce that any B-spline curve defined on this knot vector is a Bézier curve of degree d.

8.16. Implement the de Boor algorithm and verify your solutions to the above exercises.

8.1.6 Derivatives of a B-spline

The next aim is to determine the derivative of a B-spline curve of degree d as a B-spline of degree $d-1$. The first step is to determine the derivatives of the basis functions of degree d in terms of the basis functions of degree $d-1$.

Lemma 8.14

The derivative of the B-spline basis functions $N_{i,d}(t)$ of degree d may be obtained in terms of the basis functions of degree $d-1$ as follows:

$$N'_{i,d}(t) = \frac{d}{t_{i+d} - t_i} N_{i,d-1}(t) - \frac{d}{t_{i+d+1} - t_{i+1}} N_{i+1,d-1}(t) \,. \tag{8.11}$$

Proof

The proof is by induction on d. The initial induction step $(d = 1)$

$$N'_{i,1}(t) = \tfrac{1}{t_{i+1}-t_i} N_{i,0}(t) - \tfrac{1}{t_{i+2}-t_{i+1}} N_{i+1,0}(t) \,,$$

is left as an exercise to the reader. Next, suppose that (8.11) is true for all B-splines of degree d. It is necessary to show that (8.11) is true for B-splines of degree $d+1$. The recursive definition of the B-spline basis functions gives

$$N_{i,d+1} = \frac{t-t_i}{t_{i+d+1}-t_i} N_{i,d}(t) + \frac{t_{i+d+2}-t}{t_{i+d+2}-t_{i+1}} N_{i+1,d}(t) \, .$$

The derivative is obtained by applying the product rule,

$$\begin{aligned} N'_{i,d+1}(t) &= \frac{1}{t_{i+d+1}-t_i} N_{i,d}(t) + \frac{t-t_i}{t_{i+d+1}-t_i} N'_{i,d}(t) \\ &\quad - \frac{1}{t_{i+d+2}-t_{i+1}} N_{i+1,d}(t) + \frac{t_{i+d+2}-t}{t_{i+d+2}-t_{i+1}} N'_{i+1,d}(t) \, . \end{aligned}$$

Since $N'_{i,d}(t)$ and $N'_{i+1,d}(t)$ are derivatives of basis functions of degree d, the induction hypothesis (8.11) can be applied to give

$$\begin{aligned} & N'_{i,d+1}(t) \\ &= \frac{1}{t_{i+d+1}-t_i} N_{i,d}(t) + \frac{t-t_i}{t_{i+d+1}-t_i} \left(\frac{d}{t_{i+d}-t_i} N_{i,d-1}(t) - \frac{d}{t_{i+d+1}-t_{i+1}} N_{i+1,d-1}(t) \right) \\ &\quad - \frac{1}{t_{i+d+2}-t_{i+1}} N_{i+1,d}(t) + \frac{t_{i+d+2}-t}{t_{i+d+2}-t_{i+1}} \left(\frac{d}{t_{i+d+1}-t_{i+1}} N_{i+1,d-1}(t) \right. \\ &\quad \left. - \frac{d}{t_{i+d+2}-t_{i+2}} N_{i+2,d-1}(t) \right) \\ &= \frac{1}{t_{i+d+1}-t_i} N_{i,d}(t) - \frac{1}{t_{i+d+2}-t_{i+1}} N_{i+1,d}(t) + \frac{d(t-t_i)}{(t_{i+d+1}-t_i)(t_{i+d}-t_i)} N_{i,d-1} \\ &\quad + d \left(\frac{(t_{i+d+2}-t)}{(t_{i+d+2}-t_{i+1})(t_{i+d+1}-t_{i+1})} - \frac{(t-t_i)}{(t_{i+d+1}-t_i)(t_{i+d+1}-t_{i+1})} \right) N_{i+1,d-1} \\ &\quad - d \frac{(t_{i+d+2}-t)}{(t_{i+d+2}-t_{i+1})(t_{i+d+2}-t_{i+1})} N_{i+2,d-1} \, . \end{aligned}$$

But

$$\frac{t_{i+d+2}-t}{t_{i+d+2}-t_{i+1}} - \frac{t-t_i}{t_{i+d+2}-t_i} = \frac{t_{i+d+1}-t}{t_{i+d+1}-t_i} - \frac{t-t_{i+1}}{t_{i+d+2}-t_{i+1}} .$$

Hence,

$$\begin{aligned} N'_{i,d+1}(t) &= \frac{1}{t_{i+d+1}-t_i} N_{i,d}(t) - \frac{1}{t_{i+d+2}-t_{i+1}} N_{i+1,d}(t) \\ &\quad + \frac{d}{t_{i+d+1}-t_i} \left(\frac{t-t_i}{t_{i+d}-t_i} N_{i,d-1}(t) - \frac{t_{i+d+1}-t}{t_{i+d+1}-t_i} N_{i+1,d-1}(t) \right) \\ &\quad - \frac{d}{t_{i+d+2}-t_{i+1}} \left(\frac{t-t_{i+1}}{t_{i+d+1}-t_{i+1}} N_{i+1,d-1}(t) - \frac{t_{i+d+2}-t}{t_{i+d+2}-t_{i+1}} N_{i+2,d-1}(t) \right) \\ &= \frac{1}{t_{i+d+1}-t_i} N_{i,d}(t) - \frac{1}{t_{i+d+2}-t_{i+1}} N_{i+1,d}(t) + \frac{d}{t_{i+d+1}-t_i} N_{i,d}(t) \\ &\quad - \frac{d}{t_{i+d+2}-t_{i+1}} N_{i+1,d}(t) \\ &= \frac{d+1}{t_{i+d+1}-t_i} N_{i,d}(t) - \frac{d+1}{t_{i+d+2}-t_{i+1}} N_{i+1,d}(t). \end{aligned}$$

The final equation has the desired form. Hence by induction the hypothesis (8.11) is true.

□

It is now possible to prove the continuity property of Theorem 8.5.

Lemma 8.15

If the interior knot t_i has multiplicity p_i, then $N_{i,k}(t)$ is C^{k-p_i} at $t = t_i$, and C^∞ elsewhere.

Proof

Since the basis functions are piecewise polynomial of degree k, they are C^∞ everywhere except at the joins of the segments which occur at the interior knots. Suppose t_i has multiplicity p_i $(1 \leq p_i \leq k)$. The proof is by induction. For the initial induction step, $k = 1$, $p_i = 1$ and

$$N_{i,1}(t) = \frac{t-t_i}{t_{i+1}-t_i} N_{i,0}(t) + \frac{t_{i+2}-t}{t_{i+2}-t_{i+1}} N_{i+1,0}(t) = \begin{cases} \frac{t-t_i}{t_{i+1}-t_i}, & \text{if } t \in [t_i, t_{i+1}) \\ \frac{t_{i+2}-t}{t_{i+2}-t_{i+1}}, & \text{if } t \in [t_{i+1}, t_{i+2}) \\ 0, & \text{otherwise.} \end{cases}$$

For $t \neq t_i$, $N_{i,1}(t)$ is C^∞ (and hence also C^0), and since

$$\lim_{t \to t_i^+} N_{i,1}(t) = \lim_{t \to t_i^-} N_{i,1}(t) = N_{i,1}(t_i) = 0 \, ,$$
$$\lim_{t \to t_{i+1}^+} N_{i,1}(t) = \lim_{t \to t_{i+1}^-} N_{i,1}(t) = N_{i,1}(t_{i+1}) = 1 \, ,$$
$$\lim_{t \to t_{i+2}^+} N_{i,1}(t) = \lim_{t \to t_{i+2}^-} N_{i,1}(t) = N_{i,1}(t_{i+2}) = 0 \, ,$$

it follows that $N_{i,1}(t)$ is C^0.

The induction hypothesis is that all basis functions of degree $k-1$ are C^{k-1-p_i}. Then since

$$N'_{i,k}(t) = \frac{k}{t_{i+k} - t_i} N_{i,k-1}(t) - \frac{k}{t_{i+k+1} - t_{i+1}} N_{i+1,k-1}(t) \, ,$$

it follows that the derivatives $N'_{i,k}(t)$ are expressible as sums and products of C^{k-1-p_i} functions, and therefore $N'_{i,k}(t)$ is C^{k-1-p_i} at $t = t_i$. Hence, $N'_{i,k}(t)$ and its first $k - 1 - p_i$ derivatives are continuous at $t = t_i$. Thus the first $k - p_i$ derivatives of $N_{i,k}(t)$ are continuous at $t = t_i$ and, since $N_{i,k}(t)$ is itself continuous, it is deduced that $N_{i,k}(t)$ is C^{k-p_i} as required.

□

Theorem 8.16

The derivative of $\mathbf{B}(t) = \sum_{i=0}^{n} \mathbf{b}_i N_{i,d}(t)$ is

$$\mathbf{B}'(t) = \sum_{i=0}^{n-1} \mathbf{b}_i^{(1)} N_{i,d-1}^{(1)}(t)$$

where

$$\mathbf{b}_i^{(1)} = d\frac{\mathbf{b}_{i+1} - \mathbf{b}_i}{t_{i+d+1} - t_{i+1}} ,$$

and $N_{i,d-1}^{(1)}(t)$ are the degree $d-1$ basis functions defined on the knot vector $t_1, \ldots, t_{m-1}$.

Proof

Let $\mathbf{B}(t) = \sum_{i=0}^{n} \mathbf{b}_i N_{i,d}(t)$, $t \in [t_d, t_{m-d}]$, then

$$\mathbf{B}'(t) = \left(\sum_{i=0}^{n} \mathbf{b}_i N_{i,d}(t) \right)' = \sum_{i=0}^{n} \mathbf{b}_i N'_{i,d}(t) .$$

Thus (8.11) implies

$$\mathbf{B}'(t) = \sum_{i=0}^{n} \mathbf{b}_i \tfrac{d}{t_{i+d}-t_i} N_{i,d-1}(t) - \sum_{i=0}^{n} \mathbf{b}_i \tfrac{d}{t_{i+d+1}-t_{i+1}} N_{i+1,d-1}(t) .$$

Then, since $N_{0,d-1}(t) = N_{n+1,d-1}(t) = 0$ for $t \in [t_d, t_{m-d}]$, it follows that

$$\mathbf{B}'(t) = \sum_{i=1}^{n} \tfrac{d}{t_{i+d}-t_i} \left(\mathbf{b}_i - \mathbf{b}_{i-1} \right) N_{i,d-1}(t) . \tag{8.12}$$

Replacing i by $i+1$ in the summation, gives

$$\mathbf{B}'(t) = \sum_{i=0}^{n-1} \mathbf{b}_i^{(1)} N_{i,d-1}^{(1)}(t) .$$

□

As a corollary, the higher order derivatives can be obtained by repeated applications of the lemma.

Corollary 8.17

The rth derivative of $\mathbf{B}(t)$ is given by

$$\mathbf{B}^{(r)}(t) = \sum_{i=0}^{n-r} \mathbf{b}_i^{(r)} N_{i,d-r}^{(r)}(t)$$

where $\mathbf{b}_i^0 = \mathbf{b}_i$,

$$\mathbf{b}_i^{(r)} = (d-r+1)\frac{\mathbf{b}_{i+1}^{(r-1)} - \mathbf{b}_i^{(r-1)}}{t_{i+d+1} - t_{i+r}} ,$$

and $N_{i,d-r}^{(r)}(t)$ are the basis functions defined on the knot vector $t_r, \ldots, t_{m-r}$.

□

Example 8.18

Consider the B-spline $\mathbf{B}(t)$ of degree 3 defined on the knot vector $t_0 = 1.2, t_1 = 1.4, t_2 = 1.5, t_3 = 2.0, t_4 = 2.4, t_5 = 3.1, t_6 = 5.0, t_7 = 6.4, t_8 = 7.3$, with control points $\mathbf{b}_0(2,1)$, $\mathbf{b}_1(4,8)$, $\mathbf{b}_2(5,-1)$, $\mathbf{b}_3(3,-2)$, and $\mathbf{b}_4(2,-4)$. Then the control points of the derivative of $\mathbf{B}(t)$ are

$$\begin{aligned}
\mathbf{b}_0^{(1)} &= 3\frac{\mathbf{b}_1 - \mathbf{b}_0}{t_4 - t_1} = 3\frac{(4,8)-(2,1)}{1.0} = (6.0, 21.0)\ , \\
\mathbf{b}_1^{(1)} &= 3\frac{\mathbf{b}_2 - \mathbf{b}_1}{t_5 - t_2} = 3\frac{(5,-1)-(4,8)}{1.6} = (1.875, -16.875)\ , \\
\mathbf{b}_2^{(1)} &= 3\frac{\mathbf{b}_3 - \mathbf{b}_2}{t_6 - t_3} = 3\frac{(3,-2)-(5,-1)}{3.0} = (-2.0, -1.0)\ , \text{ and} \\
\mathbf{b}_3^{(1)} &= 3\frac{\mathbf{b}_4 - \mathbf{b}_3}{t_7 - t_4} = 3\frac{(2,-4)-(3,-2)}{4.0} = (-0.75, -1.5)\ .
\end{aligned}$$

The derivative has degree $d = 2$, and is defined on the knot vector $t_0 = 1.4$, $t_1 = 1.5$, $t_2 = 2.0$, $t_3 = 2.4$, $t_4 = 3.1$, $t_5 = 5.0$, $t_6 = 6.4$.

Example 8.19

The derivatives at the endpoints of an open B-spline of degree d are obtained from (8.12). Set $t_0 = t_1 = \cdots = t_d$ and $t_{m-d} = t_{m-d+1} = \cdots = t_m$ to give

$$\begin{aligned}
\mathbf{B}'(t_d) &= \sum_{i=1}^{n} d\frac{\mathbf{b}_i - \mathbf{b}_{i-1}}{t_{i+d} - t_i} N_{i,d-1}(t_d) = d\frac{\mathbf{b}_1 - \mathbf{b}_0}{t_{d+1} - t_1} N_{1,d-1}(t_d) = d\frac{\mathbf{b}_1 - \mathbf{b}_0}{t_{d+1} - t_1}\ , \\
\mathbf{B}'(t_{m-d}) &= \sum_{i=1}^{n} d\frac{\mathbf{b}_i - \mathbf{b}_{i-1}}{t_{i+d} - t_i} N_{i,d-1}(t_{m-d}) = d\frac{\mathbf{b}_n - \mathbf{b}_{n-1}}{t_{d+n} - t_n} N_{n,d-1}(t_{m-d}) \\
&= d\frac{\mathbf{b}_n - \mathbf{b}_{n-1}}{t_{m-1} - t_{m-d-1}}\ ,
\end{aligned}$$

thus verifying Equation (8.5).

EXERCISES

8.17. Determine the basis functions of the B-spline and its derivative of Example 8.18.

8.18. Let a B-spline curve $\mathbf{B}(t)$ of degree 3 be defined on the knot vector $t_0 = 0.5$, $t_1 = 0.8$, $t_2 = 1.4$, $t_3 = 2.1$, $t_4 = 2.4$, $t_5 = 2.9$, $t_6 = 4.0$, $t_7 = 4.5$, $t_8 = 4.9$ with control points $\mathbf{b}_0(-2,-3)$, $\mathbf{b}_1(-1,2)$, $\mathbf{b}_2(2,2)$, $\mathbf{b}_3(3,0)$, $\mathbf{b}_4(1,-3)$. Determine the control points of $\mathbf{B}'(t)$. Determine $\mathbf{B}'(2.8)$ in the following ways.

(a) Determine $N^{(1)}_{i,d-1}(t)$ and evaluate $\mathbf{B}'(t) = \sum_{i=0}^{n-1} \mathbf{b}_i^{(1)} N^{(1)}_{i,d-1}(t)$ at $t = 2.8$.

(b) Apply the de Boor algorithm with $t = 2.8$ to the derivative.

8.19. Determine the control points and knots of the derivative of the B-spline of Example 8.3. Evaluate $\mathbf{B}'(6.2)$ and $\mathbf{B}'(7.4)$ (use de Boor).

8.20. Determine the control points and knots of the derivative of the B-spline of Example 8.7. Evaluate $\mathbf{B}'(2.5)$.

8.21. Determine an expression for the second derivatives at the endpoints of an open B-spline.

8.22. Show that for $k > 0$, the basis functions $N_{i,k}(t)$ have just one maximum value.

8.23. Implement the derivative algorithm of Theorem 8.16.

8.2 Non-uniform Rational B-Splines (NURBS)

Rational B-spline curves are obtained from (integral) B-splines in an analogous manner to the way in which rational Bézier curves are obtained from (integral) Bézier curves. They are generally referred to as *NURBS* which stands for Non-Uniform Rational B-Splines.

Definition 8.20

The NURBS curve of degree d (order $d + 1$) with control points $\mathbf{b}_0, \ldots, \mathbf{b}_n$, weights $w_0, \ldots, w_n$, and knot vector $t_0, \ldots, t_m$, is the curve defined on the interval $[a, b] = [t_d, t_{m-d}]$ given by

$$\mathbf{B}(t) = \frac{\sum_{i=0}^{n} w_i \mathbf{b}_i N_{i,d}(t)}{\sum_{i=0}^{n} w_i N_{i,d}(t)} , \tag{8.13}$$

where $N_{i,d}(t)$ are the B-spline basis functions defined on the specified knot vector, and with the understanding that if $w_i = 0$ then $w_i \mathbf{b}_i$ is to be replaced by $\mathbf{b}_i$. The curve may also be written in the form

$$\mathbf{B}(t) = \sum_{i=0}^{n} \mathbf{b}_i R_{i,d}(t) ,$$

where

$$R_{i,d}(t) = \frac{w_i N_{i,d}(t)}{\sum_{j=0}^{n} w_j N_{j,d}(t)}$$

are the *rational B-spline basis functions.*

Let $\mathbf{b}_i = (x_i, y_i, z_i)$. Define *homogeneous control points* $\hat{\mathbf{b}}_i$ by

$$\hat{\mathbf{b}}_i = \begin{cases} (w_i x_i, w_i y_i, w_i z_i, w_i), & \text{if } w_i \neq 0 \\ (x_i, y_i, z_i, 0), & \text{if } w_i = 0 \end{cases} .$$

In homogeneous coordinates the NURBS curve has the form

$$\mathbf{B}(t) = \sum_{i=0}^{n} \hat{\mathbf{b}}_i N_{i,d}(t) .$$

Appropriate choices of knot vector and control points give rise to the concepts of open or periodic rational B-splines. An open knot vector yields a NURBS curve which is endpoint interpolating. A closed periodic NURBS is obtained by choosing a periodic knot vector, repeated control points (as described in Section 8.1.2) and a set of weights for which the ratios of the first d weights equal the ratios of the last d weights.

Example 8.21 (NURBS Circle)

A NURBS representation of a circle is used in the construction of surfaces of revolution in Section 9.4.4. The unit circle centred at the origin (see Figure 8.7) can be represented by an open quadratic NURBS defined on the interval $[0, 1]$. Take the knot vector $0, 0, 0, \frac{1}{4}, \frac{1}{2}, \frac{1}{2}, \frac{3}{4}, 1, 1, 1$, control points $\mathbf{b}_0(1, 0)$, $\mathbf{b}_1(1, 1)$, $\mathbf{b}_2(-1, 1)$, $\mathbf{b}_3(-1, 0)$, $\mathbf{b}_4(-1, -1)$, $\mathbf{b}_5(1, -1)$, $\mathbf{b}_6(1, 0)$, and corresponding weights $1, \frac{1}{2}, \frac{1}{2}, 1, \frac{1}{2}, \frac{1}{2}, 1$. Arbitrary circles and ellipses may be obtained by applying transformations to the control points. Note that there are many ways of obtaining a NURBS circle.

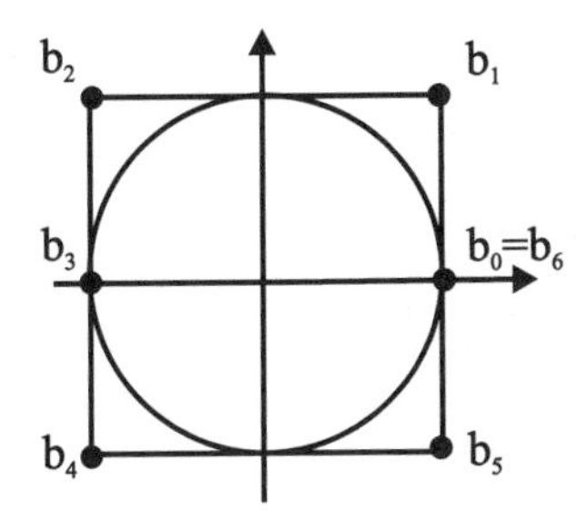

Figure 8.7 NURBS representation of a unit circle

Theorem 8.22

A NURBS curve $\mathbf{B}(t)$ given by (8.13) satisfies the following properties.

Local Control: If $t \in [t_r, t_{r+1})$ $(d \le r \le m-d-1)$ then

$$\mathbf{B}(t) = \frac{\sum_{i=r-d}^{r} w_i \mathbf{b}_i N_{i,d}(t)}{\sum_{i=r-d}^{r} w_i N_{i,d}(t)} = \sum_{i=r-d}^{r} \mathbf{b}_i R_{i,d}(t) \ .$$

Convex Hull Property: If the weights w_i are all positive and $t \in [t_r, t_{r+1})$ $(d \le r \le m-d-1)$ then $\mathbf{B}(t) \in \mathrm{CH}\{\mathbf{b}_{r-d}, ..., \mathbf{b}_r\}$.

Continuity: If p_i is the multiplicity of the breakpoint $t = u_i$, then $\mathbf{B}(t)$ is C^{d-p_i} (or greater) at $t = u_i$ and C^∞ elsewhere.

Invariance under Affine Transformations: Let T be an affine transformation. Then

$$\mathsf{T}\left(\frac{\sum_{i=r-d}^{r} w_i \mathbf{b}_i N_{i,d}(t)}{\sum_{i=r-d}^{r} w_i N_{i,d}(t)}\right) = \frac{\sum_{i=r-d}^{r} w_i \mathsf{T}\left(\mathbf{b}_i\right) N_{i,d}(t)}{\sum_{i=r-d}^{r} w_i N_{i,d}(t)} \ .$$

Invariance under Projective Transformations: Let T be a projective transformation. Then

$$\mathsf{T}\left(\sum_{i=0}^{n} \hat{\mathbf{b}}_i N_{i,d}(t)\right) = \sum_{i=0}^{n} \mathsf{T}\left(\hat{\mathbf{b}}_i\right) N_{i,d}(t)$$

where $\hat{\mathbf{b}}_i$ are the homogeneous control points. See Section 8.2.1.

The analogous result to Theorem 7.25 concerning the effect of changing a weight is the following theorem. The proof is similar.

Theorem 8.23

The effect of changing a weight from w_k to $w_k^* = w_k + \delta w_k$ is that any point $\mathbf{b} = \mathbf{B}(t)$ on the curve moves in the direction of the line $\overrightarrow{\mathbf{b}\mathbf{b}_k}$ (where $\mathbf{b}_k$ is the k-th control point).

8.2.1 Projections of NURBS Curves

The property of projective invariance is useful for the computer display of spatial NURBS curves. In order to apply a projective transformation to a NURBS curve

$$\mathbf{B}(t) = \frac{\sum_{i=0}^{n} w_i \mathbf{b}_i N_{i,n}(t)}{\sum_{i=0}^{n} w_i N_{i,n}(t)} ,$$

it is sufficient to apply the projective transformation to the homogeneous control points $\hat{\mathbf{b}}_i$, where $\hat{\mathbf{b}}_i = (w_i \mathbf{b}_i, w_i)$ if $w_i \neq 0$, and $\hat{\mathbf{b}}_i = (\mathbf{b}_i, 0)$ if $w_i = 0$. The transformed images of $\hat{\mathbf{b}}_i$ define a NURBS curve which is the transformation of $\mathbf{B}(t)$.

The proof is analogous to the equivalent result for rational Bézier curves given in Section 7.5.3. Suppose the projective transformation matrix M is applied to $\mathbf{B}(t) = \sum_{i=0}^{n} \hat{\mathbf{b}}_i N_{i,n}(t)$ (expressed in homogeneous coordinates). Then

$$\mathbf{B}(t)\mathsf{M} = \left(\sum_{i=0}^{n} \hat{\mathbf{b}}_i N_{i,n}(t) \right) \mathsf{M} = \sum_{i=0}^{n} N_{i,n}(t) \left(\hat{\mathbf{b}}_i \mathsf{M} \right) = \sum_{i=0}^{n} \hat{\mathbf{c}}_i N_{i,n}(t) ,$$

defining a NURBS curve with control points and weights given by $\hat{\mathbf{c}}_i = \hat{\mathbf{b}}_i \mathsf{M}$ from which the Cartesian control points and weights can be obtained. In particular, if the transformation is a perspective or parallel projection then the projected image of a NURBS curve onto a viewplane can be executed by applying the projection to the homogeneous control points.

As for the case of rational Bézier curves, the above argument can be adapted to show that NURBS curves are invariant under the viewplane coordinate mapping VC and the device coordinate transformation DC. It follows that the whole process of viewing a rational Bézier curve can be executed by applying the complete viewing pipeline matrix $\mathsf{VP} = \mathsf{M} \cdot \mathsf{VC} \cdot \mathsf{DC}$ to the control points.

Example 8.24

Consider the perspective projection of Examples 4.7 and 7.27 onto the xy-plane with viewpoint $\mathbf{V}(1, 5, 3)$. The projection matrix M and viewplane coordinate matrix VC are determined in Example 7.27. The quadratic NURBS curve, defined on the knot vector $t_0 = 0$, $t_1 = 0$, $t_2 = 1$, $t_3 = 2$, $t_4 = 3$, $t_5 = 3$, with control points $\mathbf{b}_0(0, 0, 0)$, $\mathbf{b}_1(1, 0, 0)$, $\mathbf{b}_2(1, 0, 1)$, $\mathbf{b}_3(1, 1, 1)$, and weights $1, 2, 2, 1$, has homogeneous control points $\hat{\mathbf{b}}_0(0, 0, 0, 1)$, $\hat{\mathbf{b}}_1(2, 0, 0, 2)$, $\hat{\mathbf{b}}_2(2, 0, 2, 2)$, $\hat{\mathbf{b}}_3(1, 1, 1, 1)$. Thus

$$\begin{pmatrix} \hat{\mathbf{b}}_0 \\ \hat{\mathbf{b}}_1 \\ \hat{\mathbf{b}}_2 \\ \hat{\mathbf{b}}_3 \end{pmatrix} \mathsf{M} \cdot \mathsf{VC} = \begin{pmatrix} 0 & 0 & 0 & 1 \\ 2 & 0 & 0 & 2 \\ 2 & 0 & 2 & 2 \\ 1 & 1 & 1 & 1 \end{pmatrix} \mathsf{M} \cdot \mathsf{VC} = \begin{pmatrix} 6.6 & -1.2 & -3.0 \\ 9.6 & -7.2 & -6.0 \\ 14.4 & -10.8 & -4.0 \\ 4.8 & -3.6 & -2.0 \end{pmatrix} .$$

Multiply the homogeneous control points through by -1 to give positive weights. Then the image of the curve is the planar quadratic NURBS curve with control points $(-2.2, 0.4)$, $(-1.6, 1.2)$, $(-3.6, 2.7)$, and $(-2.4, 1.8)$, and weights $3, 6, 4$, and 2 defined on the same knot vector. Note that the working is essentially the same as for the projection of the rational Bézier curve in Example 7.27.

8.2.2 Derivatives of NURBS

A recursive formula to determine the derivative of a NURBS is obtained from Equation (7.15) which determines the derivatives of a rational function. For a NURBS

$$\mathbf{B}(t) = \frac{\sum_{i=0}^{n} w_i \mathbf{b}_i N_{i,d}(t)}{\sum_{i=0}^{n} w_i N_{i,d}(t)} \,,$$

let $f(t) = \sum_{i=0}^{n} w_i \mathbf{b}_i N_{i,d}(t)$ and $g(t) = \sum_{i=0}^{n} w_i N_{i,d}(t)$ in (7.15). The derivatives of $f(t)$ and $g(t)$ are obtained by applying the algorithm for computing the derivatives of B-splines (Section 8.1.6) where the $w_i \mathbf{b}_i$ are considered to be the control points of $f(t)$, and the w_i are considered to be the control points of $g(t)$.

Example 8.25

Consider the NURBS of degree 3 defined on the knot vector $t_0 = 1.2$, $t_1 = 1.4$, $t_2 = 1.5$, $t_3 = 2.0$, $t_4 = 2.4$, $t_5 = 3.1$, $t_6 = 5.0$, $t_7 = 6.4$, $t_8 = 7.3$, with control points $\mathbf{b}_0(2,1)$, $\mathbf{b}_1(4,8)$, $\mathbf{b}_2(5,-1)$, $\mathbf{b}_3(3,-2)$, $\mathbf{b}_4(2,-4)$, and weights $w_0 = 1.0$, $w_1 = 1.5$, $w_2 = 2.0$, $w_3 = 1.5$, $w_4 = 1.0$. Then $f(t) = \sum_{i=0}^{n} w_i \mathbf{b}_i N_{i,d}(t)$ has control points $w_0\mathbf{b}_0 = (2,1)$, $w_1\mathbf{b}_1 = (6,12)$, $w_2\mathbf{b}_2 = (10,-2)$, $w_3\mathbf{b}_3 = (4.5,-3)$, and $w_4\mathbf{b}_4 = (2,-4)$. Thus $f'(t)$ is defined by control points

$$\begin{aligned}
\mathbf{b}_0^{(1)} &= 3\frac{w_1\mathbf{b}_1 - w_0\mathbf{b}_0}{t_4 - t_1} = \frac{3\left((6,12) - (2,1)\right)}{2.4 - 1.4} = (12.0, 33.0)\,,\\
\mathbf{b}_1^{(1)} &= 3\frac{w_2\mathbf{b}_2 - w_1\mathbf{b}_1}{t_5 - t_2} = \frac{3\left((10,-2) - (6,12)\right)}{3.1 - 1.5} = (7.5, -26.25)\,,\\
\mathbf{b}_2^{(1)} &= 3\frac{w_3\mathbf{b}_3 - w_2\mathbf{b}_2}{t_6 - t_3} = \frac{3\left((4.5,-3) - (10,-2)\right)}{5.0 - 2.0} = (-5.5, -1.0)\,,\\
\mathbf{b}_3^{(1)} &= 3\frac{w_4\mathbf{b}_4 - w_3\mathbf{b}_3}{t_7 - t_4} = \frac{3\left((2,-4) - (4.5,-3)\right)}{6.4 - 2.4} = (-1.875, -0.75)\;.
\end{aligned}$$

The function $g(t) = \sum_{i=0}^{n} w_i N_{i,d}(t)$ has a derivative with control points

$$\begin{aligned}
w_0^{(1)} &= 3\frac{w_1 - w_0}{t_4 - t_1} = \frac{3\,(1.5-1.0)}{2.4-1.4} = 1.5, \\
w_1^{(1)} &= 3\frac{w_2 - w_1}{t_5 - t_2} = \frac{3\,(2.0-1.5)}{3.1-1.5} = 0.9375, \\
w_2^{(1)} &= 3\frac{w_3 - w_2}{t_6 - t_3} = \frac{3\,(1.5-2.0)}{5.0-2.0} = -0.5, \\
w_3^{(1)} &= 3\frac{w_4 - w_3}{t_7 - t_4} = \frac{3\,(1.0-1.5)}{6.4-2.4} = -0.375\,.
\end{aligned}$$

Then $\mathbf{B}'(t)$ is computed by determining the values of $f(t), g(t), f'(t), g'(t)$ and substituting into Equation (7.15). For instance, $\mathbf{B}'(2.7)$ is computed as follows:

$$\begin{aligned}
f(2.7) &= (2,1)N_{0,3}(2.7) + (6,12)N_{1,3}(2.7) + (10,-2)N_{2,3}(2.7) \\
&\quad + (4.5,-3)N_{3,3}(2.7) + (2,-4)N_{4,3}(2.7)\,, \\
g(2.7) &= 1.0N_{0,3}(2.7) + 1.5N_{1,3}(2.7) + 2.0N_{2,3}(2.7) \\
&\quad + 1.5N_{3,3}(2.7) + 1.0N_{4,3}(2.7)\,, \\
f'(2.7) &= (12.0,33.0)\,N_{0,2}^{(1)}(2.7) + (7.5,-26.25)\,N_{1,2}^{(1)}(2.7) \\
&\quad + (-5.5,-1.0)\,N_{2,2}^{(1)}(2.7) + (-1.875,-0.75)\,N_{3,2}^{(1)}(2.7)\,, \\
g'(2.7) &= 1.5N_{0,2}^{(1)}(2.7) + 0.9375N_{1,2}^{(1)}(2.7) - 0.5N_{2,2}^{(1)}(2.7) - 0.375N_{3,2}^{(1)}(2.7)\,.
\end{aligned}$$

The basis functions were determined and evaluated in Exercise 8.17 (though the exercise was for a non-rational B-spline). At $t = 2.7$, $N_{0,3} = 0.0$, $N_{1,3} = 0.05195$, $N_{2,3} = 0.72529$, $N_{3,3} = 0.21905$, $N_{4,3} = 0.00371$, and $N_{0,2}^{(1)} = 0.0$, $N_{1,2}^{(1)} = 0.20779$, $N_{2,2}^{(1)} = 0.74276$, $N_{3,2}^{(1)} = 0.04945$. Hence

$$\begin{aligned}
f(2.7) &= (2,1)0.0 + (6,12)0.05195 + (10,-2)0.72529 \\
&\quad + (4.5,-3)0.21905 + (2,-4)0.00371 \\
&= (8.5577,-1.4992)\,, \\
g(2.7) &= (1.0)\,0.0 + (1.5)\,0.05195 + (2.0)\,0.72529 \\
&\quad + (1.5)\,0.21905 + (1.0)\,0.00371 \\
&= 1.8608\,, \\
f'(2.7) &= (12.0,33.0)\,0.0 + (7.5,-26.25)\,0.20779 \\
&\quad + (-5.5,-1.0)\,0.74276 + (-1.875,-0.75)\,0.04945 \\
&= (-2.6195,-6.2343)\,, \\
g'(2.7) &= (1.5)\,0.0 + (0.9375)\,0.20779 - (0.5)\,0.74276 - (0.375)\,0.04945 \\
&= -0.19512\,.
\end{aligned}$$

Thus $\mathbf{B}(2.7) = (8.5577, -1.4992)/\,1.8608 = (4.59894, -0.80568)$ and

$$\begin{aligned}\mathbf{B}'(2.7) &= \frac{f'(2.7) - g'(2.7)\mathbf{B}(2.7)}{g(2.7)} \\ &= \frac{(-2.6195, -6.2343) - (-0.19512)\,(4.59894, -0.80568)}{1.8608} \\ &= (-0.92549, -3.4348)\ .\end{aligned}$$

EXERCISES

8.24. Show that $\sum_{i=0}^{n} R_{i,d}(t) = 1$ and $\sum_{i=r-d}^{r} R_{i,d}(t) = 1$.

8.25. Determine the basis functions and the polynomial curve segments of the NURBS circle (Example 8.21).

8.26. Another NURBS unit circle can be obtained with a seven-point triangular control polygon $(\sqrt{3}/2, 1/2)$, $(0, 2)$, $(-\sqrt{3}/2, 1/2)$, $(-\sqrt{3}, -1)$, $(0, -1)$, $(\sqrt{3}, -1)$, $(\sqrt{3}/2, 1/2)$, knots $0, 0, 0, \frac{1}{3}, \frac{1}{3}, \frac{2}{3}, \frac{2}{3}, 1, 1, 1$, and weights $1, \frac{1}{2}, 1, \frac{1}{2}, 1, \frac{1}{2}, 1$. Determine the basis functions and the three polynomial curve segments of the curve.

8.27. For the NURBS of Example 8.25, determine $\mathbf{B}'(2.2)$.

8.28. Determine $\mathbf{B}'(0.5)$ and $\mathbf{B}'(0.8)$ for the NURBS circle (Example 8.21).

8.29. Show that for an open rational B-spline the derivatives at the end of the curve are

$$\begin{aligned}\mathbf{B}'(t_d) &= \left(\tfrac{d}{t_{d+1}-t_1}\right) \tfrac{w_1}{w_0} \left(\mathbf{b}_1 - \mathbf{b}_0\right), \quad \text{and} \\ \mathbf{B}'(t_{m-d}) &= \left(\tfrac{d}{t_{m-1}-t_{m-d-1}}\right) \tfrac{w_n}{w_{n-1}} \left(\mathbf{b}_n - \mathbf{b}_{n-1}\right)\ .\end{aligned}$$

8.30. Write a computer program (or use a computer package) to draw NURBS curves.

8.2.3 Rational de Boor Algorithm

The rational de Boor algorithm is obtained from the de Boor algorithm in a similar manner to the derivation of the rational de Casteljau algorithm from the de Casteljau algorithm. Set $\mathbf{b}_i^0 = \mathbf{b}_i$ and $w_i^0 = w_i$ and suppose $t \in [t_r, t_{r+1})$.

The rational de Boor algorithm is

$$\left.\begin{aligned} \alpha_i^j &= \frac{t - t_i}{t_{i+d-j+1} - t_i} , \\ w_i^j &= (1-\alpha_i^j) w_{i-1}^{j-1} + \alpha_i^j w_i^{j-1} , \\ w_i^j \mathbf{b}_i^j &= (1-\alpha_i^j) w_{i-1}^{j-1} \mathbf{b}_{i-1}^{j-1} + \alpha_i^j w_i^{j-1} \mathbf{b}_i^{j-1}, \text{ for } j > 0 , \end{aligned}\right\} \tag{8.14}$$

for $i = 0, \ldots, d$ and $j = r - d + i, \ldots, r$. The algorithm yields $\mathbf{B}(t) = \mathbf{b}_r^d$.

In addition to point evaluation, the de Boor or rational de Boor algorithms can be used to subdivide a B-spline or NURBS curve. Subdivision is not only a means of splitting a curve, but also a way of creating extra control points (and weights) in order to give additional freedoms for curve design. The intersection algorithms for Bézier curves described in Section 6.10, which employ the de Casteljau algorithm, can be extended to B-spline and NURBS curves using the de Boor algorithm.

Example 8.26

Consider the NURBS of degree 3 defined on the knot vector $t_0 = 1.2$, $t_1 = 1.4$, $t_2 = 1.5$, $t_3 = 2.0$, $t_4 = 2.4$, $t_5 = 3.1$, $t_6 = 5.0$, $t_7 = 6.4$, $t_8 = 7.3$, control points $\mathbf{b}_0(2,1)$, $\mathbf{b}_1(4,8)$, $\mathbf{b}_2(5,-1)$, $\mathbf{b}_3(3,-2)$, $\mathbf{b}_4(2,-4)$, and weights $w_0 = 1.0$, $w_1 = 1.5$, $w_2 = 2.0$, $w_3 = 1.5$, $w_4 = 1.0$. Determine the point $\mathbf{B}(2.7)$. Since $2.7 \in [2.4, 3.1) = [t_4, t_5)$, it follows that $r = 4$. Then

$$\alpha_2^1 = \tfrac{t-t_2}{t_5-t_2} = \tfrac{2.7-1.5}{3.1-1.5} = 0.75, \ \alpha_3^1 = \tfrac{t-t_3}{t_6-t_3} = \tfrac{2.7-2.0}{5.0-2.0} = 0.23333,$$
$$\alpha_4^1 = \tfrac{t-t_4}{t_7-t_4} = \tfrac{2.7-2.4}{6.4-2.4} = 0.075 .$$

Then

$$\begin{aligned} w_2^1 &= (1-\alpha_2^1) w_1^0 + \alpha_2^1 w_2^0 = (1-0.75)1.5 + (0.75)\,2.0 = 1.875 , \\ w_3^1 &= (1-\alpha_3^1) w_2^0 + \alpha_3^1 w_3^0 = (1-0.23333)2.0 + (0.23333)\,1.5 = 1.8833 , \\ w_4^1 &= (1-\alpha_4^1) w_3^0 + \alpha_4^1 w_4^0 = (1-0.075)1.5 + (0.075)\,1.0 = 1.4625 . \end{aligned}$$

The new row of control points is

$$\begin{aligned}
\mathbf{b}_2^1 &= \tfrac{(1-\alpha_2^1)w_1^0\mathbf{b}_1^0+\alpha_2^1 w_2^0\mathbf{b}_2^0}{w_2^1} \\
&= \tfrac{(1-0.75)1.5(4,8)+(0.75)2.0(5,-1)}{1.875} = (4.8, 0.8) \ , \\
\mathbf{b}_3^1 &= \tfrac{(1-\alpha_3^1)w_2^0\mathbf{b}_2^0+\alpha_3^1 w_3^0\mathbf{b}_3^0}{w_3^1} \\
&= \tfrac{(1-0.23333)2.0(5,-1)+(0.23333)1.5(3,-2)}{1.8833} = (4.6284, -1.1858) \ , \\
\mathbf{b}_4^1 &= \tfrac{(1-\alpha_4^1)w_3^0\mathbf{b}_3^0+\alpha_4^1 w_4^0\mathbf{b}_4^0}{w_4^1} \\
&= \tfrac{(1-0.075)1.5(3,-2)+(0.075)1.0(2,-4)}{1.4625} = (2.9487, -2.1026) \ .
\end{aligned}$$

$$\alpha_3^2 = \tfrac{t-t_3}{t_5-t_3} = \tfrac{2.7-2.0}{3.1-2.0} = 0.63636, \ \alpha_4^2 = \tfrac{t-t_4}{t_6-t_4} = \tfrac{2.7-2.4}{5.0-2.4} = 0.11538 \ .$$

Then

$$\begin{aligned}
w_3^2 &= (1-\alpha_3^2)w_2^1 + \alpha_3^2 w_3^1 \\
&= (1-0.63636)1.875 + (0.63636)\,1.8833 = 1.8803 \ , \\
w_4^2 &= (1-\alpha_4^2)w_3^1 + \alpha_4^2 w_4^1 \\
&= (1-0.11538)1.8833 + (0.11538)\,1.4625 = 1.8347 \ .
\end{aligned}$$

The new row of control points is

$$\begin{aligned}
\mathbf{b}_3^2 &= \tfrac{(1-\alpha_3^2)w_2^1\mathbf{b}_2^1+\alpha_3^2 w_3^1\mathbf{b}_3^1}{w_3^2} \\
&= \tfrac{(1-0.63636)1.875(4.8,0.8)+(0.63636)1.8833(4.6284,-1.1858)}{1.8803} \\
&= (4.6906, -0.46571) \ , \\
\mathbf{b}_4^2 &= \tfrac{(1-\alpha_4^2)w_3^1\mathbf{b}_3^1+\alpha_4^2 w_4^1\mathbf{b}_4^1}{w_4^2} \\
&= \tfrac{(1-0.11538)1.8833(4.6284,-1.1858)+(0.11538)1.4625(2.9487,-2.1026)}{1.8347} \\
&= (4.474, -1.2701) \ , \\
\alpha_4^3 &= \tfrac{t-t_4}{t_5-t_4} = \tfrac{2.7-2.4}{3.1-2.4} = 0.42857 \ .
\end{aligned}$$

Then

$$w_4^3 = (1-\alpha_4^3)w_3^2 + \alpha_4^3 w_4^2 = (1-0.42857)1.8803 + (0.42857)\,1.8347 = 1.8608 \ .$$

The final control point is

$$\begin{aligned}
\mathbf{b}_4^3 &= \tfrac{(1-\alpha_4^3)w_3^2\mathbf{b}_3^2+\alpha_4^3 w_4^2\mathbf{b}_4^2}{w_4^3} \\
&= \tfrac{(1-0.42857)1.8803(4.6906,-0.46571)+(0.42857)1.8347(4.474,-1.2701)}{1.8608} \\
&= (4.599, -0.80562) \ .
\end{aligned}$$

Hence $\mathbf{B}(2.7) = (4.599, -0.80562)$.

EXERCISES

8.31. Apply the rational de Boor algorithm to the NURBS circle to determine $\mathbf{B}(0.65)$.

8.32. Write a program (or use a computer package) which performs the rational de Boor algorithm, and use it to verify your answer to the previous question.

8.3 Knot Insertion

Knot insertion is the operation of obtaining a new representation of a B-spline curve by introducing additional knot values to the defining knot vector. The new curve has control points consisting of the original control points and additional new control points corresponding to the number of new knot values. So knot insertions give additional control points which provide extra shape control without necessarily subdividing the curve. However, if following a knot insertion operation a knot has multiplicity equal to the degree, then the B-spline is split into two B-splines at that knot value.

Definition 8.27

Let $\mathbf{B}(t) = \sum_{i=0}^{n} \mathbf{b}_i N_{i,d}(t)$ be a B-spline defined on a knot vector $t_0, \ldots, t_m$, and let $\mathbf{C}(t) = \sum_{i=0}^{q} \mathbf{c}_i \hat{N}_{i,d}(t)$ be defined on the knot vector $s_0, \ldots, s_p$. If $s_0, \ldots, s_p$ is obtained from $t_0, \ldots, t_m$ by performing knot insertions so that $\mathbf{B}(t) = \mathbf{C}(t)$ for $t \in [t_d, t_m]$, then $\mathbf{C}(t)$ is said to be a *refinement* of $\mathbf{B}(t)$.

Lemma 8.28

Let $N_{i,d}(t)$ be the B-spline basis functions of degree d defined on the knot vector $t_0, \ldots, t_m$. Suppose $\hat{t} \in [t_s, t_{s+1})$ and let $\hat{N}_{i,d}(t)$ be the basis functions defined on $\hat{t}_0 = t_0, \ldots, \hat{t}_s = t_s, \hat{t}_{s+1} = \hat{t}, \hat{t}_{s+2} = t_{s+1}, \ldots, \hat{t}_{m+1} = t_m$. Then $N_{i,d}(t) = \hat{N}_{i,d}(t)$ for $i = 0, \ldots, s-d-1$, $N_{i,d}(t) = \hat{N}_{i+1,d}(t)$ for $i = s+1, \ldots, n$, and for $i = s-d, \ldots, s$

$$N_{i,d}(t) = \frac{\hat{t} - \hat{t}_i}{\hat{t}_{i+d+1} - \hat{t}_i} \hat{N}_{i,d}(t) + \frac{\hat{t}_{i+d+2} - \hat{t}}{\hat{t}_{i+d+2} - \hat{t}_{i+1}} \hat{N}_{i+1,d}(t) \,. \qquad (8.15)$$

Theorem 8.29 (Boehm's Algorithm)

Let $\mathbf{B}(t) = \sum_{i=0}^{n} \mathbf{b}_i N_{i,d}(t)$ be a B-spline with knots $t_0, \ldots, t_m$, and let $\hat{t} \in [t_s, t_{s+1})$. Then the representation of $\mathbf{B}(t)$ with knot vector $t_0, \ldots, t_s, \hat{t}, t_{s+1}, \ldots, t_{m-1}, t_m$, is $\mathbf{B}(t) = \sum_{i=0}^{n} \hat{\mathbf{b}}_i \hat{N}_{i,d}(t)$

$$\begin{aligned} \hat{\mathbf{b}}_i &= \begin{cases} \mathbf{b}_i, & 0 \le i \le s-d \\ (1-\alpha_i)\mathbf{b}_{i-1} + \alpha_i \mathbf{b}_i, & s-d+1 \le i \le s \\ \mathbf{b}_{i-1}, & s+1 \le i \le n+1 \end{cases}, \\ \alpha_i &= \frac{\hat{t} - t_i}{t_{i+d} - t_i} = \frac{\hat{t} - \hat{t}_i}{\hat{t}_{i+d+1} - \hat{t}_i} . \end{aligned} \tag{8.16}$$

Proof

Using Lemma 8.28, and the fact that $\hat{t} = \hat{t}_{s+1}$

$$\begin{aligned} \mathbf{B}(t) &= \sum_{i=0}^{n} \mathbf{b}_i N_{i,d}(t) \\ &= \sum_{i=0}^{s-d-1} \mathbf{b}_i \hat{N}_{i,d}(t) + \sum_{i=s-d}^{s} \mathbf{b}_i \frac{\hat{t} - \hat{t}_i}{\hat{t}_{i+d+1} - \hat{t}_i} \hat{N}_{i,d}(t) \\ &\quad + \sum_{i=s-d}^{s} \mathbf{b}_i \frac{\hat{t}_{i+d+2} - \hat{t}}{\hat{t}_{i+d+2} - \hat{t}_{i+1}} \hat{N}_{i+1,d}(t) + \sum_{i=s+1}^{n} \mathbf{b}_i \hat{N}_{i+1,d}(t) \\ &= \sum_{i=0}^{s-d-1} \mathbf{b}_i \hat{N}_{i,d}(t) + \mathbf{b}_{s-d} \frac{\hat{t} - \hat{t}_{s-d}}{\hat{t}_{s+1} - \hat{t}_{s-d}} \hat{N}_{s-d,d}(t) \\ &\quad + \sum_{i=s-d+1}^{s} \left(\mathbf{b}_i \frac{\hat{t} - \hat{t}_i}{\hat{t}_{i+d+1} - \hat{t}_i} + \mathbf{b}_{i-1} \frac{\hat{t}_{i+d+1} - \hat{t}}{\hat{t}_{i+d+1} - \hat{t}_i} \right) \hat{N}_{i,d}(t) \\ &\quad + \mathbf{b}_s \frac{\hat{t}_{s+d+2} - \hat{t}}{\hat{t}_{s+d+2} - \hat{t}_{s+1}} \hat{N}_{s+1,d}(t) + \sum_{i=s+1}^{n} \mathbf{b}_i \hat{N}_{i+1,d}(t) \\ &\quad \text{(renumbering indices)} \\ &= \sum_{i=0}^{n+1} \hat{\mathbf{b}}_i \hat{N}_{i,d}(t) . \end{aligned}$$

□

Boehm's algorithm can be compared with the de Boor algorithm. The de Boor algorithm is equivalent to d insertions of the knot t. Boehm's algorithm inserts just the one knot $\hat{t}$, but several knots can be inserted by repeated applications of the algorithm, or more efficiently, by using a generalized Boehm's algorithm [3]. Multiple knots can also be inserted using the Oslo algorithm [13]. The algorithms can be generalized to NURBS curves.

Example 8.30

Consider a B-spline defined on the knot vector $t_0 = 0$, $t_1 = 0$, $t_2 = 0$, $t_3 = 1$, $t_4 = 3$, $t_5 = 3$, $t_6 = 3$, with control points $\mathbf{b}_0, \ldots, \mathbf{b}_3$. The knot $\hat{t} = 2$ is inserted as follows. Since $2 \in [t_3, t_4) = [1, 3)$, $s = 3$. So $\hat{\mathbf{b}}_0 = \mathbf{b}_0$, $\hat{\mathbf{b}}_1 = \mathbf{b}_1$, $\hat{\mathbf{b}}_4 = \mathbf{b}_3$, and

$$\alpha_2 = \frac{\hat{t} - t_2}{t_4 - t_2} = \frac{2-0}{3-0} = \frac{2}{3}, \quad \alpha_3 = \frac{\hat{t} - t_3}{t_4 - t_3} = \frac{2-1}{3-1} = \frac{1}{2},$$

$$\begin{aligned} \hat{\mathbf{b}}_2 &= (1-\alpha_2)\,\mathbf{b}_1 + \alpha_2 \mathbf{b}_2 = \frac{1}{3}\mathbf{b}_1 + \frac{2}{3}\mathbf{b}_2 , \\ \hat{\mathbf{b}}_3 &= (1-\alpha_3)\,\mathbf{b}_2 + \alpha_3 \mathbf{b}_3 = \frac{1}{2}\mathbf{b}_2 + \frac{1}{2}\mathbf{b}_3 . \end{aligned}$$

For $\mathbf{b}_0(0,0)$, $\mathbf{b}_1(6,12)$, $\mathbf{b}_2(12,12)$, $\mathbf{b}_3(16,4)$ the knot insertion yields $\hat{\mathbf{b}}_0(0,0)$, $\hat{\mathbf{b}}_1(6,12)$, $\hat{\mathbf{b}}_2\,(10,12)$, $\hat{\mathbf{b}}_3\,(14,8)$, $\hat{\mathbf{b}}_4(16,4)$, as illustrated in Figure 8.8.

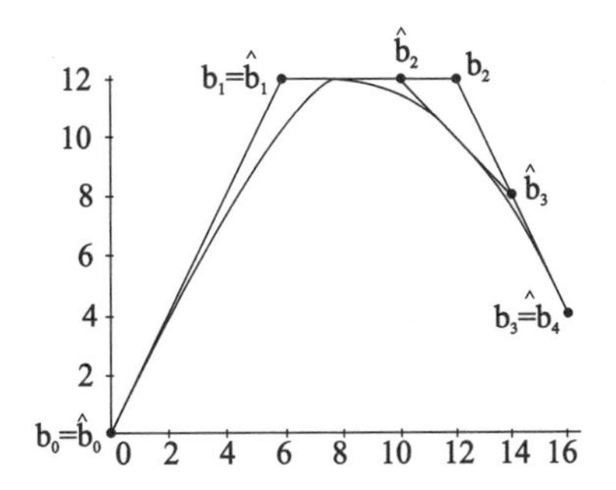

Figure 8.8 Knot insertion of a quadratic B-spline

Exercise 8.33

Using Boehm's algorithm, insert the knots $t = 1$ and $t = 2$ twice (insert one at a time) into the B-spline of Example 8.30 and show that the resulting segments are Bézier curves defined on the intervals $[0, 1]$, $[1, 2]$, and $[2, 3]$.

Exercise 8.33 exemplifies a general result that a B-spline of degree d can be converted into piecewise Bézier form by inserting sufficient knots so that each knot has multiplicity d. This fact proves that a B-spline is indeed a piecewise polynomial curve.

9
Surfaces

9.1 Introduction

Surfaces have a fundamental role in applications such as computer graphics, virtual reality, computer games, and in the computer-aided design of cars, ships, aircraft, and buildings. The earlier discussion of curves naturally leads to the study of surfaces. Conics, Bézier curves, and B-spline curves, the key curve types, have corresponding surface forms, namely, quadric surfaces, Bézier surfaces, and B-spline surfaces. Quadric surfaces are introduced in Section 9.2, and appear again in Sections 9.3 and 9.4 in Bézier and B-spline form. In some applications, surfaces occur as "surface constructs" such as extruded surfaces, ruled surfaces, and surfaces of revolution. These constructs are considered in Section 9.4. Sections 9.2.1 and 9.6 consider three other important CAD surfaces, namely, offset, skin and loft surfaces.

Definition 9.1

A subset of $\mathbb{R}^3$ of the form $\{(x, y, z) : F(x, y, z) = 0\}$ for some function $F : \mathbb{R}^3 \to \mathbb{R}$ is called an *implicit surface*. When F is a polynomial in x, y, and z, the surface is called an *algebraic surface*. If the partial derivatives of F exist, then the points of the surface satisfying

$$F(x, y, z) = \frac{\partial F}{\partial x}(x, y, z) = \frac{\partial F}{\partial y}(x, y, z) = \frac{\partial F}{\partial z}(x, y, z) = 0$$

are called *singular* points, and all other points are called *non-singular* or *reg-*

ular points. A surface with no singular points is called a *non-singular* surface. Implicity defined surfaces are important in CAD applications and provide the basis for CSG modellers discussed in Section 9.7.3.

Example 9.2

1. The implicit surfaces $ax + by + cz + d = 0$ for constants $a, b, c, d \in \mathbb{R}$ are planes.
2. The implicit surface $x^2 + y^2 + z^2 - 1 = 0$ is the unit sphere centred at the origin.

Definition 9.3

Let U be an open subset of $\mathbb{R}^2$. A *parametric surface* is a mapping $\mathbf{S} : U \longrightarrow \mathbb{R}^3$. A mapping $\mathbf{S} : V \longrightarrow \mathbb{R}^3$, defined on a closed subset V of $\mathbb{R}^2$ is said to be a parametric surface whenever there exists an open subset U containing V, and a parametric surface $\mathbf{S}_1 : U \longrightarrow \mathbb{R}^3$, such that $\mathbf{S}(s,t) = \mathbf{S}_1(s,t)$ for all $(s,t) \in V$. $\mathbf{S}_1$ is said to extend $\mathbf{S}$. The subset $S = \mathbf{S}(U)$ or $S = \mathbf{S}(V)$ of $\mathbb{R}^3$ is referred to as the *surface* S or the *trace* of $\mathbf{S}$, and $\mathbf{S}$ is said to *parametrize* S.

The coordinates of an arbitrary point of a parametric surface S can be expressed as functions of two variables, for instance,

$$\mathbf{S}(s,t) = (x(s,t), y(s,t), z(s,t)) \ .$$

The curves $\mathbf{c}_{t_0}(s) = \mathbf{S}(s, t_0)$ and $\mathbf{c}_{s_0}(t) = \mathbf{S}(s_0, t)$, obtained by fixing the value of one of the variables, are called the *s-parameter* and *t-parameter* (or *s-* and *t-coordinate*) *curves* respectively.

The parametric surface $\mathbf{S} : U \longrightarrow \mathbb{R}^3$ is said to be C^k-continuous (or just C^k) whenever the coordinate functions $x(s,t)$, $y(s,t)$, and $z(s,t)$ are C^k-continuous on U. If $|\mathbf{S}_s(s,t) \times \mathbf{S}_t(s,t)| \neq 0$, then the surface is said to be *regular* at $\mathbf{S}(s,t)$, and $\mathbf{S}(s,t)$ is said to be a *regular point*. If $\mathbf{S}(s,t)$ is regular for all $(s,t) \in U$, then the surface is said to be *regular*. If $|\mathbf{S}_s(s,t) \times \mathbf{S}_t(s,t)| = 0$, then $\mathbf{S}$ is said to be *singular* at $\mathbf{S}(s,t)$, and $\mathbf{S}(s,t)$ is said to be a *singular point*.

A parametric surface $\mathbf{S}$ defined on a closed set V is said to be C^k whenever there exists an open set U containing V, and a C^k parametric surface $\mathbf{S}_1$ defined on U, such that $\mathbf{S}(s,t) = \mathbf{S}_1(s,t)$ for all $(s,t) \in V$. The partial derivatives of $\mathbf{S}(s,t)$ at boundary points of V are obtained by taking the derivatives of the extension mapping. Then $\mathbf{S}(s,t)$ is a regular/singular point if it is a regular/singular point of $\mathbf{S}_1(s,t)$.

At a point $\mathbf{p} = \mathbf{S}(s,t)$, $\mathbf{S}_s(s,t)$ and $\mathbf{S}_t(s,t)$ are the tangent vectors to the s- and t-parameter curves. If $\mathbf{p}$ is a regular point of the surface then

$|\mathbf{S}_s(s,t) \times \mathbf{S}_t(s,t)| \neq 0$. Hence $\mathbf{S}_s(s,t)$ and $\mathbf{S}_t(s,t)$ are non-parallel vectors, and a vector perpendicular to them both is the *unit normal vector* to the surface, given by

$$\mathbf{N}(s,t) = \frac{\mathbf{S}_s(s,t) \times \mathbf{S}_t(s,t)}{|\mathbf{S}_s(s,t) \times \mathbf{S}_t(s,t)|}\,, \tag{9.1}$$

as shown in Figure 9.1. (It is also possible to take minus this vector.) Any vector $\mathbf{v}$ perpendicular to $\mathbf{N}$ is called a *tangent vector* to S at $\mathbf{p}$. The vector subspace of $\mathbb{R}^3$, consisting of all the tangent vectors to S at $\mathbf{p}$, is called the *tangent plane* at $\mathbf{p}$, and denoted $T_{\mathbf{p}}(S)$. Intuitively, $T_{\mathbf{p}}(S)$ can be visualized as the plane through $\mathbf{p}$ which is tangent to the surface at $\mathbf{p}$ (that is, perpendicular to $\mathbf{N}$), as shown in Figure 9.1.

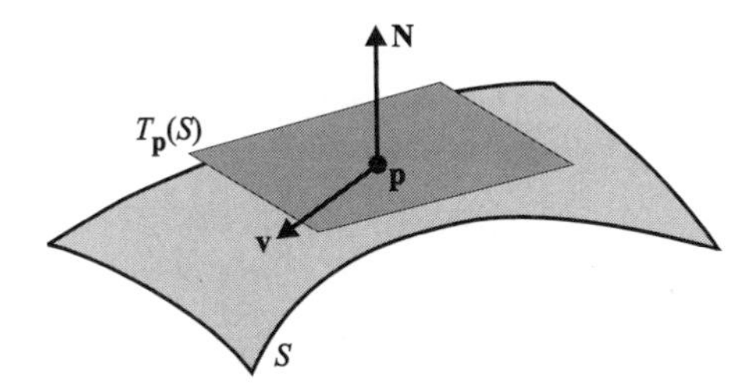

Figure 9.1 Tangent plane $T_{\mathbf{p}}(s)$

Example 9.4

Parametric surfaces of the form $\mathbf{S}(s,t) = (s,t,f(s,t))$ (or similarly, $\mathbf{S}(s,t) = (s,f(s,t),t)$ or $\mathbf{S}(s,t) = (f(s,t),s,t)$) are called non-parametric explicit surfaces or *Monge patches*. If the partial derivatives of f exist, then

$$\mathbf{S}_s(s,t) \times \mathbf{S}_t(s,t) = (1,0,f_s(s,t)) \times (0,1,f_t(s,t)) = (-f_s(s,t), -f_t(s,t), 1)\,.$$

Hence $|\mathbf{S}_s(s,t) \times \mathbf{S}_t(s,t)| = \sqrt{1 + (f_s(s,t))^2 + (f_t(s,t))^2} \neq 0$, and so the surface is regular. The normal vector is

$$\mathbf{N}(s,t) = \frac{1}{\sqrt{1 + (f_s(s,t))^2 + (f_t(s,t))^2}}(-f_s(s,t), -f_t(s,t), 1)\,.$$

The plane tangent to the surface at $\mathbf{S}(s,t)$ is

$$-f_s(s,t)\,(x-s) - f_t(s,t)\,(y-t) + (z - f(s,t)) = 0\,.$$

Example 9.5

The saddle surface $\mathbf{S}(s,t) = (s-t, s+t, s^2-t^2)$, for $(s,t) \in \mathbb{R}^2$, is the parametric surface illustrated in Figure 9.2. The curves drawn on the surface are a number of its parameter curves. Then, $\mathbf{S}_s(s,t) = (1,1,2s)$, $\mathbf{S}_t(s,t) = (-1,1,-2t)$, and

$$\mathbf{S}_s(s,t) \times \mathbf{S}_t(s,t) = (1,1,2s) \times (-1,1,-2t) = (-2(s+t), -2(s-t), 2) \ ,$$

$$|\mathbf{S}_s(s,t) \times \mathbf{S}_t(s,t)| = 2\sqrt{1+2t^2+2s^2} \ ,$$

$$\mathbf{N}(s,t) = \frac{1}{\sqrt{1+2t^2+2s^2}}(-s-t, -s+t, 1) \ .$$

The saddle surface can also be expressed in the implicit form $xy - z = 0$.

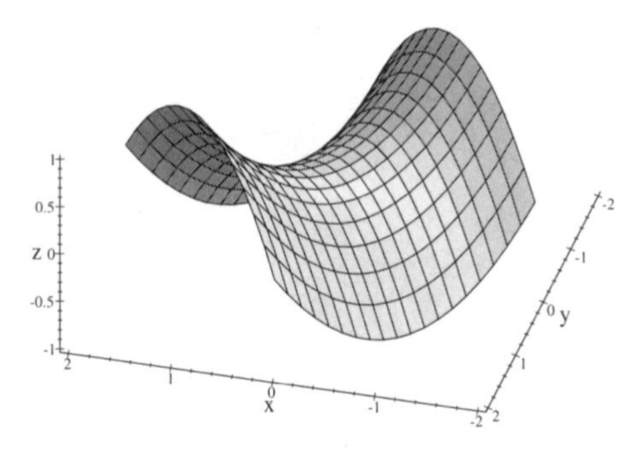

Figure 9.2 Saddle surface $(s-t, s+t, s^2-t^2)$

9.2 Quadric Surfaces

A *quadric* is an implicit surface defined by a quadratic polynomial

$$\begin{aligned}\mathbf{Q}(x,y,z) &= ax^2 + 2bxy + 2cxz + dy^2 + 2eyz + fz^2 \\ &\quad + 2gx + 2hy + 2jz + k = 0 \ ,\end{aligned} \tag{9.2}$$

for constants $a, b, c, d, e, f, g, h, j$, and k. All planar sections of a quadric are conics. Let $\mathbf{p} = (x, y, z, 1)$. The quadric surface (9.2) may be represented in the matrix form $\mathbf{Q}(x,y,z) = \mathbf{p}\mathbf{Q}\mathbf{p}^T = 0$,

$$\mathbf{Q}(x,y,z) = \begin{pmatrix} x & y & z & 1 \end{pmatrix} \begin{pmatrix} a & b & c & g \\ b & d & e & h \\ c & e & f & j \\ g & h & j & k \end{pmatrix} \begin{pmatrix} x \\ y \\ z \\ 1 \end{pmatrix} = 0 \ .$$

A point (x, y, z) of the quadric is singular if and only if $\mathbf{Q}(x, y, z) = 0$ and

$$\frac{\partial \mathbf{Q}}{\partial x}(x, y, z) = ax + by + cz + g = 0 , \tag{9.3}$$

$$\frac{\partial \mathbf{Q}}{\partial y}(x, y, z) = bx + dy + ez + h = 0 , \tag{9.4}$$

$$\frac{\partial \mathbf{Q}}{\partial z}(x, y, z) = cx + ey + fz + j = 0 . \tag{9.5}$$

Equation (9.2) can be expressed in the form

$$\begin{aligned} \mathbf{Q}(x, y, z) &= (ax + by + cz + g)\, x + (bx + dy + ez + h)\, y \\ &\quad + (cx + ey + fz + j)\, z + (gx + hy + jz + k) = 0 , \end{aligned}$$

and it follows from (9.3)–(9.5) that a singular point also satisfies

$$gx + hy + jz + k = 0 . \tag{9.6}$$

Thus a point of a quadric is singular if and only if Equations (9.3)–(9.6) are satisfied simultaneously, which occurs if and only if $\det(\mathsf{Q}) = 0$. A quadric is said to be *singular* whenever $\det(\mathsf{Q}) = 0$, and *non-singular* otherwise. Singular quadrics are cones, cylinders, or a union of planes. Quadrics which are a union of planes are called *reducible*, and those which are not are called *irreducible.*

The determinant

$$\Delta = \begin{vmatrix} a & b & c \\ b & d & e \\ c & e & f \end{vmatrix}$$

is called the *discriminant* of the quadric, and plays a similar role to the discriminant of a conic by distinguishing the types of quadric. A non-singular quadric is called a *paraboloid, hyperboloid,* or *ellipsoid* according to whether $\Delta = 0$, $\Delta > 0$, or $\Delta < 0$, respectively. The types are further distinguished as hyperboloids of one or two sheets, and hyperbolic and elliptic paraboloids.

Quadrics for which the axes of rotational symmetry or planes of reflectional symmetry are aligned with the axes are said to be in normal or standard form. Any quadric can be mapped to a quadric in normal form by applying three-dimensional rotations and translations. Space does not permit a detailed discussion of quadrics. Table 9.1 lists an implicit and a parametric normal form for each type of irreducible quadric, and the conditions on $D = \det \mathsf{Q}$ and Δ which determine the type. The quadrics are illustrated in the figures on page 231. A number of quadrics will emerge later in Bézier and B-spline form in the guise of surface constructs.

Techniques such as finding the intersection of a quadric with a line, applying transformations, and converting between parametric and implicit forms are similar to the corresponding methods for conics. The conversion problem

Table 9.1 Table of irreducible quadrics

Name	Implicit form	Parametric form
Ellipsoid $D \neq 0, \Delta < 0$	$\frac{x^2}{a^2} + \frac{y^2}{b^2} + \frac{z^2}{c^2} = 1$	$(a\cos\theta\sin\phi, b\sin\theta\sin\phi, c\cos\phi)$ $\theta \in [0, 2\pi], \phi \in [0, \pi]$
Hyperboloid (1 sheet) $D \neq 0, \Delta > 0$	$\frac{x^2}{a^2} + \frac{y^2}{b^2} - \frac{z^2}{c^2} = 1$	$(a\cos\theta\cosh t, b\sin\theta\cosh t, c\sinh t)$ $\theta \in [0, 2\pi], t \in (-\infty, \infty)$
Hyperboloid (2 sheets) $D \neq 0, \Delta > 0$	$\frac{x^2}{a^2} + \frac{y^2}{b^2} - \frac{z^2}{c^2} = -1$	$(a\cos\theta\sinh t, b\sin\theta\sinh t, \pm c\cosh t)$ $\theta \in [0, 2\pi], t \in (-\infty, \infty)$
Elliptic paraboloid $D \neq 0, \Delta = 0$	$z = \frac{x^2}{a^2} + \frac{y^2}{b^2}$	$(at\cos\theta, bt\sin\theta, t^2)$ $\theta \in [0, 2\pi], t \in (-\infty, \infty)$
Hyperbolic paraboloid $D \neq 0, \Delta = 0$	$z = -\frac{x^2}{a^2} + \frac{y^2}{b^2}$	$(at\cosh s, bt\sinh s, t^2)$ $s \in (-\infty, \infty), t \in (-\infty, \infty)$
Elliptic cone $D = 0$	$\frac{x^2}{a^2} + \frac{y^2}{b^2} - \frac{z^2}{c^2} = 0$	$(at\cos\theta, bt\sin\theta, ct)$ $\theta \in [0, 2\pi], t \in (-\infty, \infty)$
Elliptic cylinder $D = 0$	$\frac{x^2}{a^2} + \frac{y^2}{b^2} = 1$	$(a\cos\theta, b\sin\theta, t)$ $\theta \in [0, 2\pi], t \in (-\infty, \infty)$
Parabolic cylinder $D = 0$	$4ax - y^2 = 0$	$(as^2, 2as, t)$ $s \in (-\infty, \infty), t \in (-\infty, \infty)$

requires more space than is available in this text, so the reader is referred to [23] and [1]. The simpler problems are exemplified below.

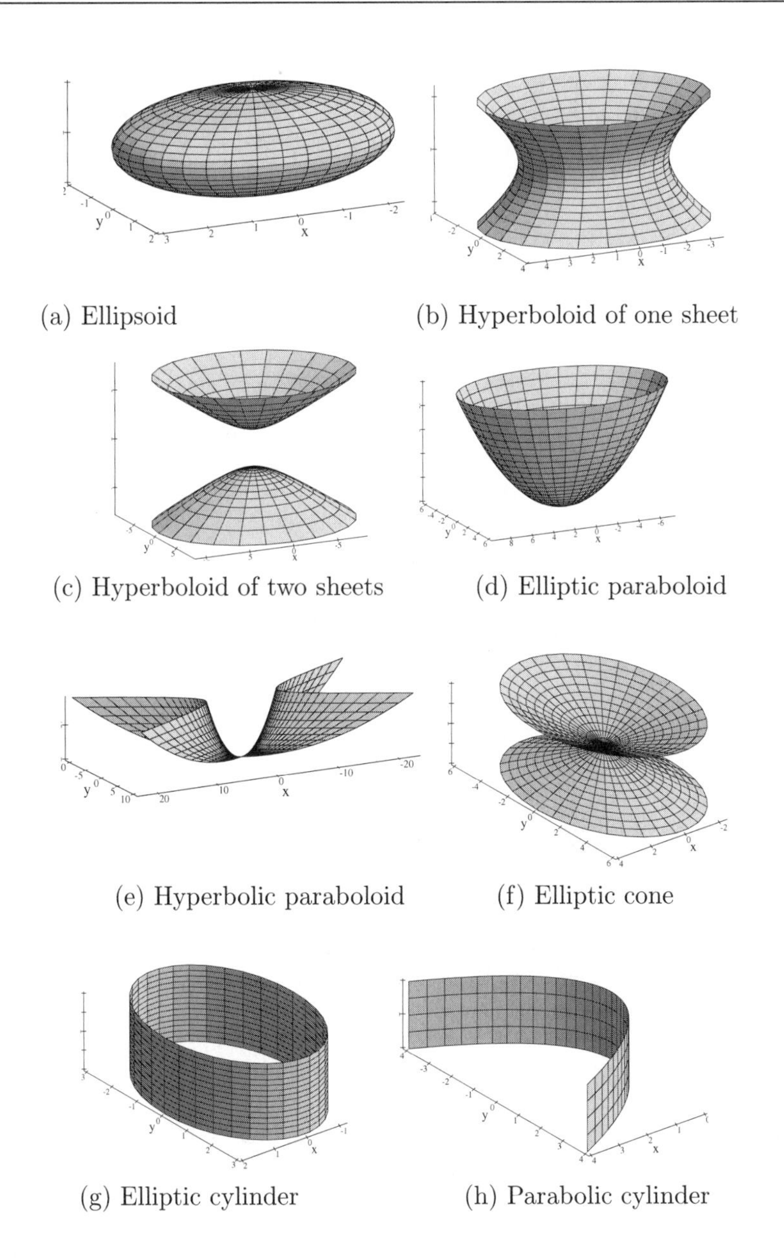

(a) Ellipsoid (b) Hyperboloid of one sheet

(c) Hyperboloid of two sheets (d) Elliptic paraboloid

(e) Hyperbolic paraboloid (f) Elliptic cone

(g) Elliptic cylinder (h) Parabolic cylinder

Example 9.6

The points of intersection of the hyperboloid $\frac{1}{4}x^2 + \frac{1}{9}y^2 - z^2 = -1$ and the line $(2t, 3t-2, t+3)$ may be obtained by substituting $x = 2t$, $y = 3t - 2$, $z = t + 3$ into the equation of the hyperboloid. This gives

$$\tfrac{1}{4}\left(2t\right)^2 + \tfrac{1}{9}\left(3t-2\right)^2 - \left(t+3\right)^2 = -1$$

which simplifies to $t^2 - \frac{22}{3}t - \frac{68}{9} = 0$. The solutions are $t = 8.2492$ and $t = -0.9159$. Substituting for t in $(2t, 3t-2, t+3)$ yields two points of intersection $(16.4984, 22.7476, 11.2492)$ and $(-1.8318, -4.7477, 2.0841)$.

Example 9.7

The parametric equation of the quadric obtained when a translation $\mathsf{T}(3,5,4)$, followed by a rotation $\mathsf{Rot}_z(\pi/2)$ about the z-axis, is applied to the elliptic cylinder $\mathbf{S}(s,t) = \big(as^2, 2as, t\big)$ is determined by

$$\begin{pmatrix} as^2 & 2as & t & 1 \end{pmatrix} \begin{pmatrix} 1 & 0 & 0 & 0 \\ 0 & 1 & 0 & 0 \\ 0 & 0 & 1 & 0 \\ 3 & 5 & 4 & 1 \end{pmatrix} \begin{pmatrix} \cos\frac{\pi}{2} & \sin\frac{\pi}{2} & 0 & 0 \\ -\sin\frac{\pi}{2} & \cos\frac{\pi}{2} & 0 & 0 \\ 0 & 0 & 1 & 0 \\ 0 & 0 & 0 & 1 \end{pmatrix}$$
$$= \begin{pmatrix} -2as-5 & as^2+3 & t+4 & 1 \end{pmatrix} .$$

The transformed quadric is $\big(-2as-5, as^2+3, t+4\big)$.

9.2.1 Offset Surfaces

Offset curves were introduced in Section 5.5 in the context of numerical controlled machining. Given a regular surface $\mathbf{S}(s,t) = (x(s,t), y(s,t), z(s,t))$ with unit normal $\mathbf{N}(s,t)$, the *offset surface* $\mathbf{O}_d(s,t)$ of $\mathbf{S}$ at a distance d is given by

$$\mathbf{O}_d(s,t) = \mathbf{S}(s,t) + d\,\mathbf{N}(s,t) .$$

Example 9.8

The offset at a distance d of the saddle surface of Example 9.5 is

$$\mathbf{O}_d(s,t) = (s-t, s+t, s^2-t^2) + \frac{d}{\sqrt{1+2s^2+2t^2}}(-s-t, -s+t, 1) ,$$

as shown in Figure 9.3.

Offset surfaces have several applications in CAD. First, offset surfaces are used to obtain paths for NC machining in a similar manner to curves. Second, two important CAD operations *thickening* and *shelling* are achieved by generating offset surfaces.

Shelling is a hollowing-out operation performed on a solid to give a new solid that has a thickness of d units. Figure 9.4(a) illustrates a solid bounded by two circular disks and half of a doughnut-shaped surface called a *torus* (see

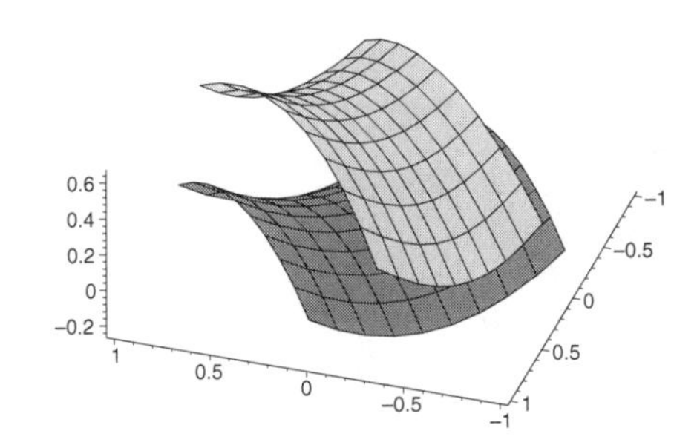

Figure 9.3 A surface and an offset

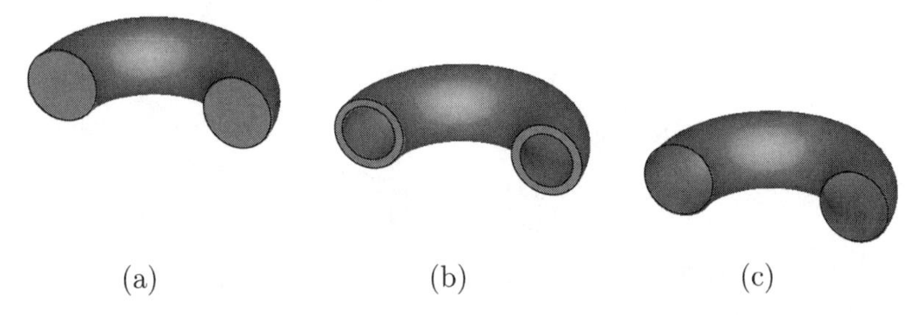

(a) (b) (c)

Figure 9.4 Shelling and thickening operations

Example 9.20). The solid is shelled to give the solid in Figure 9.4(b). The inner surface bounding the hollow of the solid is an offset surface at a distance d of the outer torus.

Thickening is the process of transforming a surface into a solid of thickness d units. Applying the thickening operation to the half-torus of Figure 9.4(c) results in a solid similar to the the one illustrated in Figure 9.4(b). Note again that the operation requires the computation of the offset of the torus.

Offset surfaces also arise in the construction of certain types of blend surfaces. Blending operations are applied to an object in order to smooth out sharp edges and vertices. In Figure 9.5(a) a *rolling-ball blend* smooths the neighbourhood of a sharp edge of a cube with a *pipe* or *canal* surface: that is, a tubular surface that is the locus of a spherical ball moving along a *spine curve.* The radius r of the ball determines the size of the blend. In Figure 9.5(c) a rolling-ball blend results in material being added to the original model shown in Figure 9.5(b). The ball is constrained to touch both surfaces during the motion as shown in Figure 9.6(a). This implies that the centre of the ball is a distance r from each surface. The spine curve is determined by computing the offset at a distance r to each of the two surfaces involved in the blend. The spine is the curve of intersection of the offset surfaces as shown in Figure 9.6(b). In Figure 9.5 the surfaces and their offsets are planes and so the spine is a line. The blend surface is obtained by rolling a ball along the line to give a cylindrical surface. A further example of a blend can be found in Example 9.20.

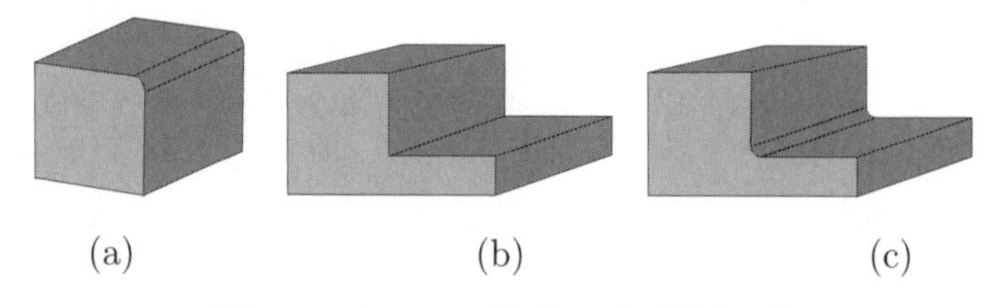

(a) (b) (c)

Figure 9.5 Rolling-ball blend

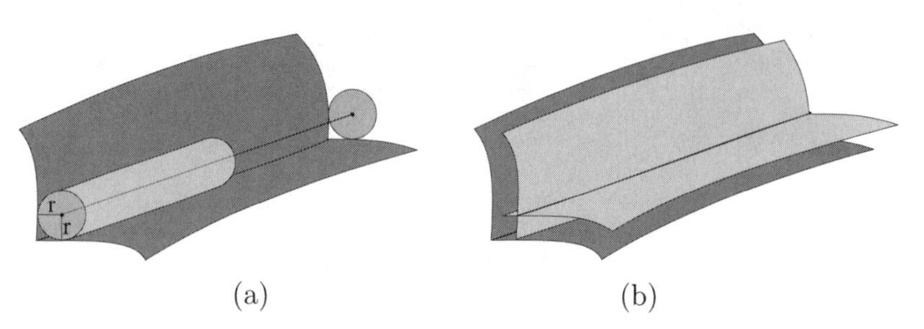

(a) (b)

Figure 9.6 Construction of the spine for a rolling-ball blend

9.3 Bézier and B-spline Surfaces

Let $B_{i,n}(s)$ and $B_{j,p}(t)$ be the Bernstein basis functions of degrees n and p in the variables s and t, respectively. A *Bézier surface* with control points $\mathbf{p}_{i,j}$ $(0 \le i \le n,\ 0 \le j \le p)$ is the parametric surface defined by

$$\mathbf{S}(s,t) = \sum_{i=0}^{n} \sum_{j=0}^{p} \mathbf{p}_{i,j} B_{i,n}(s) B_{j,p}(t), \text{ for } (s,t) \in [0,1] \times [0,1] . \tag{9.7}$$

The parameter curves of a Bézier surface are spatial Bézier curves. In particular, the parameter curves $\mathbf{S}(s,0)$, $\mathbf{S}(s,1)$, $\mathbf{S}(0,t)$, $\mathbf{S}(1,t)$, are Bézier curves which form the four edges of the Bézier surface as illustrated in Figure 9.7. A *rational Bézier surface* with control points $\mathbf{p}_{i,j}$ and weights $w_{i,j}$ $(0 \le i \le n,\ 0 \le j \le p)$ is defined by

$$\mathbf{S}(s,t) = \frac{\sum_{i=0}^{n} \sum_{j=0}^{p} w_{i,j} \mathbf{p}_{i,j} B_{i,n}(s) B_{j,p}(t)}{\sum_{i=0}^{n} \sum_{j=0}^{p} w_{i,j} B_{i,n}(s) B_{j,p}(t)}, \text{ for } (s,t) \in [0,1] \times [0,1] . \tag{9.8}$$

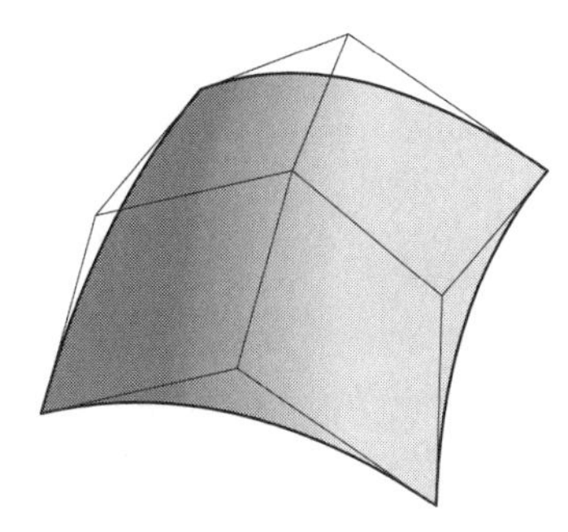

Figure 9.7 A Bézier surface and its control polyhedron

The parameter curves are rational Bézier curves. The $(n+1) \times (p+1)$ control points of a Bézier or rational Bézier surface form a *control point polyhedron.*

Let $N_{i,d}(s)$ be the B-spline basis functions of degree d with knot vector $s_0, s_1, \ldots, s_m$, and let $N_{j,e}(t)$ be the B-spline basis functions of degree e with knot vector $t_0, t_1, \ldots, t_q$. A *B-spline surface* with control points $\mathbf{p}_{i,j}$ $(0 \leq i \leq n = m - d - 1,\ 0 \leq j \leq p = q - e - 1)$ is defined by

$$\mathbf{S}(s,t) = \sum_{i=0}^{n} \sum_{j=0}^{p} \mathbf{p}_{i,j} N_{i,d}(s) N_{j,e}(t), \quad \text{for } (s,t) \in [s_d, s_{m-d}] \times [t_e, t_{q-e}]\,. \tag{9.9}$$

A *NURBS surface* with control points $\mathbf{p}_{i,j}$ and weights $w_{i,j}$ is defined by

$$\mathbf{S}(s,t) = \frac{\sum_{i=0}^{n} \sum_{j=0}^{p} w_{i,j} \mathbf{p}_{i,j} N_{i,d}(s) N_{j,e}(t)}{\sum_{i=0}^{n} \sum_{j=0}^{p} w_{i,j} N_{i,d}(s) N_{j,e}(t)}, \quad \text{for } (s,t) \in [s_d, s_{m-d}] \times [t_e, t_{q-e}]\,. \tag{9.10}$$

As for Bézier surfaces, the $(n+1) \times (p+1)$ control points of a B-spline or NURBS surface form a *control point polyhedron.* A B-spline surface is said to be *open* (respectively, *periodic, closed*) if the basis functions in both s and t are defined on *open* (respectively, *periodic, closed*) knot vectors.

Bézier or B-spline surfaces are said to be *bilinear, biquadratic, bicubic,* etc., whenever $n = p = 1$, $n = p = 2$, $n = p = 3$, etc.

9.3.1 Properties of Bézier and B-spline Surfaces

A number of the properties of Bézier and B-spline surfaces can be deduced in a similar manner to the corresponding properties for curves. The details are omitted.

Theorem 9.9

A Bézier surface (9.7) satisfies the following properties.

Endpoint Interpolation: $\mathbf{S}(0,0) = \mathbf{p}_{0,0}$, $\mathbf{S}(1,0) = \mathbf{p}_{n,0}$, $\mathbf{S}(0,1) = \mathbf{p}_{0,p}$, $\mathbf{S}(1,1) = \mathbf{p}_{n,p}$.

Convex Hull: $\mathbf{S}(s,t) \in \text{CH}\{\mathbf{p}_{0,0}, ..., \mathbf{p}_{n,p}\}$, for all $(s,t) \in [0,1] \times [0,1]$.

Invariance under Affine Transformations: Let T be a three-dimensional affine transformation. Then

$$\mathsf{T}\left(\sum_{i=0}^{n}\sum_{j=0}^{p} \mathbf{p}_{i,j} B_{i,n}(s) B_{j,p}(t)\right) = \sum_{i=0}^{n}\sum_{j=0}^{p} \mathsf{T}\left(\mathbf{p}_{i,j}\right) B_{i,n}(s) B_{j,p}(t) \, .$$

Theorem 9.10

A rational Bézier surface (9.8) satisfies the following properties.

Endpoint Interpolation: as for Theorem 9.9.

Convex Hull: If the weights are all positive, then as for Theorem 9.9.

Invariance under Affine Transformations: Let T be a three-dimensional affine transformation. Then

$$\mathsf{T}\left(\frac{\sum_{i=0}^{n}\sum_{j=0}^{p} w_{i,j}\mathbf{p}_{i,j} B_{i,n}(s) B_{j,p}(t)}{\sum_{i=0}^{n}\sum_{j=0}^{p} w_{i,j} B_{i,n}(s) B_{j,p}(t)}\right) = \frac{\sum_{i=0}^{n}\sum_{j=0}^{p} w_{i,j}\mathsf{T}\left(\mathbf{p}_{i,j}\right) B_{i,n}(s) B_{j,p}(t)}{\sum_{i=0}^{n}\sum_{j=0}^{p} w_{i,j} B_{i,n}(s) B_{j,p}(t)} \, .$$

Invariance under Projective Transformations:
Let T be a three- dimensional projective transformation, and let

$$\hat{\mathbf{p}}_{i,j} = (w_{i,j}x_{i,j}, w_{i,j}y_{i,j}, w_{i,j}z_{i,j}, w_{i,j})$$

be the homogeneous control points of $\mathbf{p}_{i,j} = (x_{i,j}, y_{i,j}, z_{i,j})$. Then

$$\mathsf{T}\left(\sum_{i=0}^{n}\sum_{j=0}^{p} \hat{\mathbf{p}}_{i,j} B_{i,n}(s) B_{j,p}(t)\right) = \sum_{i=0}^{n}\sum_{j=0}^{p} \mathsf{T}\left(\hat{\mathbf{p}}_{i,j}\right) B_{i,n}(s) B_{j,p}(t) \, .$$

Theorem 9.11

A B-spline surface (9.9) satisfies the following properties.

Local Control: Each segment is determined by a $(d+1) \times (e+1)$ mesh of control points. If $s \in [s_\sigma, s_{\sigma+1})$ and $t \in [t_\tau, t_{\tau+1})$ ($d \le \sigma \le m-d-1$, $e \le \tau \le n-e-1$), then

$$\mathbf{S}(s,t) = \sum_{i=\sigma-d}^{\sigma} \sum_{j=\tau-e}^{\tau} \mathbf{p}_{i,j} N_{i,d}(s) N_{j,e}(t), \text{ for } (s,t) \in [s_d, s_{m-d}] \times [t_e, t_{q-e}] .$$

Convex Hull: If $s \in [s_\sigma, s_{\sigma+1})$ and $t \in [t_\tau, t_{\tau+1})$ ($d \le \sigma \le m-d-1$, $e \le \tau \le n-e-1$), then $\mathbf{S}(s,t) \in \mathrm{CH}\{\mathbf{p}_{\sigma-d,\tau-e}, \ldots, \mathbf{p}_{\sigma,\tau}\}$.

Invariance under Affine Transformations: Let T be a three-dimensional affine transformation. Then

$$\mathsf{T}\left(\sum_{i=0}^{n}\sum_{j=0}^{p} \mathbf{p}_{i,j} N_{i,d}(s) N_{j,e}(t)\right) = \sum_{i=0}^{n}\sum_{j=0}^{p} \mathsf{T}(\mathbf{p}_{i,j}) N_{i,d}(s) N_{j,e}(t) .$$

Theorem 9.12

A NURBS surface (9.10) satisfies the following properties.

Local Control: If $s \in [s_\sigma, s_{\sigma+1})$ and $t \in [t_\tau, t_{\tau+1})$ ($d \le \sigma \le m-d-1$, $e \le \tau \le n-e-1$), then

$$\mathbf{S}(s,t) = \frac{\sum_{i=\sigma-d}^{\sigma} \sum_{j=\tau-e}^{\tau} w_{i,j} \mathbf{p}_{i,j} N_{i,d}(s) N_{j,e}(t)}{\sum_{i=\sigma-d}^{\sigma} \sum_{j=\tau-e}^{\tau} w_{i,j} N_{i,d}(s) N_{j,e}(t)} .$$

Convex Hull: If the weights w_i are all positive, then as for Theorem 9.11.

Invariance under Affine Transformations:
Let T be a three-dimensional affine transformation. Then

$$\mathsf{T}\left(\frac{\sum_{i=0}^{n} \sum_{j=0}^{p} w_{i,j} \mathbf{p}_{i,j} N_{i,d}(s) N_{j,e}(t)}{\sum_{i=0}^{n} \sum_{j=0}^{p} w_{i,j} N_{i,d}(s) N_{j,e}(t)}\right) = \frac{\sum_{i=0}^{n} \sum_{j=0}^{p} w_{i,j} \mathsf{T}(\mathbf{p}_{i,j}) N_{i,d}(s) N_{j,e}(t)}{\sum_{i=0}^{n} \sum_{j=0}^{p} w_{i,j} N_{i,d}(s) N_{j,e}(t)} .$$

Invariance under Projective Transformations:
Let T be a three-dimensional projective transformation, and let $\hat{\mathbf{p}}_{i,j} = (w_{i,j}x_{i,j}, w_{i,j}y_{i,j}, w_{i,j}z_{i,j}, w_{i,j})$ be the homogeneous control points of $\mathbf{p}_{i,j} = (x_{i,j}, y_{i,j}, z_{i,j})$. Then

$$\mathsf{T}\left(\sum_{i=0}^{n}\sum_{j=0}^{p} \hat{\mathbf{p}}_{i,j} N_{i,d}(s) N_{j,e}(t)\right) = \sum_{i=0}^{n}\sum_{j=0}^{p} \mathsf{T}(\hat{\mathbf{p}}_{i,j}) N_{i,d}(s) N_{j,e}(t) .$$

9.3.2 Derivatives of Bézier and B-spline Surfaces

The partial derivatives $\mathbf{S}_s(s,t)$ and $\mathbf{S}_t(s,t)$ of a Bézier surface (9.7) are obtained from the derivative formulae for Bézier curves expressed in Theorem 7.3. Then

$$\mathbf{S}(s,t) = \sum_{i=0}^{n}\sum_{j=0}^{p} \mathbf{p}_{i,j} B_{i,n}(s) B_{j,p}(t) = \sum_{j=0}^{p}\left(\sum_{i=0}^{n} \mathbf{p}_{i,j} B_{i,n}(s)\right) B_{j,p}(t) ,$$

and differentiation of the term within the bracket with respect to s gives

$$\begin{aligned}\mathbf{S}_s(s,t) &= \sum_{j=0}^{p}\left(n\sum_{i=0}^{n-1}\left(\mathbf{p}_{i+1,j}-\mathbf{p}_{i,j}\right) B_{i,n-1}(s)\right) B_{j,p}(t) \\ &= \sum_{i=0}^{n-1}\sum_{j=0}^{p} \mathbf{p}_{i,j}^{(1,0)} B_{i,n-1}(s) B_{j,p}(t), \qquad (9.11)\end{aligned}$$

where $\mathbf{p}_{i,j}^{(1,0)} = n\left(\mathbf{p}_{i+1,j}-\mathbf{p}_{i,j}\right)$. Likewise, letting $\mathbf{p}_{i,j}^{(0,1)} = p\left(\mathbf{p}_{i,j+1}-\mathbf{p}_{i,j}\right)$,

$$\mathbf{S}_t(s,t) = \sum_{i=0}^{n}\sum_{j=0}^{p-1} \mathbf{p}_{i,j}^{(0,1)} B_{i,n}(s) B_{j,p-1}(t) . \qquad (9.12)$$

Example 9.13

The partial derivative with respect to s of the biquadratic Bézier surface ($n = 2$, $p = 2$) with control points $\mathbf{p}_{0,0}(7,-3,-5)$, $\mathbf{p}_{0,1}(7,-2,-6)$, $\mathbf{p}_{0,2}(8,-1,-4)$, $\mathbf{p}_{1,0}(4,-3,-2)$, $\mathbf{p}_{1,1}(5,-1,-4)$, $\mathbf{p}_{1,2}(4,0,-3)$, $\mathbf{p}_{2,0}(1,-4,1)$, $\mathbf{p}_{2,1}(0,-2,0)$, and $\mathbf{p}_{2,2}(1,-3,1)$ has a Bézier representation ($n = 1, p = 2$), with control points

$$\begin{aligned}\mathbf{p}_{0,0}^{(1,0)} &= n\left(\mathbf{p}_{1,0}-\mathbf{p}_{0,0}\right) = 2\left((4,-3,-2)-(7,-3,-5)\right) = (-6,0,6) , \\ \mathbf{p}_{1,0}^{(1,0)} &= n\left(\mathbf{p}_{2,0}-\mathbf{p}_{1,0}\right) = 2\left((1,-4,1)-(4,-3,-2)\right) = (-6,-2,6) , \\ \mathbf{p}_{0,1}^{(1,0)} &= n\left(\mathbf{p}_{1,1}-\mathbf{p}_{0,1}\right) = 2\left((5,-1,-4)-(7,-2,-6)\right) = (-4,2,4) , \\ \mathbf{p}_{1,1}^{(1,0)} &= n\left(\mathbf{p}_{2,1}-\mathbf{p}_{1,1}\right) = 2\left((0,-2,0)-(5,-1,-4)\right) = (-10,-2,8) , \\ \mathbf{p}_{0,2}^{(1,0)} &= n\left(\mathbf{p}_{1,2}-\mathbf{p}_{0,2}\right) = 2\left((4,0,-3)-(8,-1,-4)\right) = (-8,2,2) , \\ \mathbf{p}_{1,2}^{(1,0)} &= n\left(\mathbf{p}_{2,2}-\mathbf{p}_{1,2}\right) = 2\left((1,-3,1)-(4,0,-3)\right) = (-6,-6,8) .\end{aligned}$$

Higher order partial derivatives $\mathbf{S}^{(\alpha,\beta)}(s,t) = \frac{\partial^{\alpha+\beta}}{\partial s^{\alpha}\partial t^{\beta}}\mathbf{S}(s,t)$ (the notation means αth derivative with respect to s, and βth derivative with respect to t) are obtained by repeated applications of (9.11) and (9.12) to give

$$\mathbf{S}^{(\alpha,\beta)}(s,t) = \sum_{i=0}^{n-\alpha}\sum_{j=0}^{p-\beta} \mathbf{p}_{i,j}^{(\alpha,\beta)} B_{i,n-\alpha}(s) B_{j,p-\beta}(t) ,$$

where

$$\mathbf{p}_{i,j}^{(\alpha,\beta)} = \frac{n!}{(n-\alpha)!}\frac{p!}{(p-\beta)!}\sum_{k=0}^{\alpha}\sum_{\ell=0}^{\beta}(-1)^k(-1)^\ell\binom{\alpha}{k}\binom{\beta}{\ell}\mathbf{p}_{i+\alpha-k,j+\beta-\ell}\,.$$

The derivative of a B-spline curve $\mathbf{B}(s) = \sum_{i=0}^{n}\mathbf{b}_i N_{i,d}(s)$, defined on a knot vector $s_0, \ldots, s_m$, was determined in Theorem 8.16 to be

$$\mathbf{B}'(s) = \sum_{i=0}^{n-1}\mathbf{b}_i^{(1)}N_{i,d-1}^{(1)}(s), \tag{9.13}$$

where $\mathbf{b}_i^{(1)} = d\,(\mathbf{b}_{i+1} - \mathbf{b}_i)\,/\,(t_{i+d+1} - t_{i+1})$ and $N_{i,d-1}^{(1)}(s)$ are the degree $d-1$ basis functions defined on the knot vector $s_1, \ldots, s_{m-1}$. Following the method of the derivative of a Bézier surface, the derivative with respect to s of a B-spline surface (9.9) is

$$\mathbf{S}_s(s,t) = \sum_{j=0}^{p} d\left(\sum_{i=0}^{n-1}\mathbf{p}_{i,j}^{(1)}N_{i,d-1}^{(1)}(s)\right)N_{j,e}(t) = \sum_{i=0}^{n-1}\sum_{j=0}^{p}\mathbf{p}_{i,j}^{(1,0)}N_{i,d-1}^{(1)}(s)N_{j,e}(t)$$

where

$$\mathbf{p}_{i,j}^{(1,0)} = d\frac{\mathbf{p}_{i+1,j} - \mathbf{p}_{i,j}}{s_{i+d+1} - s_{i+1}}\,,$$

and $N_{i,d-1}^{(1)}(s)$ are the degree $d-1$ basis functions defined on the knot vector $s_1, \ldots, s_{m-1}$. Likewise,

$$\mathbf{S}_t(s,t) = \sum_{i=0}^{n}\sum_{j=0}^{p-1}\mathbf{p}_{i,j}^{(0,1)}N_{i,d}(s)N_{j,e-1}^{(1)}(t)$$

where

$$\mathbf{p}_{i,j}^{(0,1)} = p\frac{\mathbf{p}_{i,j+1} - \mathbf{p}_{i,j}}{t_{j+e+1} - t_{j+1}}$$

and $N_{j,e-1}^{(1)}(t)$ are the degree $e-1$ basis functions defined on the knot vector $t_1, \ldots, t_{q-1}$.

Remark 9.14

Computation of the derivatives of rational Bézier and NURBS surfaces can be performed by combining the above formulae for integral Bézier and B-spline surfaces with the procedure for computing the derivatives of rational functions given in Section 7.5.4.

EXERCISES

9.1. Verify that the biquadratic Bézier surface with control points

$$\begin{array}{lll} \mathbf{p}_{0,0}(0,0,0), & \mathbf{p}_{0,1}(0,1/2,0), & \mathbf{p}_{0,2}(0,1,1/b), \\ \mathbf{p}_{1,0}(1/2,0,0), & \mathbf{p}_{1,1}(1/2,1/2,0), & \mathbf{p}_{1,2}(1/2,1,1/b), \\ \mathbf{p}_{2,0}(1,0,1/a), & \mathbf{p}_{2,1}(1,1/2,1/a), & \mathbf{p}_{2,2}(1,1,1/a,1/b) \end{array}$$

for non-zero constants a and b yields the quadratic surface $\mathbf{S}(s,t) = \left(s, t, \frac{1}{a}s^2 + \frac{1}{b}t^2\right)$. When a and b have the same sign the surface is an elliptic paraboloid, and when a and b have opposite signs the surface is a hyperbolic paraboloid.

9.2. Determine the control points of the first order partial derivatives with respect to s and t of the biquadratic Bézier surface with control points $\mathbf{p}_{0,0}(2,2,0)$, $\mathbf{p}_{0,1}(2,4,1)$, $\mathbf{p}_{0,2}(2,6,0)$, $\mathbf{p}_{1,0}(4,3,1)$, $\mathbf{p}_{1,1}(4,5,3)$, $\mathbf{p}_{1,2}(4,6,1)$, $\mathbf{p}_{2,0}(6,2,0)$, $\mathbf{p}_{2,1}(6,3,1)$, and $\mathbf{p}_{2,2}(6,5,0)$.

9.3. Determine the control points of the first order partial derivatives with respect to s and t of the biquadratic B-spline surface with control points

$$\begin{array}{llll} \mathbf{p}_{0,0}(1,2,1), & \mathbf{p}_{0,1}(0,4,3), & \mathbf{p}_{0,2}(1,6,2), & \mathbf{p}_{0,3}(2,9,1), \\ \mathbf{p}_{1,0}(4,1,1), & \mathbf{p}_{1,1}(4,4,5), & \mathbf{p}_{1,2}(3,6,3), & \mathbf{p}_{1,3}(3,8,1), \\ \mathbf{p}_{2,0}(6,2,0), & \mathbf{p}_{2,1}(6,5,3), & \mathbf{p}_{2,2}(7,7,2), & \mathbf{p}_{2,3}(6,8,0), \end{array}$$

weights $w_{0,0} = w_{0,1} = w_{0,2} = w_{0,3} = w_{2,0} = w_{2,1} = w_{2,2} = w_{2,3} = 1$, $w_{1,0} = w_{1,1} = w_{1,2} = w_{1,3} = 2$, knot vector $0, 1, 2, 3, 4, 5$ in the s-direction and $0, 2, 4, 6, 8, 10, 12$ in the t-direction.

9.4. (a) Express, in terms of the control points, the tangent vectors to the parameter curves at the endpoints of a Bézier surface of degree (n,p).

(b) The *endpoint normal vectors* of a Bézier surface are the normal vectors of the surface at its endpoints $\mathbf{S}(0,0)$, $\mathbf{S}(0,1)$, $\mathbf{S}(1,0)$, $\mathbf{S}(1,1)$. Express, in terms of the control points, the endpoint normal vectors of a Bézier surface of degree (n,p). (Use the fact that the normal to a surface at a point is perpendicular to the tangent directions of the parameter curves through that point.)

9.4 Surface Constructions

Consider a non-singular three-dimensional affine transformation $\mathsf{T}(t)$ depending continuously on a parameter t, that is, the entries of the transformation matrix are continuous functions of t. A surface construction is obtained by applying $\mathsf{T}(t)$ to a specified *generating* curve so that, as the parameter t varies, the curve moves through space and thereby "sweeps" out a surface. In general, a surface constructed in this manner is twisted and self-intersecting, and serves no practical purpose. However, particular choices of the generating curve and transformation do give rise to useful surface shapes. The following sections describe a number of constructions which have Bézier or B-spline representations. Such constructions are fundamental to many CAD systems.

9.4.1 Extruded Surfaces

An *extruded surface* is obtained when a spatial generating curve $\mathbf{B}(s)$ is translated in the direction of a trajectory line. A generating NURBS curve $\mathbf{B}(s) = \sum_{i=0}^{n} \mathbf{b}_i u_i N_{i,d}(s) / \sum_{i=0}^{n} u_i N_{i,d}(s)$ (knot vector $s_0, \ldots, s_m$) sweeping in the direction of the unit vector $\mathbf{n}$, through a distance δ, results in the extruded NURBS surface

$$\mathbf{S}(s,t) = \sum_{i=0}^{n} \sum_{j=0}^{1} w_{i,j} \mathbf{p}_{i,j} N_{i,d}(s) N_{j,1}(t) \Big/ \sum_{i=0}^{n} \sum_{j=0}^{1} w_{i,j} N_{i,d}(s) N_{j,1}(t) \, ,$$

with knot vector $s_0, \ldots, s_m$ in the s-direction, $t_0 = 0$, $t_1 = 0$, $t_2 = 1$, $t_3 = 1$ in the t-direction, control points $\mathbf{p}_{i,0} = \mathbf{b}_i$ and $\mathbf{p}_{i,1} = \mathbf{b}_i + \delta \mathbf{n}$, and weights $w_{i,0} = w_{i,1} = u_i$ (for $i = 1, \ldots, n$). Similar representations can be obtained for the extruded surface of a Bézier, rational Bézier, or B-spline generating curve.

Example 9.15

Let $\mathbf{B}(s)$ be the quadratic NURBS with control points $\mathbf{b}_0(0,0,0)$, $\mathbf{b}_1(1,0,0)$, $\mathbf{b}_2(1,0,1)$, $\mathbf{b}_3(1,1,1)$, weights $u_0 = 1$, $u_1 = 2$, $u_2 = 2$, $u_3 = 1$, and knot vector $s_0 = 0$, $s_1 = 0$, $s_2 = 0$, $s_3 = 1$, $s_4 = 2$, $s_5 = 2$, $s_6 = 2$. Let $\mathbf{n} = \left(\frac{2}{3}, -\frac{2}{3}, \frac{1}{3}\right)$, and $\delta = 3$. Then the extruded surface has control points $\mathbf{p}_{0,0}(0,0,0)$, $\mathbf{p}_{1,0}(1,0,0)$, $\mathbf{p}_{2,0}(1,0,1)$, $\mathbf{p}_{3,0}(1,1,1)$, $\mathbf{p}_{0,1}(2,-2,1)$, $\mathbf{p}_{1,1}(3,-2,1)$, $\mathbf{p}_{2,1}(3,-2,2)$, $\mathbf{p}_{3,1}(3,-1,2)$, weights $w_{0,0} = 1$, $w_{1,0} = 2$, $w_{2,0} = 2$, $w_{3,0} = 1$, $w_{0,1} = 1$, $w_{1,1} = 2$, $w_{2,1} = 2$, $w_{3,1} = 1$, and knots $s_0 = 0$, $s_1 = 0$, $s_2 = 0$, $s_3 = 1$, $s_4 = 2$, $s_5 = 2$, $s_6 = 2$, $t_0 = 0$, $t_1 = 0$, $t_2 = 1$, $t_3 = 1$. The surface is illustrated in Figure 9.8.

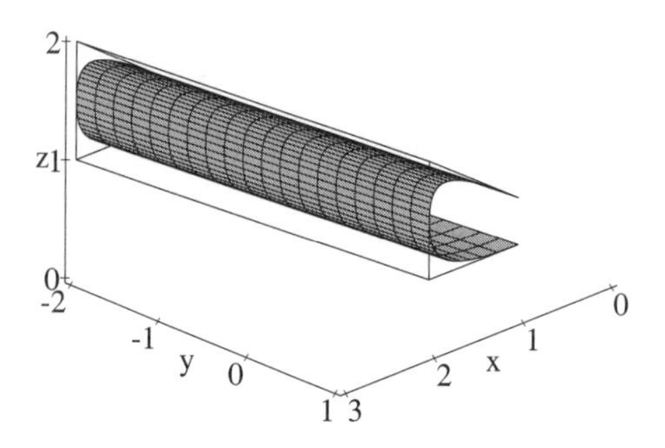

Figure 9.8 A NURBS extruded surface

9.4.2 Ruled Surfaces

A *ruled surface* is formed from two spatial curves $\mathbf{B}(s)$ and $\mathbf{C}(s)$ when points on each curve corresponding to the parameter s are joined by a line. Consider two NURBS curves $\mathbf{B}(s) = \sum_{i=0}^{n} \mathbf{b}_i u_i N_{i,d}(s) / \sum_{i=0}^{n} u_i N_{i,d}(s)$, and $\mathbf{C}(s) = \sum_{j=0}^{n} \mathbf{c}_j v_j N_{j,d}(s) \Big/ \sum_{j=0}^{n} v_j N_{j,d}(s)$. The curves are assumed to have the same degree and to be defined on the knot vector $s_0, \ldots, s_m$. The constructed NURBS ruled surface, linear in the t-direction, is given by

$$\mathbf{S}(s,t) = \sum_{i=0}^{n} \sum_{j=0}^{1} w_{i,j} \mathbf{p}_{i,j} N_{i,d}(s) N_{j,1}(t) \Bigg/ \sum_{i=0}^{n} \sum_{j=0}^{1} w_{i,j} N_{i,d}(s) N_{j,1}(t) . \quad (9.14)$$

The surface has knot vector $s_0, \ldots, s_m$ in the s-direction, and $0, 0, 1, 1$ in the t-direction. The control points are $\mathbf{p}_{i,0} = \mathbf{b}_i$ and $\mathbf{p}_{i,1} = \mathbf{c}_i$, and the weights are $w_{i,0} = u_i$, $w_{i,1} = v_i$ $(i = 0, \ldots, n)$. Clearly, an extruded surface is a special case of a ruled surface. If the specified curves do not have the same degree then it is necessary to apply a "degree raising algorithm" before the above procedure can be applied (see Exercise 6.21). If the curves have different knot vectors then a knot insertion algorithm (see Section 8.3) can be applied to obtain curves defined on identical knot vectors.

Example 9.16

Let $\mathbf{B}(s)$ be the quadratic NURBS curve with control points $\mathbf{b}_0(0,0,0)$, $\mathbf{b}_1(3,0,0)$, $\mathbf{b}_2(3,3,0)$, $\mathbf{b}_3(0,3,0)$ and weights $w_0 = 1$, $w_1 = 2$, $w_2 = 2$, $w_3 = 1$, and let $\mathbf{C}(s)$ have control points $\mathbf{c}_0(0,1,4)$, $\mathbf{c}_1(1,1,4)$, $\mathbf{c}_2(1,2,4)$, $\mathbf{c}_3(0,2,4)$ and weights $w_0 = 2$, $w_1 = 3$, $w_2 = 3$, $w_3 = 2$. Both curves are assumed to be defined on the knot vector $0, 1, 2, 3, 4, 5, 6$. Then the ruled surface is given by (9.14) with control points $\mathbf{p}_{0,0}(0,0,0)$, $\mathbf{p}_{1,0}(3,0,0)$, $\mathbf{p}_{2,0}(3,3,0)$, $\mathbf{p}_{3,0}(0,3,0)$, $\mathbf{p}_{0,1}(0,1,4)$, $\mathbf{p}_{1,1}(1,1,4)$, $\mathbf{p}_{2,1}(1,2,4)$, $\mathbf{p}_{3,1}(0,2,4)$, weights $w_{0,0} = 1$, $w_{1,0} = 2$,

$w_{2,0} = 2$, $w_{3,0} = 1$, $w_{0,1} = 2$, $w_{1,1} = 3$, $w_{2,1} = 3$, $w_{3,1} = 2$, and knots $s_0 = 0$, $s_1 = 1$, $s_2 = 2$, $s_3 = 3$, $s_4 = 4$, $s_5 = 5$, $t_0 = 0$, $t_1 = 0$, $t_2 = 1$, $t_3 = 1$. The surface is illustrated in Figure 9.9.

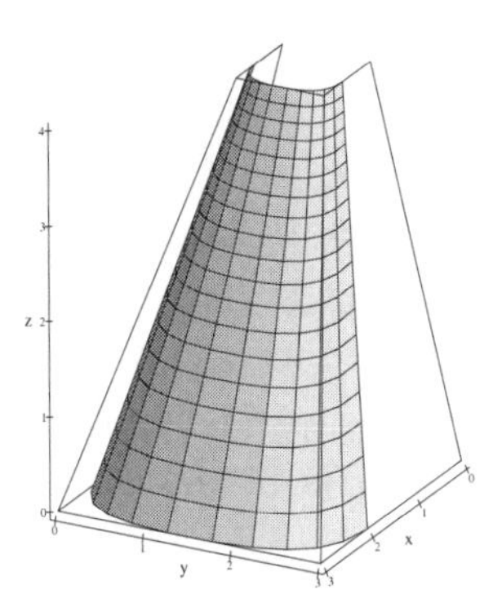

Figure 9.9 A NURBS ruled surface

Example 9.17

Given four points $\mathbf{p}_{0,0}$, $\mathbf{p}_{1,0}$, $\mathbf{p}_{0,1}$, $\mathbf{p}_{1,1}$, the ruled surface defined by the line segment joining $\mathbf{p}_{0,0}$ and $\mathbf{p}_{1,0}$, and the line segment joining $\mathbf{p}_{0,1}$ and $\mathbf{p}_{1,1}$ is a *bilinear surface.* In B-spline form the surface is

$$\begin{aligned}\mathbf{S}(s,t) &= \sum_{i=0}^{1}\sum_{j=0}^{1} \mathbf{p}_{i,j} N_{i,1}(s) N_{j,1}(t) \\ &= \mathbf{p}_{0,0}(1-s)(1-t) + \mathbf{p}_{0,1}(1-s)t + \mathbf{p}_{1,0}s(1-t) + \mathbf{p}_{1,1}st\ ,\end{aligned}$$

with knot vector $0, 0, 1, 1$ in both directions. The surface can be obtained in Bézier form by replacing the B-spline basis functions by the linear Bernstein basis functions. If the four points are coplanar then the surface is a planar quadrilateral region. A non-planar example is defined by control points $\mathbf{p}_{0,0}(0,0,0)$, $\mathbf{p}_{0,1}(0,1,1)$, $\mathbf{p}_{1,0}(1,0,1)$, $\mathbf{p}_{1,1}(1,1,0)$, which gives

$$\begin{aligned}\mathbf{S}(s,t) &= (0,0,0)(1-s)(1-t) + (0,1,1)(1-s)t \\ &\quad + (1,0,1)s(1-t) + (1,1,0)st \\ &= (s, t, s - 2st + t)\ .\end{aligned}$$

The surface is the hyperbolic paraboloid defined implicitly by $x - 2xy + y - z = 0$.

9.4.3 Translationally Swept Surfaces

The extruded surface construction can be generalized to give a *translationally swept surface* obtained by translating a generating curve $\mathbf{B}(s)$ along a trajectory curve $\mathbf{C}(t)$. When $\mathbf{B}(s)$ and $\mathbf{C}(t)$ are both Bézier, rational Bézier, B-spline, or NURBS curves, then correspondingly, the resulting translational swept surface is a Bézier, rational Bézier, B-spline or NURBS surface. For instance, let $\mathbf{B}(s) = \sum_{i=0}^{n} \mathbf{b}_i u_i N_{i,d}(s) / \sum_{i=0}^{n} u_i N_{i,d}(s)$ (knot vector $s_0, \ldots, s_m$), and $\mathbf{C}(t) = \sum_{j=0}^{p} \mathbf{c}_j v_j N_{j,e}(t) \Big/ \sum_{j=0}^{p} v_j N_{j,e}(t)$ (knot vector $t_0, \ldots, t_q$). The NURBS swept surface constructed from $\mathbf{B}(s)$ and $\mathbf{C}(t)$ is

$$\mathbf{S}(s,t) = \sum_{i=0}^{n} \sum_{j=0}^{p} w_{i,j} \mathbf{p}_{i,j} N_{i,d}(s) N_{j,e}(t) \Big/ \sum_{i=0}^{n} \sum_{j=0}^{p} w_{i,j} N_{i,d}(s) N_{j,e}(t) ,$$

with control points $\mathbf{p}_{i,j} = \mathbf{b}_i + \mathbf{c}_j$, weights $w_{i,j} = u_i v_j$, and knot vectors $s_0, \ldots, s_m$ and $t_0, \ldots, t_q$ in the s- and t-directions.

Example 9.18

Let $\mathbf{B}(s)$ be a quadratic B-spline with control points $\mathbf{b}_0(0,0,0)$, $\mathbf{b}_1(5,0,0)$, $\mathbf{b}_2(5,5,0)$, $\mathbf{b}_3(0,5,0)$, weights $u_0 = 1$, $u_1 = 2$, $u_2 = 2$, $u_3 = 1$, and knot vector $s_0 = 0$, $s_1 = 0$, $s_2 = 0$, $s_3 = 1$, $s_4 = 2$, $s_5 = 2$, $s_6 = 2$. Let $\mathbf{C}(t)$ be the cubic B-spline control points $\mathbf{c}_0(1,0,2)$, $\mathbf{c}_1(2,4,4)$, $\mathbf{c}_2(2,7,6)$, $\mathbf{c}_3(1,2,8)$, weights $v_0 = 2$, $v_1 = 3$, $v_2 = 4$, $v_3 = 1$, and knot vector $t_0 = 0$, $t_1 = 0$, $t_2 = 0$, $t_3 = 0$, $t_4 = 1$, $t_5 = 1$, $t_6 = 1, t_7 = 1$. The control points of the translational swept surface are

$$\begin{aligned}
\mathbf{p}_{0,0} &= \mathbf{b}_0 + \mathbf{c}_0 = (0,0,0) + (1,0,2) = (1,0,2), \\
\mathbf{p}_{1,0} &= \mathbf{b}_1 + \mathbf{c}_0 = (5,0,0) + (1,0,2) = (6,0,2), \\
\mathbf{p}_{2,0} &= \mathbf{b}_2 + \mathbf{c}_0 = (5,5,0) + (1,0,2) = (6,5,2), \\
\mathbf{p}_{3,0} &= \mathbf{b}_3 + \mathbf{c}_0 = (0,5,0) + (1,0,2) = (1,5,2), \\
\mathbf{p}_{0,1} &= \mathbf{b}_0 + \mathbf{c}_1 = (0,0,0) + (2,4,4) = (2,4,4),
\end{aligned}$$

and likewise $\mathbf{p}_{1,1} = (7,4,4)$, $\mathbf{p}_{2,1} = (7,9,4)$, $\mathbf{p}_{3,1} = (2,9,4)$, $\mathbf{p}_{0,2} = (2,7,6)$, $\mathbf{p}_{1,2} = (7,7,6)$, $\mathbf{p}_{2,2} = (7,12,6)$, $\mathbf{p}_{3,2} = (2,12,6)$, $\mathbf{p}_{0,3} = (1,2,8)$, $\mathbf{p}_{1,3} = (6,2,8)$, $\mathbf{p}_{2,3} = (6,7,8)$, $\mathbf{p}_{3,3} = (1,7,8)$. The weights are $w_{0,0} = u_0 v_0 = 2$, $w_{1,0} = u_1 v_0 = 4$, $w_{2,0} = 4$, $w_{3,0} = 2$, $w_{0,1} = 3$, $w_{1,1} = 6$, $w_{2,1} = 6$, $w_{3,1} = 3$, $w_{0,2} = 4$, $w_{1,2} = 8$, $w_{2,2} = 8$, $w_{3,2} = 4$, $w_{0,3} = 1$, $w_{1,3} = 2$, $w_{2,3} = 2$, $w_{3,3} = 1$. The surface is illustrated in Figure 9.10.

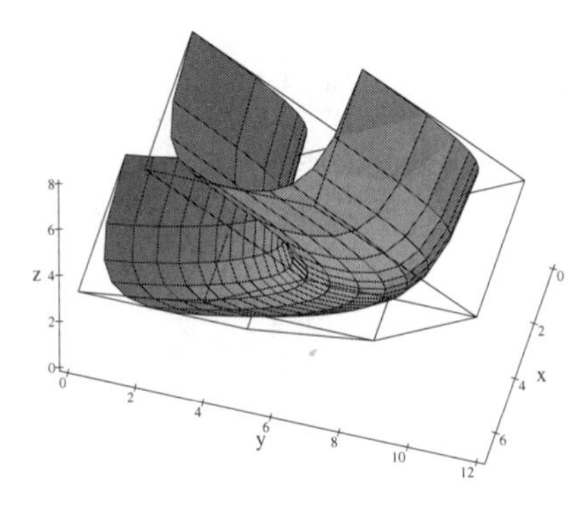

Figure 9.10 A NURBS translational swept surface

9.4.4 Surfaces of Revolution

A surface obtained by rotating a generating curve $\mathbf{B}(s)$ about a fixed axis is called a *surface of revolution.* It is assumed that the curve lies in a plane containing the axis and, to avoid self-intersections of the surface, that the axis does not intersect the curve. Let $\mathbf{B}(s) = \sum_{i=0}^{n} \mathbf{b}_i u_i N_{i,d}(s) / \sum_{i=0}^{n} u_i N_{i,d}(s)$ (knot vector $s_0, \ldots, s_m$) be a generating NURBS curve in the xz-plane. Rotating $\mathbf{B}(s)$ about the z-axis results in a NURBS surface of revolution $\mathbf{S}(s,t)$. The points in the ith row of the control polyhedron of $\mathbf{S}$ lie in a plane perpendicular to the z-axis, and consist of a copy of the control polygon of a NURBS circle scaled by a factor f_i and translated by d_i units in the z-direction. The NURBS circle (Example 8.21) has control points $(1,0)$, $(1,1)$, $(-1,1)$, $(-1,0)$, $(-1,-1)$, $(1,-1)$, $(1,0)$, weights $1, \frac{1}{2}, \frac{1}{2}, 1, \frac{1}{2}, \frac{1}{2}, 1$, and knot vector $0, 0, 0, \frac{1}{4}, \frac{1}{2}, \frac{1}{2}, \frac{3}{4}, 1, 1, 1$. These control points are expressed as three-dimensional coordinates in the $z = 0$ plane: $\mathbf{c}_0(1,0,0)$, $\mathbf{c}_1(1,1,0)$, $\mathbf{c}_2(-1,1,0)$, $\mathbf{c}_3(-1,0,0)$, $\mathbf{c}_4(-1,-1,0)$, $\mathbf{c}_5(1,-1,0)$, $\mathbf{c}_6(1,0,0)$. The scale factor f_i is equal to the distance of $\mathbf{b}_i$ from the z-axis (equal to the x-coordinate of $\mathbf{b}_i$), and the distance d_i of translation is the distance of $\mathbf{b}_i$ from the x-axis (equal to the z-coordinate of $\mathbf{b}_i$). The knots in the s-direction are $s_0, \ldots, s_m$, and the knots in the t-direction are inherited from the NURBS circle. The rows of weights $w_{i,j} = \{u_i, \frac{1}{2}u_i, \frac{1}{2}u_i, u_i, \frac{1}{2}u_i, \frac{1}{2}u_i, u_i\}$ for $i = 0, \ldots, n$, are the weights of the NURBS circle scaled by a factor u_j.

Example 9.19

Let $\mathbf{B}(s)$ be a NURBS curve of degree $d = 3$ with control points $\mathbf{b}_0(2,0,1)$, $\mathbf{b}_1(1,0,2)$, $\mathbf{b}_2(3,0,3)$, $\mathbf{b}_3(1,0,4)$, $\mathbf{b}_4(1,0,5)$, weights $u_0 = 1$, $u_1 = 2$, $u_2 = 3$, $u_3 = 4$, $u_4 = 2$, and knots $0, 0, 0, 0, 1, 2, 2, 2, 2$. The surface of revolution is given

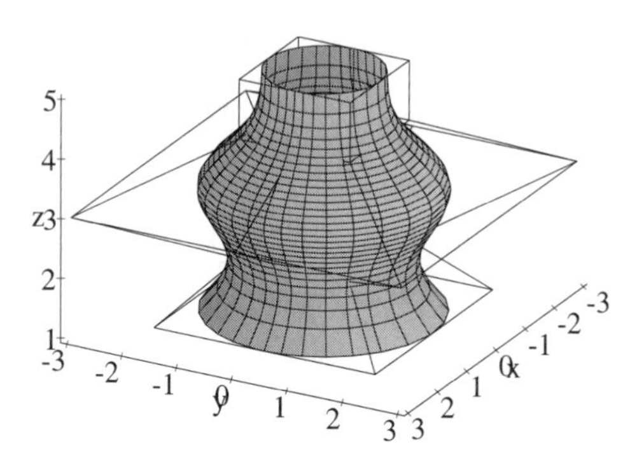

Figure 9.11 A NURBS surface of revolution

by

$$\mathbf{S}(s,t) = \sum_{i=0}^{4}\sum_{j=0}^{6} w_{i,j}\mathbf{p}_{i,j}N_{i,3}(s)N_{j,2}(t) \Bigg/ \sum_{i=0}^{4}\sum_{j=0}^{6} w_{i,j}N_{i,3}(s)N_{j,2}(t) \ .$$

The control points are computed as follows. The scale factor of the $i = 0$ row is the x-coordinate of $\mathbf{b}_0(2,0,1)$, and the translation distance is the z-coordinate. Thus $f_0 = 2$ and $d_0 = 1$. Then $\mathbf{p}_{0,j} = f_0\mathbf{c}_j + (0,0,d_0)$, giving $\mathbf{p}_{0,0}\,(2,0,1)$, $\mathbf{p}_{0,1}\,(2,2,1)$, $\mathbf{p}_{0,2}\,(-2,2,1)$, $\mathbf{p}_{0,3}\,(-2,0,1)$, $\mathbf{p}_{0,4}\,(-2,-2,1)$, $\mathbf{p}_{0,5}\,(2,-2,1)$, $\mathbf{p}_{0,6}\,(2,0,1)$. The scale factor of the $i = 1$ row is the x-coordinate of $\mathbf{b}_1(1,0,2)$, and the translation distance is the z-coordinate. Thus $f_1 = 1$ and $d_1 = 2$. Then $\mathbf{p}_{1,j} = f_1\mathbf{c}_j + (0,0,d_1)$, giving $\mathbf{p}_{1,0}\,(1,0,2)$, $\mathbf{p}_{1,1}\,(1,1,2)$, $\mathbf{p}_{1,2}\,(-1,1,2)$, $\mathbf{p}_{1,3}\,(-1,0,2)$, $\mathbf{p}_{1,4}\,(-1,-1,2)$, $\mathbf{p}_{1,5}\,(1,-1,2)$, $\mathbf{p}_{1,6}\,(1,0,2)$, etc. The rows of weights are $w_{0,j} = \{1,\frac{1}{2},\frac{1}{2},1,\frac{1}{2},\frac{1}{2},1\}$, $w_{1,j} = \{2,1,1,2,1,1,2\}$, $w_{2,j} = \{3,\frac{3}{2},\frac{3}{2},3,\frac{3}{2},\frac{3}{2},3\}$, $w_{3,j} = \{4,2,2,4,2,2,4\}$, $w_{4,j} = \{2,1,1,2,1,1,2\}$. The knots are $0,0,0,0,1,2,2,2,2$ in the s-direction and $0,0,0,\frac{1}{4},\frac{1}{2},\frac{1}{2},\frac{3}{4},1,1,1$ in the t-direction. The surface of revolution is illustrated in Figure 9.11.

Example 9.20

A *torus* is the surface obtained when a circle is swept about an axis that lies in the plane of the circle (see Section 9.2.1). A torus can be obtained in NURBS form or by the parametric equation

$$\mathbf{S}(u,v) = ((r\cos u + R)\cos v, (r\cos u + R)\sin v, r\sin u) \ ,$$

where $r > 0$ is the radius of the circle, and $R > 0$ is the distance from the axis to the circle centre. The torus with $R = 2$ and $r = 1$ is shown in Figure 9.12(a). Toroidal surfaces arise as rolling-ball blends (see Section 9.2.1). For instance,

Figures 9.12(c) and (d) show the effect of blend operations involving a cylindrical solid (b). When $R > r$ the torus has the shape of a doughnut. When $r > R$ the torus self-intersects in two points on the axis. Removing these two points gives two surfaces: the outer surface is called an *apple* torus and the inner one is referred to as a *lemon* torus (make a sketch to see why the torii have these names). The case $R = r$ is called a *vortex* torus. All the torus surface types arise in CAD applications.

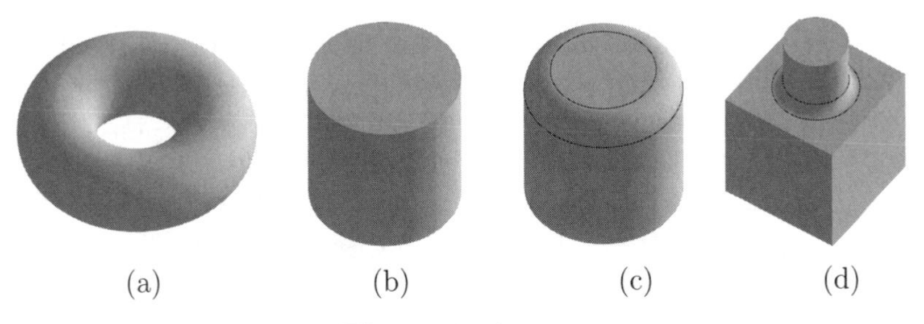

Figure 9.12

EXERCISES

9.5. Determine the remaining rows of control points for Example 9.19.

9.6. Determine the control points of the Bézier surface obtained by extruding the cubic Bézier curve with control points $\mathbf{b}_0(2,3,0)$, $\mathbf{b}_1(1,5,2)$, $\mathbf{b}_2(1,7,-1)$, $\mathbf{b}_3(2,9,-3)$, in the direction $\mathbf{n} = \left(\frac{1}{3}, \frac{2}{3}, -\frac{2}{3}\right)$ through $\delta = 6$ units.

9.7. Determine the control points and weights of the rational Bézier surface obtained by extruding the cubic rational Bézier curve with control points $\mathbf{b}_0(2,3,0)$, $\mathbf{b}_1(1,5,2)$, $\mathbf{b}_2(1,7,-1)$, $\mathbf{b}_3(2,9,-3)$, and weights $u_0 = 1$, $u_1 = 2$, $u_2 = 3$, $u_3 = 1$, in the direction $\mathbf{n} = \left(\frac{1}{3}, \frac{2}{3}, -\frac{2}{3}\right)$ through $\delta = 6$ units.

9.8. Determine the control points and knots of the B-spline surface obtained by extruding the quadratic open B-spline curve with control points $\mathbf{b}_0(4,7,2)$, $\mathbf{b}_1(4,7,4)$, $\mathbf{b}_2(4,9,2)$, $\mathbf{b}_3(4,9,4)$, and knots $s_0 = 0$, $s_1 = 0$, $s_2 = 0$, $s_3 = 1$, $s_4 = 2$, $s_5 = 2$, $s_6 = 2$ in the direction $\mathbf{n} = \left(\frac{5}{13}, \frac{1}{13}, \frac{12}{13}\right)$ through $\delta = 13$ units.

9.9. Extrude the quadratic Bézier curve with control points $\mathbf{b}_0(0,0,0)$, $\mathbf{b}_1(0,a,0)$, $\mathbf{b}_2(a,2a,0)$ through 1 unit in the direction of the z-axis to give a parabolic cylinder. List the control points. Describe how a parabolic cylinder can be obtained as a translationally swept surface.

9.10. Determine a NURBS sphere by rotating the NURBS circle of Example 8.21 about the z-axis. Apply a scaling of 3 units in the x-direction and 2 units in the y-direction to the NURBS sphere to obtain an ellipsoid.

9.11. Determine a NURBS representation for a torus.

9.12. (a) Apply a scaling of a units in the x-direction and b units in the y-direction to the NURBS circle to obtain an ellipse.

(b) Assume that the ellipse lies in the $z = 0$ plane (by adding a zero third coordinate to the control points). Determine a NURBS representation of an elliptic cylinder by (i) extruding the ellipse through 1 unit in the direction of the z-axis, and (ii) translationally sweeping the line segment $(1 - t)(0, 0, 0) + t(0, 0, 1)$ along the trajectory curve defined by the ellipse.

(c) Assume that the ellipse lies in the $y = 0$ plane (by adding a zero second coordinate to the control points). Determine a NURBS representation of an ellipsoid (of revolution) by rotating the ellipse about the z-axis.

9.13. Let a hyperbola be defined in NURBS form by control points $\mathbf{b}_0(-1, 0, 1)$, $\mathbf{b}_1(0, 0, 0)$, $\mathbf{b}_2(1, 0, 1)$, weights $w_0 = 1$, $w_1 = 3$, $w_2 = 1$, and knot vector $-1, -1, -1, 1, 1, 1$. Determine a NURBS representation of a hyperboloid (of one sheet) by rotating the hyperbola about the z-axis.

9.14. Determine the ruled surface defined by the NURBS circle $\mathbf{B}(t)$ (assumed to be in the $z = 0$ plane), and the quadratic NURBS curve with control points $\mathbf{b}_0(4, 0, 3)$, $\mathbf{b}_1(4, 2, 3)$, $\mathbf{b}_2(-4, 2, 3)$, $\mathbf{b}_3(-4, 0, 3)$, $\mathbf{b}_4(-4, -6, 3)$, $\mathbf{b}_5(4, -6, 3)$, $\mathbf{b}_6(4, 0, 3)$, weights $1, 1, 1/2, 1, 1/2, 1, 1$, and knot vector $0, 0, 0, 1/4, 3/4, 1, 1, 1$.

9.5 Surface Subdivision

The de Casteljau, de Boor, and knot insertion algorithms for integral and rational Bézier and B-spline curves can be applied to surfaces. For instance, to subdivide (or to evaluate the coordinates of a point of) a Bézier surface $\mathbf{S}(s, t)$ at the parameter value (s_0, t_0), the de Casteljau algorithm is applied first in the t direction, and then again in the s direction, or vice versa. To apply the algorithm in the t direction, each row of the control polyhedron (that is, the control points $\mathbf{p}_{i,j}$ with fixed i) is treated as the control polygon of a Bézier curve in the parameter t, and the de Casteljau algorithm is executed with $t = t_0$. This yields a subdivision of $\mathbf{S}(s,t)$ into two Bézier subsurfaces along the parameter curve $\mathbf{S}(s,t_0)$. Similarly, the de Casteljau algorithm is executed with $s = s_0$ to each column of the control polygon (that is, the control points $\mathbf{p}_{i,j}$ with fixed j). The result is that the surfaces are subdivided along the parameter curve $\mathbf{S}(s_0,t)$, giving a subdivision of the original surface into four surface patches. The algorithm also yields an evaluation of the point $\mathbf{S}(s_0,t_0)$.

In a similar manner, the rational de Casteljau algorithm can be applied to a rational Bézier surface, and the de Boor and knot insertion algorithms can be applied to a B-spline or NURBS surface. The following example elucidates the method further.

Example 9.21

A Bézier surface $\mathbf{S}(s,t)$ has control points

$$\begin{array}{lll} \mathbf{p}_{0,0}(2,3,0), & \mathbf{p}_{0,1}(2,6,3), & \mathbf{p}_{0,2}(2,10,0)\,, \\ \mathbf{p}_{1,0}(6,2,1), & \mathbf{p}_{1,1}(6,6,4), & \mathbf{p}_{1,2}(6,9,1)\,, \\ \mathbf{p}_{2,0}(10,2,0), & \mathbf{p}_{2,1}(10,6,3), & \mathbf{p}_{2,2}(10,10,0)\,. \end{array}$$

Apply the de Casteljau algorithm to subdivide the surface along the coordinate curves $\mathbf{S}(0.5,t)$ and $\mathbf{S}(s,0.25)$, and to evaluate the point $\mathbf{S}(0.5,0.25)$. Applying the de Casteljau algorithm with $t = 0.25$ to each row of control points yields

$$\begin{array}{lll} & (2,3,0) & (2,6,3) & (2,10,0) \\ \text{row } i = 0 & (2.0,3.75,0.75) & (2.0,7.0,2.25) & \\ & (2.0,4.5625,1.125) & & \\ & (6,2,1) & (6,6,4) & (6,9,1) \\ \text{row } i = 1 & (6.0,3.0,1.75) & (6.0,6.75,3.25) & \\ & (6.0,3.9375,2.125) & & \\ & (10,2,0) & (10,6,3) & (10,10,0) \\ \text{row } i = 2 & (10.0,3.0,0.75) & (10.0,7.0,2.25) & \\ & (10.0,4.0,1.125) & & \end{array}$$

The result is two surfaces: the first with control points

$$\begin{array}{lll} \mathbf{p}_{0,0}(2,3,0), & \mathbf{p}_{0,1}\,(2.0,3.75,0.75)\,, & \mathbf{p}_{0,2}\,(2.0,4.5625,1.125)\,, \\ \mathbf{p}_{1,0}(6,2,1), & \mathbf{p}_{1,1}\,(6.0,3.0,1.75)\,, & \mathbf{p}_{1,2}\,(6.0,3.9375,2.125)\,, \\ \mathbf{p}_{2,0}(10,2,0), & \mathbf{p}_{2,1}\,(10.0,3.0,0.75)\,, & \mathbf{p}_{2,2}\,(10.0,4.0,1.125)\,, \end{array}$$

and the second surface with control points

$$\begin{array}{lll} \mathbf{p}_{0,0}\,(2.0,4.5625,1.125)\,, & \mathbf{p}_{0,1}\,(2.0,7.0,2.25)\,, & \mathbf{p}_{0,2}(2,10,0)\,, \\ \mathbf{p}_{1,0}\,(6.0,3.9375,2.125)\,, & \mathbf{p}_{1,1}\,(6.0,6.75,3.25)\,, & \mathbf{p}_{1,2}(6,9,1)\,, \\ \mathbf{p}_{2,0}\,(10.0,4.0,1.125)\,, & \mathbf{p}_{2,1}\,(10.0,7.0,2.25)\,, & \mathbf{p}_{2,2}(10,10,0)\,. \end{array}$$

Next, apply the de Casteljau algorithm with $s = 0.5$ to the first subsurface

$$\begin{array}{lll} (2,3,0) & (6,2,1) & (10,2,0) \\ (4,2.5,0.5) & (8,2,0.5) & \\ (6,2.25,0.5) & & \end{array}$$

$$\begin{array}{lll} (2,3.75,0.75) & (6,3,1.75) & (10,3,0.75) \\ (4,3.375,1.25) & (8,3,1.25) & \\ (6,3.1875,1.25) & & \end{array}$$

$$\begin{array}{lll} (2.0,4.5625,1.125) & (6.0,3.9375,2.125) & (10.0,4.0,1.125) \\ (4.0,4.25,1.625) & (8.0,3.96875,1.625) & \\ (6.0,4.109375,1.625) & & \end{array}$$

and then to the second subsurface

$$\begin{array}{lll} (2.0,4.5625,1.125) & (6.0,3.9375,2.125) & (10.0,4.0,1.125) \\ (4.0,4.25,1.625) & (8.0,3.96875,1.625) & \\ (6.0,4.109375,1.625) & & \end{array}$$

$$\begin{array}{lll} (2,7,2.25) & (6.0,6.75,3.25) & (10,7,2.25) \\ (4.0,6.875,2.75) & (8.0,6.875,2.75) & \\ (6.0,6.875,2.75) & & \end{array}$$

$$\begin{array}{lll} (2,10,0) & (6,9,1) & (10,10,0) \\ (4.0,9.5,0.5) & (8.0,9.5,0.5) & \\ (6.0,9.5,0.5) & & \end{array}$$

The last triangle of points of the first subsurface is identical to the first triangle of the second subsurface since the subsurfaces have a common row of control points. In practice the triangle would only be computed once, but it is included here twice to illustrate that the surface is divided first into two and then into four. As before the edges of the computed triangle give the control points of the four surfaces resulting from the subdivision. The point $\mathbf{S}(0.5, 0.25)$ is $(6.0, 4.109375, 1.625)$. Note that only one of the $t = 0.25$ subsurfaces needs to be subdivided in order to evaluate this point. The subdivided surface is shown in Figure 9.13.

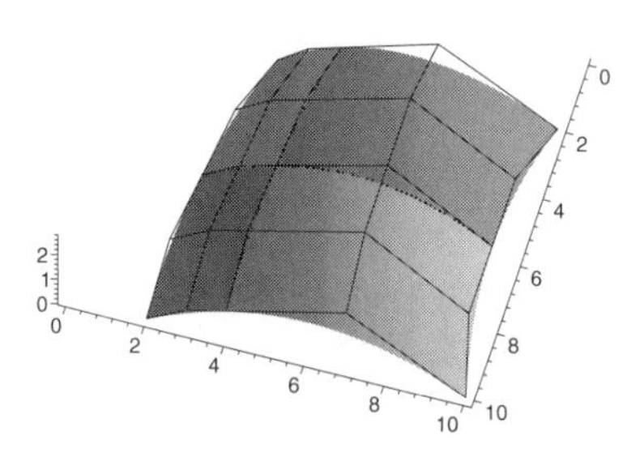

Figure 9.13 Subdivided biquadratic Bézier surface of Example 9.21

Subdivision of rational Bézier, B-spline, and NURBS surfaces is performed by applying the rational de Casteljau, the de Boor, or a knot insertion algorithm in a similar manner.

Remark 9.22

The number of linear interpolations computed in a subdivision of a Bézier surface into four is easily determined. Suppose the surface has degree m in the s direction and degree n in the t direction. The de Casteljau algorithm in the s direction requires $\frac{1}{2}m(m+1)$ interpolations for each column giving a total of $\frac{1}{2}m(m+1)(n+1)$ interpolations. The algorithm in the t direction (remembering that there are now $2m+1$ rows of n control points) requires $\frac{1}{2}n(n+1)$ interpolations per row giving $\frac{1}{2}(2m+1)\,n(n+1)$ interpolations in all. The total number of interpolations is $\frac{1}{2}m(m+1)(n+1)+\frac{1}{2}(2m+1)\,n(n+1) = \frac{1}{2}(n+1)\left(2mn+n+m^2+m\right)$. Thus, depending on the degrees m and n, it matters which direction is subdivided first. For instance, if $m=3$, $n=4$, then the number of computations is 100. But if $m=4$, $n=3$, then the number of computations is 94. Thus it is most efficient to subdivide first in the variable which has the largest degree.

EXERCISES

9.15. In Section 6.10.3, a method to determine the intersection of two Bézier curves using subdivision was described. Describe how the operation of subdivision together with the convex hull property for surfaces can be used to determine the curve of intersection of two Bézier surfaces.

9.16. Subdivide the Bézier surface of Example 9.21 at $s=0.25$ and $t=0.5$.

9.17. Subdivide the Bézier surface of Example 9.13 at $s=0.4$ and $t=0.2$.

9.18. Use the de Boor algorithm to subdivide the B-spline surface of Exercise 9.3 at $s=2.2$ and $t=6.4$.

9.6 Skin and Loft Surfaces

Skinning is the operation of constructing a surface that interpolates a number of user specified curve sections. Clearly, there are an infinite number of surfaces passing through two curves $\mathbf{c}(t)$ and $\mathbf{d}(t)$. One solution is to linearly interpolate the two curves to give the ruled surface

$$\mathbf{x}(s,t) = (1-s)\mathbf{c}(t) + s\mathbf{d}(t) \ . \tag{9.15}$$

A sequence of curve sections can be skinned by computing ruled surfaces between adjacent pairs of curves as shown in Figure 9.14.

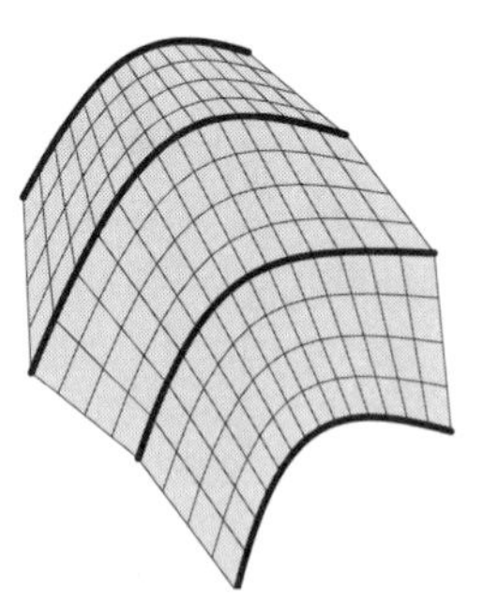

Figure 9.14 Skinning through curve sections

Example 9.23

Let two curve sections be $\mathbf{c}(t) = (t, t^2, 0)$ and $\mathbf{d}(t) = (t, t^4 - t^2, 10)$. Then the skinning surface given by Equation (9.15) is

$$\begin{aligned} \mathbf{x}(s,t) &= \big((1-s)t + st, (1-s)t^2 + s(t^4 - t^2), (1-s)0 + s10\big) \\ &= (t, t^2 - 2st^2 + st^4, 10s) \ . \end{aligned}$$

Figure 9.15(a) shows the skin surface and the intermediate parameter curves corresponding to $s = 0.25$, $s = 0.5$ and $s = 0.75$.

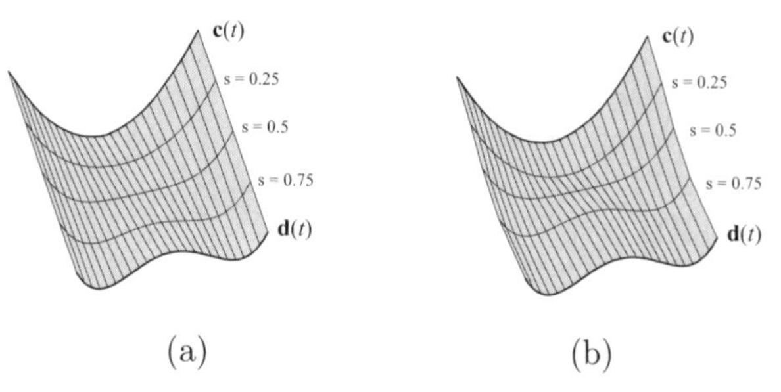

Figure 9.15 Skin and loft surfaces of Examples 9.23 and 9.25

More general skinning surfaces can be achieved by replacing the blending functions $1-s$ and s in (9.15) by continuous functions $f_0(s)$ and $f_1(s)$ that satisfy $f_0(0) = f_1(1) = 1$ and $f_0(1) = f_1(0) = 0$. Additionally, the blending functions can be required to satisfy $f_0(s) + f_1(s) = 1$ in order to obtain skin surfaces that are planar whenever the boundary curves are coplanar (Exercise 9.23).

To proceed further, it is necessary to introduce the notion of *geometric continuity* for surfaces in a similar manner to curves in Definition 7.14.

Definition 9.24

Two regular surfaces $\mathbf{x}(s,t)$ and $\mathbf{y}(u,v)$ are said to *meet with (parametric) C^k-continuity at a point* $\mathbf{P} = \mathbf{x}(s_0,t_0) = \mathbf{y}(u_0,v_0)$ whenever

$$\frac{\partial^{i+j}\mathbf{x}(s_0,t_0)}{\partial^i s \partial^j t} = \frac{\partial^{i+j}\mathbf{y}(u_0,v_0)}{\partial^i u \partial^j v},$$

for $0 \leq i+j \leq k$.

Two regular surfaces $\mathbf{x}(s,t)$ and $\mathbf{y}(u,v)$ are said to *meet with G^k-continuity at a point* $\mathbf{P} = \mathbf{x}(s_0,t_0) = \mathbf{y}(u_0,v_0)$ whenever there is an invertible mapping (called a *reparametrization*) $\mathbf{h} : (\tilde{u},\tilde{v}) \mapsto (u(\tilde{u},\tilde{v}), v(\tilde{u},\tilde{v}))$ such that $\mathbf{x}(s,t)$ and $\mathbf{y}(u(\tilde{u},\tilde{v}), v(\tilde{u},\tilde{v}))$ meet with C^k-continuity at $\mathbf{P}$. Two surfaces are said to *meet with G^k-continuity along a curve* if they meet with G^k-continuity at every point of that curve.

Suppose that $\mathbf{x}(s,t)$ and $\mathbf{y}(u,v)$ meet with G^1-continuity at $\mathbf{P} = \mathbf{x}(s_0,t_0) = \mathbf{y}(u_0,v_0)$. Then there is a reparametrization $u = u(\tilde{u},\tilde{v})$, $v = v(\tilde{u},\tilde{v})$ for which the chain rule gives

$$\mathbf{x}_s = \frac{\partial}{\partial \tilde{u}}\mathbf{y}(u(\tilde{u},\tilde{v}), v(\tilde{u},\tilde{v})) = \frac{\partial u}{\partial \tilde{u}}\mathbf{y}_u + \frac{\partial v}{\partial \tilde{u}}\mathbf{y}_v, \tag{9.16}$$

$$\mathbf{x}_t = \frac{\partial}{\partial \tilde{v}}\mathbf{y}(u(\tilde{u},\tilde{v}), v(\tilde{u},\tilde{v})) = \frac{\partial u}{\partial \tilde{v}}\mathbf{y}_u + \frac{\partial v}{\partial \tilde{v}}\mathbf{y}_v. \tag{9.17}$$

Hence

$$\begin{aligned}\mathbf{x}_s \times \mathbf{x}_t &= \left(\frac{\partial u}{\partial \tilde{u}}\mathbf{y}_u + \frac{\partial v}{\partial \tilde{u}}\mathbf{y}_v\right) \times \left(\frac{\partial u}{\partial \tilde{v}}\mathbf{y}_u + \frac{\partial v}{\partial \tilde{v}}\mathbf{y}_v\right) \\ &= \left(\frac{\partial u}{\partial \tilde{u}}\frac{\partial v}{\partial \tilde{v}} - \frac{\partial v}{\partial \tilde{u}}\frac{\partial u}{\partial \tilde{v}}\right)(\mathbf{y}_u \times \mathbf{y}_v).\end{aligned}$$

Therefore, when two surfaces meet with G^1-continuity at a point $\mathbf{P}$, they have the same normal directions, and hence the same tangent planes. The converse is also true: if two surfaces meet at a point $\mathbf{P}$, and have common surface normals at $\mathbf{P}$, then the surfaces meet with G^1-continuity (Exercise 9.22).

Skinning is often referred to as *lofting*, a term that arises from the ship-building industry where some aspects of ship design took place in the lofts of hangers. Some authors reserve the term "lofting" to mean a skinning operation where the interpolating surfaces satisfy specified derivative conditions along the curve sections. When a skinning operation is applied to a sequence of curve sections, the resulting surfaces meet (in general) with only C^0-continuity. Lofting, however, can yield surfaces that have G^1-continuity along the curve sections.

Suppose two curve sections $\mathbf{c}(t)$ and $\mathbf{d}(t)$ and two derivative functions $\mathbf{c}_s(t)$ and $\mathbf{d}_s(t)$ are specified. There is no unique loft surface but a commonly used one is

$$\mathbf{x}(s,t) = H_0(s)\mathbf{c}(t) + H_1(s)\mathbf{d}(t) + \hat{H}_0(s)\mathbf{c}_s(t) + \hat{H}_1(s)\mathbf{d}_s(t) , \qquad (9.18)$$

where $H_i(s)$ and $\hat{H}_i(s)$ are the Hermite polynomials:

$$\begin{aligned} H_0(s) &= 1 - 3s^2 + 2s^3 , & \hat{H}_0(s) &= s - 2s^2 + s^3 , \\ H_1(s) &= 3s^2 - 2s^3 , & \hat{H}_1(s) &= -s^2 + s^3 . \end{aligned}$$

Example 9.25

Let the curve sections be $\mathbf{c}(t) = (t, 0, t^2)$ and $\mathbf{d}(t) = (t, 2, t^4 - t^2)$, and the derivative conditions be $\mathbf{c}_s(t) = (0, 2, -1)$ and $\mathbf{d}_s(t) = (0, 4, -0.5)$. (Note that the derivative conditions can be non-constant.) Then the lofting surface given by Equation (9.18) is

$$\begin{aligned} \mathbf{x}(s,t) &= (1 - 3s^2 + 2s^3)(t, 0, t^2) + (3s^2 - 2s^3)(t, 2, t^4 - t^2) \\ &\quad + (s - 2s^2 + s^3)(0, 2, -1) + (-s^2 + s^3)(0, 4, -0.5) . \end{aligned}$$

The coordinate functions simplify to

$$\begin{aligned} x(t) &= (1 - 3s^2 + 2s^3)t + (3s^2 - 2s^3)t = t , \\ y(t) &= 2(3s^2 - 2s^3) + 2(s - 2s^2 + s^3) + 4(-s^2 + s^3) \\ &= 2s - 2s^2 + 2s^3 , \\ z(t) &= (1 - 3s^2 + 2s^3)t^2 + (3s^2 - 2s^3)(t^4 - t^2) \\ &\quad - (s - 2s^2 + s^3) - 0.5(-s^2 + s^3) \\ &= -s + 2.5s^2 - 1.5s^3 + t^2 - 6s^2t^2 + 4s^3t^2 + 3s^2t^4 - 2s^3t^4 . \end{aligned}$$

The surface is shown in Figure 9.15(b).

More general lofting surfaces can be obtained by replacing the blending functions H_0, H_1, $\hat{H}_0$ and $\hat{H}_1$ by functions $f_0(s)$, $f_1(s)$, $g_0(t)$, $g_1(t)$ that satisfy

$$\begin{aligned} f_0(0) &= f_1(1) = 1\,, & f_0(1) &= f_1(0) = 0\,, \\ f_0'(0) &= f_0'(1) = 0\,, & f_1'(0) &= f_1'(1) = 0\,, \\ g_0(0) &= g_0(1) = 0\,, & g_1(0) &= g_1(1) = 0\,, \\ g_0'(0) &= g_1'(1) = 1\,, & g_0'(1) &= g_1'(0) = 0\,, \end{aligned}$$

to give the surface

$$\mathbf{x}(s,t) = f_0(s)\mathbf{c}(t) + f_1(s)\mathbf{d}(t) + g_0(s)\mathbf{c}_s(t) + g_1(s)\mathbf{d}_s(t)\,. \tag{9.19}$$

Next consider the task of constructing a surface through four boundary curves: $\mathbf{c}(t)$, $\mathbf{d}(t)$, $\mathbf{e}(s)$ and $\mathbf{f}(s)$, as showed in Figure 9.16. Consider the skin

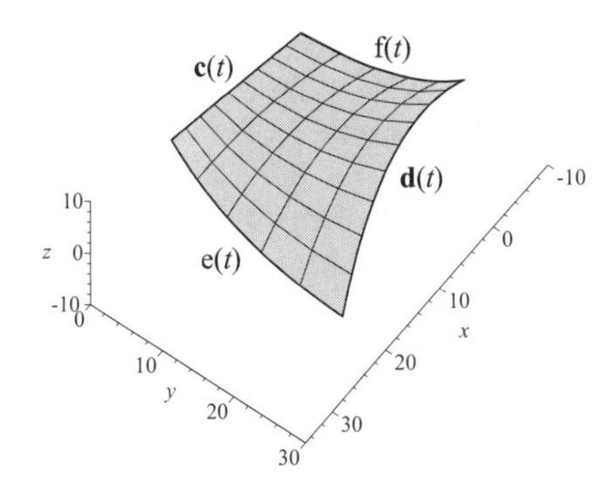

Figure 9.16 Gordon–Coons surface interpolating four boundary curves

surface $\mathbf{x}(s,t)$ that interpolates $\mathbf{c}(t)$ and $\mathbf{d}(t)$ given by (9.15). Substituting $t = 0$ into (9.15) gives

$$\mathbf{x}(s,0) = (1-s)\mathbf{c}(0) + s\mathbf{d}(0)\,.$$

Therefore, in order for $\mathbf{x}(s,t)$ to interpolate $\mathbf{e}(s)$ when $t = 0$ it is necessary to modify (9.15) by adding

$$\mathbf{e}(s) - (1-s)\mathbf{c}(0) - s\mathbf{d}(0)\,.$$

Similarly, substituting $t = 1$ into (9.15) gives

$$\mathbf{x}(s,1) = (1-s)\mathbf{c}(1) + s\mathbf{d}(1)\,,$$

and in order for $\mathbf{x}(s,t)$ to interpolate $\mathbf{f}(s)$ when $t = 1$ it is necessary to modify (9.15) by adding

$$\mathbf{f}(s) - (1-s)\mathbf{c}(1) - s\mathbf{d}(1)\,.$$

The necessary correction across the entire surface is obtained by linearly interpolating the two correction terms

$$(1-t)\,(\mathbf{e}(s)-(1-s)\mathbf{c}(0)-s\mathbf{d}(0))+t\,(\mathbf{f}(s)-(1-s)\mathbf{c}(1)-s\mathbf{d}(1))\,. \quad (9.20)$$

Subtracting (9.20) from (9.15) yields the *Gordon–Coons surface*

$$\begin{aligned}\hat{\mathbf{x}}(s,t) &= (1-s)\mathbf{c}(t)+s\mathbf{d}(t)+(1-t)\mathbf{e}(s)+t\mathbf{f}(s)\\ &\quad -(1-s)(1-t)\mathbf{x}_{0,0}-(1-s)t\mathbf{x}_{0,1}-s(1-t)\mathbf{x}_{1,0}-st\mathbf{x}_{1,1}\,,\end{aligned} \quad (9.21)$$

where

$$\begin{aligned}\mathbf{x}_{0,0}&=\mathbf{c}(0)=\mathbf{e}(0)\,, & \mathbf{x}_{0,1}&=\mathbf{c}(1)=\mathbf{f}(0)\,,\\ \mathbf{x}_{1,0}&=\mathbf{d}(0)=\mathbf{e}(1)\,, & \mathbf{x}_{1,1}&=\mathbf{d}(1)=\mathbf{f}(1)\,.\end{aligned}$$

The Gordon–Coons surface can be expressed in matrix form

$$\mathbf{x}(s,t)=\begin{pmatrix}1-s & s & 1\end{pmatrix}\begin{pmatrix}-\mathbf{x}_{0,0} & -\mathbf{x}_{0,1} & \mathbf{c}(t)\\ -\mathbf{x}_{1,0} & -\mathbf{x}_{1,1} & \mathbf{d}(t)\\ \mathbf{e}(s) & \mathbf{f}(s) & \mathbf{0}\end{pmatrix}\begin{pmatrix}1-t\\ t\\ 1\end{pmatrix}\,. \quad (9.22)$$

The reader should verify that an identical formula is obtained if the above method is applied to the interpolant of $\mathbf{e}(s)$ and $\mathbf{f}(s)$.

Example 9.26

Consider four boundary Bézier curves that meet at vertices $\mathbf{x}_{0,0}(-4,0,-4)$, $\mathbf{x}_{0,1}(20,0,4)$, $\mathbf{x}_{1,0}(-8,20,0)$ and $\mathbf{x}_{1,1}(28,30,6)$ given by

$$\begin{aligned}\mathbf{c}(t) &= (-4,0,-4)(1-t)+(20,0,4)t\,,\\ \mathbf{d}(t) &= (-8,20,0)(1-t)^2+(10,22,16)2(1-t)t+(28,30,6)t^2\,,\\ \mathbf{e}(s) &= (-4,0,-4)(1-s)^2+(-7,12,-8)2(1-s)s+(-8,20,0)s^2\,,\\ \mathbf{f}(s) &= (20,0,4)(1-s)^2+(25,12,1)2(1-s)s+(28,30,6)s^2\,.\end{aligned}$$

Using (9.21), the Gordon–Coons surface is $\mathbf{x}(s,t) = (x(s,t),y(s,t),z(s,t))$

where

$$\begin{aligned}
x(s,t) &= 4\,(1-s)\,(1-t) - 20\,(1-s)\,t + 8\,s\,(1-t) - 28\,st \\
&\quad + (1-s)\,(-4+24\,t) + s\left(-8\,(1-t)^2 + 20\,(1-t)\,t + 28\,t^2\right) \\
&\quad + (1-t)\left(-4\,(1-s)^2 - 14\,(1-s)\,s - 8\,s^2\right) \\
&\quad + t\left(20\,(1-s)^2 + 50\,(1-s)\,s + 28\,s^2\right) \\
&= -4 + 24\,t - 6\,s + 16\,st + 2\,s^2 - 4\,ts^2\,, \\
y(s,t) &= -20\,s\,(1-t) - 30\,st + s\left(20\,(1-t)^2 + 44\,(1-t)\,t + 30\,t^2\right) \\
&\quad + (1-t)\left(24\,(1-s)\,s + 20\,s^2\right) + t\left(24\,(1-s)\,s + 30\,s^2\right) \\
&= 24\,s - 6\,st + 6\,st^2 - 4\,s^2 + 10\,ts^2\,, \\
z(s,t) &= 4\,(1-s)\,(1-t) - 4\,(1-s)\,t - 6\,st + (1-s)\,(-4+8\,t) \\
&\quad + s\left(32\,(1-t)\,t + 6\,t^2\right) + (1-t)\left(-4\,(1-s)^2 - 16\,(1-s)\,s\right) \\
&\quad + t\left(4\,(1-s)^2 + 2\,(1-s)\,s + 6\,s^2\right) \\
&= -4 + 8\,t - 8\,s + 28\,st - 26\,st^2 + 12\,s^2 - 4\,ts^2\,.
\end{aligned}$$

The surface is shown in Figure 9.16.

Given a grid or network of curves, as showed in Figure 9.17, Gordon–Coons surfaces can be used to skin between each set of four boundary curve segments to give a C^0 *network surface.* A general Gordon–Coons surface can be obtained by replacing the blending functions $1-s$, s, $1-t$, t, that are used in (9.22), by functions $f_0(s)$, $f_1(s)$, $g_0(t)$, $g_1(t)$ that satisfy

$$\begin{aligned}
f_0(0) = g_0(0) = 1, &\qquad f_0(s) + f_1(s) = 1, \\
f_1(1) = g_1(1) = 0, &\qquad g_0(t) + g_1(t) = 1.
\end{aligned}$$

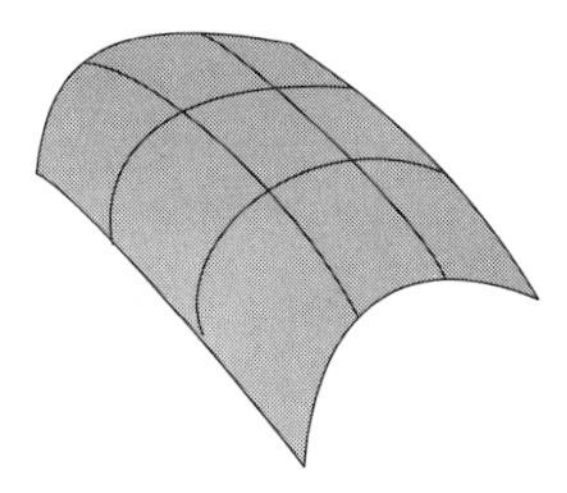

Figure 9.17 Network of curves interpolated by Gordon–Coons surfaces

The derivation of the Gordon–Coons surface can be generalised to give a loft surface that interpolates four boundary curves $\mathbf{c}(t)$, $\mathbf{d}(t)$, $\mathbf{e}(s)$ and $\mathbf{f}(s)$ and satisfies four derivative conditions $\mathbf{c}_s(t)$, $\mathbf{d}_s(t)$, $\mathbf{e}_t(s)$ and $\mathbf{f}_t(s)$. The surface is given by

$$\mathbf{x}(s,t) = \mathbf{H}(s)\begin{pmatrix} -\mathbf{x}_{0,0} & -\mathbf{x}_{0,1} & -\mathbf{e}_t(0) & -\mathbf{f}_t(0) & \mathbf{c}(t) \\ -\mathbf{x}_{1,0} & -\mathbf{x}_{1,1} & -\mathbf{e}_t(1) & -\mathbf{f}_t(1) & \mathbf{d}(t) \\ -\mathbf{c}_s(0) & -\mathbf{c}_s(1) & -\mathbf{x}_{st,0,0} & -\mathbf{x}_{st,0,1} & \mathbf{c}_s(t) \\ -\mathbf{d}_s(0) & -\mathbf{d}_s(1) & -\mathbf{x}_{st,1,0} & -\mathbf{x}_{st,1,1} & \mathbf{d}_s(t) \\ \mathbf{e}(s) & \mathbf{f}(s) & \mathbf{e}_t(s) & \mathbf{f}_t(s) & \mathbf{0} \end{pmatrix}\mathbf{H}(t)^T, \quad (9.23)$$

where

$$\mathbf{H} = \begin{pmatrix} H_0 & H_1 & \hat{H}_0 & \hat{H}_1 & 1 \end{pmatrix},$$

and

$$\mathbf{x}_{0,0} = \mathbf{c}(0) = \mathbf{e}(0)\,, \quad \mathbf{x}_{0,1} = \mathbf{c}(1) = \mathbf{f}(0)\,,$$
$$\mathbf{x}_{1,0} = \mathbf{d}(0) = \mathbf{e}(1)\,, \quad \mathbf{x}_{1,1} = \mathbf{d}(1) = \mathbf{f}(1)\,.$$

Note that the left hand $\mathbf{H}$ of Equation (9.23) is a function of s and the right hand $\mathbf{H}^T$ is a function of t. All the entries in the matrix can be obtained from the specified curve and derivative data except for those involving $\mathbf{x}_{st}$. The four values $\mathbf{x}_{st,0,0}$, $\mathbf{x}_{st,0,1}$, $\mathbf{x}_{st,1,0}$ and $\mathbf{x}_{st,1,1}$ are called *twist* vectors, and specify the second order partial derivatives $\frac{\partial^2 \mathbf{x}}{\partial s \partial t}(s,t)$ at the corners of the surface. They can be used to control the shape of the interior of the surface without changing the shape of the boundary curves.

The loft surface (9.23) can be used to obtain a G^1 network surface defined by a grid of curves in a similar manner to the skin surface. In the case of a network surface, the second order derivative terms $\mathbf{x}_{st,i,j}$ must satisfy additional *compatibility conditions* in order to achieve continuity at the corners:

$$\mathbf{x}_{st,0,0} = \frac{d\mathbf{e}_t}{ds}(0) = \frac{d\mathbf{c}_s}{dt}(0)\,, \quad \mathbf{x}_{st,0,1} = \frac{d\mathbf{f}_t}{ds}(0) = \frac{d\mathbf{c}_s}{dt}(1)\,,$$
$$\mathbf{x}_{st,1,0} = \frac{d\mathbf{e}_t}{ds}(1) = \frac{d\mathbf{d}_s}{dt}(0)\,, \quad \mathbf{x}_{st,1,1} = \frac{d\mathbf{f}_t}{ds}(1) = \frac{d\mathbf{d}_s}{dt}(1)\,.$$

Example 9.27

Consider a Gordon–Coons surface given by linear boundary curves in the $z = 0$ plane: $\mathbf{c}(t) = (0, t, 0)$, $\mathbf{d}(t) = (1, t, 0)$, $\mathbf{e}(s) = (s, 0, 0)$, and $\mathbf{f}(s) = (s, 1, 0)$. Let the derivative conditions be $\mathbf{c}_s(t) = (1, 0, 0)$, $\mathbf{d}_s(t) = (1, 0, 0)$, $\mathbf{e}_t(s) = (0, 1, 0)$, and $\mathbf{f}_t(s) = (0,1,0)$. Then

$$\mathbf{x}(s,t) = \mathbf{H}(s)\begin{pmatrix} -(0,0,0) & -(0,1,0) & -(1,0,0) & -(1,0,0) & (0,t,0) \\ -(1,0,0) & -(1,1,0) & -(1,0,0) & -(1,0,0) & (1,t,0) \\ -(1,0,0) & -(0,1,0) & -\mathbf{x}_{st,0,0} & -\mathbf{x}_{st,0,1} & (0,1,0) \\ -(1,0,0) & -(0,1,0) & -\mathbf{x}_{st,1,0} & -\mathbf{x}_{st,1,1} & (0,1,0) \\ (s,0,0) & (s,1,0) & (1,0,0) & (1,0,0) & (0,0,0) \end{pmatrix}\mathbf{H}(t)^T.$$

If the twist vectors have zero z-component, then the Gordon–Coons surface has zero z-component and so the surface is planar. However, if the twist vectors are chosen so that they point out of the $z = 0$ plane, then the surface is no longer planar despite the fact that the boundary curves and the derivatives lie in the plane. The twist vectors provide a potentially powerful tool for surface design.

EXERCISES

9.19. Determine a skin surface that interpolates the curves $\mathbf{c}(t) = (3t^2 + 4, 2t^2, -t)$ and $\mathbf{d}(t) = (2t, -t^4, 2t+4)$, for $0 \le t \le 1$.

9.20. Let $\mathbf{c}(t) = (3t^2, 2t^2, t)$ and $\mathbf{d}(t) = (2t+10, 3t, 2t^2)$. Determine the loft surfaces that interpolate $\mathbf{c}(t)$ and $\mathbf{d}(t)$, and satisfy the derivative conditions

a) $\mathbf{c}_s(t) = (1,1,0)$ and $\mathbf{d}_s(t) = (0,-1,0)$,

b) $\mathbf{c}_s(t) = (0,-t^2,0)$ and $\mathbf{d}_s(t) = (-1,t^2,-1)$.

9.21. Determine the Coons surface interpolating the four boundary Bézier curves

$$\begin{aligned} \mathbf{c}(t) &= (-5,-5,0)B_{0,1}(t) + (20,3,0)B_{1,1}(t), \\ \mathbf{d}(t) &= (-8,0,10)B_{0,2}(t) + (10,16,4)B_{1,2}(t) + (27,7,10)B_{2,2}(t), \\ \mathbf{e}(s) &= (-5,-5,0)B_{0,3}(s) + (-6,-4,3)B_{1,3}(s) \\ &\quad + (-7,-2,7)B_{2,3}(s) + (-8,0,10)B_{3,3}(s), \\ \mathbf{f}(s) &= (20,3,0)B_{0,2}(s) + (22,5,4)B_{1,2}(s) + (27,7,10)B_{2,2}(s). \end{aligned}$$

9.22. Show that if two surfaces meet at a point $\mathbf{P}$ and the surfaces have a common normal direction at $\mathbf{P}$, then the surfaces meet with G^1-continuity.

9.23. Consider the skinning operation through two section curves with general blending functions functions $f_0(s)$ and $f_1(s)$ that satisfy

$f_0(0) = f_1(1) = 1$, $f_0(1) = f_1(0) = 0$ and $f_0(s) + f_1(s) = 1$. Show that the skin surface

$$\mathbf{x}(s,t) = f_0(s)\mathbf{c}(t) + f_1(s)\mathbf{d}(t)$$

is planar whenever the boundary curves are coplanar.

9.24. Verify that the partial derivatives of the loft surface $\mathbf{x}(s,t)$ given by (9.18), and evaluated at $s = 0$ and $s = 1$, agree with the specified derivative conditions $\mathbf{c}_s(t)$ and $\mathbf{d}_s(t)$ respectively.

9.25. Suppose two Bézier surfaces $\mathbf{B}(s,t) = \sum_{i=0}^{n} \sum_{j=0}^{p} \mathbf{b}_{i,j} B_{i,n}(s) B_{j,p}(t)$ and $\mathbf{C}(s,t) = \sum_{i=0}^{n} \sum_{j=0}^{p} \mathbf{c}_{i,j} B_{i,n}(s) B_{j,p}(t)$ (for $(s,t) \in [0,1] \times [0,1]$) meet along a common boundary $\mathbf{B}(1,t) = \mathbf{C}(0,t)$. Determine conditions on the control points for the surfaces to have the same tangent planes along the boundary.

9.7 Geometric Modelling and CAD

Geometric modelling is concerned with developing tools to create, represent, and manipulate geometric shapes. Every commercial CAD system incorporates a *geometric modeller*, an implementation of modelling tools that enable the user to specify curves, surfaces and solids. Most systems are able to represent the curve and surface types that were introduced in the earlier chapters such as lines, conics, planes, quadrics, and Bézier and B-spline curves and surfaces.

Geometric modellers make a careful distinction between surfaces and solids. The term "sphere", for instance, can refer to either the outer boundary surface or to the solid consisting of the boundary and all the points inside. A *solid* is a finite volume bounded by a finite number of surfaces. The intersections of the bounding surfaces give the edges of the solid, and the intersections of the edges give the vertices. For example, a solid cube is bounded by six planes. The planes intersect to give twelve linear edges, and the edges intersect in eight vertices. Each face of a cube is a planar region bounded by four of the edges.

In addition to the mathematical definitions of the surfaces and edges, a specification of a solid requires information about the *topology* of the model, that is, how the surfaces, edges and vertices interconnect. Not all modellers can represent solids, and so those that do are referred to as *solid modellers*. A number of different representations are used in both industrial and academic modellers and some of these are described in the following sections.

9.7.1 Wireframe Modeller

A wireframe modeller uses curves to represent both surfaces and solids. Typically, the curves comprise the boundaries of each surface, and additional (parameter) curves to indicate the shape of each surface. The surfaces and solids are not fully represented and no topological information is stored. Wireframe modellers are very easy to implement and are useful for obtaining fast renderings of surfaces.

9.7.2 Surface Modeller

A surface modeller represents each surface mathematically, but no topological information is stored and, therefore, there is no true concept of a solid. The simplest type of surface modeller is the *polyhedral* modeller for which the surfaces are all planar. More sophisticated surface modellers are able to represent Bézier, B-spline and NURBS surfaces, and general parametric surfaces. Surface modellers are useful for product concept, specification and styling, and may be sufficient for some applications such as numerical controlled (NC) machining and computer-aided manufacture (CAM).

9.7.3 Constructive Solid Geometry (CSG) Modellers

A constructive solid geometry (CSG) modeller uses implicit surface definitions (see Section 9.1). An implicitly defined surface $f(x, y, z) = 0$ partitions the three-dimensional workspace into two regions or *half-spaces* consisting of points satisfying $f(x, y, z) \geq 0$ and $f(x, y, z) \leq 0$ respectively. (Half-spaces can also be constructed using $>$ and $<$.) For instance, the unit sphere with the implicit equation $x^2 + y^2 + z^2 - 1 = 0$ yields two half-spaces: the inside of the sphere, which is the set of points (x, y, z) satisfying $x^2 + y^2 + z^2 - 1 \leq 0$, and the outside which is the set of points satisfying $x^2 + y^2 + z^2 - 1 > 0$. Likewise, an infinite plane divides the workspace into the two regions on either side of the plane. Solids are defined in terms of half-spaces. For example, a hemisphere comprises the points satisfying both $x^2 + y^2 + z^2 - 1 \leq 0$ and $z \geq 0$, that is, the set of points inside the sphere and to one side of the plane $z = 0$.

The constructive solid geometry (CSG) modeller pre-defines a number of solids called primitives. Common primitives include solids derived from planar and quadric geometries such as spheres, cubes, cylinders and cones. A solid can be represented by the modeller if it is one of the primitives or can be constructed from the primitives by applying one or more modelling operations, such as linear transformations (see Chapters 1 and 2), and Boolean operations.

Three *Boolean operations* are used: union, intersection and difference. Let A and B be two sets of points representing solids or half-spaces. Then the *union* of A and B, denoted $A \cup B$, is the set of all points contained in A or B (including those points in both A and B). The *intersection* of A and B, denoted $A \cap B$, is the set of all points contained in both A and B. The *difference* of A and B, denoted $A \setminus B$, is the set of all points contained in A but not contained in B. Figure 9.18 exemplifies the three operations for two solid blocks. Boolean operations are a more general concept than presented here and can be defined for sets of any objects.

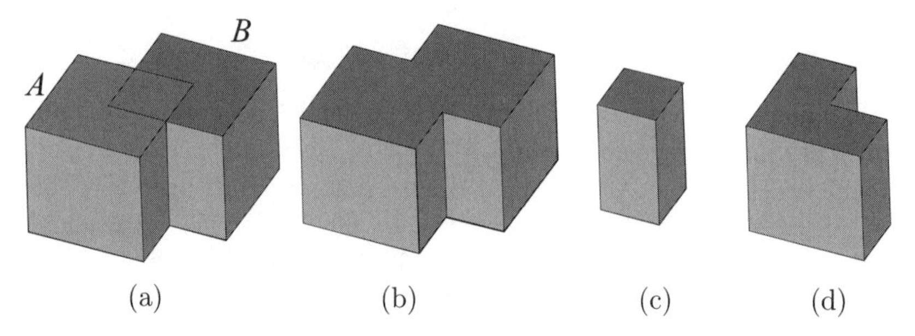

Figure 9.18 (a) Blocks A and B, (b) union $A \cup B$, (c) intersection $A \cap B$, and (d) difference $A \setminus B$

Example 9.28 (Boolean Operations)

In order to cut a hole in a block of material A (called the *blank*) it is necessary to construct an object (called the *tool*) that has the shape of the required hole. Let B be a cylindrical tool. A transformation is applied to B so that it overlaps the region of A where the hole is to be cut. Note that the tool can be larger than the required hole. The material that is to be removed from the block is obtained by performing the difference operation $A \setminus B$ as shown in Figure 9.19.

Whenever modelling operations are applied to objects there is a potential problem that the result is invalid, that is, it cannot be represented by the modeller. For instance, a modeller that can only represent curves and surfaces in Bézier form cannot represent the intersection of two Bézier surfaces since the curve of intersection is not (in general) a Bézier curve. One of the strengths of the CSG representation is that Boolean operations are guaranteed to give valid results. Further, the implicit surface definitions make it straightforward to per-

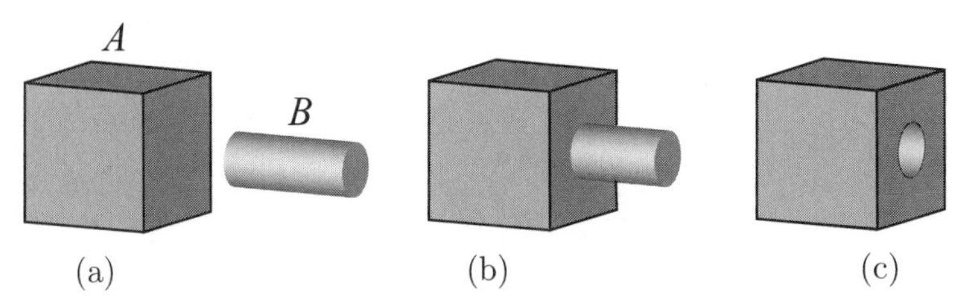

Figure 9.19 Cutting a hole using a Boolean operation

form interrogations such as determining whether a point lies inside or outside of a solid or on a boundary surface. However, the implicit representations and the lack of topological information makes rendering computationally expensive since the edge and boundary information needs to be computed before it can be drawn.

9.7.4 Boundary Representations (B-rep)

B-rep modellers distinguish the topology of a model from its geometry. A solid has three main topological entities, namely, faces, edges and vertices corresponding to the geometric entities surfaces, curves and points. The geometric entities store information about shape such as the coordinates of a point, the defining function of a parametric curve or surface, or the control points and knots of a B-spline curve or surface. The topological entities store information about their relationship to the other entities together with a reference to the corresponding geometric entity. B-rep modellers commonly use the following terminology.

Vertex: A vertex represents a point that specifies where edges meet.

Edge: An edge represents a finite arc of a curve bounded by two vertices, and specifies where two faces intersect.

Loop: A loop is a connected sequence of edges.

Face: A face represents a finite region of a surface bounded by one outer loop and a number of inner loops. The inner loops cannot be nested and must not intersect.

Shell: A shell is a collection of connected faces.

Solid: A solid is a union of volumes enclosed by shells.

B-rep modellers may also represent non-solid objects such as a *sheet* which is defined by a face and does not necessarily bound a finite volume, and a *wire* which is defined by a connected sequence of edges.

Example 9.29

Consider the tetrahedron of Figure 9.20(a). There are four vertices $\mathbf{V}_1$, $\mathbf{V}_2$, $\mathbf{V}_3$, $\mathbf{V}_4$, six edges $\mathbf{E}_1$, $\mathbf{E}_2$, $\mathbf{E}_3$, $\mathbf{E}_4$, $\mathbf{E}_5$, $\mathbf{E}_6$, and four faces $\mathbf{F}_1$, $\mathbf{F}_2$, $\mathbf{F}_3$, $\mathbf{F}_4$. Face $\mathbf{F}_1$ has one loop of edges $\mathbf{E}_1 \rightarrow \mathbf{E}_2 \rightarrow \mathbf{E}_3 \rightarrow \mathbf{E}_1$. Edge $\mathbf{E}_1$ has vertices $\mathbf{V}_1$ and $\mathbf{V}_2$.

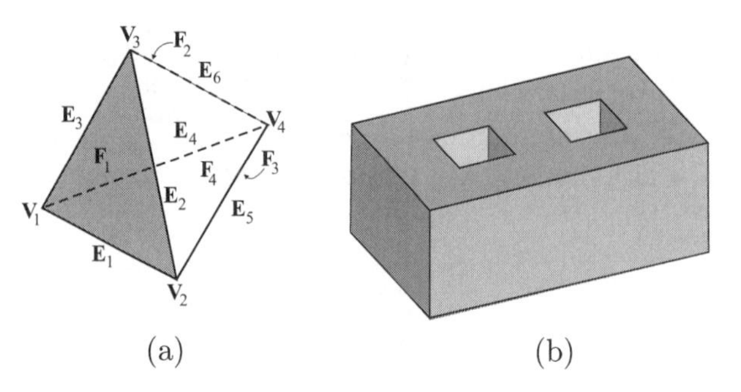

(a) (b)

Figure 9.20

Example 9.30

Consider Figure 9.20(b) showing a block with two holes. The solid is defined by one shell with 14 faces. Twelve of the faces are rectangular and each has one loop of four edges and four vertices. The two remaining faces, forming the top and bottom of the object, have three loops each: one outer and two inner. The solid is defined by 24 vertices, 36 edges, 14 faces and one shell.

The separation of the topology and the geometry is an important aspect of the B-rep. The topology of two instances of a tetrahedron, for example, is identical, whereas the underlying geometry, such as the coordinates of the vertices, may differ. An application of a non-singular transform to a solid has no effect on the topology, but it does change the geometry.

Solids with more than one shell arise, for instance, when the solid contains a void. Voids are obtained by the Boolean subtraction $A \setminus B$ of two solids whenever B is contained in the interior of A. Multiple shells also arise when a Boolean operation causes a volume to be cut into disjoint volumes called *lumps*.

In CAD applications a solid object is said to be *manifold* whenever the following conditions are satisfied:

1. All edges and faces are bounded.
2. Edges only intersect at vertices.
3. Faces only intersect in edges.
4. Edges are contained in exactly two faces.

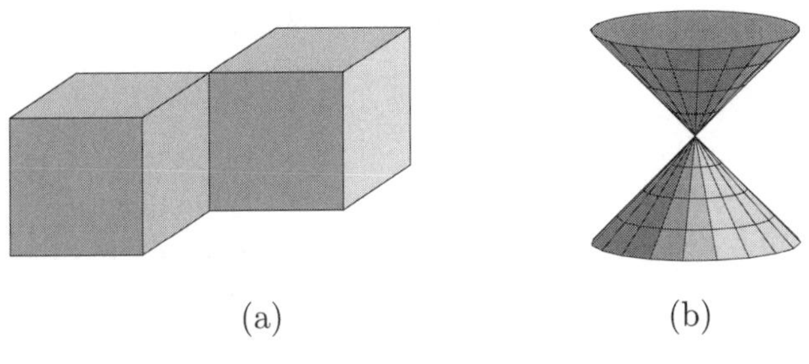

(a) (b)

Figure 9.21 Non-manifold bodies

Figure 9.21 shows examples of non-manifold objects. In the process of applying Boolean operations to manifold objects it is possible to obtain objects that are no longer manifold. For instance the object of Figure 9.21(a), which violates condition (4) above, can be obtained by applying the union operation to two blocks. The solid of Figure 9.21(b) violates condition (3).

As a sanity check, solids must satisfy the Euler–Poincaré Formula

$$V - E + F - I = 2(S - H) \tag{9.24}$$

where V, E, F, I, S, and H are the numbers of vertices, edges, faces, inner loops, shells, and holes, respectively. The solid of Example 9.30 satisfies the Euler–Poincaré formula since there are 24 vertices, 36 edges, 14 faces, and two holes. Further, there are two faces, each with two inner loops, giving 4 loops in total, and therefore yielding the identity $24 - 36 + 14 - 4 = 2(1 - 2)$.

A B-rep modeller is considered to be more versatile than a CSG modeller because it is possible to define objects with complex boundary faces using B-splines or general parametric surfaces, whereas a CSG modeller is limited to solids obtained from the primitives. The operation of determining whether a point lies inside or outside of a solid, or on a boundary surface is often computationally expensive for a B-rep modeller than for a CSG modeller which can yield the information from the half-spaces defining the solid.

CSG modellers retain information about the Boolean operations used to construct a solid. So if a user wishes to make a change to a model parameter such as the position of a vertex or the radius of a sphere, then the solid can easily be reconstructed by repeating the Boolean operations on the objects with the new dimensions. To emulate the CSG data structure, some B-rep modellers have a model design history that retains information about how each solid is created so that the system can repeat the design process if a user modifies a parameter, or wishes to "undo" an operation to rectify an error. The danger for a B-rep system is that some modifications can result in self-intersections or other types of invalid model. In view of the pros and cons, the reader should not be surprised to learn that all the leading commercial CAD modellers use the B-rep method or a hybrid of B-rep and CSG representations.

EXERCISES

9.26. The B-rep data structure is complex and a number of similar solutions have been proposed. Investigate Baumgart's winged-edge B-rep structure and two improved representations: Mäntylä's half-edge structure [16] and Braid et al. [4].

9.27. Euler operations are a set of primitive operations that can be applied to a solid to give a new solid that satisfies the Euler–Poincaré formula (9.24) and therefore maintains validity. Investigate. See Toriya [16].

10

Curve and Surface Curvatures

The aim of this chapter is to discuss the local geometry of a curve or surface. In particular, to determine measures of how much a curve or surface "bends", and to describe the shape of a curve or surface in the vicinity of a point on that curve or surface. These measures or "curvatures" have applications in determining the quality, or isolating imperfections of, the curves and surfaces produced by a designer using a CAD package.

10.1 Curvature of a Plane Curve

Let $\mathbf{C}(t) = (x(t), y(t))$ be a regular parametric plane curve defined on an interval I (open or closed). It is assumed that $\mathbf{C}(t)$ is C^1-continuous (that is, the derivatives of $x(t)$ and $y(t)$ exist and are continuous) and that higher order derivatives exist whenever the context suggests that they are required. In Section 5.4 it was shown that a regular curve can be reparametrized with respect to the arclength parameter s to give a unit speed curve. The curve $\mathbf{C}(t)$ and its unit speed reparametrization $\mathbf{C}(t(s))$ are both denoted by $\mathbf{C}$. Differentiation with respect to a general parameter will be denoted by a " $\cdot$ " and differentiation with respect to the unit speed parameter will be denoted by a " $'$ ". For instance, $\dot{\mathbf{C}} = \frac{d\mathbf{C}}{dt}$ and $\mathbf{C}' = \frac{d\mathbf{C}}{ds}$. Recall that

$$s(t) = \int_{t_0}^{t} \left((\dot{x}(u))^2 + (\dot{y}(u))^2 \right)^{1/2} du$$

and

$$\nu(t) = \dot{s}(t) = \left((\dot{x}(t))^2 + (\dot{y}(t))^2 \right)^{1/2} .$$

Let $\mathbf{t}$ and $\mathbf{n}$ denote the unit tangent and normal of $\mathbf{C}$. Since $\mathbf{t}(s)\cdot\mathbf{t}(s) = 1$, differentiation yields $\mathbf{t}(s)\cdot\mathbf{t}'(s) = 0$, implying $\mathbf{t}'(s)$ is perpendicular to $\mathbf{t}(s)$. Hence $\mathbf{t}'(s)$ is parallel to $\mathbf{n}(s)$, and so

$$\mathbf{t}'(s) = \kappa(s)\mathbf{n}(s) \tag{10.1}$$

for some $\kappa(s)$ called the *curvature* of $\mathbf{C}$. Since $\dot{s}(t) = \nu(t)$, the chain rule gives $\dot{\mathbf{t}} = \nu\mathbf{t}'$, and hence in terms of a general parameter the curvature is given by

$$\dot{\mathbf{t}}(t) = \kappa(t)\nu(t)\mathbf{n}(t) . \tag{10.2}$$

It follows that

$$\kappa = |\mathbf{t}'| = \left|\dot{\mathbf{t}}\right| / \nu .$$

Further, differentiation of $\mathbf{n}(s) \cdot \mathbf{n}(s) = 1$ gives $\mathbf{n}(s) \cdot \mathbf{n}'(s) = 0$, implying $\mathbf{n}'(s)$ is perpendicular to $\mathbf{n}(s)$. Hence $\mathbf{n}'(s)$ is parallel to $\mathbf{t}(s)$, and so $\mathbf{n}'(s) = \mu(s)\mathbf{t}(s)$ for some $\mu(s)$. To determine $\mu(s)$, differentiate $\mathbf{t}(s) \cdot \mathbf{n}(s) = 0$ to give $\mathbf{t}'(s)\cdot\mathbf{n}(\mathbf{s}) + \mathbf{t}(s) \cdot \mathbf{n}'(s) = 0$. Thus

$$(\kappa(s)\mathbf{n}(s)) \cdot\mathbf{n}(s) + \mathbf{t}(s) \cdot (\mu(s)\mathbf{t}(s)) = \kappa(s) + \mu(s) = 0 .$$

Hence $\mu(s) = -\kappa(s)$, and

$$\mathbf{n}'(s) = -\kappa(s)\mathbf{t}(s) .$$

The chain rule gives $\dot{\mathbf{n}} = \nu\mathbf{n}'$, and hence in terms of a general parameter

$$\dot{\mathbf{n}}(t) = -\kappa(t)\nu(t)\mathbf{t}(t) . \tag{10.3}$$

Equations (10.2) and (10.3) are known as the *Frenet formulae* for plane curves.

Theorem 10.1

The curvature of a regular plane curve $\mathbf{C}(t) = (x(t), y(t))$ is

$$\kappa(t) = \frac{\dot{x}(t)\ddot{y}(t) - \ddot{x}(t)\dot{y}(t)}{(\dot{x}(t)^2 + \dot{y}(t)^2)^{3/2}} .$$

Proof

The derivatives of $\mathbf{t}(t)= \dot{\mathbf{C}}(t) / \nu(t)$ and $\nu(t) = \left(\dot{x}(t)^2 + \dot{y}(t)^2\right)^{1/2}$ are

$$\begin{aligned} \dot{\mathbf{t}}(t) &= \left(\nu(t)\ddot{\mathbf{C}}(t) - \dot{\nu}(t)\dot{\mathbf{C}}(t)\right) \Big/ \nu(t)^2 , \text{ and} \\ \dot{\nu}(t) &= (\dot{x}(t)\ddot{x}(t) + \dot{y}(t)\ddot{y}(t))/ \nu(t) . \end{aligned}$$

So

$$\dot{\mathbf{t}}(t) = \frac{(\dot{x}(t)\ddot{y}(t) - \dot{y}(t)\ddot{x}(t))\,(-\dot{y}(t), \dot{x}(t))}{(\dot{x}(t)^2 + \dot{y}(t)^2)^{3/2}} .$$

Then $\dot{\mathbf{t}}(t) = \kappa(t)\nu(t)\mathbf{n}(t)$ implies that $\dot{\mathbf{t}}(t) \cdot \mathbf{n}(t) = \kappa(t)\nu(t)$, and hence

$$\kappa(t) = \frac{\dot{\mathbf{t}}(t) \cdot \mathbf{n}(t)}{\nu(t)} = \frac{\dot{x}(t)\ddot{y}(t) - \dot{y}(t)\ddot{x}(t)}{(\dot{x}(t)^2 + \dot{y}(t)^2)^{3/2}} .$$

□

The next aim is to show that $\kappa(t)$ is a measure of the "bendiness" of a regular plane curve. This is accomplished by showing that, near a given point on a curve, the curve is well approximated by a circle. The radius of that circle measures the extent to which the curve bends. It will be shown that $\kappa(t)$ is the reciprocal of the radius.

To this end, consider three neighbouring points $\mathbf{C}(t-\delta t)$, $\mathbf{C}(t)$, and $\mathbf{C}(t+\delta t)$ of a regular curve $\mathbf{C}$. Let $\mathbf{x} = (x, y)$. Suppose $(\mathbf{x}-\mathbf{c})\cdot(\mathbf{x}-\mathbf{c})-r^2 = 0$ ($\mathbf{c}$ constant) is the unique circle through the points, and let

$$\sigma(t) = (\mathbf{C}(t) - \mathbf{c}) \cdot (\mathbf{C}(t) - \mathbf{c}) - r^2 .$$

Then $\sigma(t - \delta t) = \sigma(t) = \sigma(t + \delta t) = 0$, and by Rolle's theorem, there exist $t_0 \in (t - \delta t, t)$ and $t_1 \in (t, t + \delta t)$ such that $\dot{\sigma}(t_0) = \dot{\sigma}(t_1) = 0$. A further application of Rolle's theorem implies that there exists a $t_2 \in (t_0, t_1)$ such that $\ddot{\sigma}(t_2) = 0$. Thus

$$\begin{aligned}
\dot{\sigma}(t_0) &= 2(\mathbf{C}(t_0) - \mathbf{c}) \cdot \dot{\mathbf{C}}(t_0) = 0 , \\
\dot{\sigma}(t_1) &= 2(\mathbf{C}(t_1) - \mathbf{c}) \cdot \dot{\mathbf{C}}(t_1) = 0 , \\
\ddot{\sigma}(t_2) &= 2(\mathbf{C}(t_2) - \mathbf{c}) \cdot \ddot{\mathbf{C}}(t_2) + 2\dot{\mathbf{C}}(t_2) \cdot \dot{\mathbf{C}}(t_2) = 0 .
\end{aligned}$$

Letting $\delta t \to 0$, then the three points converge to the point $\mathbf{C}(t)$, and t_0, t_1, t_2 all converge to t so that

$$2(\mathbf{C}(t) - \mathbf{c}) \cdot \dot{\mathbf{C}}(t) = 0 , \tag{10.4}$$

$$2(\mathbf{C}(t) - \mathbf{c}) \cdot \ddot{\mathbf{C}}(t) + 2\dot{\mathbf{C}}(t) \cdot \dot{\mathbf{C}}(t) = 0 . \tag{10.5}$$

The circle converges to the circle, known as the *osculating circle*, which best fits $\mathbf{C}$ at the point $\mathbf{C}(t)$. Equation (10.4) implies that $\mathbf{C}(t) - \mathbf{c}$ is perpendicular to the tangent vector and hence parallel to the normal vector. Thus $\mathbf{C}(t) - \mathbf{c} = \mu\mathbf{n}(t)$ for some μ. Substituting in (10.5) gives

$$2(\mu\mathbf{n}(t)) \cdot \ddot{\mathbf{C}}(t) + 2\dot{\mathbf{C}}(t) \cdot \dot{\mathbf{C}}(t) = 0 . \tag{10.6}$$

Since $\dot{\mathbf{C}} = \nu\mathbf{t}$ and $\dot{\mathbf{t}} = \nu\kappa\mathbf{n}$, it follows that $\ddot{\mathbf{C}} = \dot{\nu}\mathbf{t} + \nu\dot{\mathbf{t}} = \dot{\nu}\mathbf{t} + \nu^2\kappa\mathbf{n}$. Substituting for $\dot{\mathbf{C}}$ and $\ddot{\mathbf{C}}$ in (10.6) yields

$$2(\mu\mathbf{n}) \cdot \left(\dot{\nu}\mathbf{t} + \nu^2\kappa\mathbf{n}\right) + 2\,(\nu\mathbf{t}) \cdot (\nu\mathbf{t}) = 2\left(\mu\nu^2\kappa + \nu^2\right) = 0\,.$$

Hence $\mu = -\frac{1}{\kappa}$ and the centre of the osculating circle is $\mathbf{c} = \mathbf{C}(t) + \frac{1}{\kappa(t)}\mathbf{n}(t)$, called the *centre of curvature*. The osculating circle has radius $\rho(t) = \frac{1}{|\kappa(t)|}$, called the *radius of curvature*. Since the curve is well approximated by the osculating circle, the curvature $\kappa(t)$ measures the bendiness of the curve at $\mathbf{C}(t)$. When $|\kappa(t)|$ is small, $\rho(t)$ is large and therefore the curve is fairly flat, whereas when $|\kappa(t)|$ is large, $\rho(t)$ is small and the curve bends a fair amount. When $\kappa(t) > 0$, $\mathbf{c}$ lies on the same side of the curve as $\mathbf{n}$, and when $\kappa(t) < 0$, $\mathbf{c}$ lies on the opposite side of the curve to $\mathbf{n}$, as shown later in Figure 10.1.

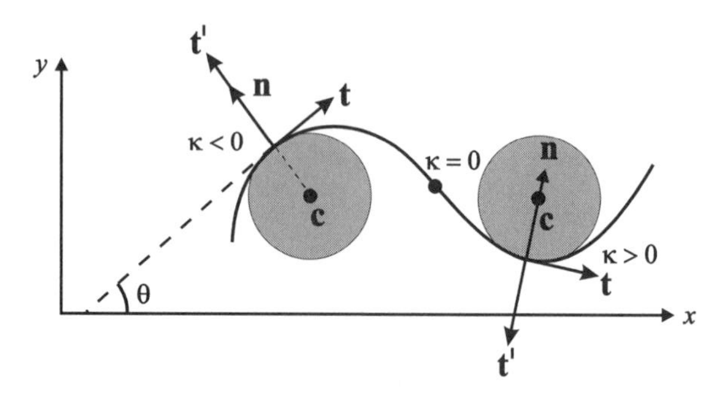

Figure 10.1 Curvature of a plane curve

Example 10.2

Consider the ellipse $\mathbf{C}(\theta) = (a\cos\theta, b\sin\theta)$. Then $\dot{\mathbf{C}}(\theta) = (-a\sin\theta, b\cos\theta)$ and $\ddot{\mathbf{C}}(\theta) = (-a\cos\theta, -b\sin\theta)$, giving

$$\kappa(\theta) = \frac{(-a\sin\theta)\,(-b\sin\theta) - (b\cos\theta)\,(-a\cos\theta)}{\left((-a\sin\theta)^2 + (b\cos\theta)^2\right)^{3/2}} = \frac{ab}{\left(a^2\sin^2\theta + b^2\cos^2\theta\right)^{3/2}}\,.$$

The graph of the curvature for $a = 3$, $b = 2$, and $0 \le \theta \le 2\pi$ is shown in Figure 10.2. The curvature has maximum values at $\theta = 0$ and $\theta = \pi$, and minimum values at $\theta = \pi/2$ and $\theta = 3\pi/2$. The corresponding points on the ellipse are easily identified.

Suppose $\kappa(s)$ is the curvature function of a unit speed curve $\mathbf{C}(s) = (x(s), y(s))$. Then since $\mathbf{t}(s)$ is a unit vector, $\mathbf{t}(s) = (\cos\theta(s), \sin\theta(s))$ where

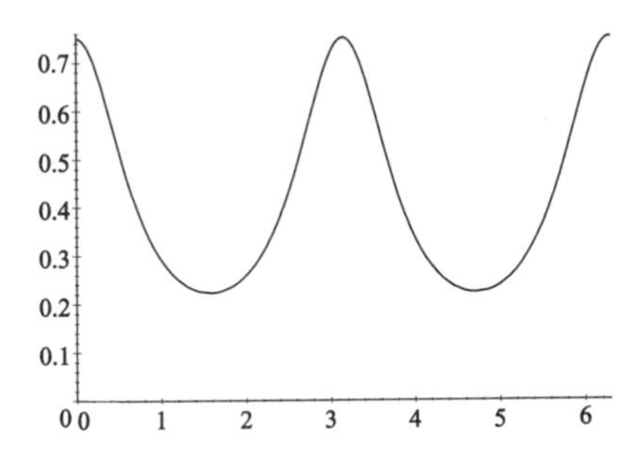

Figure 10.2 Curvature function for the ellipse $\mathbf{C}(\theta) = (3\cos\theta, 2\sin\theta)$

$\theta(s)$ is the angle the tangent vector makes with the x-axis (see Figure 10.1). Then $\mathbf{n}(s) = (-\sin\theta(s), \cos\theta(s))$, $\mathbf{t}'(s) = \theta'(s)\,(-\sin\theta(s), \cos\theta(s)) = \theta'(s)\mathbf{n}(s)$, and comparison with (10.1) gives

$$\kappa(s) = \theta'(s)\ . \tag{10.7}$$

So curvature is the rate of change of the tangent (when the curve is unit speed). This fact is used in Theorem 10.3. When $\kappa > 0$, $\mathbf{n}$ and $\mathbf{t}'$ have the same direction, and when $\kappa < 0$, $\mathbf{n}$ and $\mathbf{t}'$ have opposite directions, as shown in Figure 10.1.

Theorem 10.3

Let $\kappa(s)$ be a continuous function. Then there exists a planar unit speed curve $\mathbf{C}(s) = (x(s), y(s))$ with curvature $\kappa(s)$. The curve is unique up to its position and orientation in the plane.

Proof

Suppose $\kappa(s)$ is the curvature function of a unit speed curve $\mathbf{C}(s) = (x(s), y(s))$, and suppose $(x(s_0), y(s_0)) = (x_0, y_0)$ and $(x'(s_0), y'(s_0)) = (x'_0, y'_0)$. Let $\mathbf{t}(s) = (x'(s), y'(s)) = (\cos\theta(s), \sin\theta(s))$. Then (10.7) gives

$$\theta(s) = \theta(s_0) + \int_{s_0}^{s} \kappa(u)\ du\ ,$$

where $\theta(s_0) = \tan^{-1}(y'_0/x'_0)$ if $x'_0 \neq 0$, and $\theta(s_0) = \pi/2$ if $x'_0 = 0$. Thus

$$x(s) = x_0 + \int_{s_0}^{s} \cos\theta(u)\ du, \quad y(s) = y_0 + \int_{s_0}^{s} \sin\theta(u)\ du\ . \tag{10.8}$$

Equations (10.8) show the existence of a unit speed curve with curvature $\kappa(s)$. For a given initial point (x_0, y_0) and tangent direction (x'_0, y'_0), the constructed curve is uniquely determined. So all unit speed curves with curvature $\kappa(s)$ can be mapped to one another by a planar transformation consisting of a rotation which aligns the initial tangent directions of the curves, and a translation which maps the initial point of one curve to the initial point of the other.

□

The curvature $\kappa(s)$ is often referred to as the *natural* or *intrinsic equation* of a curve.

Remark 10.4

An alternative proof of Theorem 10.3 is obtained by noting that the identity $\mathbf{t}'(s) = \kappa(s)\mathbf{n}(s)$ yields a system of ordinary differential equations

$$\begin{cases} x''(s) = -\kappa(s)y'(s) \\ y''(s) = \kappa(s)x'(s) \end{cases} . \tag{10.9}$$

The existence of solutions to this system is a result in the theory of differential equations. The system can be converted to a second order differential equation. Let $X(s) = x'(s)$ and $Y(s) = y'(s)$, then (10.9) gives the system of first order differential equations

$$\begin{cases} X'(s) = -\kappa(s)Y(s) \\ Y'(s) = \kappa(s)X(s) \end{cases} . \tag{10.10}$$

Differentiating the first equation gives $X''(s) = -\kappa'(s)Y(s) - \kappa(s)Y'(s)$ and substituting $Y' = \kappa(s)X(s)$ and $Y(s) = -X'(s)/\kappa(s)$ yields

$$X''(s) - \frac{\kappa'(s)}{\kappa(s)}X'(s) + \kappa(s)^2X(s) = 0 . \tag{10.11}$$

Equation (10.11) can be solved for $X(s)$ and the result used to determine $Y(s)$. Finally, $X(s)$ and $Y(s)$ are integrated to obtain $x(s)$ and $y(s)$.

For fairly simple choices of $\kappa(s)$, Equations (10.11), (10.9), or (10.8) can be solved to give analytical solutions for $x(s)$ and $y(s)$ using standard techniques. For more complicated $\kappa(s)$, numerical integration methods can be used to evaluate the integrals (10.8).

Example 10.5

Let $\kappa(s) = 1/R$, $(x(0), y(0)) = (R, 0)$, and $(x'(0), y'(0)) = (0, 1)$. Following the proof of Theorem 10.3, $s_0 = 0$, $\alpha = \pi/2$, and $\theta(s) = \pi/2 + \int_0^s 1/R \, du =$

$\pi/2 + s/R$. So

$$x(s) = R + \int_0^s \cos(\pi/2 + u/R)\ du = R + \int_0^s -\sin(u/R)\ du$$
$$= R + (R\cos(s/R) - R) = R\cos(s/R)\ .$$

Similarly,

$$y(s) = \int_0^s \sin(\pi/2 + u/R)\ du = \int_0^s \cos(u/R)\ du = R\sin(s/R)\ .$$

Thus $(x(s), y(s)) = (R\cos(s/R), R\sin(s/R))$. The curve is a circle radius $|R|$.

Alternatively, $\kappa(s) = 1/R$, $\kappa'(s) = 0$, and (10.11) gives

$$X''(s) + \frac{1}{R^2}X(s) = 0\ ,$$

which has solutions of the form $X(s) = A\cos(s/R) + B\sin(s/R)$. The initial condition $X(0) = x'(0) = 0$ implies $A = 0$. Thus $x'(s) = B\sin(s/R)$, and integrating gives $x(s) = -BR\cos(s/R)$. The initial condition $x(0) = R$ yields $B = -1$ and $x(s) = R\cos(s/R)$. Finally, $Y(s) = -X'(s)/\kappa(s) = \frac{1}{R}\cos(s/R)$, and so $y(s) = \sin(s/R)$.

Example 10.6

Let $\kappa(s) = as$, $(x(0), y(0)) = (0,0)$, and $(x'(0), y'(0)) = (1,0)$. Then, $s_0 = 0$, $\alpha = 0$, $\theta(s) = \int_0^s au\ du = \frac{1}{2}as^2$, and

$$x(s) = \int_0^s \cos\left(\frac{1}{2}as^2\right)\ du, \quad y(s) = \int_0^s \sin\left(\frac{1}{2}au^2\right)\ du\ .$$

The integrals in the above expressions are known as Fresnel integrals. The curve obtained is called a clothoid or Cornu spiral, and is illustrated in Figure 10.3.

EXERCISES

10.1. Compute the unit tangent vector, unit normal vector, and curvature for each of the following curves:

(a) *catenary:* $\mathbf{C}(t) = (t, c\cosh(t/c))$;

(b) *cycloid:* $\mathbf{C}(t) = (t - \sin t, 1 - \cos t)$, $t \in [-\pi, \pi]$;

(c) *logarithmic spiral:* $\mathbf{C}(t) = \left(ae^{bt}\cos t, ae^{bt}\sin t\right)$.

10.2. Determine the curvature $\kappa(t)$ of the curve (t, t^3). Sketch the curve, and indicate the parts of the curve where $\kappa > 0$ and $\kappa < 0$.

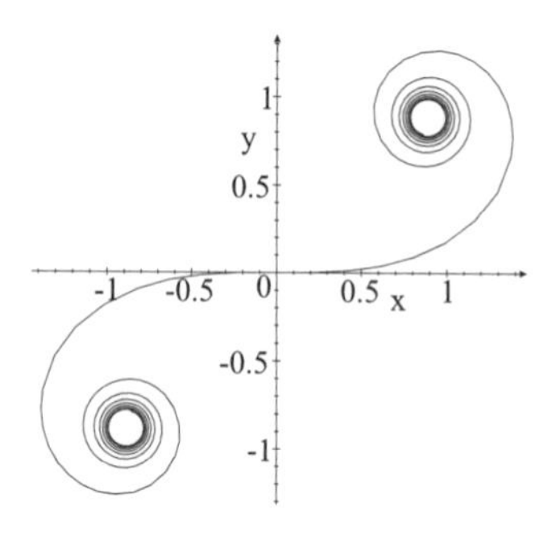

Figure 10.3 Clothoid or Cornu spiral

10.3. Determine the parametrization of the plane unit speed curve with curvature

(a) $\kappa(s) = 1/\sqrt{1-s^2}$;

(b) $\kappa(s) = 1/\sqrt{s}$;

(c) $\kappa(s) = -a/(a^2+s^2)$ where a is a positive real number.

10.4. Show that a regular plane curve with curvature $\kappa = 0$ is a straight line.

10.5. Show that a plane curve with polar coordinates $r = r(\theta)$ for $\theta \in [a, b]$ (so $(x(\theta), y(\theta)) = (r(\theta)\cos\theta, r(\theta)\sin\theta)$) has arclength

$$\int_a^b \sqrt{r(\theta)^2 + r'(\theta)^2}\, d\theta\,,$$

and curvature

$$\kappa(\theta) = \frac{2r'(\theta)^2 - r(\theta)\, r'(\theta) + r(\theta)^2}{\left(r(\theta)^2 + r'(\theta)^2\right)^{3/2}}\,.$$

10.6. Let $\mathbf{C}(t)$ be a regular plane curve such that $\kappa(t) \neq 0$. The curve $\mathbf{E}(t) = \mathbf{C}(t) + \frac{1}{\kappa(t)}\mathbf{n}(t)$ is called the *evolute* of $\mathbf{C}(t)$. The evolute is the locus of the centres of curvature of $\mathbf{C}(t)$. Determine the evolute of the following curves:

(a) *cycloid*: $\mathbf{C}(t) = (t + \sin t, 1 - \cos t)$.

(b) *ellipse*: $\mathbf{C}(t) = (a\cos t, b\sin t)$.

10.7. Consider the ellipse $\mathbf{C}(t) = (3\cos t, 2\sin t)$, $t \in [0, 2\pi]$.

(a) Determine the parametric equation of the offset of the ellipse at a distance d.

(b) Determine κ and $\dot{\kappa}$. Hence calculate the parameter values, and the corresponding points on the ellipse, where the curvature is at a maximum or a minimum.

(c) Obtain the maximum and minimum values of curvature. Deduce the maximum radius d that a ball cutter can be in order to cut the shape of the ellipse (assuming the cutter is in the interior of the ellipse).

10.2 Curvature and Torsion of a Space Curve

Let $\mathbf{C}(t) = (x(t), y(t), z(t))$ be a regular parametric space curve defined on an interval I (open or closed). As for the case of plane curves, the curve $\mathbf{C}(t)$ and its unit speed reparametrization $\mathbf{C}(t(s))$ are both denoted by $\mathbf{C}$, and differentiation with respect to a general and unit speed parameter are distinguished by $\cdot$ and $'$ respectively. The speed of a space curve is $\nu(t) = \left(\dot{x}(t)^2 + \dot{y}(t)^2 + \dot{z}(t)^2\right)^{1/2}$, and the chain rule for differentiation yields $\dot{\mathbf{C}} = \nu\mathbf{C}'$. The *unit tangent vector* is defined to be

$$\mathbf{t} = \mathbf{C}' = \dot{\mathbf{C}}/\nu \,. \tag{10.12}$$

The vector $\mathbf{k} = \mathbf{t}' = \dot{\mathbf{t}}/\nu$ is called the *curvature vector*, and its magnitude

$$\kappa = |\mathbf{k}| = |\mathbf{t}'| = \left|\dot{\mathbf{t}}\right|/\,\nu,$$

is called the *curvature* of $\mathbf{C}$. The curvature measures the rate of change of the tangent $\mathbf{t}$ along the curve with respect to arclength. At a given point of a space curve, there are infinitely many vectors which are perpendicular to $\mathbf{t}$, and therefore normal to the curve. Since $\mathbf{t}$ is a unit vector, $\mathbf{t}\cdot\mathbf{t} = 1$ and $\mathbf{t}\cdot\mathbf{t}' = 0$. Hence $\mathbf{k} = \mathbf{t}'$ is perpendicular to $\mathbf{t}$. At every point of the curve for which $\kappa \neq 0$ there is a well-defined unit vector

$$\mathbf{n} = \mathbf{t}'\,/|\mathbf{t}'| = \dot{\mathbf{t}}\,/\left|\dot{\mathbf{t}}\right|$$

called the *principal normal.* It follows that

$$\mathbf{t}' = \kappa\mathbf{n} \quad \text{and} \quad \dot{\mathbf{t}} = \kappa\nu\mathbf{n}\,. \tag{10.13}$$

If $\kappa = 0$ then the principal normal is not well defined.

At a point $\mathbf{p}$ on the curve $\mathbf{C}$, the plane containing point $\mathbf{p}$, and directions $\mathbf{t}$ and $\mathbf{n}$ is called the *osculating plane.* The unit vector $\mathbf{b} = \mathbf{t}\times\mathbf{n}$, which is perpendicular to the osculating plane, is called the *binormal* vector. The plane containing $\mathbf{p}$, $\mathbf{n}$, and $\mathbf{b}$ is called the *normal plane*, and the plane containing $\mathbf{p}$, $\mathbf{t}$, and $\mathbf{b}$ is called the *rectifying plane*. The planes are depicted in Figure 10.4. The

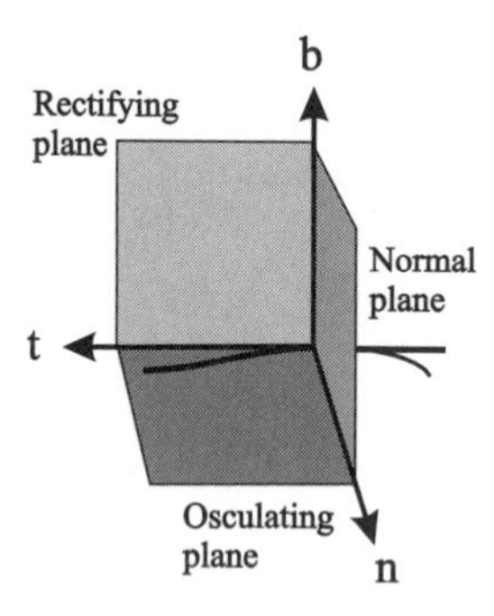

Figure 10.4 Osculating, normal and rectifying planes of a space curve

mutually perpendicular unit vectors $\mathbf{t}$, $\mathbf{n}$, and $\mathbf{b}$ are called the *Frenet frame*, and they satisfy

$$\begin{aligned} \mathbf{t}\cdot\mathbf{t}=1, \quad \mathbf{n}\cdot\mathbf{n}=1, \quad \mathbf{b}\cdot\mathbf{b}=1\,, \\ \mathbf{t}\cdot\mathbf{n}=0, \quad \mathbf{t}\cdot\mathbf{b}=0, \quad \mathbf{n}\cdot\mathbf{b}=0\,. \end{aligned} \tag{10.14}$$

Any vector $\mathbf{v}$ can be expressed in terms of the frame: $\mathbf{v} = v_1\mathbf{t}+v_2\mathbf{n}+v_3\mathbf{b}$. Then $\mathbf{v}\cdot\mathbf{t} = (v_1\mathbf{t}+v_2\mathbf{n}+v_3\mathbf{b})\cdot\mathbf{t} = v_1(\mathbf{t}\cdot\mathbf{t}) + v_2(\mathbf{n}\cdot\mathbf{t}) + v_3(\mathbf{b}\cdot\mathbf{t}) = v_1$. Similarly, $v_2 = \mathbf{v}\cdot\mathbf{n}$ and $v_3 = \mathbf{v}\cdot\mathbf{b}$. So $\mathbf{v} = (\mathbf{v}\cdot\mathbf{t})\,\mathbf{t} + (\mathbf{v}\cdot\mathbf{n})\,\mathbf{n} + (\mathbf{v}\cdot\mathbf{b})\,\mathbf{b}$. The expression is called the orthonormal expansion of $\mathbf{v}$ with respect to the Frenet frame.

Differentiating the first row of equations of (10.14) yields

$$\mathbf{t}\cdot\dot{\mathbf{t}}=0, \quad \mathbf{n}\cdot\dot{\mathbf{n}}=0, \quad \mathbf{b}\cdot\dot{\mathbf{b}}=0\,. \tag{10.15}$$

The orthonormal expansion of $\dot{\mathbf{b}}$ with respect to the Frenet frame gives

$$\dot{\mathbf{b}} = (\dot{\mathbf{b}}\cdot\mathbf{t})\mathbf{t} + (\dot{\mathbf{b}}\cdot\mathbf{n})\mathbf{n} + (\dot{\mathbf{b}}\cdot\mathbf{b})\mathbf{b} = (\dot{\mathbf{b}}\cdot\mathbf{t})\mathbf{t} + (\dot{\mathbf{b}}\cdot\mathbf{n})\mathbf{n}\,.$$

Differentiating $\mathbf{b}\cdot\mathbf{t}=0$ gives $\dot{\mathbf{b}}\cdot\mathbf{t}+\mathbf{b}\cdot\dot{\mathbf{t}}=0$, and (10.13) implies

$$\dot{\mathbf{b}}\cdot\mathbf{t} = -\mathbf{b}\cdot\dot{\mathbf{t}} = -\mathbf{b}\cdot(\kappa\nu\mathbf{n}) = 0\,.$$

Hence $\dot{\mathbf{b}} = (\dot{\mathbf{b}}\cdot\mathbf{n})\mathbf{n}$ and therefore $\dot{\mathbf{b}}$ and $\mathbf{b}'$ are parallel to $\mathbf{n}$. Thus $\mathbf{b}' = -\tau\mathbf{n}$ for some τ called the *torsion*, and

$$\dot{\mathbf{b}} = \mathbf{b}'\nu = -\tau\nu\mathbf{n}\,. \tag{10.16}$$

$\mathbf{b}'$ is called the *torsion vector*. Torsion measures the bending of the curve out of the osculating plane, and can be computed using $\tau = -\left(\dot{\mathbf{b}} \cdot \mathbf{n}\right)/\nu$ (also see Theorem 10.9).

Differentiating $\mathbf{n} \cdot \mathbf{b} = 0$ yields $\mathbf{n} \cdot \dot{\mathbf{b}} + \dot{\mathbf{n}} \cdot \mathbf{b} = 0$, and (10.16) implies

$$\dot{\mathbf{n}} \cdot \mathbf{b} = -\mathbf{n} \cdot \dot{\mathbf{b}} = -\mathbf{n} \cdot (-\tau\nu\mathbf{n}) = \tau\nu \,. \tag{10.17}$$

Likewise, differentiating $\mathbf{n} \cdot \mathbf{t} = 0$ gives $\mathbf{n} \cdot \dot{\mathbf{t}} + \dot{\mathbf{n}} \cdot \mathbf{t} = 0$, and (10.13) gives

$$\dot{\mathbf{n}} \cdot \mathbf{t} = -\mathbf{n} \cdot \dot{\mathbf{t}} = -\mathbf{n} \cdot (\kappa\nu\mathbf{n}) = -\kappa\nu \,. \tag{10.18}$$

The orthonormal expansion for $\dot{\mathbf{n}}$ is

$$\dot{\mathbf{n}} = (\dot{\mathbf{n}} \cdot \mathbf{t})\mathbf{t} + (\dot{\mathbf{n}} \cdot \mathbf{n})\mathbf{n} + (\dot{\mathbf{n}} \cdot \mathbf{b})\mathbf{b} \,,$$

and it follows from (10.15), (10.17), and (10.18) that

$$\dot{\mathbf{n}} = -\kappa\nu\mathbf{t} + \tau\nu\mathbf{b} \,. \tag{10.19}$$

Together Equations (10.13), (10.16), and (10.19) yield the following theorem which expresses the vectors $\dot{\mathbf{t}}$, $\dot{\mathbf{n}}$, $\dot{\mathbf{b}}$ in terms of the Frenet frame.

Theorem 10.7 (Frenet–Serret Formulae)

Let $\mathbf{C}(t)$ be a regular curve with $\kappa(t) \neq 0$. Then

$$\begin{aligned} \dot{\mathbf{t}}(t) &= \kappa(t)\nu(t)\mathbf{n}(t) \,, \\ \dot{\mathbf{n}}(t) &= -\kappa(t)\nu(t)\mathbf{t}(t) + \tau(t)\nu(t)\mathbf{b}(t) \,, \\ \dot{\mathbf{b}}(t) &= -\tau(t)\nu(t)\mathbf{n}(t) \,. \end{aligned}$$

□

Example 10.8

Consider the *twisted cubic* $\mathbf{C}(t) = (t, \frac{1}{2}t^2, \frac{1}{6}t^3)$. Then $\dot{\mathbf{C}}(t) = (1, t, \frac{1}{2}t^2)$ and $\nu(t) = \left(1^2 + t^2 + \left(\frac{1}{2}t^2\right)\right)^{1/2} = \frac{2+t^2}{2}$. Hence

$$\mathbf{t} = \dot{\mathbf{C}}(t) \Big/ \left|\dot{\mathbf{C}}(t)\right| = \left(\frac{2}{2+t^2}, \frac{2t}{2+t^2}, \frac{t^2}{2+t^2}\right) .$$

Thus $\dot{\mathbf{t}} = \left(-\frac{4t}{(2+t^2)^2}, \frac{4-2t^2}{(2+t^2)^2}, \frac{4t}{(2+t^2)^2}\right)$, $\left|\dot{\mathbf{t}}\right| = \frac{2}{2+t^2}$, and

$$\mathbf{n} = \dot{\mathbf{t}} \,/ \left|\dot{\mathbf{t}}\right| = \left(-\frac{2t}{2+t^2}, \frac{2-t^2}{2+t^2}, \frac{2t}{2+t^2}\right) .$$

Further,

$$\mathbf{b} = \mathbf{t} \times \mathbf{n} = \left(\frac{t^2}{2+t^2}, -\frac{2t}{2+t^2}, \frac{2}{2+t^2} \right) .$$

The curvature is

$$\kappa = \left| \dot{\mathbf{t}} \right| / \nu = \frac{4}{(2+t^2)^2} .$$

Further,

$$\begin{aligned} \mathbf{n} \cdot \dot{\mathbf{b}} &= \left(-\frac{2t}{2+t^2}, \frac{2-t^2}{2+t^2}, \frac{2t}{2+t^2} \right) \cdot \left(\frac{4t}{(2+t^2)^2}, -\frac{4-2t^2}{(2+t^2)^2}, -\frac{4t}{(2+t^2)^2} \right) \\ &= -\frac{2}{2+t^2} . \end{aligned}$$

Thus, the torsion is

$$\tau = -\left(\mathbf{n} \cdot \dot{\mathbf{b}} \right) \Big/ \nu = \frac{4}{(2+t^2)^2} .$$

EXERCISES

10.8. Show that $\ddot{\mathbf{C}} = \dot{\nu}\mathbf{t} + \kappa\nu^2\mathbf{n}$ and $\dddot{\mathbf{C}} = \left(\ddot{\nu} - \kappa^2\nu^3 \right) \mathbf{t} + \left(3\kappa\nu\dot{\nu} + \dot{\kappa}\nu^2 \right) \mathbf{n} + \kappa\tau\nu^3\mathbf{b}$.

10.9. Show that for a unit speed curve $\mathbf{C}(s)$, $\kappa(s) = |\mathbf{C}''(s)|$.

10.10. Determine the Frenet frame, curvature, and torsion of the curve $\mathbf{C}(t) = (4\cos t, 5 - 5\sin t, -3\cos t)$. Describe the curve.

10.11. Determine the Frenet frame, curvature, and torsion of the curve $\mathbf{C}(t) = \left(3\cos t - 4\sin t, 3\sin t + 4\cos t, 5\sqrt{3}t \right)$.

The next theorem gives one of the most direct methods of computing the curvature, torsion, and Frenet frame. In particular, the curvature and torsion can be obtained without having to compute the Frenet frame.

Theorem 10.9

Let $\mathbf{C}$ be a regular curve. Then

$$\kappa = \frac{|\dot{\mathbf{C}} \times \ddot{\mathbf{C}}|}{|\dot{\mathbf{C}}|^3}, \quad \tau = \frac{(\dot{\mathbf{C}} \times \ddot{\mathbf{C}}) \cdot \dddot{\mathbf{C}}}{|\dot{\mathbf{C}} \times \ddot{\mathbf{C}}|^2}, \quad \mathbf{t} = \frac{\dot{\mathbf{C}}}{|\dot{\mathbf{C}}|}, \quad \mathbf{b} = \frac{\dot{\mathbf{C}} \times \ddot{\mathbf{C}}}{|\dot{\mathbf{C}} \times \ddot{\mathbf{C}}|}, \quad \mathbf{n} = \mathbf{b} \times \mathbf{t} .$$

Proof

Using $\dot{\mathbf{C}} = \nu\mathbf{t}$ and the results of Exercise 10.8,

$$\dot{\mathbf{C}} \times \ddot{\mathbf{C}} = \nu\mathbf{t} \times \left(\dot{\nu}\mathbf{t} + \kappa\nu^2\mathbf{n}\right) = \nu\dot{\nu}\left(\mathbf{t} \times \mathbf{t}\right) + \kappa\nu^3\left(\mathbf{t} \times \mathbf{n}\right) .$$

Then, since $\mathbf{t} \times \mathbf{t} = \mathbf{0}$ and $\mathbf{t} \times \mathbf{n} = \mathbf{b}$ it follows that

$$\dot{\mathbf{C}} \times \ddot{\mathbf{C}} = \kappa\nu^3\mathbf{b} . \tag{10.20}$$

Hence $\left|\dot{\mathbf{C}} \times \ddot{\mathbf{C}}\right| = \left|\kappa\nu^3\mathbf{b}\right| = \kappa\nu^3$, so that

$$\kappa = \frac{\left|\dot{\mathbf{C}}\times\ddot{\mathbf{C}}\right|}{\nu^3} = \frac{\left|\dot{\mathbf{C}}\times\ddot{\mathbf{C}}\right|}{\left|\dot{\mathbf{C}}\right|^3} \quad \text{and} \quad \mathbf{b} = \frac{\dot{\mathbf{C}}\times\ddot{\mathbf{C}}}{\left|\dot{\mathbf{C}}\times\ddot{\mathbf{C}}\right|} .$$

Equation (10.12) gives $\mathbf{t} = \dot{\mathbf{C}} \Big/ \left|\dot{\mathbf{C}}\right|$, and it follows that $\mathbf{n} = \mathbf{b} \times \mathbf{t}$. Equations (10.20) and (10.14), and Exercise 10.8, imply

$$\left(\dot{\mathbf{C}} \times \ddot{\mathbf{C}}\right) \cdot \dddot{\mathbf{C}} = \left(\kappa\nu^3\mathbf{b}\right) \cdot \left(\left(\ddot{\nu} - \kappa^2\nu^3\right)\mathbf{t} + \left(3\kappa\nu\dot{\nu} + \dot{\kappa}\nu^2\right)\mathbf{n} + \kappa\tau\nu^3\mathbf{b}\right) = \kappa^2\tau\nu^6 .$$

Hence,

$$\tau = \frac{\left(\dot{\mathbf{C}}\times\ddot{\mathbf{C}}\right)\cdot\dddot{\mathbf{C}}}{\kappa^2\nu^6} = \frac{\left(\dot{\mathbf{C}}\times\ddot{\mathbf{C}}\right)\cdot\dddot{\mathbf{C}}}{\left|\dot{\mathbf{C}}\times\ddot{\mathbf{C}}\right|^2} .$$

□

Example 10.10

Consider the helix $\mathbf{C}(t) = \left(a\cos\frac{t}{c}, a\sin\frac{t}{c}, \frac{bt}{c}\right)$ where $c = (a^2 + b^2)^{1/2}$ and $a > 0$. Then $\dot{\mathbf{C}} = \left(-\frac{a}{c}\sin\frac{t}{c}, \frac{a}{c}\cos\frac{t}{c}, \frac{b}{c}\right)$, $\ddot{\mathbf{C}} = \left(-\frac{a}{c^2}\cos\frac{t}{c}, -\frac{a}{c^2}\sin\frac{t}{c}, 0\right)$, $\dddot{\mathbf{C}} = \left(\frac{a}{c^3}\sin\frac{t}{c}, -\frac{a}{c^3}\cos\frac{t}{c}, 0\right)$. The curve is unit speed since

$$\left|\dot{\mathbf{C}}\right| = \sqrt{\left(-\frac{a}{c}\sin\frac{t}{c}\right)^2 + \left(\frac{a}{c}\cos\frac{t}{c}\right)^2 + \left(\frac{bt}{c}\right)^2} = 1 .$$

Then $\dot{\mathbf{C}} \times \ddot{\mathbf{C}} = \left(\frac{b}{c^3}a\sin\frac{t}{c}, -\frac{b}{c^3}a\cos\frac{t}{c}, \frac{a^2}{c^3}\right)$ and $\left|\dot{\mathbf{C}} \times \ddot{\mathbf{C}}\right| = \frac{a}{a^2+b^2}$. Hence

$$\kappa = \frac{\left|\dot{\mathbf{C}}\times\ddot{\mathbf{C}}\right|}{\left|\dot{\mathbf{C}}\right|^3} = \frac{a}{a^2 + b^2} .$$

(Since the curve is unit speed the curvature can also be computed using $\kappa = |\mathbf{t}'| = \left|\ddot{\mathbf{C}}\right|$.) Further, $\left(\dot{\mathbf{C}} \times \ddot{\mathbf{C}}\right) \cdot \dddot{\mathbf{C}} = \frac{a^2 b}{(a^2+b^2)^3}$, and hence

$$\tau = \frac{\left(\dot{\mathbf{C}}\times\ddot{\mathbf{C}}\right)\cdot\dddot{\mathbf{C}}}{\left|\dot{\mathbf{C}}\times\ddot{\mathbf{C}}\right|^2} = \frac{b}{a^2 + b^2} .$$

The Frenet frame is

$$\begin{aligned}
\mathbf{t} &= \dot{\mathbf{C}} \Big/ \left|\dot{\mathbf{C}}\right| = \left(-\frac{a}{c}\sin\frac{t}{c}, \frac{a}{c}\cos\frac{t}{c}, \frac{b}{c}\right), \\
\mathbf{b} &= \left(\dot{\mathbf{C}} \times \ddot{\mathbf{C}}\right) \Big/ \left|\dot{\mathbf{C}} \times \ddot{\mathbf{C}}\right| = \left(\frac{b}{c}\sin\frac{t}{c}, -\frac{b}{c}\cos\frac{t}{c}, \frac{a}{c}\right), \\
\mathbf{n} &= \mathbf{b} \times \mathbf{t} = \left(-\cos\frac{t}{c}, -\sin\frac{t}{c}, 0\right).
\end{aligned}$$

Example 10.11

Consider the curve $\mathbf{C}(t) = \left(t, \frac{1}{2}t^2, \frac{1}{6}t^3\right)$. Then $\dot{\mathbf{C}} = \left(1, t, \frac{1}{2}t^2\right)$, $\ddot{\mathbf{C}} = (0, 1, t)$, and $\dddot{\mathbf{C}} = (0, 0, 1)$. Then $\nu = \left|\dot{\mathbf{C}}\right| = \frac{2+t^2}{2}$, and

$$\left|\dot{\mathbf{C}} \times \ddot{\mathbf{C}}\right| = \left|\left(1, t, \frac{1}{2}t^2\right) \times (0, 1, t)\right| = \left|\left(\frac{1}{2}t^2, -t, 1\right)\right| = \frac{2+t^2}{2}.$$

Hence

$$\kappa = \frac{\left|\dot{\mathbf{C}} \times \ddot{\mathbf{C}}\right|}{\left|\dot{\mathbf{C}}\right|^3} = \frac{4}{(2+t^2)^2}.$$

Further, $\left(\dot{\mathbf{C}} \times \ddot{\mathbf{C}}\right) \cdot \dddot{\mathbf{C}} = \left(\frac{1}{2}t^2, -t, 1\right) \cdot (0, 0, 1) = 1$. So

$$\tau = \frac{\left(\dot{\mathbf{C}} \times \ddot{\mathbf{C}}\right) \cdot \dddot{\mathbf{C}}}{\left|\dot{\mathbf{C}} \times \ddot{\mathbf{C}}\right|^2} = \frac{4}{(2+t^2)^2}.$$

Example 10.12 (Application to Rigid Body Motion)

Consider the motion of a rigid body. Suppose that a reference point on the body moves along a unit speed curve $\mathbf{C}(s)$. Instantaneously, a line of points of the rigid body are stationary, and the body rotates about that line. The line has the direction of the *Darboux vector* ω which satisfies

$$\mathbf{t}' = \omega \times \mathbf{t}, \quad \mathbf{n}' = \omega \times \mathbf{n}, \quad \mathbf{b}' = \omega \times \mathbf{b}.$$

The magnitude ω of ω is the angular velocity of the motion at that instant. Suppose the orthonormal expansion of ω with respect to the Frenet frame is $\omega = \alpha\mathbf{t} + \beta\mathbf{n} + \gamma\mathbf{b}$. Then

$$\begin{aligned}
\mathbf{t}' &= (\alpha\mathbf{t} + \beta\mathbf{n} + \gamma\mathbf{b}) \times \mathbf{t} = \gamma\mathbf{n} - \beta\mathbf{b} \\
\mathbf{n}' &= (\alpha\mathbf{t} + \beta\mathbf{n} + \gamma\mathbf{b}) \times \mathbf{n} = -\gamma\mathbf{t} + \alpha\mathbf{b} \\
\mathbf{b}' &= (\alpha\mathbf{t} + \beta\mathbf{n} + \gamma\mathbf{b}) \times \mathbf{b} = \beta\mathbf{t} - \alpha\mathbf{n}
\end{aligned}$$

and the Frenet formulae give $\alpha = \tau$, $\beta = 0$, and $\gamma = \kappa$. Hence $\omega = \tau\mathbf{t} + \kappa\mathbf{b}$.

Theorem 10.13

Let $\mathbf{C}(t)$ be a regular curve, defined on an interval I, with curvature $\kappa(t)$ and torsion $\tau(t)$.

1. If $\kappa(t) = 0$ for all $t \in I$, then $\mathbf{C}(t)$ is a line segment.
2. If $\tau(t) = 0$ and $\kappa(t) \neq 0$ for all $t \in I$, then $\mathbf{C}(t)$ is a planar curve.
3. If $\tau(t) = 0$ for all $t \in I$, and κ is a non-zero constant, then $\mathbf{C}(t)$ is an arc of a circle of radius $1/\kappa$.

Proof

Reparametrize the curve so that the curve is unit speed.

1. Since $\mathbf{t}' = \kappa\mathbf{n}$, the assumption $\kappa = 0$ implies that $\mathbf{t}' = \mathbf{0}$. Thus $\mathbf{t}$ is a constant vector and

$$(\mathbf{C} \times \mathbf{t})' = (\mathbf{C}' \times \mathbf{t}) + (\mathbf{C} \times \mathbf{t}') = (\mathbf{t} \times \mathbf{t}) + (\mathbf{C} \times \mathbf{0}) = \mathbf{0} .$$

   Hence $\mathbf{C} \times \mathbf{t} = \mathbf{v}$ for some constant vector $\mathbf{v}$. The identity imposes two independent linear constraints on the points of $\mathbf{C}$, and hence the curve is a line.

2. Since $\mathbf{b}' = \tau\mathbf{n}$, the assumption $\tau = 0$ implies that $\mathbf{b}' = \mathbf{0}$. Thus $\mathbf{b}$ is a constant vector (non-zero since $\kappa \neq 0$). Then

$$(\mathbf{C} \cdot \mathbf{b})' = (\mathbf{C}' \cdot \mathbf{b}) + (\mathbf{C} \cdot \mathbf{b}') = (\mathbf{t} \cdot \mathbf{b}) + (\mathbf{C} \cdot \mathbf{0}) = \mathbf{0} .$$

   Hence $\mathbf{C} \cdot \mathbf{b} = \alpha$ for some constant α, and therefore $\mathbf{C}$ is planar.

3. Since $\tau = 0$, the curve is planar and it is sufficient to show that every point of $\mathbf{C}$ has a constant distance from a fixed point. Since $\mathbf{n}' = -\kappa\mathbf{t} + \tau\mathbf{b} = -\kappa\mathbf{t}$ and $1/\kappa$ is constant,

$$\left(\mathbf{C} + \frac{1}{\kappa}\mathbf{n}\right)' = \mathbf{C}' + \frac{1}{\kappa}\mathbf{n}' = \mathbf{t} + \frac{1}{\kappa}(-\kappa\mathbf{t}) = \mathbf{0} .$$

   Hence $\mathbf{C} + 1/\kappa\mathbf{n} = \mathbf{v}$ for some constant vector $\mathbf{v}$. Then $|\mathbf{C} - \mathbf{v}| = 1/|\kappa|$ implying every point of the curve has a constant distance $1/|\kappa|$ from $\mathbf{v}$.

□

The discussion on curvature of space curves is concluded with two theorems (without proofs) which are generalizations of Theorem 10.3 to space curves.

Theorem 10.14

Let $\kappa(s) > 0$ and $\tau(s)$ be continuous functions. Then there exists a curve for which s is the arclength parameter, and κ and τ are the curvature and torsion functions.

Theorem 10.15

Any two space curves, parametrized with respect to arclength, with identical curvature $\kappa(s)$ and torsion $\tau(s)$ functions, differ only by a translation and rotation. If two space curves have the same curvature function $\kappa(s)$, but have torsion functions of opposite sign, they differ only by a translation, a rotation, and a reflection.

EXERCISES

10.12. Determine the curvature and torsion of the curve

$\mathbf{C}(t) = (5\cos t, 3\cos t - 4\sin t, 4\cos t + 3\sin t)$. Deduce that the curve is planar, and identify the type of curve.

10.13. Determine the curvature and torsion of the following curves:

(a) $\mathbf{C}(t) = \left(3t - t^3, 3t^2, 3t + t^3\right)$;

(b) $\mathbf{C}(t) = ((t + \sin t), (1 - \cos t), t)$;

(c) $\mathbf{C}(t) = \left(t, \frac{1}{t} + 1, \frac{1}{t} - t\right)$;

(d) $\mathbf{C}(t) = (1 - \cos(t), t - \sin(t), 4\sin(t/2))$.

10.14. Determine the Frenet frame, curvature, and torsion of the curve $\mathbf{C}(t) = \left(\frac{1}{\sqrt{2}}t, \frac{1}{3}(1+t)^{3/2}, \frac{1}{3}(1-t)^{3/2}\right)$, $t \in (-1, 1)$.

10.15. Show that the Darboux vector satisfies $\mathbf{t}' \times \mathbf{t}'' = \kappa^2 \omega$.

10.16. Let $\mathbf{C} : \mathbb{R} \to \mathbb{R}^n$ be a smooth function at $t = t_0$. Then Taylor's theorem gives

$$\mathbf{C}(t) = \mathbf{C}(t_0) + (t - t_0)\dot{\mathbf{C}}(t_0) + \tfrac{1}{2}(t - t_0)^2\ddot{\mathbf{C}}(t_0) + \tfrac{1}{6}(t - t_0)^3\dddot{\mathbf{C}}(t_0) + \cdots .$$

(a) Apply the formulae which express the derivatives $\dot{\mathbf{C}}, \ddot{\mathbf{C}}, \dddot{\mathbf{C}}$ in terms of the Frenet frame to give the *Frenet approximation* of

$\mathbf{C}(t)$ of the form

$$\begin{aligned}&\mathbf{C}(t_0) + \mathbf{t}_0\left((t-t_0)\nu_0 + \tfrac{1}{2}(t-t_0)^2\dot{\nu}_0 + \tfrac{1}{6}(t-t_0)^3\left(\ddot{\nu}_0 - \kappa_0^2\nu_0^3\right) + \cdots\right)\\ &+ \mathbf{n}_0\left(\tfrac{1}{2}(t-t_0)^2\kappa_0\nu_0^2 + \tfrac{1}{6}(t-t_0)^3\left(3\kappa_0\nu_0\dot{\nu}_0 + \dot{\kappa}_0\nu_0^2\right) + \cdots\right)\\ &+ \mathbf{b}_0\left(\tfrac{1}{6}(t-t_0)^3\kappa\tau_0\nu_0^3 + \cdots\right)\end{aligned} \tag{10.21}$$

where $\nu_0, \dot{\nu}_0, \ddot{\nu}_0, \kappa_0, \tau_0$ denote the speed, the derivatives of speed, the curvature, and the torsion at the point $\mathbf{C}(t_0)$.

(b) Suppose $\mathbf{C}(t_0)$ is the origin. Show that the orthographic projections of the curve onto the osculating, rectifying and normal planes are approximated by the following curves:

(i) $\left((t-t_0)\nu_0, \tfrac{1}{2}(t-t_0)^2\kappa_0\nu_0^2\right)$, i.e. $y = \tfrac{1}{2}\kappa_0 x^2$,

(ii) $\left((t-t_0)\nu_0, \tfrac{1}{6}(t-t_0)^3\kappa_0\tau_0\nu_0^3\right)$, i.e. $z = \tfrac{1}{6}\kappa_0\tau_0 x^3$, and

(iii) $\left(\tfrac{1}{2}(t-t_0)^2\kappa_0\nu_0^2, \tfrac{1}{6}(t-t_0)^3\kappa_0\tau_0\nu_0^3\right)$, i.e. $z^2 = \frac{2\tau_0^2}{9\kappa_0}x^3$, respectively.

10.3 Curvature of Bézier Curves

Theorem 10.16

The curvature and torsion of a Bézier curve $\mathbf{B}(t)$ at $t = 0$ are

$$\kappa = \frac{n-1}{n}\frac{b}{a^3}, \qquad \tau = \frac{n-2}{n}\frac{c}{b^2}$$

where $a = |\mathbf{b}_1 - \mathbf{b}_0|$, $b = |(\mathbf{b}_1 - \mathbf{b}_0) \times (\mathbf{b}_2 - \mathbf{b}_1)|$, and

$$c = ((\mathbf{b}_1 - \mathbf{b}_0) \times (\mathbf{b}_2 - \mathbf{b}_1)) \cdot (\mathbf{b}_3 - \mathbf{b}_2) \ .$$

Proof

The derivatives of $\mathbf{B}(t)$ are obtained using Theorem 7.3 and its corollaries,

$$\begin{aligned}\dot{\mathbf{B}}(0) &= n\,(\mathbf{b}_1 - \mathbf{b}_0) \ ,\\ \ddot{\mathbf{B}}(0) &= n\,(n-1)\,(\mathbf{b}_2 - 2\mathbf{b}_1 + \mathbf{b}_0) = n\,(n-1)\,((\mathbf{b}_2 - \mathbf{b}_1) - (\mathbf{b}_1 - \mathbf{b}_0)) \ ,\\ \dddot{\mathbf{B}}(0) &= n\,(n-1)\,(n-2)\,(\mathbf{b}_3 - 3\mathbf{b}_2 + 3\mathbf{b}_1 - \mathbf{b}_0) \ .\end{aligned}$$

Then,

$$\begin{aligned}\left|\dot{\mathbf{B}}\right| &= n\left|\mathbf{b}_1-\mathbf{b}_0\right| , \\ \left|\dot{\mathbf{B}}\times\ddot{\mathbf{B}}\right| &= n^2(n-1)\left|(\mathbf{b}_1-\mathbf{b}_0)\times((\mathbf{b}_2-\mathbf{b}_1)-(\mathbf{b}_1-\mathbf{b}_0))\right| \\ &= n^2(n-1)\left|(\mathbf{b}_1-\mathbf{b}_0)\times(\mathbf{b}_2-\mathbf{b}_1)\right| ,\end{aligned}$$

and

$$\kappa = \frac{\left|\dot{\mathbf{B}}\times\ddot{\mathbf{B}}\right|}{\left|\dot{\mathbf{B}}\right|^3} = \frac{(n-1)\left|(\mathbf{b}_1-\mathbf{b}_0)\times(\mathbf{b}_2-\mathbf{b}_1)\right|}{n\left|\mathbf{b}_1-\mathbf{b}_0\right|^3} .$$

Further,

$$\begin{aligned}&\left(\dot{\mathbf{B}}\times\ddot{\mathbf{B}}\right)\cdot\dddot{\mathbf{B}} \\ &= n^3(n-1)^2(n-2)\left((\mathbf{b}_1-\mathbf{b}_0)\times(\mathbf{b}_2-\mathbf{b}_1)\right)\cdot(\mathbf{b}_3-3\mathbf{b}_2+3\mathbf{b}_1-\mathbf{b}_0) .\end{aligned}$$

Expressing $\mathbf{b}_3-3\mathbf{b}_2+3\mathbf{b}_1-\mathbf{b}_0 = (\mathbf{b}_3-\mathbf{b}_2)-(\mathbf{b}_2-\mathbf{b}_1)-(\mathbf{b}_2-\mathbf{b}_1)+(\mathbf{b}_1-\mathbf{b}_0)$ in the previous equation, expanding the cross product, and simplifying gives

$$\left(\dot{\mathbf{B}}\times\ddot{\mathbf{B}}\right)\cdot\dddot{\mathbf{B}} = n^3(n-1)^2(n-2)\left((\mathbf{b}_1-\mathbf{b}_0)\times(\mathbf{b}_2-\mathbf{b}_1)\right)\cdot(\mathbf{b}_3-\mathbf{b}_2) .$$

Hence

$$\tau = \frac{\left(\dot{\mathbf{B}}\times\ddot{\mathbf{B}}\right)\cdot\dddot{\mathbf{B}}}{\left|\dot{\mathbf{B}}\times\ddot{\mathbf{B}}\right|^2} = \frac{(n-2)\left(\left((\mathbf{b}_1-\mathbf{b}_0)\times(\mathbf{b}_2-\mathbf{b}_1)\right)\cdot(\mathbf{b}_3-\mathbf{b}_2)\right)}{n\left|(\mathbf{b}_1-\mathbf{b}_0)\times(\mathbf{b}_2-\mathbf{b}_1)\right|^2} .$$

□

The reader is left the exercise of proving the next theorem.

Theorem 10.17

The curvature and torsion of a rational Bézier curve $\mathbf{B}(t)$ at $t=0$ are

$$\kappa = \frac{n-1}{n}\frac{w_0w_2}{w_1^2}\frac{b}{a^2}, \qquad \tau = \frac{n-2}{n}\frac{w_0w_3}{w_1w_2}\frac{c}{b^2}$$

where a, b, and c are as above.

The curvature and torsion of an integral or rational Bézier curve $\mathbf{B}(t)$ at the point $\mathbf{B}(t_0)$, $t_0 \in [0,1]$ can be computed by applying the integral or rational de Casteljau algorithm to subdivide the curve at $t=t_0$ into the two segments $\mathbf{B}_{\text{left}}(t)$ and $\mathbf{B}_{\text{right}}(t)$. Applying Theorem 10.16 or 10.17 to $\mathbf{B}_{\text{right}}(t)$ at $t=0$ gives the curvature and torsion at $\mathbf{B}(t_0)$.

Example 10.18

Consider the cubic Bézier curve $\mathbf{B}(t)$ with control points $\mathbf{b}_0(0,1,4)$, $\mathbf{b}_1(2,-1,3)$, $\mathbf{b}_2(3,2,7)$, $\mathbf{b}_3(5,2,2)$. Then

$$\begin{aligned} a &= |(2,-1,3)-(0,1,4)| = 3\,, \\ b &= |((2,-1,3)-(0,1,4)) \times ((3,2,7)-(2,-1,3))| = |(-5,-9,8)| = \sqrt{170}\,, \\ c &= (-5,-9,8)\cdot((5,2,2)-(3,2,7)) = -50\,. \end{aligned}$$

Hence the curvature and torsion of $\mathbf{B}(t)$ at $\mathbf{B}(0)$ are $\kappa(0) = \frac{3-1}{3}\frac{\sqrt{170}}{(3)^3} = \frac{2}{81}\sqrt{170}$, and $\tau(0) = \frac{3-2}{3}\frac{-50}{170} = -\frac{5}{51}$.

EXERCISES

10.17. Determine the curvature and torsion of the cubic Bézier curve with control points $\mathbf{b}_0(1,2,1)$, $\mathbf{b}_1(3,0,4)$, $\mathbf{b}_2(6,-3,2)$, and $\mathbf{b}_3(4,2,3)$

(a) at the point $\mathbf{B}(0)$, and

(b) at the point $\mathbf{B}(0.3)$. (Hint: Example 7.1 may help!)

10.18. Determine the curvature of the cubic rational Bézier curve with control points $\mathbf{b}_0(3,2,7)$, $\mathbf{b}_1(5,4,3)$, $\mathbf{b}_2(8,3,3)$, $\mathbf{b}_3(5,2,4)$, and weights $1,2,2,1$ at the points $\mathbf{B}(0)$ and $\mathbf{B}(0.6)$.

10.19. Determine the control points of the planar Bézier curve $\mathbf{B}(t)$ of degree 5 satisfying $\mathbf{B}(0) = (0,0)$, $\mathbf{B}(1) = (5,0)$, $\mathbf{B}'(0) = (1,4)$, $\mathbf{B}'(1) = (-1,2)$, $\kappa(0) = 1$, and $\kappa(1) = 2$.

10.20. Determine expressions for the Frenet frame $\mathbf{t}, \mathbf{n}, \mathbf{b}$ of a Bézier curve $\mathbf{B}(t)$ at $t = 0$.

10.21. Prove Theorem 10.17.

10.4 Surface Curvatures

In this section surfaces are parametrized using the variables u and v (s and t were used in the chapter on surfaces). Let U be an open or closed subset of $\mathbb{R}^2$ and let $\mathbf{S} : U \to \mathbb{R}^3$, be a parametrized surface with unit normal $\mathbf{N}(u,v)$. If $(u(t), v(t))$ is a regular curve in U, then it is mapped by $\mathbf{S}$ to the curve $\mathbf{C}(t) = \mathbf{S}(u(t), v(t))$ on the surface. The chain rule gives $\dot{\mathbf{C}} = \dot{u}\mathbf{S}_u + \dot{v}\mathbf{S}_v$ and $\dot{\mathbf{C}}\cdot\mathbf{N} = (\dot{u}\mathbf{S}_u + \dot{v}\mathbf{S}_v)\cdot\mathbf{N} = \dot{u}\,(\mathbf{S}_u\cdot\mathbf{N}) + \dot{v}\,(\mathbf{S}_v\cdot\mathbf{N}) = 0$. So every tangent vector to the curve $\mathbf{C}$ is a tangent vector to the surface. The converse, that every tangent vector of the surface is a tangent vector to some curve on the surface, is proved in the following lemma.

Lemma 10.19

Let $\mathbf{S} : U \to \mathbb{R}^3$ be a regular surface. If $\mathbf{v}$ is a tangent vector to the surface at a point $\mathbf{p}$ then there exists a curve $\mathbf{C}(t)$, $t \in (-a, a)$ (some $a > 0$), such that $\mathbf{C}(0) = \mathbf{p}$ and $\dot{\mathbf{C}}(0) = \mathbf{v}$.

Proof

Since $\mathbf{S}$ is regular, $\mathbf{S}_u$ and $\mathbf{S}_v$ are linearly independent vectors, and therefore $\mathbf{v} = \alpha\mathbf{S}_u(u,v)+\beta\mathbf{S}_v(u,v)$ for some α and β. Let $\mathbf{C}(t) = \mathbf{S}(u+\alpha t, v+\beta t)$ be defined on $(-a, a)$ where $a > 0$ is chosen so that $(-a, a)$ is contained in U. Then $\mathbf{C}(0) = \mathbf{p}$, and the chain rule yields $\dot{\mathbf{C}}(t) = \alpha\mathbf{S}_u(u+\alpha t, v+\beta t) + \beta\mathbf{S}_v(u+\alpha t, v+\beta t)$ and $\dot{\mathbf{C}}(0) = \alpha\mathbf{S}_u(u,v) + \beta\mathbf{S}_v(u,v) = \mathbf{v}$.

□

In view of the lemma, the tangent vectors at a point $\mathbf{p}$ will often be expressed in the form $\dot{u}\mathbf{S}_u + \dot{v}\mathbf{S}_v$.

Suppose $\mathbf{p} = \mathbf{S}(u,v)$ is a regular point, and let $\mathbf{v}$ be a tangent vector to the surface at $\mathbf{p}$. The plane through $\mathbf{p}$ containing the directions $\mathbf{v}$ and $\mathbf{N}$ intersects $\mathbf{S}$ in a curve $\mathbf{C}(t)$ as shown in Figure 10.5. The curve can be parametrized so that $\mathbf{C}(0) = \mathbf{p}$ and $\dot{\mathbf{C}}(0) = \mathbf{v}$. The curvature $\kappa(t)$ of $\mathbf{C}(t)$ at $t = 0$ is called the *normal curvature* of $\mathbf{S}$ in the direction $\mathbf{v}$ at $\mathbf{p}$, and denoted $\kappa_{\mathbf{p}}(\mathbf{v})$. Theorem 10.21 will prove that $\kappa_{\mathbf{p}}(\mathbf{v})$ has a maximum and a minimum value, denoted $\kappa_{\max}$ and $\kappa_{\min}$ respectively, called the *principal curvatures*. The tangent vectors which give rise to the principal curvatures are called the *principal directions*. A curve $\mathbf{C}(t)$ on $\mathbf{S}$ for which every tangent vector $\dot{\mathbf{C}}(t)$ is a principal direction of the surface is called a *line of principal curvature*.

Let $\mathbf{C}(t) = \mathbf{S}(u(t), v(t))$ be a curve on $\mathbf{S}$ with unit tangent $\mathbf{t}$ and unit normal $\mathbf{n}$. Then $\dot{\mathbf{C}} = \nu\mathbf{t}$, $\ddot{\mathbf{C}} = \dot{\nu}\mathbf{t}+\nu^2\kappa\mathbf{n}$ and

$$\ddot{\mathbf{C}} \cdot \mathbf{N} = \left(\dot{\nu}\mathbf{t}+\nu^2\kappa\mathbf{n}\right) \cdot\mathbf{N} =\nu^2\kappa\mathbf{n} \cdot \mathbf{N} \,. \tag{10.22}$$

The chain rule applied to $\mathbf{C}(t) = \mathbf{S}(u(t), v(t))$ gives

$$\begin{aligned} \dot{\mathbf{C}} &= \dot{u}\mathbf{S}_u + \dot{v}\mathbf{S}_v \,, \\ \ddot{\mathbf{C}} &= \dot{u}\left(\dot{u}\mathbf{S}_{uu} + \dot{v}\mathbf{S}_{vu}\right) + \ddot{u}\mathbf{S}_u + \dot{v}\left(\dot{u}\mathbf{S}_{uv} + \dot{v}\mathbf{S}_{vv}\right) + \ddot{v}\mathbf{S}_v \\ &= \dot{u}^2\mathbf{S}_{uu} + 2\dot{u}\dot{v}\mathbf{S}_{uv} + \dot{v}^2\mathbf{S}_{vv} + \ddot{u}\mathbf{S}_u + \ddot{v}\mathbf{S}_v \,. \end{aligned}$$

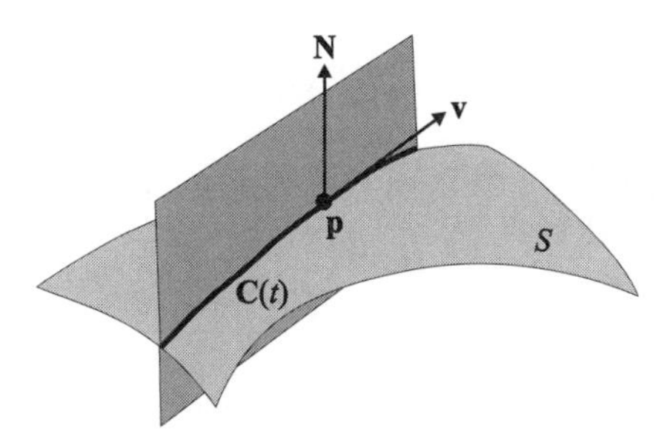

Figure 10.5

Let $E = \mathbf{S}_u \cdot \mathbf{S}_u$, $F = \mathbf{S}_u \cdot \mathbf{S}_v$, and $G = \mathbf{S}_v \cdot \mathbf{S}_v$, then

$$\begin{aligned}\nu^2 &= \dot{\mathbf{C}} \cdot \dot{\mathbf{C}} = (\dot{u}\mathbf{S}_u + \dot{v}\mathbf{S}_v) \cdot (\dot{u}\mathbf{S}_u + \dot{v}\mathbf{S}_v) \\ &= \dot{u}^2\mathbf{S}_u \cdot \mathbf{S}_u + 2\dot{u}\dot{v}\mathbf{S}_u \cdot \mathbf{S}_v + \dot{v}^2\mathbf{S}_v \cdot \mathbf{S}_v \\ &= E\dot{u}^2 + 2F\dot{u}\dot{v} + G\dot{v}^2 \,.\end{aligned}$$

Further, let $L = \mathbf{S}_{uu}{\cdot}\mathbf{N}$, $M = \mathbf{S}_{uv}{\cdot}\mathbf{N}$, and $N = \mathbf{S}_{vv}{\cdot}\mathbf{N}$, then

$$\begin{aligned}\ddot{\mathbf{C}} \cdot \mathbf{N} &= \left(\dot{u}^2\mathbf{S}_{uu} + 2\dot{u}\dot{v}\mathbf{S}_{uv} + \dot{v}^2\mathbf{S}_{vv} + \ddot{u}\mathbf{S}_u + \ddot{v}\mathbf{S}_v\right) \cdot \mathbf{N} \\ &= \dot{u}^2\mathbf{S}_{uu} \cdot \mathbf{N} + 2\dot{u}\dot{v}\mathbf{S}_{uv} \cdot \mathbf{N} + \dot{v}^2\mathbf{S}_{vv} \cdot \mathbf{N} \\ &= L\dot{u}^2 + 2M\dot{u}\dot{v} + N\dot{v}^2 \,. \end{aligned} \tag{10.23}$$

The expressions $E\dot{u}^2 + 2F\dot{u}\dot{v} + G\dot{v}^2$ and $L\dot{u}^2 + 2M\dot{u}\dot{v} + N\dot{v}^2$ are called the *first and second fundamental forms* of the surface. Equations (10.22) and (10.23) give the curvature κ of $\mathbf{C}$ at $\mathbf{p}$

$$\kappa = \frac{L\dot{u}^2 + 2M\dot{u}\dot{v} + N\dot{v}^2}{\nu^2 \mathbf{n} \cdot \mathbf{N}} \,.$$

Suppose $\mathbf{C}$ is a curve, through the point $\mathbf{p}$, in the plane containing the unit normal $\mathbf{N}$ to the surface at $\mathbf{p}$. Then $\mathbf{n} \cdot \mathbf{N} = 1$ and the formula yields the normal curvature of the surface in the direction $\mathbf{v} = \dot{\mathbf{C}}$

$$\kappa_{\mathbf{p}}(\dot{\mathbf{C}}) = \frac{L\dot{u}^2 + 2M\dot{u}\dot{v} + N\dot{v}^2}{\nu^2} = \frac{L\dot{u}^2 + 2M\dot{u}\dot{v} + N\dot{v}^2}{E\dot{u}^2 + 2F\dot{u}\dot{v} + G\dot{v}^2} \,. \tag{10.24}$$

The sign of $\kappa_{\mathbf{p}}$ indicates whether, near $\mathbf{p}$, the curve (and the surface in the direction $\mathbf{v}$) is bending towards or away from the normal, as illustrated in Figure 10.6.

Example 10.20

Consider the surface $\mathbf{S}(u,v) = (u, v, u^2 - v^2)$. Then $\mathbf{S}_u = (1, 0, 2u)$, $\mathbf{S}_v = (0, 1, -2v)$, $\mathbf{S}_{uu} = (0, 0, 2)$, $\mathbf{S}_{uv} = (0, 0, 0)$, $\mathbf{S}_{vv} = (0, 0, -2)$, $\mathbf{S}_u \times \mathbf{S}_v =$

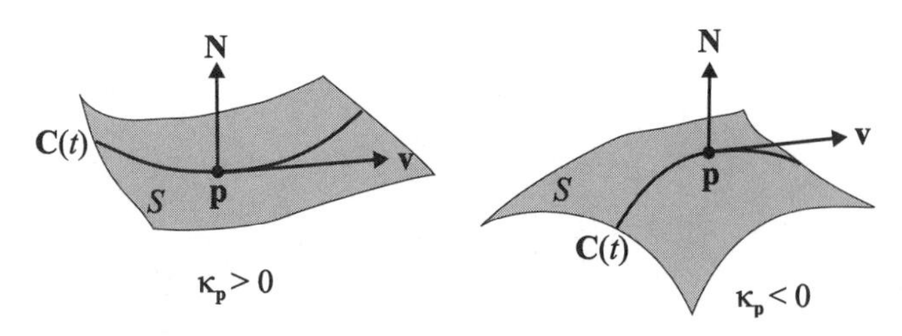

Figure 10.6 Geometric interpretation of the sign of the normal curvature

$(-2u, 2v, 1)$,

$$\mathbf{N} = (-2u, 2v, 1) / |(-2u, 2v, 1)| = \left(1 + 4u^2 + 4v^2\right)^{-1/2} (-2u, 2v, 1),$$

$E = \mathbf{S}_u \cdot \mathbf{S}_u = 1 + 4u^2$, $F = \mathbf{S}_u \cdot \mathbf{S}_v = -4uv$, $G = \mathbf{S}_v \cdot \mathbf{S}_v = 1 + 4v^2$, $L = \mathbf{S}_{uu} \cdot \mathbf{N} = 2/\left(1 + 4u^2 + 4v^2\right)^{1/2}$, $M = \mathbf{S}_{uv} \cdot \mathbf{N} = 0$, $N = \mathbf{S}_{vv} \cdot \mathbf{N} = -2/\left(1 + 4u^2 + 4v^2\right)^{1/2}$. The curve $(u(t), v(t)) = (t, t^2)$ is mapped by $\mathbf{S}$ to the surface curve $\mathbf{C}(t) = \mathbf{S}(t, t^2) = (t, t^2, t^2 - t^4)$. Then $\dot{u}(t) = 1$ and $\dot{v}(t) = 2t$. At $t = 0$ the curve passes through the origin and has tangent vector $\dot{\mathbf{C}}(0) = (1, 0, 0)$, $u = v = 0$, $\dot{u} = 1$, $\dot{v} = 0$, $E = 1$, $F = 0$, $G = 1$, $L = 2$, $M = 0$, and $N = -2$. Hence, the normal curvature at the origin in the direction $(1, 0, 0)$ is $\kappa_{\mathbf{p}}(1, 0, 0) = 2$.

The curve $(u(t), v(t)) = (\cos t, \sin t)$ is mapped to the surface curve $\mathbf{C}(t) = (\cos t, \sin t, \cos^2 t - \sin^2 t)$. Then $\dot{u}(t) = -\sin t$, $\dot{v}(t) = \cos t$. At $t = \pi/2$ the curve passes through the point $(0, 1, -1)$ and has tangent vector $\dot{\mathbf{C}}(\pi/2) = (-1, 0, 0)$, $u = 0$, $v = 1$, $\dot{u} = -1$, $\dot{v} = 0$, $E = 1$, $F = 0$, $G = 5$, $L = 2/\sqrt{5}$, $M = 0$, and $N = -2/\sqrt{5}$. Hence the normal curvature at $(0, 1, -1)$ in the direction $(-1, 0, 0)$ is $\kappa_{\mathbf{p}}(-1, 0, 0) = 2/\sqrt{5}$.

Theorem 10.21 (Euler)

Let $\mathbf{p}$ be a regular point of a surface $\mathbf{S}(u, v)$. Suppose the normal curvature $\kappa_{\mathbf{p}}(\mathbf{v})$ is a non-constant function of $\mathbf{v}$. Then there are unique unit tangent vectors $\mathbf{v}_{\max}$ and $\mathbf{v}_{\min}$ such that the normal curvature $\kappa_{\mathbf{p}}(\mathbf{v}_{\max}) = \kappa_{\max}$ is maximal, and $\kappa_{\mathbf{p}}(\mathbf{v}_{\min}) = \kappa_{\min}$ is minimal. Further, $\mathbf{v}_{\max}$ and $\mathbf{v}_{\min}$ are perpendicular.

Proof

The surface curve $\dot{\mathbf{C}}(t) = \mathbf{S}(u(t), v(t))$ has tangent vector $\dot{\mathbf{C}} = \dot{u}\mathbf{S}_u + \dot{v}\mathbf{S}_v$, and by (10.24) the normal curvature in this direction is

$$\kappa_{\mathbf{p}}(\dot{u}\mathbf{S}_u + \dot{v}\mathbf{S}_v) = \frac{L\dot{u}^2 + 2M\dot{u}\dot{v} + N\dot{v}^2}{E\dot{u}^2 + 2F\dot{u}\dot{v} + G\dot{v}^2} \,. \tag{10.25}$$

The maximum and minimum normal curvatures are the extrema of (10.25) for all tangent vectors $\dot{u}\mathbf{S}_u + \dot{v}\mathbf{S}_v$. By reparametrizing $(u(t), v(t))$, the tangent vectors $\dot{u}\mathbf{S}_u + \dot{v}\mathbf{S}_v$ can be assumed to have unit length. Then $E\dot{u}^2 + 2F\dot{u}\dot{v} + G\dot{v}^2 = (\dot{u}\mathbf{S}_u + \dot{v}\mathbf{S}_v) \cdot (\dot{u}\mathbf{S}_u + \dot{v}\mathbf{S}_v) = 1$. Therefore the problem is to find the extrema of $\kappa_{\mathbf{p}}(\dot{u}\mathbf{S}_u + \dot{v}\mathbf{S}_v) = L\dot{u}^2 + 2M\dot{u}\dot{v} + N\dot{v}^2$ subject to the constraint $E\dot{u}^2 + 2F\dot{u}\dot{v} + G\dot{v}^2 = 1$. The solution can be found by applying the method of Lagrange multipliers (treating $\dot{u}$ and $\dot{v}$ as variables). Let

$$\mathcal{L}(\dot{u},\dot{v}) = L\dot{u}^2 + 2M\dot{u}\dot{v} + N\dot{v}^2 - \lambda\left(E\dot{u}^2 + 2F\dot{u}\dot{v} + G\dot{v}^2 - 1\right) \,.$$

The conditions for the extrema are

$$\begin{aligned} \frac{\partial \mathcal{L}}{\partial \dot{u}} &= 2L\dot{u} + 2M\dot{v} - 2\lambda\left(2E\dot{u} + 2F\dot{v}\right) = 0 \,, \\ \frac{\partial \mathcal{L}}{\partial \dot{v}} &= 2M\dot{u} + 2N\dot{v} - 2\lambda\left(2G\dot{v} + 2F\dot{u}\right) = 0 \,, \end{aligned}$$

giving

$$(L - \lambda E)\dot{u} + (M - \lambda F)\dot{v} = 0 \,, \tag{10.26}$$

$$(M - \lambda F)\dot{u} + (N - \lambda G)\dot{v} = 0 \,, \tag{10.27}$$

which can be expressed in the matrix form

$$(\mathsf{S} - \lambda \mathsf{F})\begin{pmatrix} \dot{u} \\ \dot{v} \end{pmatrix} = 0 \,, \tag{10.28}$$

where

$$\mathsf{F} = \begin{pmatrix} E & F \\ F & G \end{pmatrix} \quad \text{and} \quad \mathsf{S} = \begin{pmatrix} L & M \\ M & N \end{pmatrix} .$$

F and S are called the first and second fundamental matrices, respectively. Since $EG - F^2 \neq 0$ at a regular point (Exercise 10.22), F is non-singular and (10.28) gives

$$(\mathsf{F}^{-1}\mathsf{S} - \lambda I)\begin{pmatrix} \dot{u} \\ \dot{v} \end{pmatrix} = 0 \,.$$

Solving (10.28) for λ is equivalent to solving for the eigenvalues of $\mathsf{F}^{-1}\mathsf{S}$. The eigenvalues λ satisfy

$$\begin{aligned} 0 &= \begin{pmatrix} \dot{u} & \dot{v} \end{pmatrix}(\mathsf{S} - \lambda \mathsf{F})\begin{pmatrix} \dot{u} \\ \dot{v} \end{pmatrix} = \begin{pmatrix} \dot{u} & \dot{v} \end{pmatrix}\mathsf{S}\begin{pmatrix} \dot{u} \\ \dot{v} \end{pmatrix} - \lambda\begin{pmatrix} \dot{u} & \dot{v} \end{pmatrix}\mathsf{F}\begin{pmatrix} \dot{u} \\ \dot{v} \end{pmatrix} \\ &= L\dot{u}^2 + 2M\dot{u}\dot{v} + N\dot{v}^2 - \lambda\left(E\dot{u}^2 + 2F\dot{u}\dot{v} + G\dot{v}^2\right) \,. \end{aligned}$$

Hence

$$\lambda = \frac{L\dot{u}^2 + 2M\dot{u}\dot{v} + N\dot{v}^2}{E\dot{u}^2 + 2F\dot{u}\dot{v} + G\dot{v}^2} .$$

Thus the eigenvalues of $\mathsf{F}^{-1}\mathsf{S}$ are the principal curvatures $\kappa_{\max}$ and $\kappa_{\min}$, and the corresponding principal directions $\mathbf{w}_1=\binom{\dot{u}_1}{\dot{v}_1}$ and $\mathbf{w}_2=\binom{\dot{u}_2}{\dot{v}_2}$ yield the principal directions $\mathbf{v}_{\max} = \dot{u}_1\mathbf{S}_u + \dot{v}_1\mathbf{S}_v$ and $\mathbf{v}_{\min} = \dot{u}_2\mathbf{S}_u + \dot{v}_2\mathbf{S}_v$.

Then (10.28) implies $(\mathsf{S} - \mathsf{F}\kappa_{\max})\,\mathbf{w}_1 = 0$, and the fact that F and S are symmetric gives

$$\mathbf{w}_1^T (\mathsf{S} - \mathsf{F}\kappa_{\max})^T = \mathbf{w}_1^T \left(\mathsf{S}^T - \mathsf{F}^T\kappa_{\max}\right) = \mathbf{w}_1^T (\mathsf{S} - \mathsf{F}\kappa_{\max}) = 0 .$$

So $\mathbf{w}_1^T\mathsf{S} = \mathbf{w}_1^T\mathsf{F}\kappa_{\max}$ and multiplying on the right by $\mathbf{w}_2$ gives

$$\mathbf{w}_1^T\mathsf{S}\mathbf{w}_2 = \kappa_{\max}\mathbf{w}_1^T\mathsf{F}\mathbf{w}_2 . \tag{10.29}$$

Further, (10.28) implies $(\mathsf{F} - \mathsf{S}\kappa_{\min})\,\mathbf{w}_2 = 0$, so

$$\mathbf{w}_1^T\mathsf{S}\mathbf{w}_2 = \kappa_{\min}\mathbf{w}_1^T\mathsf{F}\mathbf{w}_2 . \tag{10.30}$$

Then subtracting Equation (10.30) from (10.29) gives $(\kappa_{\max} - \kappa_{\min})\,\mathbf{w}_1^T\mathsf{F}\mathbf{w}_2 = 0$, and since $\kappa_{\max} \neq \kappa_{\min}$ it follows that $\mathbf{w}_1^T\mathsf{F}\mathbf{w}_2 = 0$. Further, since

$$\begin{aligned} \mathbf{w}_1^T\mathsf{F}\mathbf{w}_2 &= (\dot{u}_1\mathbf{S}_u + \dot{v}_1\mathbf{S}_v)\cdot(\dot{u}_2\mathbf{S}_u + \dot{v}_2\mathbf{S}_v) \\ &= \mathbf{v}_{\max}\cdot\mathbf{v}_{\min} \\ &= 0 \end{aligned}$$

it follows that $\mathbf{v}_{\max}$ and $\mathbf{v}_{\min}$ are perpendicular.

□

Eliminating λ from Equations (10.26) and (10.27) gives

$$(EM - FL)\dot{u}^2 + (EN - GL)\,\dot{u}\dot{v} + (FN - GM)\dot{v}^2 = 0 . \tag{10.31}$$

Equation (10.31) is a necessary and sufficient condition for a curve $\mathbf{S}(u(t), v(t))$ to be a line of curvature. In particular, the u-parameter curve $(u(t), v(t)) = (t, v_0)$, for which $(\dot{u}(t), \dot{v}(t)) = (1, 0)$, is a line of curvature if and only if $EM - FL = 0$. Likewise, the v-parameter curve is a line of curvature if and only if $FN - GM = 0$. Since $EG - F^2 \neq 0$, the conditions $EM - FL = FN - GM = 0$ can be satisfied if and only $F = M = 0$. Therefore the parameter curves are the lines of curvature if and only if $F = M = 0$.

Let $\mathbf{v} = \dot{u}\mathbf{S}_u + \dot{v}\mathbf{S}_v$ and $\mathbf{w} = \binom{\dot{u}}{\dot{v}}$. The formula for normal curvature (10.25) can be expressed in terms of the fundamental matrices

$$\kappa_{\mathbf{p}}(\mathbf{v}) = \frac{\mathbf{w}^T\mathsf{S}\mathbf{w}}{\mathbf{w}^T\mathsf{F}\mathbf{w}} .$$

A second theorem due to Euler states that the normal curvature $\kappa_{\mathbf{p}}(\mathbf{v})$ in an arbitrary direction $\mathbf{v}$ can be obtained from the principal curvatures [8].

Theorem 10.22 (Euler)

If θ is the angle between a tangent vector $\mathbf{v}$ and $\mathbf{v}_{\max}$, then

$$\kappa_{\mathbf{p}}(\mathbf{v}) = \kappa_{\max} \cos^2 \theta + \kappa_{\min} \sin^2 \theta \, .$$

□

The curvatures of a surface most commonly used are not the principal curvatures but the *Gaussian curvature* $K = \kappa_{\max}\kappa_{\min}$ and the *mean curvature* $H = \frac{1}{2}(\kappa_{\max} + \kappa_{\min})$. The Gaussian and mean curvatures can be computed without computing the principal curvatures. The principal curvatures are the roots of the quadratic

$$\det(\mathsf{S} - \lambda \mathsf{F}) = (EG - F^2)\lambda^2 - (EN + GL - 2FM)\lambda + (LN - M^2) = 0 \, ,$$

and the fact that the sum and product of the roots of a quadratic $ax^2 + bx + c$ (with $a \neq 0$) are $-b/a$ and c/a, yields

$$K = \frac{LN - M^2}{EG - F^2}, \qquad H = \frac{1}{2}\left(\frac{EN + GL - 2FM}{EG - F^2}\right) .$$

The principal, mean, and Gaussian curvatures distinguish the local geometry of a surface at a point $\mathbf{p}$ as follows.

Elliptic Point: $H \neq 0, K > 0$. At an elliptic point $\kappa_{\min}$ and $\kappa_{\max}$ have the same sign. Therefore the normal sections have the same profile, implying the surface near $\mathbf{p}$ has the shape of an ellipsoid.

Hyperbolic Point: $H \neq 0, K < 0$. At a hyperbolic point $\kappa_{\min}$ and $\kappa_{\max}$ have opposite signs. So the surface near $\mathbf{p}$ has the shape of a saddle.

Parabolic Point: $H \neq 0, K = 0$. So either $\kappa_{\min} = 0$ or $\kappa_{\max} = 0$. Therefore the surface is linear in one principal direction, and near $\mathbf{p}$ the surface has the shape of a parabolic cylinder. In computer vision applications the surface is said to be a *ridge* or a *trough*.

Umbilic Point: $\kappa_{\min} = \kappa_{\max} \neq 0$ $(H \neq 0, K > 0)$. An umbilic point is a special case of an elliptic point. The normal curvature is constant (non-zero) and near $\mathbf{p}$ the surface has the shape of a sphere.

Flat or Planar Point: $\kappa_{\min} = \kappa_{\max} = 0$ $(H = K = 0)$. The normal curvature is identically zero and the surface near $\mathbf{p}$ is flat.

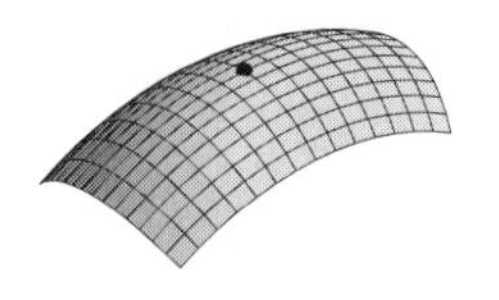

(a) Elliptic point $K > 0, H \neq 0$

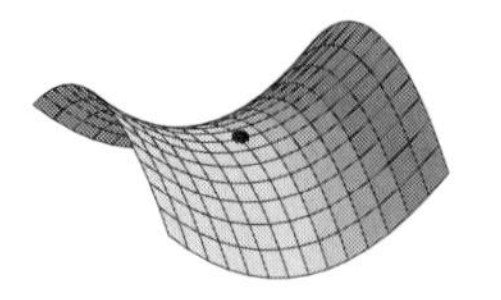

(b) Hyperbolic point $K < 0, H \neq 0$

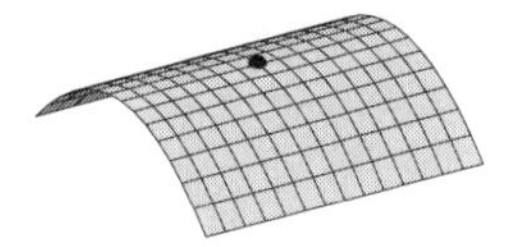

(c) Parabolic point $K = 0, H \neq 0$

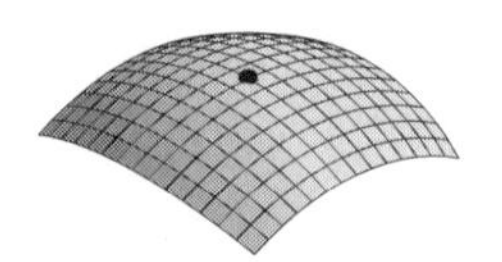

(d) Umbilic point $\kappa_{\text{max}} = \kappa_{\text{min}} \neq 0$

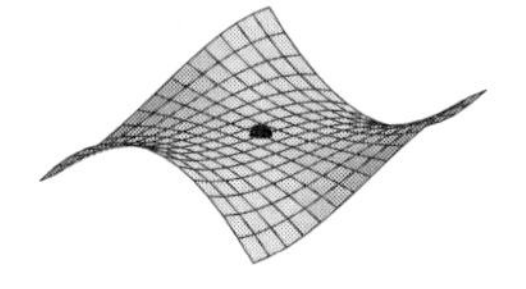

(e) Planar point $\kappa_{\text{max}} = \kappa_{\text{min}} = 0$

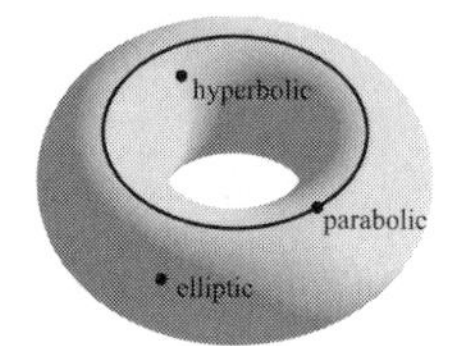

(f) Elliptic, hyperbolic, and parabolic points on a torus

Figure 10.7

Example 10.23

Let $\mathbf{S}(u,v) = (u\cos v, u\sin v, v)$, $0 < v < 2\pi$, $u > 0$. Then $\mathbf{S}_u = (\cos v, \sin v, 0)$, $\mathbf{S}_v = (-u\sin v, u\cos v, 1)$, and hence $E = 1$, $F = 0$ and $G = 1 + u^2$. Further $\mathbf{S}_{uu} = (0,0,0)$, $\mathbf{S}_{uv} = (-\sin v, \cos v, 0)$, $\mathbf{S}_{vv} = (-u\cos v, -u\sin v, 0)$ and $\mathbf{N} = \left(1+u^2\right)^{-1/2}(\sin v, -\cos v, u)$, and hence $L = 0$, $M = -\left(1+u^2\right)^{-1/2}$ and $N = 0$. Then

$$
\begin{aligned}
&\det(\mathsf{S} - \lambda\mathsf{F}) \\
&= \det\left(\begin{pmatrix} 0 & -\left(1+u^2\right)^{-1/2} \\ -\left(1+u^2\right)^{-1/2} & 0 \end{pmatrix} - \lambda\begin{pmatrix} 1 & 0 \\ 0 & 1+u^2 \end{pmatrix}\right) \\
&= \det\begin{pmatrix} -\lambda & -\left(1+u^2\right)^{-1/2} \\ -\left(1+u^2\right)^{-1/2} & -\lambda\left(1+u^2\right) \end{pmatrix} = \frac{\lambda^2\left(1+u^2\right)^2 - 1}{1+u^2}\,.
\end{aligned}
$$

Solving $\det(\mathsf{S} - \lambda\mathsf{F}) = 0$ gives $\lambda = \left(1+u^2\right)^{-1}$ and $\lambda = -\left(1+u^2\right)^{-1}$. The Gaussian curvature is $K = -\left(1+u^2\right)^{-2}$, and the mean curvature is $H = 0$. Since $K < 0$, every point of the surface is hyperbolic.

Example 10.24

Consider the torus $\mathbf{S}(u,v) = ((r\cos u + R)\cos v, (r\cos u + R)\sin v, r\sin u)$, for $R > r > 0$. Then

$$\begin{aligned}
\mathbf{S}_u(u,v) &= (-r\sin u\cos v, -r\sin u\sin v, r\cos u),\\
\mathbf{S}_v(u,v) &= \left(-\left(r\cos u + R\right)\sin v, \left(r\cos u + R\right)\cos v, 0\right),\\
\mathbf{S}_{uu}(u,v) &= (-r\cos u\cos v, -r\cos u\sin v, -r\sin u),\\
\mathbf{S}_{uv}(u,v) &= (r\sin u\sin v, -r\sin u\cos v, 0),\\
\mathbf{S}_{vv}(u,v) &= \left(-\left(r\cos u + R\right)\cos v, -\left(r\cos u + R\right)\sin v, 0\right).
\end{aligned}$$

Thus $E = \mathbf{S}_u\cdot\mathbf{S}_u = r^2$, $F = \mathbf{S}_u\cdot\mathbf{S}_v = 0$, $G = \mathbf{S}_v\cdot\mathbf{S}_v = \left(r\cos u + R\right)^2$. The surface normal is $\mathbf{n} = (\mathbf{S}_u \times \mathbf{S}_v)/\left|\mathbf{S}_u \times \mathbf{S}_v\right| = (-\cos u\cos v, -\cos u\sin v, -\sin u)$. Hence $L = \mathbf{n}\cdot\mathbf{S}_{uu} = r$, $M = \mathbf{n}\cdot\mathbf{S}_{uv} = 0$, $N = \mathbf{n}\cdot\mathbf{S}_{vv} = \cos u\left(r\cos u + R\right)$. Therefore, $LN - M^2 = r\cos u\left(r\cos u + R\right)$, $EG - F^2 = r^2\left(r\cos u + R\right)^2$,

$$K = \frac{r\cos u\left(r\cos u + R\right)}{r^2\left(r\cos u + R\right)^2} = \frac{\cos u}{r\left(r\cos u + R\right)}, \text{ and}$$

$$H = \frac{1}{2}\frac{r^2\cos u\left(r\cos u + R\right) + r\left(r\cos u + R\right)^2}{r^2\left(r\cos u + R\right)^2} = \frac{r\cos u + \frac{1}{2}R}{r\left(r\cos u + R\right)}.$$

Since $R > r$, the denominator of K is positive. Thus, when $0 \le u < \pi/2$ or $3\pi/2 \le u < 2\pi$, then $\cos u > 0$ and $K > 0$. When $\pi/2 < u \le \pi$ or $\pi/2 \le u < 3\pi/2$, then $\cos u < 0$ and $K < 0$. When $u = \pi/2$ or $3\pi/2$, then $K = 0$. Thus the torus has regions of elliptic and hyperbolic points separated by two circles of parabolic points parametrized by $\mathbf{S}(\pi/2, v) = (R\cos v, R\sin v, r)$, $\mathbf{S}(3\pi/2, v) = (R\cos v, R\sin v, -r)$. The partition of the torus according to the type of point is illustrated in Figure 10.7.

Example 10.25 (Developable Surfaces)

Aircraft wings are constructed from a special honeycomb material which cannot be shaped by the methods used for plate metal. The wing shape is obtained by rolling the material. The resulting surface shapes are a special type of ruled surface known as *developable surfaces* for which the Gaussian curvature is zero at every point of the surface. See Exercise 10.31.

Example 10.26 (Minimal Surfaces)

A surface for which the mean curvature is zero at every point is called a *minimal surface*. Minimal surfaces arise in the study of soap films which form on a closed curve. The surface of a soap film is such that the surface tension is minimized. The resulting surface is a minimal surface.

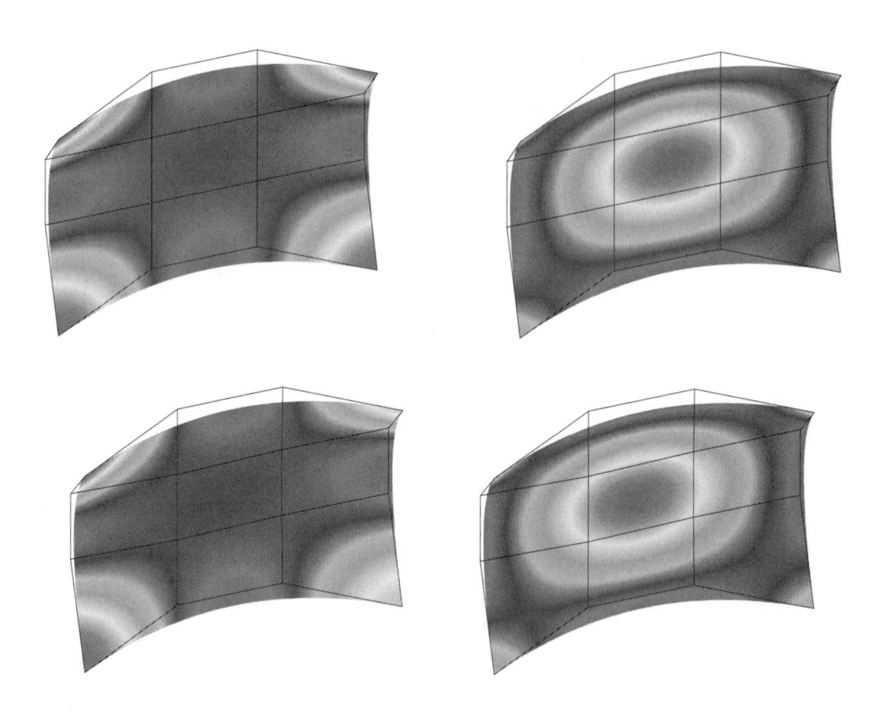

Figure 10.8 Gaussian and mean curvatures for Bézier surfaces

Example 10.27 (Curvatures of Bezier and B-spline Surfaces)

Curvatures are used to assess the quality of manufactured surfaces. Applications to the car and ship building industries can be found in [9], [17], [18], [7]. Surfaces can be coloured to indicate the value of a particular curvature at a point on the surface. For a Bézier surface $\mathbf{B}(s,t)$, the first and second order partial derivatives at $(s,t) = (0,0)$ are easily determined in terms of the control points using the formulae derived in Section 9.3.2, and the surface curvatures are easily determined from the derivatives. Curvatures at other parameter values can be obtained by subdivision in a manner similar to computing the curvature of a Bézier curve. A similar method applies to B-spline surfaces.

To illustrate how curvatures can be used, the plots of two similarly shaped Bézier surfaces are shown in Figure 10.8. The darker shades indicate high values of Gaussian curvature and the lighter shades indicate low values. Note that although the surfaces look very similar the curvatures show areas of difference. Such shading techniques (which work better in colour) can be used to highlight imperfections or potentially troublesome areas such as flat spots (when $K = H = 0$).

EXERCISES

10.22. Using the notation of Section 10.4, show that a surface $\mathbf{S}(u, v)$ is regular if and only if $EG - F^2 \neq 0$.

10.23. Consider a curve $\mathbf{C}(t)$ on a regular parametric surface $\mathbf{S}(u, v)$, and suppose $\mathbf{p} = \mathbf{C}(t) = \mathbf{S}(u, v)$ is a point on the curve. Let $\mathbf{N_p}$ be the principal normal of $\mathbf{C}$ at $\mathbf{p}$, $\mathbf{n_p}$ be the surface normal at $\mathbf{p}$, $\kappa_{\mathbf{p}}$ be the normal curvature of $\mathbf{S}$ in the direction $\dot{\mathbf{C}}$, and κ be the curvature of $\mathbf{C}$ at $\mathbf{p}$. Show that if θ is the angle between $\mathbf{N_p}$ and $\mathbf{n_p}$, then $\kappa_{\mathbf{p}} = \kappa \cos \theta$.

10.24. Determine the principal curvatures of the following surfaces:

(a) $\mathbf{S}(u, v) = \left(u, v, u^2 + v^2\right)$,

(b) $\mathbf{S}(u, v) = (u, \sin v, u + \cos v)$,

(c) *Torus*: $\mathbf{S}(u, v) = ((R + r \cos u) \cos v, (R + r \cos u) \sin v, r \sin u)$.

10.25. Determine the Gaussian and mean curvatures of the following surfaces:

(a) *Saddle surface*: $\mathbf{S}(u, v) = (u, v, uv)$,

(b) $\mathbf{S}(u, v) = \left(u, v, u^2 - v^2\right)$,

(c) $\mathbf{S}(u, v) = \left(u, v, u^3 + v^3\right)$.

10.26. Determine the umbilics of the following surfaces:

(a) *Ellipsoid*: $\mathbf{S}(u, v) = (4 \cos u \cos v, 2 \cos u \sin v, \sin u)$, $0 \leq u \leq 2\pi$, $0 \leq v \leq 2\pi$,

(b) $\mathbf{S}(u, v) = \left(u, v, u^2 + v^2\right)$.

10.27. Show that the following surfaces are minimal (that is, $H = 0$):

(a) *Bugle surface*: $\mathbf{S}(u,v) = (a\cosh(u/a)\cos v, a\cosh(u/a)\sin v, u)$,

(b) *Scherk's surface*: $\mathbf{S}(u,v) = (u, v, \ln(\cos(u)) - \ln(\cos(v)))$,

(c) *Enneper's surface*:

$$\mathbf{S}(u,v) = \left(u - u^3/3 + uv^2, v - v^3/3 + u^2v, u^2 - v^2\right) ,$$

(d) *Catalan's surface*:

$$\mathbf{S}(u,v) = (u - \sin u \cosh v, 1 - \cos u \cosh v, 4\sin(u/2)\sinh(v/2)) .$$

10.28. Show that the parameter curves of Enneper's surface are lines of curvature.

10.29. Show that $\mathbf{S}(u,v) = (u\cos v, u\sin v, u)$ is developable ($K = 0$).

10.30. The *offset* at a distance d of a regular surface $\mathbf{S}(u,v)$ with unit normal $\mathbf{N}(u,v)$ is $\mathbf{O}(u,v) = \mathbf{S}(u,v) + d\,\mathbf{N}(u,v)$ (see Section 9.2.1). Show that if K and H are the Gaussian and mean curvatures of $\mathbf{S}$ then the offset has Gaussian curvature $K/\left(Kd^2 - 2Hd + 1\right)$ and mean curvature $(H - Kd)/\left(Kd^2 - 2Hd + 1\right)$.

10.31. Let $\mathbf{A}(u)$ and $\mathbf{B}(u)$ be unit speed curves. Show that the ruled surface $\mathbf{S}(u,v) = \mathbf{A}(u) + v\mathbf{B}(u)$ is developable if and only if $(\mathbf{A}(u) \times \mathbf{A}'(u)) \cdot \mathbf{B}(u) = 0$. It can be shown that any developable surface is one of the following: (i) a cone, i.e $\mathbf{A}(u)$ is constant, (ii) a cylinder, i.e. $\mathbf{B}(u)$ is constant, or (iii) a *tangential developable*, that is, the surface consisting of all the tangents of a space curve, i.e. $\mathbf{B}(u) = \mathbf{A}'(u)$.

10.32. The *mean value* of a function $f(t)$ defined on an interval $[a,b]$ is $\frac{1}{b-a}\int_a^b f(t)\,dt$. By integrating $\kappa_{\mathbf{p}}(\theta) = \kappa_{\max}\cos^2\theta + \kappa_{\min}\sin^2\theta$ over the interval $[0,\pi]$, show that the mean value of the normal curvature $\kappa_{\mathbf{p}}$ at a point $\mathbf{p}$ is the mean curvature H.

10.33. Write a program or use a package to determine the curvatures of a Bézier surface $\mathbf{B}(s,t)$ (or B-spline) at a mesh of parameter values (s_i, t_j).

11
Rendering

11.1 Introduction

This chapter introduces techniques for object rendering. The two areas of CAD and computer graphics have differing opinions of what constitutes a good rendering. In the field of CAD the user needs a highly accurate and well defined line drawing that conforms to international drawing standards. Further, the objects must be drawn to scale so that they can be annotated with dimensions. In contrast, the computer graphics user desires a photographic realism of objects in a scene showing qualities such as colour, surface texture, and shadow. The position, shape, direction, and intensity of each light source play an important role. The following sections consider various elements that contribute to both accurate CAD drawings and realistic object rendering. Colour is introduced in Section 11.2, and a model for reflected light is developed in Section 11.3. Shading algorithms, which apply the light intensities obtained from the reflected light model, are discussed in Section 11.4. Section 11.5 introduces a new geometric feature of a surface, namely, the silhouette. Silhouettes are an essential feature in CAD drawings, and they are used in Section 11.6 to create shadow effects.

11.2 Colour

Colours are described in terms of *hues* that represent distinct colours such as red, blue or yellow. Artists commonly refer to tints, shades and tones. A *tint* is obtained from a hue by adding white. The amount of added white reflects the level of *saturation*. For instance a "dark blue" is highly saturated (less washed-out) while a "light blue" is unsaturated (more washed-out). Similarly, a *shade* is obtained by adding black to decrease the intensity or lightness of the hue. Addition of both white and black to a hue results in a *tone*. There are several models that are used to specify colour. These include the red, green, blue (RGB) and the hue, saturation, value (HSV) models used for monitors, and the cyan, magenta, yellow (CMY) model for printing devices.

The RGB model specifies the amounts of the three primary colours red r, green g, and blue b, as a coordinate (r, g, b) in a unit cube as shown in Figure 11.1(a). The primaries are *additive* meaning that the colours are mixed to the give the desired colour. Equal amounts of red, green and blue give a shade of grey ranging from white to black. Greys correspond to points on the diagonal of the RGB cube with white at $(0, 0, 0)$ and black at $(1, 1, 1)$.

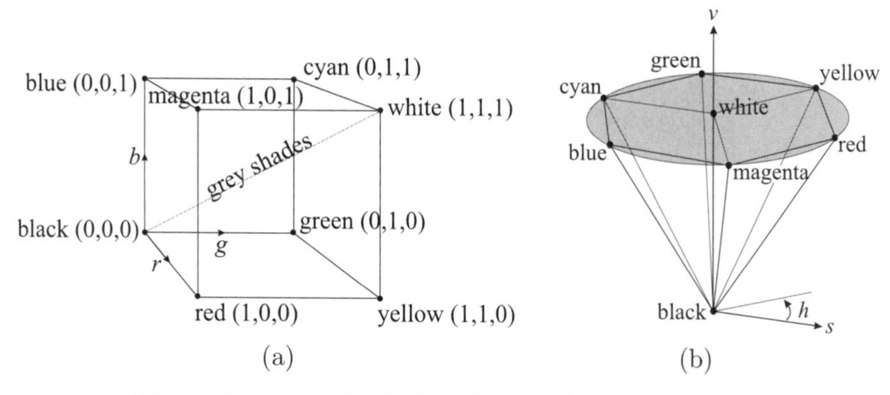

Figure 11.1 (a) RGB cube, and (b) HSV hexcone

In the RGB cube the "pure" hues are represented by a loop of vertices (omitting black and white): red-yellow-green-cyan-blue-magenta-red. This leads naturally to the HSV model for which colour is specified by a coordinate (h, s, v) to indicate the values of hue h, saturation s, and shade value v. The vertices are mapped to a plane to form a *hexcone* as shown in Figure 11.1(b). The hexcone vertex at $(0, 0, 0)$ represents black and the point $(0, 0, 1)$ represents white. The hue h specifies the colour in the loop as an angle about the cone axis with $0°$ representing red. The saturation s is the distance to the hexcone axis, and the value v denotes the darkness of the colour given by the distance in the axis direction. The tints are represented by points in the plane of the pure hues (shown shaded in Figure 11.1(b)). Points on the planar boundaries of the hexcone represent shades, and points on the axis joining white and black represent greys.

Printed colour performs differently to coloured light. For instance, when white light shines on blue paper, green and red light is absorbed and only blue light is reflected giving it a blue appearance. Therefore, printed colour is *subtractive*, and behaves as a filter removing colour components from the light that shines on it. The primary colours are cyan c, magenta m, and yellow y, and specified by the coordinate (c, m, y) in the CMY cube. Cyan, magenta and yellow absorb the *complementary* colours red, green and blue, respectively.

A variation of CMY is the CMYK model where the K stands for black. The motivation for CMYK is that, in reality, cyan, magenta and yellow inks do not mix to a true black but to a very dark brown. So many colour printers use an amount of pure black ink k as well as the three primary colours. The CMYK model represents each colour by a coordinate (c, m, y, k) which is obtained from the colour's (c, m, y) coordinate by taking $k = \min\{c, m, y\}$ and replacing c, m, and y by $c - k$, $m - k$, and $y - k$ respectively.

11.3 An Illumination Model for Reflected Light

The light in a scene can originate from either light-emitting sources such as the sun, light bulbs or a television, or light-reflecting surfaces such as mirrors, the walls of a room, and objects in a scene. In reality, light emanates from an area such as the surface of a light bulb or a television screen. This kind of light is referred to as *distributed light*. Distributed light is often simplified by assuming that the light emanates from a point to give a *point source*. This is a natural simplification to make when the light source is far away or when it is relatively small in comparison to the objects in the scene. A light source that is located at infinity produces rays of light that are parallel, and is referred to as a *directional* light source. Sunlight is often treated as a directional source.

When light falls on the surface of an object it can be (i) reflected: light bounces off the surface of the object, (ii) refracted: light passes through the object, or (iii) absorbed: light does not pass through the object and it is not reflected. The amalgamation of reflected light from several objects in a scene is called the *ambient* or *background* light. When light is reflected off a surface, it scatters to give *diffuse* reflection. The light from point sources creates a highlight or hotspot on a surface called *specular* reflection. The material properties of the object play an important role. Matte objects such as cardboard, wood, copper and some textiles give little specular reflection. In contrast, shiny surfaces such as polished metals and mirrors produce many highlights.

Let $\mathbf{P}$ be a point of a surface with unit normal $\mathbf{N}$, and let $\mathbf{L}$ be the unit vector pointing to the light source as shown in Figure 11.2(a). The incoming *incident ray* is reflected along an outgoing *reflected ray* $\mathbf{R}$, which is also assumed to be a unit vector. The angles that the incident and reflected rays make with the normal are called the angles of *incidence* and *reflection* respectively. The Laws of Reflection state:

Law 1: The angle of incidence is equal to the angle of reflection.

Law 2: The surface normal, incident ray and reflected ray lie in the same plane.

Referring to Figure 11.2(b), the reflected ray $\mathbf{R}$ can be obtained as the vector sum $\mathbf{R} = \overrightarrow{\mathbf{PC}} + \overrightarrow{\mathbf{CD}}$. Projection of the vector $\mathbf{L}$ onto $\mathbf{N}$ implies that $\overrightarrow{\mathbf{PB}}$ has length $\mathbf{N} \cdot \mathbf{L}$. By symmetry $\overrightarrow{\mathbf{PC}}$ has length $2(\mathbf{N} \cdot \mathbf{L})$ and since $\overrightarrow{\mathbf{CD}} = \overrightarrow{\mathbf{AP}} = -\mathbf{L}$ it follows that

$$\mathbf{R} = 2(\mathbf{N} \cdot \mathbf{L})\mathbf{N} - \mathbf{L} \,. \tag{11.1}$$

The vector should be normalized to obtain a unit vector.

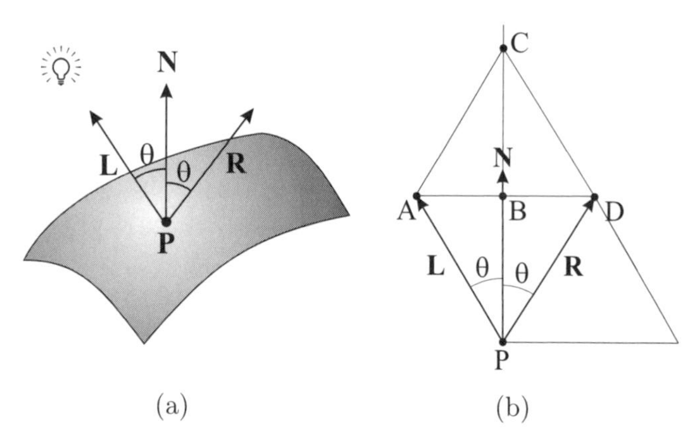

Figure 11.2 Laws of reflection

11.3.1 Diffuse Reflection

The model for diffuse reflection assumes that the surface is perfectly diffusing, that is, light is scattered equally in all directions. This is a reasonable assumption for matte surfaces, such as cardboard and chalk, which are microscopically rough so that light has an equal chance of being reflected in every direction. Such surfaces are referred to as *Lambertian surfaces* as they satisfy Lambert's Law:

Lambert's Law: Let ϕ be the angle between the surface normal and the view direction. Then the intensity of light reflected in the view direction is directly proportional to $\cos\phi$.

Note that *intensity* is defined to be the amount of light energy per unit of area.

Consider a narrow beam of light, of intensity $\hat{I}$, shining on an area of ΔA of an objects's surface. An application of elementary trigonometry, yields that the light reflected from the area ΔA is directed towards the viewer (in the direction $\mathbf{V}$) as a beam of area $\cos\phi\Delta A$ as indicated in Figure 11.3(a). Therefore, the reflected beam has light energy $\hat{I}\Delta A$ acting on an area $\cos\phi\Delta A$. Then, Lambert's Law implies that the intensity of the reflected light is

$$I = \cos\phi\left(\frac{\hat{I}\Delta A}{\cos\phi\Delta A}\right) = \hat{I} \ .$$

It can be concluded that the intensity of the reflected light is independent of the view direction.

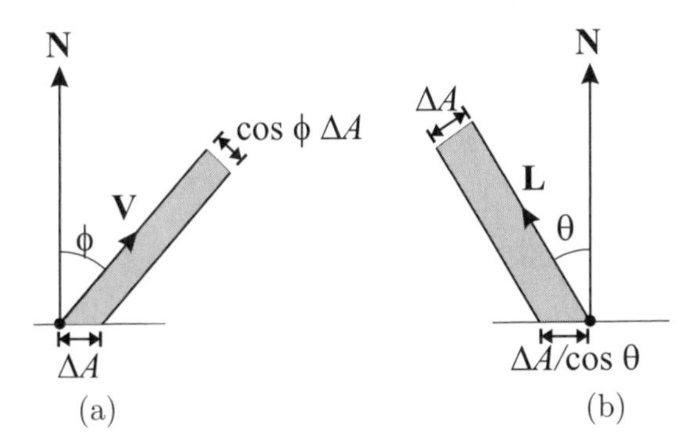

Figure 11.3

To compute the intensity I, consider a narrow beam of light with direction $\mathbf{L}$ and area ΔA. Suppose that the angle of incidence is θ, and that the intensity of the light source is I_d. Then, by elementary trigonometry, the beam covers an area $\Delta A/\cos\theta$ as shown in Figure 11.3(b). Hence, the beam has energy $I_d\Delta A$ acting on an area $\Delta A/\cos\theta$, giving

$$I = R_d\frac{I_d\Delta A}{\Delta A/\cos\theta} = I_d R_d\cos\theta \ .$$

where $0 \leq R_d \leq 1$ is the *coefficient of diffuse reflection* that specifies the proportion of light reflected by the surface material of the object. If the ray of light makes an angle of incidence that is greater than $\pi/2$, then the light is cast from behind the surface. Therefore, the ray does not illuminate the surface, and the surface is said to be *self-occluding.*

Finally, if $\mathbf{N}$ is the unit normal to the surface, and $\mathbf{L}$ is the unit vector pointing to the light source, then $\cos\theta = \mathbf{N} \cdot \mathbf{L}$. The reflected diffuse intensity I_D is given by

$$I_D = I_d R_d \cos\theta = I_d R_d (\mathbf{N} \cdot \mathbf{L}) , \tag{11.2}$$

where $0 \leq \theta \leq \pi/2$, and $0 \leq R_d \leq 1$. The colour of reflected diffuse light is the colour of the object. Maximum reflected intensity is obtained when $\mathbf{N} \cdot \mathbf{L} = 1$ which occurs when the light ray is perpendicular to the object's surface, that is, parallel to the surface normal.

11.3.2 Specular Reflection

Specular reflection is obtained when light from a point source is reflected at a certain angle from a shiny surface. The colour of specular reflection is equal to the colour of the light source. Unlike diffuse reflection, specular reflection is dependent on the position of the viewer.

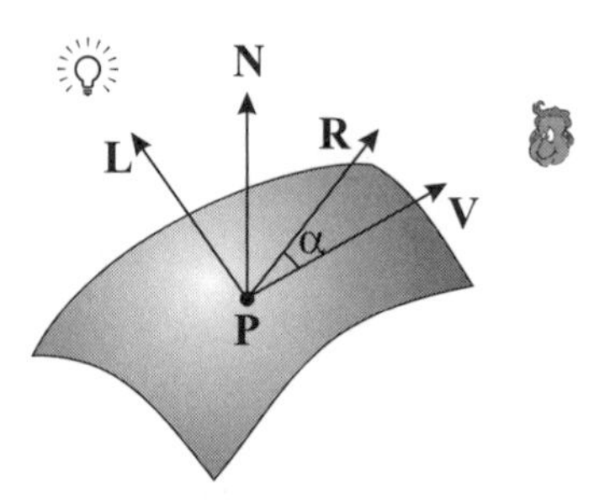

Figure 11.4

Referring to Figure 11.4, let $\mathbf{V}$ be the unit vector in the direction from a surface point $\mathbf{P}$ to the viewpoint. (Note that, for a fixed finite viewpoint, the vector $\mathbf{V}$ is dependent on $\mathbf{P}$.) Further, let α be the angle between $\mathbf{V}$ and the (unit) reflected light ray $\mathbf{R}$. Then, $\cos\alpha = \mathbf{V} \cdot \mathbf{R}$. If the object is made from a material that is perfectly reflecting, then the specular reflections have direction $\mathbf{R}$ and are visible only when $\mathbf{R} = \mathbf{V}$ and $\alpha = 0$. In reality, however, there are specular reflections in a range of angles about $\mathbf{R}$. The intensity of reflected light that the eye receives depends on α, and is such that smaller angles yield greater intensities. This results in a region of the surface with higher intensities called a specular *highlight* or *hotspot* as illustrated in Figure 11.5.

Figure 11.5 Specular highlights

A model for specular reflection, based on empirical findings, has been introduced by Phong [19] and assumes that the intensity of specular reflection is proportional to $\cos^m \alpha$, for some positive number m. The graphs of $\cos^m \alpha$ for various values of m can be seen in Figure 11.6. The cosine functions provide reasonable profiles of intensity: values close to 1 for small angles, and rapidly decreasing values as the angle increases. Let θ continue to denote the angle of incidence. Then, using the fact that $\cos \alpha = \mathbf{V} \cdot \mathbf{R}$, the specular intensity is given by

$$I_S = I_s R_s(\theta) \cos^m \alpha = I_s R_s(\theta)(\mathbf{V} \cdot \mathbf{R})^m , \qquad (11.3)$$

where $0 \leq R_s(\theta) \leq 1$ is the *coefficient of specular reflectance* for the surface. Note that $R_s(\theta)$ is dependent on the angle θ and is governed by the material properties of the surface. The function $R_s(\theta)$ should be non-decreasing with increasing θ. One obvious simplification of (11.3) is to assume that the coefficient of specular reflectance is constant: $R_s(\theta) = R_s$.

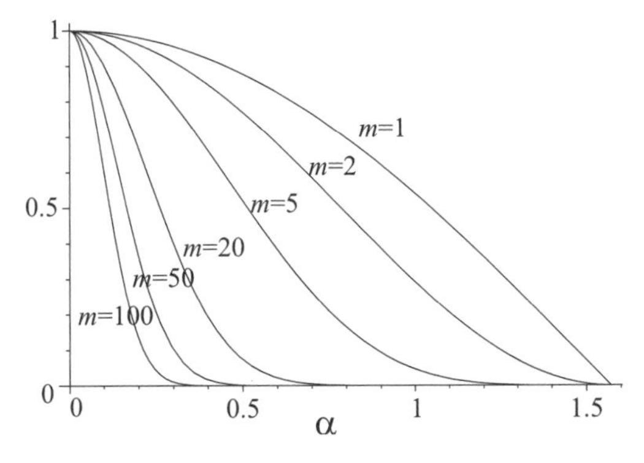

Figure 11.6 $\cos^m \alpha$ for various m

When the light source is distant, it can be assumed that the light rays are parallel, and that $\mathbf{L}$ is constant. A further simplification is to assume that the viewpoint is at infinity so that $\mathbf{V}$ is constant. With both of these assumptions, Equation (11.3) can be modified to give an alternative model for which the computation of intensities is more efficient. Let

$$\mathbf{B} = \frac{\mathbf{V} + \mathbf{L}}{|\mathbf{V} + \mathbf{L}|} \tag{11.4}$$

be the unit vector shown in Figure 11.7. Let β be the angle between $\mathbf{N}$ and $\mathbf{B}$ so that $\cos\beta = \mathbf{N} \cdot \mathbf{B}$. (When $\mathbf{V}$ lies in the plane of $\mathbf{N}$ and $\mathbf{L}$ it is easily shown that $\beta = \alpha/2$.) The modified model for specular intensities is obtained by replacing $\mathbf{V} \cdot \mathbf{R}$ by $\mathbf{N} \cdot \mathbf{B}$ in (11.3) to give

$$I_{\mathrm{S}} = I_s R_s(\theta)(\mathbf{N} \cdot \mathbf{B})^m \, . \tag{11.5}$$

The computation of $\mathbf{B}$ requires fewer arithmetic operations than the computation of $\mathbf{R}$ using (11.1). Further, $\mathbf{B}$ is constant, and so I_{D} and I_{S} are functions of $\mathbf{N}$. It should be noted that the two models (11.3) and (11.5) do not give identical results. The intensities of the second model are those of a surface with normal vector $\mathbf{B}$ and $\mathbf{R} = \mathbf{V}$, and therefore the resulting intensities are maximal in the view direction.

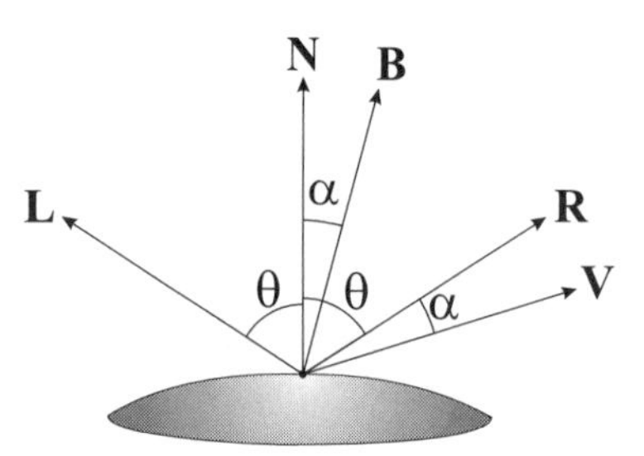

Figure 11.7 $\mathbf{B}$ is cheaper to compute than $\mathbf{R}$

11.3.3 Ambient Reflection

Ambient (or background) reflection is difficult to model due to the complexity of multiple reflections from one object onto other. A simple model of ambient lighting is obtained by assuming that every object receives the same amount of light from all directions. The colour of reflected ambient light is the same as the colour of the object. If the uniform intensity of ambient light is I_a, then the reflected ambient intensity of an object is given by

$$I_{\mathrm{A}} = I_a R_a \ , \tag{11.6}$$

where $0 \leq R_a \leq 1$ is the *coefficient of ambient reflection*. The value of R_a is dependent on the material characteristics of the object. When $R_a = 0$ the object produces no reflected light, and when $R_a = 1$ the object reflects light at full intensity.

11.3.4 Attenuation

When a light source is a finite distance from the objects in a scene, the distance between the objects and the light source should be considered. The brightness of reflected light is inversely proportional to the square of the distance d of the object to the light source. This means that each reflected intensity should be multiplied by an *attenuation factor* $\mathrm{att}(d) = 1/d^2$ giving

$$\begin{aligned} I_{\mathrm{D}} &= \mathrm{att}(d) I_d R_d \cos\theta = \mathrm{att}(d) I_d R_d (\mathbf{N} \cdot \mathbf{L}) \ , \quad \text{and} \\ I_{\mathrm{S}} &= \mathrm{att}(d) I_s R_s(\theta) (\mathbf{N} \cdot \mathbf{B})^m \ . \end{aligned}$$

In practice the formulae give poor results for non-point light sources, and an adapted multiplier

$$\mathrm{att}(d) = \frac{1}{a_2 d^2 + a_1 d + a_0}$$

is more commonly used (with $a_0 \neq 0$ to prevent a divide by zero). The presence of the quadratic term can cause a wide range of intensity values so many applications set $a_2 = 0$ to give $\mathrm{att}(d) = 1/(a_1 d + a_0)$, or ignore attenuation completely ($\mathrm{att}(d) = 1$). An alternative is to cap the attenuation at a certain value to remove the possibility of extreme values.

11.3.5 Total Intensity

The reflection model is completed by combining the ambient intensity, with the diffuse and specular intensities for each light source, to give the total intensity

$$I = I_a R_a + \mathrm{att}(d) \sum_i \left(I_d R_d (\mathbf{N} \cdot \mathbf{L}_i) + I_s R_s(\theta) (\mathbf{N} \cdot \mathbf{B}_i)^m \right) \ , \tag{11.7}$$

where $\sum_i$ denotes summing over all light sources.

For a colour rendering device, the intensity formula (11.7) is applied to each colour component and combined to give a colour intensity. For example, suppose a device uses an RGB model. Then it is necessary to specify the coefficients of reflection and intensities for each colour component. For compactness, let the red, green and blue coefficients of ambient reflectance be represented by the vector $\mathbf{R}_a = (R_{a,r}, R_{a,g}, R_{a,b})$. Similarly, let the coefficients of diffuse and specular reflectance be $\mathbf{R}_d = (R_{d,r}, R_{d,g}, R_{d,b})$ and $\mathbf{R}_s = (R_{s,r}, R_{s,g}, R_{s,b})$, respectively. Further, let the ambient, diffuse, and specular intensities be $\mathbf{I}_a = (I_{a,r}, I_{a,g}, I_{a,b})$, $\mathbf{I}_d = (I_{d,r}, I_{d,g}, I_{d,b})$, and $\mathbf{I}_s = (I_{s,r}, I_{s,g}, I_{s,b})$. Then the total intensity for red is

$$I_{\text{red}} = I_{a,r}R_{a,r} + \text{att}(d)\sum_i \left(I_{d,r}R_{d,r}(\mathbf{N}\cdot\mathbf{L}_i) + I_{s,r}R_{s,r}(\theta)(\mathbf{N}\cdot\mathbf{B}_i)^m\right) . \quad (11.8)$$

There are similar equations for green and blue.

11.4 Shading Algorithms

The illumination model discussed in Section 11.3 determines the colour intensity of a point of an object surface. Shading algorithms determine how the illumination model is applied across the entire surface. Ideally, intensities would be calculated at every visible point of every surface in the scene (and computed to the resolution of the display device). This is not feasible since each point intensity calculation requires a surface unit normal to be computed, and this entails expensive surface derivative evaluations and the application of a square root (see Equation (9.1)). Therefore it is important to implement shading algorithms in a manner that minimises the number of intensity and surface normal computations.

The shading algorithms described in the following sections require the object surfaces to be *faceted*, that is, approximated by a mesh of planar polygonal faces or *facets*. A simple way to facet a surface $\mathbf{S}(u,v)$ is to evaluate along the u and v parameter lines to give a rectangular grid of points $\mathbf{P}_{i,j} = \mathbf{S}(u_i, v_j)$ for $0 \le i \le m$ and $0 \le j \le n$. Triangular facets can be obtained by splitting the polygon with vertices $\mathbf{P}_{i,j}$, $\mathbf{P}_{i+1,j}$, $\mathbf{P}_{i,j+1}$, $\mathbf{P}_{i+1,j+1}$ into two triangles. More elaborate faceting methods take the curvature of the surface into account in order to obtain many facets in regions where the surface bends the most, and few facets where the surface is flat. Bézier and B-spline surfaces can be faceted using the subdivision methods of Section 9.5. If the surface is subdivided sufficiently, then the control polygon can be used to obtain facets.

11.4.1 Flat Shading

The flat shading algorithm assigns one colour intensity value uniformly across each facet. This gives a very cheap shading method as the intensity is computed at just one point of each facet. However, since adjacent facets have different intensities the facets are clearly distinguishable, and the lack of variation in the shading makes the facets look flat. The facets of the sphere in Figure 11.8(a) are fairly pronounced. Better shading can be obtained by taking very small facets as shown in Figure 11.8(b).

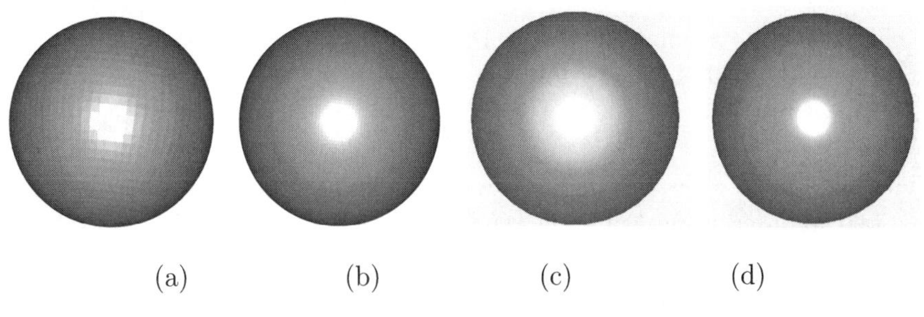

(a) (b) (c) (d)

Figure 11.8 Renderings of a sphere

11.4.2 Gouraud Shading

Gouraud shading tries to overcome the unnatural uniform intensity of the flat shading method. The illumination model (11.7) is used to determine the colour intensities of the vertices of each triangular facet of the surface. Then, the intensities of each vertex are interpolated to give approximate intensities for other points of the facet. The variation in intensities provided by the interpolation reduces the flat appearance of the facet.

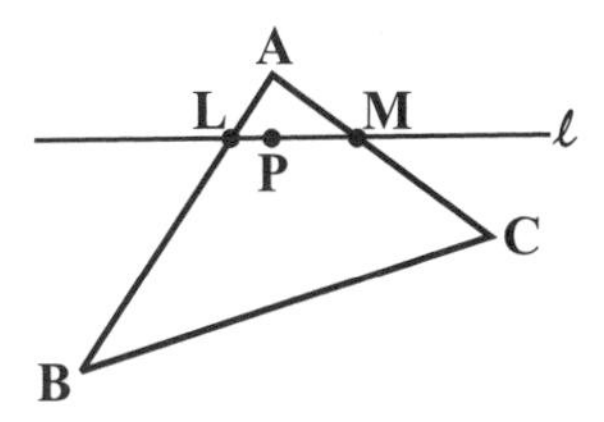

Figure 11.9 Intensity calculation using a scanline

Consider the triangular facet in Figure 11.9. Let the intensities of the vertices **A**, **B** and **C**, be $\mathbf{I}_A$, $\mathbf{I}_B$ and $\mathbf{I}_C$, respectively. Consider a line ℓ (called the *scanline*) that moves across the triangle. Suppose ℓ intersects the facet edge **AB** in the point **L**, and the edge **AC** in the point **M**. Then, by linear interpolation, $\mathbf{L} = (1-s)\mathbf{A} + s\mathbf{B}$, for some $0 \leq s \leq 1$, and an approximation for the intensity at **L** is taken to be

$$\mathbf{I}_L = (1-s)\mathbf{I}_A + s\mathbf{I}_B \ .$$

Similarly, $\mathbf{M} = (1-t)\mathbf{A} + t\mathbf{C}$, for some $0 \leq t \leq 1$, and the intensity of **M** is taken to be

$$\mathbf{I}_M = (1-t)\mathbf{I}_A + t\mathbf{I}_C \ .$$

Finally, let **P** be a point on ℓ, then $\mathbf{P} = (1-u)\mathbf{L} + u\mathbf{M}$ for some $0 \leq u \leq 1$, and an approximate intensity for **P** is

$$\mathbf{I}_P = (1-u)\mathbf{I}_L + u\mathbf{I}_M \ . \tag{11.9}$$

Since the scanline passes over the whole triangle, an intensity for every point can be computed using Equation (11.9).

The Gouraud shading method described above can be implemented using the facet normals in the intensity computations as they are less expensive to compute than true surface normals. However, this results in uneven shading since pairs of adjacent facets have different intensity values along the edge where the facets meet. Smoother shading can be obtained by averaging the normals of the facets that contain the vertex, as shown in Figure 11.10. Note that a vertex may be contained in more than three facets. The average vector is normalized to obtain a unit vector (so there is no need to divide through by the number of facets when computing the the average). Since the intensities along an edge are identical for the two facets containing the edge, the resulting shading is smooth across the facets. The Gouraud method achieves smoother shading is obtained at the additional computational cost of averaging and normalizing the facet normals at each vertex. Figure 11.8(c) shows a rendering of a sphere using the Gouraud method.

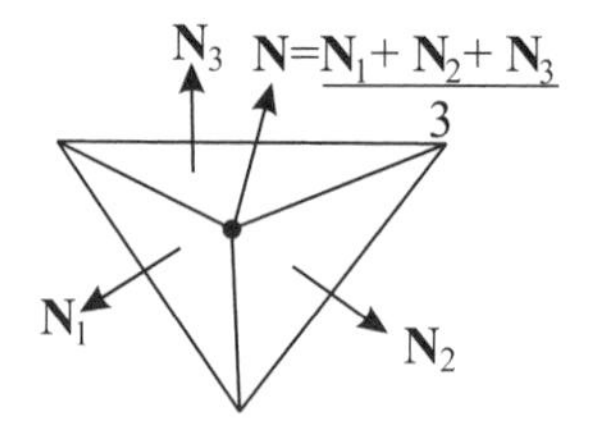

Figure 11.10 Average facet normal

11.4.3 Phong Shading

In Phong shading true surface normals are computed at the facet vertices. Approximate surface normals at points in the facet are obtained by linearly interpolating the vertex normals. If the exact surface normals of two vertices $\mathbf{A}$ and $\mathbf{B}$ are $\mathbf{N}_0(x_0, y_0, z_0)$ and $\mathbf{N}_1(x_1, y_1, z_1)$, then the interpolated normal of a point $\mathbf{L} = (1-s)\mathbf{A} + s\mathbf{B}$, for $0 \le s \le 1$, is taken to be the normalization of $\mathbf{N} = (1-s)\mathbf{N}_0 + s\mathbf{N}_1$. The approximate normal of any point in the facet can be obtained using a scanline method similar to the one used in Gouraud shading, but normals are interpolated in place of intensities. Since approximate normals avoid computing surface derivatives, they are more efficient to use than true surface normals. The Phong method then applies (11.7), using the approximate normals, to compute the intensities at points in each facet. The gradual variation of normals across each facet means that Phong shading produces a smoother and more natural looking shading than both the flat and Gouraud methods. Figure 11.8(d) shows a rendering of a sphere using the Gouraud method.

EXERCISES

11.1. Write a computer program that converts a RGB coordinate (r, g, b) to a HSV coordinate (h, s, v) and vice versa.

11.2. Consider a point light source positioned at $(0, 10, 20)$ shining on the surface $\mathbf{x}(s,t) = (s, t, -s^2 - t^2)$, $0 \le s, t \le 1$. Determine the reflected vector $\mathbf{R}$ of the incident ray that hits the surface at the point $\mathbf{x}(0.5, 0.5) = (0.5, 0.5, -0.5)$. Determine the angle of incidence.

11.3. Write a computer program to perform Flat, Gouraud or Phong shading of an object with polygonal faces.

11.5 Silhouettes

In contrast to the photorealism of computer graphics rendering, CAD drawings are predominately line drawings with little or no shading. For objects with faces that are planar, a CAD style drawing can be obtained using projections together with edge and vertex information. In Section 4.3 it was shown that a parallel or perspective projection maps a linear edge $\mathbf{AB}$ to the linear segment joining the images of $\mathbf{A}$ and $\mathbf{B}$. Projections of non-linear edges can be determined by other methods. Sections 7.5.3 and 8.2.1 described techniques for projecting Bézier and NURBS curves.

Objects with curved faces cannot be rendered by projecting edges alone. For instance, a sphere, which has no edges, is rendered by drawing a circle to represent the extremities of the sphere with respect to the view: the circle is an example of a *silhouette*. The ability to compute silhouettes is an essential tool for CAD drawings.

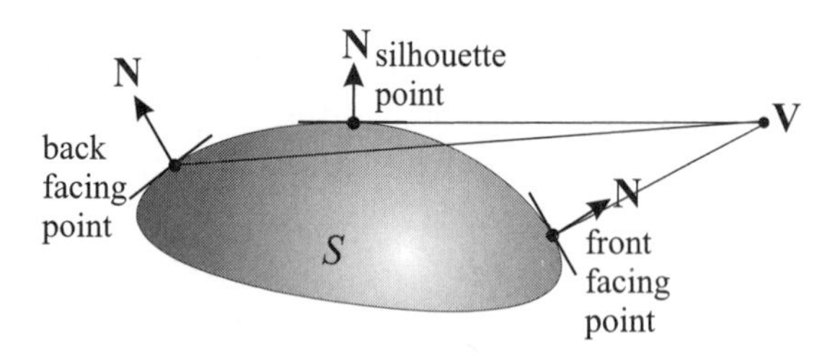

Figure 11.11

Consider a projection of a surface S from a viewpoint $\mathbf{V}$ (considered to be at infinity for a parallel projection) as shown in Figure 11.11. Let $\mathbf{P}$ be a point on S and let $\mathbf{N}$ be the surface normal at $\mathbf{P}$. In a neighbourhood of $\mathbf{P}$ the surface is approximated by the tangent plane which passes through $\mathbf{P}$ and has normal direction $\mathbf{N}$. If $\mathbf{N}$ points towards the viewpoint then a neighbourhood of $\mathbf{P}$ is *front-facing* and is visible in the given view. Similarly, if $\mathbf{N}$ points away from the viewpoint, then a neighbourhood of $\mathbf{P}$ is *back-facing* and is invisible in the given view. When $\mathbf{N}$ is neither front-facing nor back-facing then, in a neighbourhood of $\mathbf{P}$, the face is (in general) turning from front-facing to back-facing and $\mathbf{P}$ is called a *silhouette* point. For a given projection, the set of all silhouette points of a surface is called the *silhouette*. The image of the silhouette in the viewplane is called the *apparent contour*. Silhouette points are expressed more precisely in the following definition.

Definition 11.1

Let $\mathbf{P}$ be a point of a surface S, and let $\mathbf{N}$ be a surface normal at $\mathbf{P}$. Then $\mathbf{P}$ is said to be a *silhouette point of a parallel projection* in the direction $\mathbf{V}$ whenever $\mathbf{V} \cdot \mathbf{N} = 0$. $\mathbf{P}$ is said to be a *silhouette point of a perspective projection* from a viewpoint $\mathbf{V}$ whenever $(\mathbf{V} - \mathbf{P}) \cdot \mathbf{N} = 0$.

Example 11.2

Consider the parallel projection of the surface $\mathbf{x}(s,t) = (s, t, st + s^3 + t^2)$ in the direction $\mathbf{V}(0,1,0)$. Then $\mathbf{x}_s(s,t) = (1, 0, t + 3s^2)$ and $\mathbf{x}_t(s,t) = (0, 1, s + 2t)$ and the unit normal is

$$\mathbf{N} = \frac{1}{|\mathbf{x}_s \times \mathbf{x}_t|}(-t - 3s^2, -s - 2t, 1) .$$

The silhouette points satisfy $\mathbf{V} \cdot \mathbf{N} = -s - 2t = 0$. Therefore $s = -2t$ and substituting for s in $\mathbf{x}(s,t)$ gives the silhouette curve $(-2t, t, -t^2 - 8t^3)$. The surface and its silhouette are shown in Figure 11.12. Note that the silhouette calculation does not require the surface normal to be a unit vector, and so the surface normal will not be normalized in the subsequent examples.

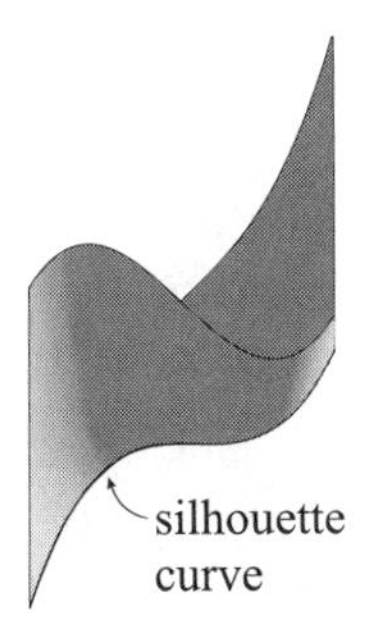

Figure 11.12 Surface and silhouette of Example 11.2

Example 11.3

Consider the parallel projection of the surface $\mathbf{x}(s,t) = \left(s, t, st^2 + t^2 - \frac{1}{3}s^3\right)$ in the direction $\mathbf{V}(1,0,0)$. Then $\mathbf{x}_s(s,t) = (1, 0, t^2 - s^2)$ and $\mathbf{x}_t(s,t) = (0, 1, 2st + 2t)$ and a normal is

$$\mathbf{N} = (s^2 - t^2, -2st - 2t, 1) .$$

The silhouette points satisfy $\mathbf{V} \cdot \mathbf{N} = s^2 - t^2 = 0$. Therefore $s = t$ or $s = -t$. Substituting $s = t$ in $\mathbf{x}(s,t)$ gives the silhouette curve $\left(t, t, t^2 + \frac{2}{3}t^3\right)$. Similarly, substituting $s = -t$ gives a second silhouette curve $\left(-t, t, t^2 - \frac{2}{3}t^3\right)$. The surface and the two silhouettes are shown in Figure 11.13(a) and viewed from above. Figures 11.13(b) and (c) show the surface and silhouettes viewed along the positive and negative x-axis.

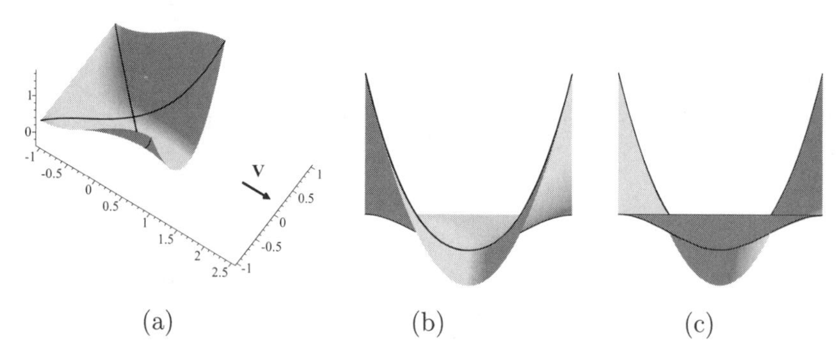

Figure 11.13 Surface and silhouettes of Example 11.3

Example 11.4

Consider the perspective projection of the surface $\mathbf{x}(s,t) = (s, t, t^2 - s^2)$ from the viewpoint $\mathbf{V}(0,5,0)$. Then $\mathbf{x}_s(s,t) = (1, 0, -2s)$ and $\mathbf{x}_t(s,t) = (0, 1, 2t)$ and a normal is

$$\mathbf{N} = (2s, -2t, 1) \ .$$

The silhouette points satisfy $(\mathbf{V} - \mathbf{x}(s,t)) \cdot \mathbf{N} = -s^2 + t^2 - 10t = 0$. Therefore $s = \pm\sqrt{t(t-10)}$ for $t \leq 0$ and $t \geq 10$. Substituting for s into $\mathbf{x}(s,t)$ gives two silhouette curves $(\sqrt{t(t-10)}, t, 10t)$ and $(-\sqrt{t(t-10)}, t, 10t)$. An alternative substitution can be obtained by parametrizing the conic $-s^2+t^2-10t = 0$ using the method of Section 5.6.4: for instance, $(s,t) = \big(10u/(1-u^2), 10/(1-u^2)\big)$. The trigonometric parametrizations of the conic can also be used: $(s,t) = (5\tan\theta, 5\sec\theta + 5)$ or $(s,t) = (5\sinh\theta, \pm 5\cosh\theta + 5)$. The surface and the two silhouette curves are shown in Figure 11.14(a) and viewed from above. Figure 11.14(b) shows the surface and silhouettes viewed from $\mathbf{V}(0,5,0)$.

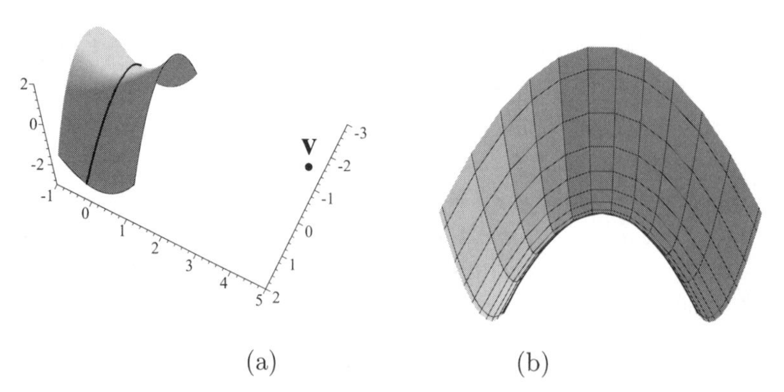

Figure 11.14 Surface and silhouettes of Example 11.4

Example 11.5 (Silhouettes of a Sphere)

To show that the silhouette of a sphere is a circle, consider the sphere with radius r and centred at $\mathbf{C}(x_0, y_0, z_0)$. In implicit form the sphere is given by

$$(x - x_0)^2 + (y - y_0)^2 + (z - z_0)^2 = r^2 \tag{11.10}$$

and a normal vector at the point $\mathbf{P}(x, y, z)$ is

$$\mathbf{N} = (x - x_0, y - y_0, z - z_0) = \mathbf{P} - \mathbf{C} \, . \tag{11.11}$$

The silhouette points of the sphere for a parallel projection with direction $\mathbf{V}(v_0, v_1, v_2)$ satisfy $\mathbf{V} \cdot \mathbf{N} = 0$ giving

$$v_0(x - x_0) + v_1(y - y_0) + v_2(z - z_0) = 0 \, . \tag{11.12}$$

Equation (11.12) defines the plane through $\mathbf{C}$ with normal direction $\mathbf{V}$. Therefore, the silhouette curve is the intersection of the sphere with the plane (11.12), that is, a circle centred at $\mathbf{C}$ with radius r.

For a perspective projection with viewpoint $\mathbf{V}(v_0, v_1, v_2)$, the silhouette points of the sphere (11.10) satisfy $(\mathbf{V} - \mathbf{P}) \cdot \mathbf{N} = 0$ and hence

$$(\mathbf{V} - \mathbf{P}) \cdot (\mathbf{P} - \mathbf{C}) = 0 \, . \tag{11.13}$$

In vector form, the sphere (11.10) is given by

$$(\mathbf{P} - \mathbf{C}) \cdot (\mathbf{P} - \mathbf{C}) = r^2 \, . \tag{11.14}$$

Adding Equations (11.13) and (11.14) gives

$$(\mathbf{V} - \mathbf{C}) \cdot (\mathbf{P} - \mathbf{C}) = r^2 \, ,$$

which, after some rearrangement of the equation, gives

$$(\mathbf{V} - \mathbf{C}) \cdot \mathbf{P} = r^2 + \mathbf{C} \cdot (\mathbf{V} - \mathbf{C}) \, . \tag{11.15}$$

Therefore, the silhouette points lie on a plane with normal direction $\mathbf{V} - \mathbf{C}$. Intersecting the plane (11.15) with the sphere (11.10) gives a silhouette circle. See Exercise 11.7 for further details.

EXERCISES

11.4. Determine the silhouette curve of the surface $\mathbf{x}(s,t) = (s, t, st + s^2 - \frac{1}{3}t^3)$ for a parallel projection in the direction $\mathbf{V}(0,1,0)$.

11.5. Determine the silhouette curve of the surface $\mathbf{x}(s,t) = (s, t, st^2 - s + \frac{1}{3}s^3)$ for a parallel projection in the direction $\mathbf{V}(1,0,0)$. (Hint: the substitution $s = \cos\theta$ and $t = \sin\theta$ might be helpful for the parametrization of the silhouette curve.)

11.6. Determine the silhouette curve of the surface $\mathbf{x}(s,t) = (s, t, 9s^2t - t - 3t^3)$ for a parallel projection in the direction $\mathbf{V}(0,1,0)$.

11.7. Show that, for a parallel or perspective projection, a plane either has no silhouette points or the entire plane is "in silhouette" (that is, every point of the plane is a silhouette point).

11.8. Generalise the method shown for a sphere to determine the silhouette plane of the ellipsoid $\frac{x^2}{a^2} + \frac{y^2}{b^2} + \frac{z^2}{c^2} = 1$ for a parallel projection in the direction $\mathbf{V}(v_0, v_1, v_2)$.

11.9. Show that, for a given projection, the silhouette of a surface is the locus of points on the surface for which there is a tangent line passing through the viewpoint (considered to be at infinity for a parallel projection).

11.10. Consider the perspective projection of the sphere of Example 11.6.

a) Show that the distance from the silhouette plane (11.12) to the sphere centre $\mathbf{C}$ is $d = r^2 / |\mathbf{V} - \mathbf{C}|$.

b) Let $\mathbf{U} = (\mathbf{V} - \mathbf{C}) / |\mathbf{V} - \mathbf{C}|$ denote the silhouette plane unit normal. Show that the silhouette circle has centre $\mathbf{C}_1$ and radius r_1 given by

$$\mathbf{C}_1 = \mathbf{C} + d\mathbf{U}, \quad \text{and} \quad r_1 = (r^2 - d^2)^{1/2} .$$

c) Let $\mathbf{X}$ be a unit vector perpendicular to $\mathbf{U}$, and let $\mathbf{Y} = (\mathbf{U} \times \mathbf{X}) / |\mathbf{U} \times \mathbf{X}|$. Using $\mathbf{C}_1$ as the origin, and $\mathbf{X}$ and $\mathbf{Y}$ as the x- and y- axes for the silhouette plane, write down a parametrization for the silhouette circle.

11.11. Show that, for a perspective projection, the sphere (11.10) has no silhouette points when the viewpoint is inside the sphere.

Silhouette curves of quadric surfaces are relatively straightforward to compute since they can be shown to lie in a *silhouette plane*. The intersection of the silhouette plane with the quadric yields silhouette curves that are conics. Consider a quadric with the *homogeneous* equation

$$\mathbf{x}\mathrm{Q}\mathbf{x}^T = 0\,, \tag{11.16}$$

and a projection with viewpoint $\mathbf{V}$ (using homogeneous coordinates). The aim is to apply Exercise 11.6 to obtain the silhouette. Let $\mathbf{P}$ be a point on the quadric so that $\mathbf{PQP}^T = 0$. The line through $\mathbf{P}$ and $\mathbf{V}$ is

$$\mathbf{x}(s) = (1-s)\mathbf{P} + s\mathbf{V}\,. \tag{11.17}$$

It follows that the point $\mathbf{x}(s)$ lies on the quadric whenever

$$\mathbf{x}(s)\mathrm{Q}\mathbf{x}(s)^T = ((1-s)\mathbf{P} + s\mathbf{V})\mathrm{Q}((1-s)\mathbf{P}^T + s\mathbf{V}^T) = 0\,. \tag{11.18}$$

Expanding (11.18), gives a quadratic equation in s

$$(1-s)^2\mathbf{PQP}^T + s(1-s)\mathbf{VQP}^T + s(1-s)\mathbf{PQV}^T + s^2\mathbf{VQV}^T = 0\,. \tag{11.19}$$

Since $\mathbf{PQP}^T = 0$ and $\mathbf{VQP}^T = \mathbf{PQV}^T$ (Exercise 11.12), it follows that Equation (11.19) simplifies to

$$2s(1-s)\mathbf{VQP}^T + s^2\mathbf{VQV}^T = 0\,. \tag{11.20}$$

Applying Exercise 11.9, $\mathbf{P}$ is a silhouette point if and only if the line (11.17) is tangent to the quadric at $\mathbf{P}$, which can occur if and only if (11.20) has a multiple root. Since $s = 0$ is a root, a multiple root can can only arise when

$$\mathbf{VQP}^T = 0\,. \tag{11.21}$$

This is the condition for $\mathbf{P}$ to be a silhouette point of the quadric (11.16).

Example 11.6

Consider a projection of the sphere $(x-x_0)^2+(y-y_0)^2+(z-z_0)^2-r^2=0$ from a viewpoint $\mathbf{V}(v_0, v_1, v_2, v_3)$. Condition (11.21) implies that $\mathbf{P}(X, Y, Z, W)$ is a silhouette point whenever

$$\begin{pmatrix} v_0 & v_1 & v_2 & v_3 \end{pmatrix} \begin{pmatrix} 1 & 0 & 0 & -x_0 \\ 0 & 1 & 0 & -y_0 \\ 0 & 0 & 1 & -z_0 \\ -x_0 & -y_0 & -z_0 & x_0^2+y_0^2+z_0^2-r^2 \end{pmatrix} \begin{pmatrix} X \\ Y \\ Z \\ W \end{pmatrix} = 0\,,$$

yielding the silhouette plane,

$$\begin{aligned} &(v_0 - v_3x_0)X + (v_1 - v_3y_0)Y + (v_2 - v_3z_0)Z \\ &\quad -(v_0x_0 + v_1y_0 + v_2z_0 + v_3(x_0^2+y_0^2+z_0^2-r^2))W = 0\,. \end{aligned} \tag{11.22}$$

For a parallel projection, $v_3 = 0$, and in affine coordinates (11.22) gives

$$v_0 x + v_1 y + v_2 z - (v_0 x_0 + v_1 y_0 + v_2 z_0) = 0 .$$

This agrees with Equation (11.12). For a perspective projection, set $v_3 = 1$, so that in affine coordinates (11.22) yields

$$(v_0 - x_0)x + (v_1 - y_0)y + (v_2 - z_0)z - (v_0 x_0 + v_1 y_0 + v_2 z_0 + (x_0^2 + y_0^2 + z_0^2 - r^2)) = 0 . \tag{11.23}$$

It left as an exercise to the reader (Exercise 11.13) to verify that this equation is equivalent to (11.15) .

Example 11.7

Consider the sphere $\left(x - \frac{1}{2}\right)^2 + \left(y - \frac{1}{2}\right)^2 + (z-1)^2 = \frac{1}{4}$ with radius $\frac{1}{2}$ and centred at $\left(\frac{1}{2}, \frac{1}{2}, 1\right)$. For a perspective projection with viewpoint $(-1, 0, 2)$, Equation (11.23) can be applied to yield the silhouette plane

$$-\frac{3}{2}x - \frac{1}{2}y + z - \frac{1}{4} = 0 .$$

The silhouette curve is the circle of intersection of the silhouette plane and the sphere. Using Exercise 11.7(a) and (b),

$$|\mathbf{V} - \mathbf{C}| = \left| \left(-\frac{3}{2}, -\frac{1}{2}, 1 \right) \right| = \frac{\sqrt{14}}{2}, \quad \text{and} \quad d = \frac{1}{2\sqrt{14}} .$$

It follows that the circle has radius $r_1 = \frac{\sqrt{13}}{2\sqrt{14}}$ and centre

$$\mathbf{C} = \left(\frac{1}{2}, \frac{1}{2}, 1 \right) + \frac{1}{14} \left(-\frac{3}{2}, -\frac{1}{2}, 1 \right) = \left(-\frac{3}{28}, -\frac{1}{28}, \frac{1}{14} \right) .$$

A parametric equation for the silhouette circle can be obtained using Exercise 11.7(c). The silhouette plane has unit normal $\mathbf{U} = \frac{1}{\sqrt{14}}(-3, -1, 2)$. The x-axis for the silhouette plane is chosen to be $\mathbf{X} = (1/\sqrt{10}, -3/\sqrt{10}, 0)$, a unit vector perpendicular to $\mathbf{U}$. The y-axis is $\mathbf{Y} = (3/\sqrt{35}, 1/\sqrt{35}, 5/\sqrt{35})$, the unit vector with direction $\mathbf{U} \times \mathbf{X}$. In this coordinate system, the silhouette circle has the

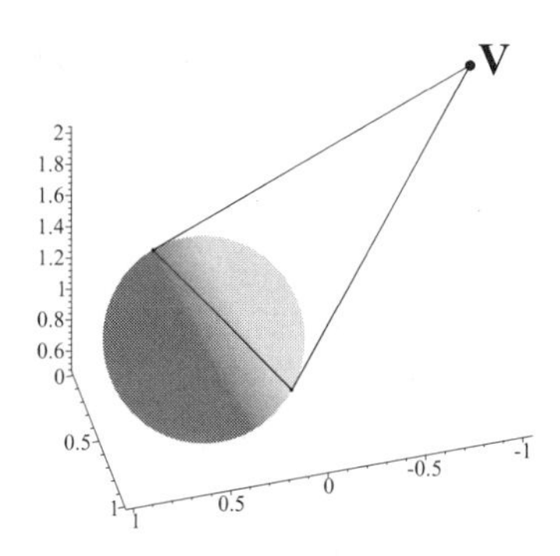

Figure 11.15 Silhouette of Example 11.7

parametric equation

$$\begin{aligned}
&\mathbf{C} + r_1 \cos\theta \mathbf{X} + r_1 \sin\theta \mathbf{Y} \\
&= \left(-\frac{3}{28}, -\frac{1}{28}, \frac{1}{14}\right) + \frac{\sqrt{13}}{2\sqrt{14}} \cos\theta \left(\frac{1}{\sqrt{10}}, -\frac{3}{\sqrt{10}}, 0\right) \\
&\quad + \frac{\sqrt{13}}{2\sqrt{14}} \sin\theta \left(\frac{3}{\sqrt{35}}, \frac{1}{\sqrt{35}}, \frac{5}{\sqrt{35}}\right), \\
&= \left(-\frac{3}{28} + \frac{\sqrt{13}\cos\theta}{2\sqrt{14}\sqrt{10}} + \frac{3\sqrt{13}\sin\theta}{2\sqrt{14}\sqrt{35}}, \right. \\
&\qquad \left. -\frac{1}{28} - \frac{3\sqrt{13}\cos\theta}{2\sqrt{14}\sqrt{10}} + \frac{\sqrt{13}\sin\theta}{2\sqrt{14}\sqrt{35}}, \frac{1}{14} + \frac{5\sqrt{13}\sin\theta}{2\sqrt{14}\sqrt{35}}\right).
\end{aligned}$$

The sphere and the circular silhouette are shown in Figure 11.15.

Example 11.8

Consider the torus $\mathbf{x}(s,t) = ((r\cos s + R)\cos t, (r\cos s + R)\sin t, r\sin s)$ for $0 \le s \le 2\pi$ and $0 \le t \le 2\pi$. The unit normal of the torus is

$$\mathbf{N}(s,t) = (-\cos s\cos t, -\cos s\sin t, -\sin s).$$

The silhouette for a parallel projection in the direction $\mathbf{V}(v_0, v_1, v_2)$ satisfies $\mathbf{V} \cdot \mathbf{N} = 0$ giving

$$v_0(\cos s\cos t) + v_1(\cos s\sin t) + v_2 \sin s = 0,$$

and therefore

$$\tan s = -\frac{v_0}{v_2}\cos t - \frac{v_1}{v_2}\sin t. \tag{11.24}$$

Equation (11.24) defines the curve in the (s,t)-parameter space of the torus that corresponds to the silhouette of the torus. A parametric equation for the silhouette can be obtained, for instance, by using (11.24) to solve for s, and then substituting for s in the parametric equation of the torus. Figures 11.16(a) and (b) show the curve in the (s,t)-parameter space of the torus, and the corresponding silhouette when $r = 5$, $R = 10$ and $\mathbf{V}(50,50,25)$. Part of the silhouette curve is drawn using a dashed line to indicate that that it is hidden by the torus in the given projection. Observe that the parameter-space curve has two branches (taking the periodicity of the domain in account). Each branch corresponds to a connected component of the silhouette curve.

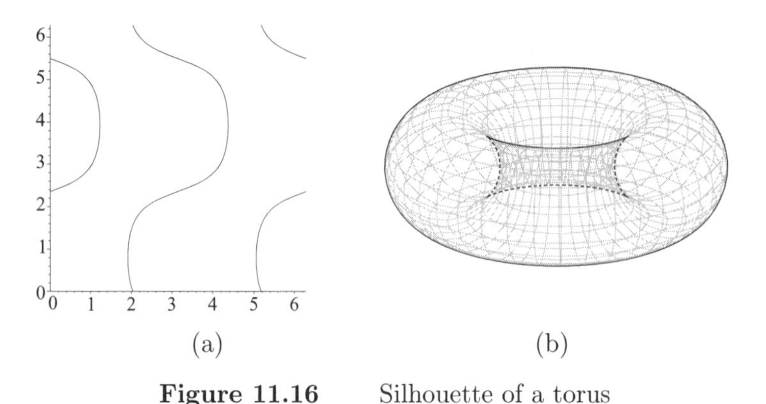

(a) (b)

Figure 11.16 Silhouette of a torus

In general, it is not possible to obtain an analytical solution for the silhouettes of a surface. Silhouettes of more complex surfaces such as Bézier and B-spline surfaces are found by numerical methods. One approach is similar to that used in Example 11.8. For a parametric surface $\mathbf{x}(s,t) = (x(s,t), y(s,t), z(s,t))$ the normal direction is $(\mathbf{x}_s \times \mathbf{x}_t)$. The silhouette for a parallel projection is given by the vector triple product

$$\mathbf{V} \cdot (\mathbf{x}_s \times \mathbf{x}_t) = 0. \tag{11.25}$$

Equation (11.25) defines a curve $\mathbf{C}(s,t) = 0$ in the (s,t) parameter space of the surface. Points of this curve are obtained numerically using, for instance, a marching method or subdivision (see [13]). The points of this curve correspond to the silhouette points of the surface. Perspective projections can be determined similarly.

The CAD drawing in Figure 11.17(a) has been obtained by projecting the edges and silhouettes of the surfaces. The CAD drawing is completed by performing a *hidden line* computation to determine the segments of the edges and silhouettes that are visible. The silhouette plays an important part in the hidden line calculation. In Figure 11.17(b), the line **AC** has a visible segment between points **A** and **B**, and an invisible segment **BC** that lies behind the hemisphere. Similarly, the line **DF** has a visible segment between points **D** and **E**, and an invisible segment **EF** that lies behind one of the blocks. Observe that changes in the visibility of an edge (or a silhouette) correspond to points in the viewplane where the projected image of the edge (or silhouette) intersects the projected images of itself, and other edges and silhouettes. However, not all such intersections give rise to a change in visibility, and other information about the surfaces must be taken into account. Changes in visibility can also arise at *apparent cusps* of a silhouette, that is, silhouette points that correspond to cusps of the apparent contour. For instance, the visibility of the torus silhouette in Figure 11.16 changes at two of the four apparent cusps.

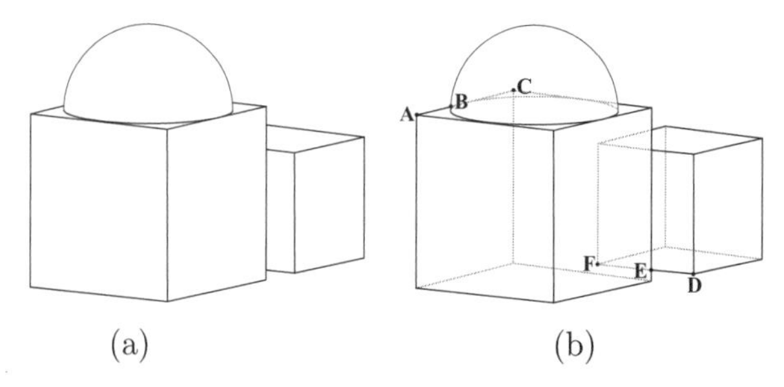

Figure 11.17 CAD drawing with (a) hidden lines removed and (b) hidden lines drawn faint

EXERCISES

11.12. Show that $\mathbf{VQP}^T = \mathbf{PQV}^T$.

11.13. Verify that Equations (11.15) and (11.23) are equivalent.

11.14. Compute the silhouette curves of the cylinder $x^2 + y^2 = 1$ for the perspective projection with viewpoint $(\lambda, 0, 0)$, for $\lambda > 1$ or $\lambda < -1$.

11.15. Compute the silhouette curve of the quadric $4x^2 - 2xz + y^2 - z^2 - 4 = 0$ for the perspective projection with viewpoint $(10, 0, 0)$.

11.16. Write a computer program (or use a computer package) that obtains the silhouettes of a quadric surface. The user should input the coefficients of the quadric and the viewpoint, and the program should determine the conic silhouette in parametric form. Care should be taken to identify the cases when there are no silhouette points; for instance, when the viewpoint lies inside a sphere or ellipsoid. The cases when the silhouettes are linear may need to be treated separately.

11.17. Consider the surface of revolution of the curve $(f(u), 0, g(u))$ $(f(u) \neq 0)$ about the z-axis given by $\mathbf{x}(u, v) = (f(u)\cos v, f(u)\sin v, g(u))$. Show that the silhouettes for a parallel view in the z-direction are circles lying in planes that are perpendicular to the z-axis.

11.18. Show that, for a parallel projection in the z-direction, the silhouette curves of the torus of Example 11.5 are circles.

11.19. Show that, for a parallel projection in the x-direction, the silhouette curves of the torus of Example 11.5 are circles.

11.6 Shadows

A shadow is a region of a object's surface where illumination is reduced due to the obstruction of light sources by other objects in the scene. The amount of diffuse light is reduced, but there is still a contribution of ambient light. When a shading algorithm is applied to a scene, shadows appear in regions where the intensities are significantly less than the surrounding intensities. Shading algorithms do not always produce a satisfactory result. For some surfaces it is possible to determine the boundary of the shadow region mathematically by applying a projection with the light source as the viewpoint and the surfaces of objects in the scene as "viewplanes". The boundaries of an object's shadow are the projected images of edges and silhouettes of the obscuring objects onto the surface.

The shadows cast by a polygonal surface onto a planar surface are straightforward to compute. A polygon is a union of triangular planar facets, and it follows from Exercise 11.4 that each facet either has no silhouette, or is entirely in silhouette and does not contribute to the shadow. Therefore, a polygon casts a shadow that is bounded by the projected images of the polygon edges: these can be found using the method of Section 4.3.

Example 11.9

The scene in Figure 11.18(a) consists of a sphere, with radius $\frac{1}{2}$ and centre $\left(\frac{1}{2}, \frac{1}{2}, 1\right)$, that has been embedded in a unit cube. The shadow cast onto the plane $z = 0$ is determined by (i) computing the silhouette of the sphere, (ii) projecting the silhouette and the edges of the cube using the light source as viewpoint and the flat surface as viewplane, and (iii) shading the bounded shadow region. The unit cube, viewpoint $(-1,0,2)$ and viewplane $z=0$ are

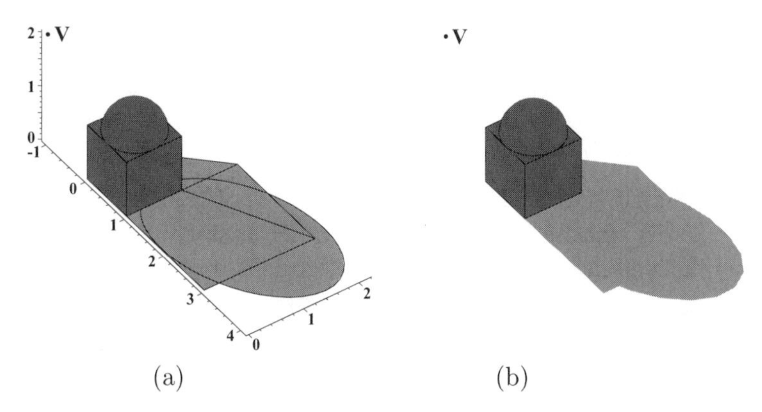

(a) (b)

Figure 11.18

as given in Example 4.18 and so the projection matrix M has already been determined. The silhouette of the sphere was obtained in Example 11.7. The projected vertices of the cube are calculated in the usual manner:

$$\begin{pmatrix} 0&0&0&1\\ 1&0&0&1\\ 1&1&0&1\\ 0&1&0&1\\ 0&0&1&1\\ 1&0&1&1\\ 1&1&1&1\\ 0&1&1&1 \end{pmatrix} \begin{pmatrix} -2&0&0&0\\ 0&-2&0&0\\ -1&0&0&1\\ 0&0&0&-2 \end{pmatrix} = \begin{pmatrix} 0&0&0&-2\\ -2&0&0&-2\\ -2&-2&0&-2\\ 0&-2&0&-2\\ -1&0&0&-1\\ -3&0&0&-1\\ -3&-2&0&-1\\ -1&-2&0&-1 \end{pmatrix},$$

giving the points with affine coordinates $(0,0,0)$, $(1,0,0)$, $(1,1,0)$, $(0,1,0)$, $(1,0,0)$, $(3,0,0)$, $(3,2,0)$ and $(1,2,0)$. The image of the silhouette can be obtained by either computing points on the circle and applying the projection matrix, or applying the projection matrix to the equation of the circle to yield a parametric equation for the silhouette image. The shadow is shown in Figure 11.18(b).

Solutions

Chapter 1

1.2. $-4x - 3y + 23 = 0.$

1.4. Use Example 1.2 with $\theta = \pi/2$, to give $A_1A_2 + B_1B_2 = 0$.

1.6. $\mathbf{PQ}$ has parametric equation $(x(t), y(t)) = ((1-t)\,p_1 + tq_1, (1-t)\,p_2 + tq_2)$ for $0 \leq t \leq 1$. Then

$$\begin{aligned} L\,(x(t), y(t)) &= (ax(t) + by(t) + c, dx(t) + ey(t) + f) \\ &= (a\,((1-t)\,p_1 + tq_1) + b\,((1-t)\,p_2 + tq_2) + c, \\ &\quad d\,((1-t)\,p_1 + tq_1) + e\,((1-t)\,p_2 + tq_2) + f) \\ &= (1-t)\,(ap_1 + bp_2 + c, dp_1 + ep_2 + f) \\ &\quad + t\,(ap_1 + bp_2 + c, dp_1 + ep_2 + f) \\ &= (1-t)\,L\,(\mathbf{P}) + tL\,(\mathbf{Q})\ . \end{aligned}$$

The final line defines the line segment with endpoints $L\,(\mathbf{P})$ and $L\,(\mathbf{Q})$.

1.7. (a) $\mathbf{A}'(4,-1)$, $\mathbf{B}'(6,-1)$, $\mathbf{C}'(5,0)$, $\mathbf{D}'(4.5,1)$. (b) $\mathsf{T}(-3,2)$.

1.8. A mirror image.

1.10. $\mathbf{A}'(1,-1)$, $\mathbf{B}'(3,-1)$, $\mathbf{C}'(2,-2)$, $\mathbf{D}'(1.5,-3)$.

1.13. For $\mathsf{Rot}\,(\pi/3) : \mathbf{P}'(-0.366, 1.366)$, $\mathbf{Q}'(0.634, 3.10)$, $\mathbf{R}'(-0.732, 2.732)$.

For $\mathsf{Rot}\,(\pi/4) : \mathbf{P}'(0, 1.414)$, $\mathbf{Q}'(1.414, 2.828)$, $\mathbf{R}'(0, 2.828)$.

1.15. No.

1.16. (a) $\mathbf{v} = (3/5, -4/5)\,,\ \mathsf{Sh}\,((3/5,-4/5)\,,4) = \begin{pmatrix} 2.92 & -2.56 \\ 1.44 & -0.92 \end{pmatrix}.$

(b) $\mathbf{v} = (8/10, 6/10)$, $\mathsf{Sh}\left((8/10, 6/10), -1\right) = \begin{pmatrix} 1.48 & 0.36 \\ -0.64 & 0.52 \end{pmatrix}$.

1.17. Bottom left window is obtained from **Square** by a scaling of 0.5 in the y-direction followed by a translation of 1.5 units in the x-direction and 1.5 units in the y-direction.

1.20. Apply Exercise 1.6 with $t = 0.5$.

Chapter 2

2.1. $(2, 6, 4)$, $(-1, -3, -2)$, $(1, 3, 2)$, $(4, 12, 8)$.

2.2. $(2, -3, 1)$, $(4, -6, 2)$ etc.

2.3. $(-5, 20, 1)$, $(10, -40, 2)$.

2.4. Reflexive: $M_1 \sim M_1$ since $M_1 = 1M_1$. Symmetric: if $M_1 \sim M_2$ then $M_1 = \mu M_2$ for some $\mu \neq 0$. Then $M_2 = \frac{1}{\mu} M_1$ and so $M_2 \sim M_1$. Transitive: if $M_1 \sim M_2$ and $M_2 \sim M_3$, then $M_1 = \mu M_2$ and $M_2 = \eta M_3$ for some $\mu \neq 0$ and $\eta \neq 0$. Then $M_1 = (\mu\eta) M_3$ and $\mu\eta \neq 0$. So $M_1 \sim M_3$. Hence $\sim$ is an equivalence relation.

2.5. $(6, -3, 0)$.

2.6. $(3, 4, 0)$.

2.7. $3X + 4Y - 5W = 0$.

2.8. $(9, 2, 0)$.

2.9. $(-b, a, 0)$.

2.10. $\mathbf{A}(1, 1)$, $\mathbf{B}(3, 1)$, $\mathbf{C}(2, 2)$, $\mathbf{D}(1.5, 3)$.

2.11. $\begin{pmatrix} 2 & 0 & 0 \\ 0 & 1.5 & 0 \\ 0 & 0 & 1 \end{pmatrix}$, $\mathbf{A}'(2, 1.5)$, $\mathbf{B}'(6, 1.5)$, $\mathbf{C}'(4, 3)$, $\mathbf{D}'(3, 4.5)$.

2.12. $\begin{pmatrix} 0 & 1 & 0 \\ -1 & 0 & 0 \\ 0 & 0 & 1 \end{pmatrix}$, $\mathbf{A}'(-1, 1)$, $\mathbf{B}'(-1, 3)$, $\mathbf{C}'(-2, 2)$, $\mathbf{D}'(1 - 3, 1.5)$.

2.13. $\begin{pmatrix} 1/2 & 0 & 0 \\ 0 & 2/3 & 0 \\ 0 & 0 & 1 \end{pmatrix}$.

2.14. $\begin{pmatrix} \cos\theta & -\sin\theta & 0 \\ \sin\theta & \cos\theta & 0 \\ 0 & 0 & 1 \end{pmatrix}$.

2.16. $\begin{pmatrix} 0 & -3 & 0 \\ 2 & 0 & 0 \\ 0 & 0 & 1 \end{pmatrix}$.

In general, changing the order of transformations results in a different transformation.

2.17. $\begin{pmatrix} -4 & 0 & 0 \\ 0 & -0.5 & 0 \\ 0 & 0 & 1 \end{pmatrix}$.

2.18. $\begin{pmatrix} 0 & -1 & 0 \\ 1 & 0 & 0 \\ -1 & 1 & 1 \end{pmatrix}$.

2.19.

$$\begin{aligned} (\mathsf{T}(-2,5)\cdot\mathsf{Rot}(-\pi/3))^{-1} &= \mathsf{Rot}(-\pi/3)^{-1}\cdot\mathsf{T}(-2,5)^{-1} \\ &= \mathsf{Rot}(\pi/3)\cdot\mathsf{T}(2,-5) \\ &= \begin{pmatrix} 0.5 & 0.866 & 0 \\ -0.866 & 0.5 & 0 \\ 2.0 & -5.0 & 1.0 \end{pmatrix}. \end{aligned}$$

2.20. $\begin{pmatrix} 0.75 & 0.3125 & 0 \\ 0.5 & -0.625 & 0 \\ -2.0 & -2.5 & 1.0 \end{pmatrix}$.

2.21. (a) $\mathsf{A} = \begin{pmatrix} \cos\theta & \sin\theta & 0 \\ -\sin\theta & \cos\theta & 0 \\ x_0 & y_0 & 1 \end{pmatrix}$.

(b) $\mathsf{A}^{-1} = \begin{pmatrix} \cos\theta & -\sin\theta & 0 \\ \sin\theta & \cos\theta & 0 \\ -x_0\cos\theta - y_0\sin\theta & x_0\sin\theta - y_0\cos\theta & 1 \end{pmatrix}$ giving

$$\begin{aligned} x' &= x\cos\theta + y\sin\theta - y_0\sin\theta - x_0\cos\theta\,, \\ y' &= -x\sin\theta + y\cos\theta - y_0\cos\theta + x_0\sin\theta\,. \end{aligned}$$

(d) The equations of the x'- and y'-axes are obtained by setting $y' = 0$ and $x' = 0$ in the previous equations.

2.22. $\mathsf{T}(a,b)$ where $a = x_0\cos\theta - x_0 + y_0\sin\theta - x_1\cos\theta - y_1\sin\theta + x_1$ and $b = -y_0 - x_0\sin\theta + y_0\cos\theta + x_1\sin\theta - y_1\cos\theta + y_1$.

2.23. $\begin{pmatrix} -21 & 20 & 0 \\ 20 & 21 & 0 \\ -80 & 32 & 29 \end{pmatrix}$.

2.26. Bottom left window: $\begin{pmatrix} 1 & 0 & 0 \\ 0 & 0.5 & 0 \\ 1.5 & 1.5 & 1 \end{pmatrix}$.

2.27.

$$\mathsf{T}(10,5)\cdot\mathsf{S}(20,20)\cdot\mathsf{T}(200,200) = \begin{pmatrix} 20 & 0 & 0 \\ 0 & 20 & 0 \\ 400 & 300 & 1 \end{pmatrix}.$$

2.28. $(1,3,1)\times(4,-2,1) = (5,3,-14)$, giving $5x + 3y - 14 = 0$.

2.29. $(1,-3,7)\times(4,3,-5) = (-6,33,15)$, giving $(-6/15, 33/15)$.

Chapter 3

3.1. $(1.5,3,2.5)$, $(0.5,1.0,1.5)$, $(0,0,2)$, $(1,0,0)$.

3.2. $(3,4,1,0)$, $(7,2,0,0)$.

3.3. (a) $\begin{pmatrix} 0.7071 & 0.7071 & 0 & 0 \\ -0.7071 & 0.7071 & 0 & 0 \\ 0 & 0 & 1 & 0 \\ 0 & 0 & 0 & 1 \end{pmatrix}$.

(b) $\begin{pmatrix} 1.0 & 0 & 0 & 0 \\ 0 & -2.5981 & -1.5 & 0 \\ 0 & 0.5 & -0.8660 & 0 \\ 2.0 & 2.5 & -4.3301 & 1.0 \end{pmatrix}$.

(c) Translate $(1,0,-11/3)$ to the origin. Plane is $6x - 6y + 3z = 0$ and $\mathbf{R} = (2/3, -2/3, 1/3)$. Then $\sin\theta_x = (-2/3)\,/\,\left(\frac{1}{3}\sqrt{5}\right) = -\frac{2}{5}\sqrt{5}$, $\cos\theta_x = (1/3)\,/\,\left(\frac{1}{3}\sqrt{5}\right) = \frac{1}{5}\sqrt{5}$, $\sin\theta_y = 2/3$, $\cos\theta_y = \frac{1}{3}\sqrt{5}$, giving

$$\begin{pmatrix} 0.1111 & 0.8889 & -0.4444 & 0 \\ 0.8889 & 0.1111 & 0.4444 & 0 \\ -0.4444 & 0.4444 & 0.7778 & 0 \\ -0.7407 & 0.7407 & -0.3704 & 1.0 \end{pmatrix}.$$

(d) Then $\mathbf{Q}-\mathbf{P}=(6,2,3)$, and hence $\mathbf{R}=(6/7,2/7,3/7)$. Then $\sin\theta_x = (2/7)/\left(\frac{1}{7}\sqrt{13}\right) = \frac{2}{13}\sqrt{13}$, $\cos\theta_x = (3/7)/\left(\frac{1}{7}\sqrt{13}\right) = \frac{3}{13}\sqrt{13}$, $\sin\theta_y = 6/7$, and $\cos\theta_y = \frac{1}{7}\sqrt{13}$, giving

$$\begin{pmatrix} 0.5049 & 0.6713 & 0.5426 & 0 \\ 0.2427 & -0.7137 & 0.6571 & 0 \\ 0.8283 & -0.2001 & -0.5233 & 0 \\ -0.9092 & 0.7713 & 1.3043 & 1.0 \end{pmatrix}.$$

3.4. $(-11,5,27)$.

3.5. $10x+16y+z-25=0$.

3.8. (a) The points are either collinear or lie on a plane through the origin.

3.9. $(p_{12},p_{20},p_{01},p_{03},p_{13},p_{23}) = (x_1y_2-x_2y_1, x_2y_0-x_0y_2, x_0y_1-x_1y_0, x_0-y_0, x_1-y_1, x_2-y_2)$, so $\omega=(x_0-y_0,x_1-y_1,x_2-y_2)=\mathbf{P}-\mathbf{Q}$ and $\mathbf{v}=(x_1y_2-x_2y_1,x_2y_0-x_0y_2,x_0y_1-x_1y_0)=\mathbf{P}\times\mathbf{Q}$. Clearly $\overrightarrow{\mathbf{QP}}=\mathbf{P}-\mathbf{Q}$ is the direction vector of the line. $\overrightarrow{\mathbf{OP}}$ and $\overrightarrow{\mathbf{OQ}}$ are vectors parallel to the plane containing $\mathbf{P}$, $\mathbf{Q}$, and $\mathbf{O}$. The normal to the plane is $\overrightarrow{\mathbf{OP}}\times\overrightarrow{\mathbf{OQ}}=\mathbf{P}\times\mathbf{Q}$.

3.10. $(p_{03},p_{13},p_{23})\cdot(p_{12},p_{20},p_{01}) = (x_1y_2-x_2y_1,x_2y_0-x_0y_2,x_0y_1-x_1y_0)\cdot(x_0-y_0,x_1-y_1,x_2-y_2) = (x_0-y_0)(x_1y_2-x_2y_1)+(x_1-y_1)(x_2y_0-x_0y_2)+(x_2-y_2)(x_0y_1-x_1y_0)=0$.

3.11. The rotation is $\mathsf{Rot}_x(\theta_x)\,\mathsf{Rot}_y(-\theta_y)\,\mathsf{Rot}_z(\theta)\,\mathsf{Rot}_y(\theta_y)\,\mathsf{Rot}_x(-\theta_x)$, which after simplification using $r_1^2+r_2^2+r_3^2=1$ (following the notation of Section 3.2.4) yields

$$\begin{pmatrix} \begin{matrix}(\cos\theta+\\ {r_1}^2(1-\cos\theta))\end{matrix} & \begin{matrix}(r_3\sin\theta+\\ r_1r_2(1-\cos\theta))\end{matrix} & \begin{matrix}(-r_2\sin\theta+\\ r_1r_3(1-\cos\theta))\end{matrix} & 0 \\ \begin{matrix}(-r_3\sin\theta+\\ r_1r_2(1-\cos\theta))\end{matrix} & \begin{matrix}(\cos\theta+\\ {r_2}^2(1-\cos\theta))\end{matrix} & \begin{matrix}(r_1\sin\theta+\\ r_2r_3(1-\cos\theta))\end{matrix} & 0 \\ \begin{matrix}(r_2\sin\theta+\\ r_1r_3(1-\cos\theta))\end{matrix} & \begin{matrix}(-r_1\sin\theta+\\ r_2r_3(1-\cos\theta))\end{matrix} & \begin{matrix}(\cos\theta+\\ {r_3}^2(1-\cos\theta))\end{matrix} & 0 \\ 0 & 0 & 0 & 1 \end{pmatrix}$$

3.12. (a) $(5+6\mathbf{i}+11\mathbf{j}-7\mathbf{k})$, (b) $(2,(3,-1,1))$, (c) $(8,(44,-22,48))$, (d) $(4,(34,0,-40))$, (e) $(4,(8,24,-46))$.

3.13. (a) $(1/27,-5/54,1/18,-2/27)$, (b) $(-3/26,-2/13,0,1/26)$.

3.14. $\mathbf{ijk} = -1$ implies $\mathbf{iijk} = -\mathbf{i}$. Since $\mathbf{i}^2 = -1$ it follows that $\mathbf{jk} = \mathbf{i}$. Then $\mathbf{jjk} = \mathbf{ji}$ and so $\mathbf{k} = -\mathbf{ji}$, etc.

3.17. Let $\mathbf{q}_i = (s_i, \mathbf{v})$ for $i = 1, 2, 3$. Then

$$\begin{aligned}
\mathbf{q}_1(\mathbf{q}_2 + \mathbf{q}_3) &= (s_1(s_2 + s_3) - \mathbf{v_1} \cdot (\mathbf{v_2} + \mathbf{v_3}), \\
&\quad s_1(\mathbf{v_2} + \mathbf{v_3}) + (s_2 + s_3)\mathbf{v}_1 + (\mathbf{v_1} \times (\mathbf{v_2} + \mathbf{v_3}))) \\
&= ((s_1 s_2 + \mathbf{v_1} \cdot \mathbf{v_2}) + (s_1 s_3 - \mathbf{v_1} \cdot \mathbf{v_3}), \\
&\quad (s_1 \mathbf{v_2} + s_2 \mathbf{v}_1 + (\mathbf{v_1} \times \mathbf{v_2})) + (s_1 \mathbf{v_3} + s_3 \mathbf{v}_1 + (\mathbf{v_1} \times \mathbf{v_3})) \\
&= \mathbf{q}_1\mathbf{q}_2 + \mathbf{q}_1\mathbf{q}_3 \,.
\end{aligned}$$

3.18. (a) $\overline{\mathbf{q}_1\mathbf{q}_2} = (s_1 s_2 - \mathbf{v_1} \cdot \mathbf{v_2}, -s_1\mathbf{v}_2 - s_2\mathbf{v}_1 - (\mathbf{v_1} \times \mathbf{v_2}))$ and $\overline{\mathbf{q}_2}\ \overline{\mathbf{q}_1} = (s_2 s_1 - (-\mathbf{v_2}) \cdot (-\mathbf{v_1}), -s_2\mathbf{v}_1 - s_1\mathbf{v}_2 + ((-\mathbf{v_2}) \times (-\mathbf{v_1})))$. The right hand sides of the two equations are equal so $\overline{\mathbf{q_1 q_2}} = \overline{\mathbf{q_2}}\ \overline{\mathbf{q_1}}$.

3.19. $\mathbf{I}^2 = (0, \mathbf{v})(0, \mathbf{v}) = ((0)(0) - \mathbf{v} \cdot \mathbf{v}, (0)\mathbf{v} + (0)\mathbf{v} + (\mathbf{v} \times \mathbf{v})) = (-|\mathbf{v}|^2, \mathbf{0})$.

3.21. (a) $|\mathbf{q}| = 5$, so $(2, (1, 2, 4)) = 5\left(\frac{2}{5}, \left(\frac{1}{5}, \frac{2}{5}, \frac{4}{5}\right)\right) = 5(s, \mathbf{v})$. Lemma 3.11 can be applied to $(s, \mathbf{v})$: $|\mathbf{v}| = \sqrt{21}/5$, and $\mathbf{I} = \frac{\mathbf{v}}{|\mathbf{v}|} = \left(\frac{1}{\sqrt{21}}, \frac{2}{\sqrt{21}}, \frac{4}{\sqrt{21}}\right)$. Let $s = \cos\theta = 2/5$ ($\theta = 1.1593$ radians). Then $(2, (1, 2, 4)) = 5(\cos\theta, \sin\theta\mathbf{I})$.

3.22. (a)

$$\begin{aligned}
\mathbf{q}_1\mathbf{q}_2 &= r_1(\cos\theta_1, \sin\theta_1\mathbf{I})\, r_2(\cos\theta_2, \sin\theta_2\mathbf{I}) \\
&= r_1 r_2(\cos\theta_1\cos\theta_2 - (\sin\theta_1\mathbf{I}) \cdot (\sin\theta_2\mathbf{I}), \\
&\quad \cos\theta_1\sin\theta_2\mathbf{I} + \cos\theta_2\sin\theta_1\mathbf{I} + ((\sin\theta_1\mathbf{I}) \times (\sin\theta_2\mathbf{I}))) \\
&= r_1 r_2(\cos\theta_1\cos\theta_2 - \sin\theta_1\sin\theta_2, (\cos\theta_1\sin\theta_2 + \cos\theta_2\sin\theta_1)\mathbf{I}) \\
&= r_1 r_2(\cos(\theta_1 + \theta_2), \sin(\theta_1 + \theta_2)\mathbf{I}) \,.
\end{aligned}$$

3.25. $\mathbf{I} = \left(-\frac{1}{3}, \frac{2}{3}, \frac{2}{3}\right)$ and $\mathbf{q} = \left(\cos\left(\frac{\pi}{4}\right), \sin\left(\frac{\pi}{4}\right)\mathbf{I}\right) = \left(\frac{\sqrt{2}}{2}, \left(-\frac{\sqrt{2}}{6}, \frac{\sqrt{2}}{3}, \frac{\sqrt{2}}{3}\right)\right)$. The rotation is given by $\mathbf{q}(0, (5, 6, 7))\mathbf{q}^{-1} = (0, (-5/3, 31/3, -2/3))$. Thus $(5, 6, 7)$ rotates to the point $(-5/3, 31/3, -2/3)$.

3.26. Using the fact that $\overline{\mathbf{q}_1\mathbf{q}_2} = \overline{\mathbf{q}_2}\,\overline{\mathbf{q}_1}$, $\mathbf{q}\overline{\mathbf{q}} = \overline{\mathbf{q}}\mathbf{q} = 1$, and $\mathbf{x}\overline{\mathbf{x}} = 1$, gives

$$|C_{\mathbf{q}}(\mathbf{x})|^2 = \mathbf{q}\mathbf{x}\overline{\mathbf{q}}\,\overline{\mathbf{q}\mathbf{x}\overline{\mathbf{q}}} = \mathbf{q}\,\mathbf{x}\,\overline{\mathbf{q}}\,\mathbf{q}\,\overline{\mathbf{x}}\,\overline{\mathbf{q}} = \mathbf{q}\,\mathbf{x}\,\overline{\mathbf{x}}\,\overline{\mathbf{q}} = \mathbf{q}\,\overline{\mathbf{q}}\,|\mathbf{x}|^2 = |\mathbf{x}|^2$$

and the result follows.

3.30. $\phi = 1.0745$, $\cos\phi = \frac{10}{21}$, and $\sin\phi = \frac{\sqrt{341}}{21}$. The motion is given by

$$\frac{\sin(\phi(1-t))}{\sqrt{341}}(6, 0, 18, -9) + \frac{\sin(\phi t)}{\sqrt{341}}(14, 14, 7, 0) \,.$$

Chapter 4

4.1. $\begin{pmatrix} -127 & -33 & -3 \\ 24 & 11 & 12 \\ -10 & -55 & -126 \end{pmatrix}$.

4.2. $\begin{pmatrix} -10 & -14 & 0 \\ -15 & -21 & 0 \\ -6 & 4 & -31 \end{pmatrix}$.

4.3. $\mathsf{M} = \begin{pmatrix} -12 & -3 & 1 \\ -7 & -16 & -1 \\ 63 & -27 & -10 \end{pmatrix}$, and $\mathbf{A}'\left(-\frac{5}{2}, \frac{13}{2}\right)$, $\mathbf{B}'\left(\frac{2}{3}, \frac{29}{3}\right)$, $\mathbf{C}'\left(\frac{2}{3}, \frac{29}{3}\right)$.

4.4. $\mathsf{M} = \begin{pmatrix} 4 & 8 & 0 \\ 1 & 2 & 0 \\ -8 & 32 & 6 \end{pmatrix}$, and $\mathbf{A}'\left(\frac{1}{3}, \frac{26}{3}\right)$, $\mathbf{B}'\left(\frac{11}{6}, \frac{35}{3}\right)$, $\mathbf{C}'\left(\frac{3}{2}, 11\right)$.

4.6. $\begin{pmatrix} 5 & 1 & -1 & -1 \\ 6 & 4 & 3 & 3 \\ 4 & -2 & 9 & 2 \\ -8 & 4 & -4 & 3 \end{pmatrix}$.

4.8. $\begin{pmatrix} -5 & 0 & 0 & 0 \\ 2 & -9 & 6 & 0 \\ 3 & -6 & 4 & 0 \\ 4 & -8 & 12 & -5 \end{pmatrix}$.

4.10. For Exercise 4.6: $\mathbf{A}'(-8/3, 4/3, -4/3)$, $\mathbf{B}'(-3/2, 5/2, -5/2)$, $\mathbf{C}'(-1/3, -4/3, -1/6)$, $\mathbf{D}'(1, 1, 1)$.

For Exercise 4.8: $\mathbf{A}'(-4/5, 8/5, -12/5)$, $\mathbf{B}'(1/5, 8/5, -12/5)$, $\mathbf{C}'(-6/5, 17/5, -18/5)$, $\mathbf{D}'(-4/5, 23/5, -22/5)$.

4.12. $\mathsf{K} = \begin{pmatrix} 12/13 & 5/13 & 0 & 0 \\ -5/13 & 12/13 & 0 & 0 \\ 4 & 3 & 0 & 1 \end{pmatrix}$, $\mathsf{VC}= \begin{pmatrix} 12/13 & -5/13 & 0 \\ 5/13 & 12/13 & 0 \\ 0 & 0 & 0 \\ -63/13 & -16/13 & 1 \end{pmatrix}$,

giving $\mathbf{A}''(-63/11, -16/13)$, $\mathbf{B}''(-3, -2)$, $\mathbf{C}''(-24/13, 10/13)$, and $\mathbf{D}''(-48/13, 20/13)$, $\mathbf{E}''(-97/26, -15/13)$, $\mathbf{F}''(-56/13, -33/13)$.

4.13.

$$K = \begin{pmatrix} 3/\sqrt{34} & 0 & 5/\sqrt{34} & 0 \\ 0 & -1 & 0 & 0 \\ -1 & 1 & 1 & 1 \end{pmatrix},$$

$$VC = \begin{pmatrix} \frac{11}{98}\sqrt{34} & -20/49 & -20/49 \\ 0 & -1 & 0 \\ \frac{13}{98}\sqrt{34} & 12/49 & 12/49 \\ -\frac{1}{49}\sqrt{34} & 17/49 & 17/49 \end{pmatrix}.$$

Images of vertices: $(-\frac{1}{17}\sqrt{34}, -\frac{32}{17})$, $(-\frac{22}{31}\sqrt{34}, \frac{80}{31})$, $(0,2)$, $(-\frac{3}{2}\sqrt{34}, \frac{103}{5})$.

4.15. $M= \begin{pmatrix} -12 & 0 & 0 & 0 \\ 0 & -12 & 0 & 0 \\ 2 & 3 & -4 & 1 \\ 8 & 12 & 32 & -8 \end{pmatrix}$, $VC= \begin{pmatrix} 0.7071 & -0.7071 & 0 \\ 0.7071 & 0.7071 & 0 \\ -0.1664 & 0.4991 & -0.2353 \\ 0.0416 & -0.1248 & 0.0588 \end{pmatrix}$,

$$DC = \begin{pmatrix} 80 & 0 & 0 \\ 0 & 80 & 0 \\ 480 & 860 & 1 \end{pmatrix}, \ VP = \begin{pmatrix} -678.8225 & 678.8225 & 0 \\ -678.8225 & -678.8225 & 0 \\ 819.4113 & 746.8629 & 1 \\ -3161.1775 & -5296.0808 & -8 \end{pmatrix}.$$

Then $\mathbf{A}''(431.51, 552.91)$, $\mathbf{B}''(625.46, 358.96)$, $\mathbf{C}''(722.44, 649.89)$, and $\mathbf{D}''(593.14, 520.59)$.

4.16. $M = \begin{pmatrix} 1 & 0 & 1 & 1 \\ -7 & -6 & -1 & -1 \\ 0 & 0 & -6 & 0 \\ -7 & 0 & -1 & -7 \end{pmatrix}$, $VC = \begin{pmatrix} 0 & -0.333 & 0.333 \\ 1.414 & 0.333 & -0.333 \\ 0 & 1 & 0 \\ -1.414 & -0.667 & 0.667 \end{pmatrix}$,

$$DC = \begin{pmatrix} 25 & 0 & 0 \\ 0 & 100/9 & 0 \\ 100 & 250/3 & 1 \end{pmatrix}, \ VP = \begin{pmatrix} 64.645 & 83.333 & 1 \\ -276.777 & -83.333 & -1 \\ 0 & -66.667 & 0 \\ -452.513 & -516.667 & -7 \end{pmatrix}.$$

$\mathbf{A}''(64.65, 83.33)$, $\mathbf{B}''(11.61, 133.33)$, $\mathbf{C}''(185.86, 102.38)$, $\mathbf{D}''(64.64, 116.67)$.

4.18. Consider the line segment through $\mathbf{P}(p_1, p_2, p_3)$ in direction (v_1, v_2, v_3), from $\mathbf{P}$ to $\mathbf{Q}(p_1 + tv_1, p_2 + tv_2, p_3 + tv_3)$. The line segment has length $|t|\,|(v_1, v_2, v_3)|$. Then

$$\mathsf{Q}M = (p_1 + tv_1, p_2 + tv_2, p_3 + tv_3, 1)M = \mathbf{P}M + t(v_1, v_2, v_3, 0)M \,.$$

The image of the segment has length $|\mathbf{Q}M - \mathbf{P}M| = |t|\,|(v_1, v_2, v_3, 0)M|$. The foreshortening ratio is $|(v_1, v_2, v_3, 0)M| / |(v_1, v_2, v_3)|$ which depends only on (v_1, v_2, v_3). Therefore all segments with the same direction have the same foreshortening ratio.

4.19. Use the solution to Exercise 4.18. The direction of the x-axis is $(v_1, v_2, v_3) = (1, 0, 0)$. Then

$$(v_1, v_2, v_3, 0)\mathsf{M} = (1\ 0\ 0\ 0)\ \mathsf{M} = \left(n_2^2 + n_3^2 \quad -n_1 n_2 \quad -n_1 n_3\ 0\right)$$

which has magnitude

$$\left(n_2^2 + n_3^2\right)^{1/2} \left(n_1^2 + n_2^2 + n_3^2\right)^{1/2} = \left(n_2^2 + n_3^2\right)^{1/2} .$$

The foreshortening ratio is $|(v_1\ v_2\ v_3\ 0)\mathsf{M}| / |(1, 0, 0)| = \left(n_2^2 + n_3^2\right)^{1/2}$. Similarly for the other principal directions.

4.20. Let $\mathbf{P}\ (x\ y\ z\ w)$ be a point in the viewplane in world coordinates and let the corresponding viewplane coordinates be $(X\ Y\ W)$. Then $(x\ y\ z\ w) = (\ Xr_1 + Ys_1 + Wq_0 \quad Xr_2 + Ys_2 + Wq_0 \quad Xr_3 + Ys_3 + Wq_0 \quad W\ )$. Then $(x\ y\ z\ w)$ is a point at infinity if and only if $(X\ Y\ W)$ is a point at infinity. Similarly for the device coordinate transformation.

Chapter 5

5.1. $(((6t - 2)\,t + 4)\,t - 5)\,t + 3$ requires 4 $\times$'s and 4 $\pm$'s. $3 - 5 \cdot t + 4 \cdot t \cdot t - 2 \cdot t \cdot t \cdot t + 6 \cdot t \cdot t \cdot t \cdot t$ requires 10 $\times$'s and 4 $\pm$'s.

5.3. Horner requires n $\times$'s and n $\pm$'s. The ordinary method requires $n(n+1)/2$ $\times$'s and n $\pm$'s. Saving $n(n-1)/2$ $\times$'s – quite a saving!

5.5. (a) $\mathbf{t}(u) = (x(t) + ux'(t), y(t) + uy'(t))$. (c) $x'(t)x + y'(t)y + y'(t)y(t) - x'(t)x(t) = 0$.

5.6. (a)(i) $(1, 1)$ corresponds to $t = 1$; unit tangent is $\mathbf{t}(t) = \left(1/\sqrt{1 + 4t^2}, 2t/\sqrt{1 + 4t^2}\right)$, and so $\mathbf{t}(1) = \left(1/\sqrt{5}, 2/\sqrt{5}\right)$. (ii) $\mathbf{n}(1) = \left(-2/\sqrt{5}, 1/\sqrt{5}\right)$.

(iii) $2/\sqrt{5}(x - 1) - 1/\sqrt{5}(y - 1) = 0$.

(c) (i) $\mathbf{t}(t) = \frac{1}{\sqrt{1+b^2}} \left(b\cos t - \sin t, b\sin t + \cos t\right)$.

(ii) $\mathbf{n}(t) = \frac{1}{\sqrt{1+b^2}} \left(-b\sin t - \cos t, b\cos t - \sin t\right)$.

(iii) $(b\sin t + \cos t)\,(x - x(t)) - (b\cos t - \sin t)\,(y - y(t)) = 0$.

(d) (i) $\mathbf{t}(t) = \frac{1}{\sqrt{2+2\cos t}} \left(1 + \cos t, \sin t\right)$.

(ii) $\mathbf{n}(t) = \frac{1}{\sqrt{2+2\cos t}} \left(-\sin t, 1 + \cos t\right)$.

(iii) $\sin t(x - x(t)) - (1 + \cos t)\,(y - y(t)) = 0$.

5.8. Rotation of $\mathbf{C}(t) = (x(t), y(t))$ is the curve $\mathbf{R}(t) = (x(t)\cos\theta - y(t)\sin\theta,$ $x\sin\theta + y\cos\theta)$. Then

$$\mathbf{R}'(t) = (x'(t)\cos\theta - y'(t)\sin\theta, x'(t)\sin\theta + y'(t)\cos\theta) .$$

Then speed of $\mathbf{R}(t)$ is

$$\begin{aligned}&\sqrt{(x'(t)\cos\theta - y'(t)\sin\theta)^2 + (x'(t)\sin\theta + y'(t)\cos\theta)^2}\\&= \sqrt{(x'(t))^2 + (y'(t))^2}\end{aligned}$$

which is the speed of $\mathbf{C}(t)$.

5.10. (a) $L_{\mathbf{C}}(t) = \int_{-\pi}^{t} \sqrt{(1+\cos u)^2 + (\sin u)^2}\; du = 4\sin\frac{1}{2}t + 4$.

(c) $L_{\mathbf{C}}(t) = \int_{t_0}^{t} \sqrt{\left(\sinh\left(\frac{u}{c}\right)\right)^2 + 1}\; du = \int_{t_0}^{t} \cosh\left(\frac{u}{c}\right)\; du = \sinh\left(\frac{t}{c}\right) - \sinh\left(\frac{t_0}{c}\right)$.

(e) $L_{\mathbf{C}}(t) = a\sqrt{1+b^2}\int_{t_0}^{t} e^{bu}\; du = \frac{a}{b}\sqrt{1+b^2}\left(e^{bt} - e^{bt_0}\right)$.

5.11. 8.

5.14. (a) Take $t_0 = 0$ so that $s(t) = L_{\mathbf{C}}(t) = 4\sin\frac{1}{2}t$. Then $t = 2\arcsin\frac{1}{4}s$ and reparametrization gives

$$\left(2\arcsin\frac{1}{4}s + \sin\left(2\arcsin\frac{1}{4}s\right), 1 - \cos\left(2\arcsin\frac{1}{4}s\right)\right) .$$

(c) Take $t_0 = 0$ so that $s(t) = \sinh\left(\frac{t}{c}\right)$. Then $t = c\,\mathrm{arcsinh}\, s$, and reparametrization gives $\left(c\cosh\left(\mathrm{arcsinh}\, s\right), c\,\mathrm{arcsinh}\, s\right)$ which simplifies to

$$\left(c\sqrt{1+s^2}, c\,\mathrm{arcsinh}\, s\right) .$$

5.15. $(x'(t), y'(t)) = \left(-3 + 6t, 6t - 6t^2\right)$, $\mathbf{n}(t) = \frac{1}{1-2t+2t^2}\left(-2t + 2t^2, -1 + 2t\right)$. Then the offset is

$$\left(1 - 3t + 3t^2, 3t^2 - 2t^3\right) + \frac{d}{1 - 2t + 2t^2}\left(-2t + 2t^2, -1 + 2t\right) .$$

The curve and an offset are shown in the following figure.

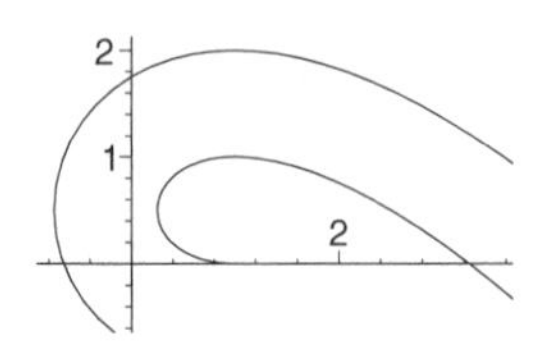

5.16. (a) $\mathbf{n}(t) = \frac{1}{\cosh(t/c)}\left(-1, \sinh(t/c)\right)$.

The offset is $\left(c\cosh(t/c), t\right) + \frac{d}{\cosh(t/c)}\left(-1, \sinh(t/c)\right)$.

(b) $\mathbf{n}(t) = \frac{1}{\sqrt{1+b^2}}\left(-b\sin t - \cos t, b\cos t - \sin t\right)$.

The offset is $\left(e^{bt}\cos t, e^{bt}\sin t\right) + \frac{d}{\sqrt{1+b^2}}\left(-b\sin t - \cos t, b\cos t - \sin t\right)$.

5.19. (d) Ellipse, (e) reducible $(2x-3)(x+y-1)=0$, (f) hyperbola,

(g) hyperbola, (h) reducible $(3x+2y)(x-y+2)=0$.

5.22. (a) $a = 13$, $b = -5$, $c = 13$, $\theta = \pi/4$, $x = \frac{1}{\sqrt{2}}X' - \frac{1}{\sqrt{2}}Y'$ and $y = \frac{1}{\sqrt{2}}X' + \frac{1}{\sqrt{2}}Y'$ puts conic in the form $8\left(X'+3\right)^2 + 18\left(Y'+2\right)^2 - 72 = 0$. Applying the translation $\mathsf{T}(-3,-2)$ puts the conic in the standard form for an ellipse $\frac{1}{9}x^2 + \frac{1}{4}y^2 = 1$. (b) $y = 6x^2$. (c) $\frac{1}{4}x^2 - \frac{1}{16}y^2 = 1$.

5.26. (d) $(1,-1)$ and $(1/3,-5/3)$. (e) $(2.92,-2.39)$ and $(1.52,-0.28)$.

5.29. (a) $\left(\frac{-2t}{1-2t+2t^2}, \frac{-2t^2}{1-2t+2t^2}\right)$. (c) Consider lines through $(1,0)$: $y = t(x-1)$. $x^2 + 2x\left(t(x-1)\right) - \left(t(x-1)\right)^2 - 1 = (x-1)\left(x - t^2x + 2xt + t^2 + 1\right) = 0$. Then $x = \frac{t^2+1}{t^2-2t-1}$, $y = t(x-1) = t\left(\frac{t^2+1}{t^2-2t-1} - 1\right) = \frac{2t+2t^2}{t^2-2t-1}$.

5.30. (a) $x = t^2 - 1$, $y = t+2$. Then $t = y-2$, $x = (y-2)^2 - 1$ giving $x - y^2 + 4y - 3 = 0$. (c) $x = 2t^2 + t - 1$, $y = t^2 - 3t + 3$. Then $x - 2y = \left(2t^2 + t - 1\right) - 2\left(t^2 - 3t + 3\right) = 7t - 7$. So $t = \frac{1}{7}(x - 2y + 7)$. Then $x = 2\left(\frac{1}{7}(x-2y+7)\right)^2 + \left(\frac{1}{7}(x-2y+7)\right) - 1$ which simplifies to $2x^2 - 8xy - 14x + 8y^2 - 70y + 98 = 0$.

(e) $\begin{pmatrix} -1 & -1 & -4 \\ 0 & 0 & -3 \\ 1 & -2 & 4 \end{pmatrix}$ has signed minors $A_0 = -6$, $A_1 = -3$, $A_2 = 0$, $B_0 = 12$, $B_1 = 0$, $B_2 = -3$, $C_0 = 3$, $C_1 = -3$, $C_2 = 0$. Then the implicit equation is $\left(-3x-3\right)^2 - \left(-6x + 12y + 3\right)\left(-3y\right) = 0$, which simplifies to $x^2 - 2xy + 4y^2 + 2x + y + 1 = 0$.

5.31. $\begin{vmatrix} \mathbf{e}_1 & \mathbf{e}_2 & \mathbf{e}_3 & \mathbf{e}_4 \\ 9 & 0 & 1 & 3 \\ 3 & 4 & 0 & 4 \\ -4 & -3 & 1 & 4 \end{vmatrix} = -16\mathbf{e}_1 + 55\mathbf{e}_2 + 273\mathbf{e}_3 - 43\mathbf{e}_4.$

Chapter 6

6.1. $(1-t)^2(-1,5)+2(1-t)t(2,0)+t^2(4,6)=\left(-1+6t-t^2,5-10t+11t^2\right)$. $\mathbf{B}(0.75)=(2.9375,3.6875)$. $\mathbf{B}(1.25)$ is not a point on the curve since $t=1.25$ is not in the interval $[0,1]$.

6.2. $(1-t)^2(p_0,q_0)+2(1-t)t(p_1,q_1)+t^2(p_2,q_2)=$

$$\left((p_0-2p_1+p_2)\,t^2+2\,(p_1-p_0)\,t+p_0,\right.$$
$$\left.(q_0-2q_1+q_2)\,t^2+2\,(q_1-q_0)\,t+q_0\right)$$

which is the parametric equation of a parabola.

6.3. $\mathbf{B}(t)=(1-t)^3\,(1,0)+3(1-t)^2t\,(2,3)+3(1-t)t^2\,(5,4)+t^3\,(2,1)=$

$\left(1+3t+6t^2-8t^3,9t-6t^2-2t^3\right)$. $\mathbf{B}(0)=(1,0)$, $\mathbf{B}(0.5)=(3.0,2.75)$, $\mathbf{B}(1)=(2,1)$.

6.5. (a) $\mathbf{b}_1-\mathbf{b}_0$ and $\mathbf{b}_1-\mathbf{b}_0$. (b) $2\mathbf{b}_1-2\mathbf{b}_0$ and $2\mathbf{b}_2-2\mathbf{b}_1$.

6.7. $\mathbf{B}(t)=(1-t)^3\mathbf{b}_0+3(1-t)^2t\left(\frac{2\mathbf{b}_0+\mathbf{b}_3}{3}\right)+3(1-t)t^2\left(\frac{\mathbf{b}_0+2\mathbf{b}_3}{3}\right)+t^3\mathbf{b}_3$. Expanding gives $\mathbf{B}(t)=\mathbf{b}_0-\mathbf{b}_0t+t\mathbf{b}_3=\mathbf{b}_0\,(1-t)+t\mathbf{b}_3$.

6.9. $\mathbf{b}_0(2,2)$, $\mathbf{b}_1(2,5)$, $\mathbf{b}_2(5,6)$, $\mathbf{b}_3(8,6)$ using endpoint interpolation and knowledge of the end tangents.

6.10. $\mathbf{b}_0(3,6)$, $\mathbf{b}_1(3,8)$, $\mathbf{b}_2(4,9)$, $\mathbf{b}_3(5,9)$ using endpoint interpolation and knowledge of the end tangents.

6.15. $\binom{n}{i}+\binom{n}{i+1}=\frac{n!}{i!(n-i)!}+\frac{n!}{(i+1)!(n-i-1)!}=\frac{n!(i+1)-n!(n-i)}{(i+1)!(n-i)!}=\frac{(n+1)!}{(i+1)!(n-i)!}=\binom{n+1}{i+1}$.

6.16. $\int_0^1(1-t)^3\,dt=\left[-\frac{1}{4}\,(1-t)^4\right]_0^1=\frac{1}{4}$, $\int_0^1 3(1-t)^2t\,dt=\left[\frac{3}{4}t^4-2t^3+\frac{3}{2}t^2\right]_0^1=\frac{1}{4}$, etc.

6.17. (c) $n=130$.

6.19. $B_{i,n}(1-t)=\binom{n}{i}(1-(1-t))^{n-i}(1-t)^i=\binom{n}{i}(1-t)^it^{n-i}$
$=\binom{n}{n-i}(1-t)^it^{n-i}=B_{n-i,n}(t)$.

6.20. $\frac{i}{n}B_{i,n}(t)=\frac{i}{n}\frac{n!}{i!(n-i)!}(1-t)^{n-i}t^i=\frac{(n-1)!}{(i-1)!(n-i)!}(1-t)^{n-i}t^i$ so that

$\sum_{i=0}^n\frac{i}{n}B_{i,n}(t)=\sum_{i=1}^n\frac{(n-1)!}{(i-1)!(n-i)!}(1-t)^{n-i}t^i$

$=t\sum_{i=0}^{n-1}\frac{(n-1)!}{i!(n-i-1)!}(1-t)^{n-i-1}t^i=t\sum_{i=0}^{n-1}B_{i,n-1}(t)=t$. Then $\mathbf{B}(t)=\sum_{i=0}^n\left(\left(1-\frac{i}{n}\right)\mathbf{a}+\frac{i}{n}\mathbf{b}\right)B_{i,n}(t)=$

$\mathbf{a}\sum_{i=0}^nB_{i,n}(t)-\mathbf{a}\sum_{i=0}^n\frac{i}{n}B_{i,n}(t)+\mathbf{b}\sum_{i=0}^n\frac{i}{n}B_{i,n}(t)=\mathbf{a}(1-t)+\mathbf{b}t$ yielding the line segment $\overline{\mathbf{ab}}$.

6.24. When the control points are collinear the convex hull is a line segment implying that the Bézier curve is contained in a line segment.

6.25. (a) $\mathbf{b}_0(3,4)$, $\mathbf{b}_1(5,5)$, $\mathbf{b}_2(6,3)$, $\mathbf{b}_3(4,2)$. (b) $\mathbf{b}_0(0,0)$, $\mathbf{b}_1(-1,2)$, $\mathbf{b}_2(1,3)$, $\mathbf{b}_3(2,1)$. (c) $\mathbf{b}_0(0,0)$, $\mathbf{b}_1(1,2)$, $\mathbf{b}_2(-1,3)$, $\mathbf{b}_3(-2,1)$.

6.27. $\mathbf{b}_0^1(1.5, 0.75)$, $\mathbf{b}_1^1(3.5, 3.5)$, $\mathbf{b}_2^1(5.5, 4.25)$, $\mathbf{b}_0^2(2, 1.4375)$, $\mathbf{b}_1^2(4, 3.6875)$, $\mathbf{b}_0^3(2.5, 2.0)$. $\mathbf{B}(0.25) = (2.5, 2.0)$.

6.28. $(3.456, 1.3776)$.

6.32. (a) $\mathbf{b}_0^1(0.4, 0.1)$, $\mathbf{b}_1^1(1.2, 0.6)$, $\mathbf{b}_2^1(2.2, 0.9)$, $\mathbf{b}_0^2(0.6, 0.225)$, $\mathbf{b}_1^2(1.45, 0.675)$, $\mathbf{b}_0^3(0.8125, 0.3375)$. $\mathbf{B}(0.25) = (0.8125, 0.3375)$.

(b) $\mathbf{B}_{\text{left}}$: $(0.2, 0.0)$, $(0.4, 0.1)$, $(0.6, 0.225)$, $(0.8125, 0.3375)$;

$\mathbf{B}_{\text{right}}$: $(0.8125, 0.3375)$, $(1.45, 0.675)$, $(2.2, 0.9)$, $(3.4, 0.0)$.

6.34. (a) $\mathbf{B}(1/3) = (0.7, 0.275)$.

Chapter 7

7.1. $\mathbf{b}_0^1(2.6, 6.7, 4.3)$, $\mathbf{b}_1^1(4.3, 6.6, 4.7)$, $\mathbf{b}_2^1(4.4, 7.1, 3.7)$, $\mathbf{b}_0^2(3.11, 6.67, 4.42)$, $\mathbf{b}_1^2(4.33, 6.75, 4.40)$, $\mathbf{b}_0^3(3.476, 6.694, 4.414)$. $\mathbf{B}(0.3) = (3.476, 6.694, 4.414)$.

7.2. First derivative: $3\,((4,3) - (6,3)) = (-6,0)$, $3\,((1,2) - (4,3)) = (-9,-3)$, $3\,((-1,2) - (1,2)) = (-6,0)$. Second derivative: $2\,((-9,-3) - (-6,0)) = (-6,-6)$.

7.4. $n\,(n-1)\,(\mathbf{b}_2 - 2\mathbf{b}_1 + \mathbf{b}_0)$ and $n\,(n-1)\,(\mathbf{b}_n - 2\mathbf{b}_{n-1} + \mathbf{b}_{n-2})$.

7.5.

$$\begin{aligned} B'_{i,n}(t) &= \binom{n}{i} i(1-t)^{n-i}t^{i-1} - \binom{n}{i}(n-i)(1-t)^{n-i-1}t^i \\ &= n\binom{n-1}{i-1}(1-t)^{n-i}t^{i-1} - n\binom{n-1}{i}(1-t)^{n-i-1}t^i \\ &= n\left(B_{i-1,n-1}(t) - B_{i,n-1}(t)\right) . \end{aligned}$$

7.7. $\mathbf{b}_0(1,4)$, $\mathbf{b}_1(1,3)$, $\mathbf{b}_2(0,5)$.

7.8. $(-3 + 8t - 6t^2, -3 + 4t + t^2)$.

7.14. C^0 since $\mathbf{c}_0 = \mathbf{b}_3 = (3,6)$. For visual continuity $\mu 3(\mathbf{b}_3 - \mathbf{b}_2) = 3(-2,2)$ and $\mu 3(\mathbf{c}_1 - \mathbf{c}_0) = 3(-1,1)$. Then take $\mu = 1/2$. Change $\mathbf{b}_2$ to $(4,5)$ to obtain C^1.

7.20. (a)–(c) follow from formulae in the text with $w_0 = w_2 = 1$.

$$\frac{\mathbf{b}_0(1-0.5)^2+w_1\mathbf{b}_1 2(0.5)(1-0.5)+\mathbf{b}_2(0.5)^2}{(1-0.5)^2+w_1 2(0.5)(1-0.5)+(0.5)^2} = \frac{\mathbf{b}_0+2w_1\mathbf{b}_1+\mathbf{b}_2}{2(1+w_1)}$$

$$= \left(1-\frac{w_1}{1+w_1}\right)\left(\frac{\mathbf{b}_0+\mathbf{b}_2}{2}\right)+\left(\frac{w_1}{1+w_1}\right)\mathbf{b}_1 = \mathbf{S}.$$

7.21. $w_1^0 = 2.6, w_1^1 = 4.2, w_1^2 = 2.6, w_2^0 = 3.56, w_2^1 = 3.24, w_3^0 = 3.38$, and $\mathbf{b}_0^1\,(5.769, 4.769)$, $\mathbf{b}_1^1\,(5.571, 3.857)$, $\mathbf{b}_2^1\,(4.538, 2.308)$, $\mathbf{b}_0^2\,(5.629, 4.124)$, $\mathbf{b}_1^2\,(5.074, 3.111)$, $\mathbf{b}_0^3\,(5.309, 3.539)$.

7.24. $\mathsf{M} = \begin{pmatrix} -33 & 21 & 15 & 3 \\ 27 & -39 & 15 & 3 \\ 0 & 0 & -60 & 0 \\ 108 & 84 & 60 & -48 \end{pmatrix}$. $\mathsf{VC} = \begin{pmatrix} -\frac{1}{18} & 0 & -\frac{2}{9} \\ \frac{17}{18} & 0 & -\frac{2}{9} \\ 0 & 1 & 0 \\ -\frac{2}{9} & 0 & \frac{1}{9} \end{pmatrix}$.

$\begin{pmatrix} 2 & 4 & -2 & 2 \\ 3/2 & 5/2 & 4/2 & 1/2 \\ -4 & 12 & 12 & 4 \\ 0 & 3 & 6 & 3 \end{pmatrix}$ $\mathsf{M}\cdot\mathsf{VC} = \begin{pmatrix} 54 & 330 & -78 \\ -24 & -30 & -12 \\ -216 & -360 & -168 \\ 135 & -135 & -135 \end{pmatrix}$.

The projected control points are $(-0.692, -4.231)$, $(2, 2.5)$, $(1.286, 2.143)$, $(-1, 1)$ and weights $-78, -12, -168, -135$.

7.25. An integral curve is obtained when weights are equal. Rewrite the expression for the weights $w_i = (n_1, n_2, n_3)\cdot\mathbf{b}_i v_4 - (n_1, n_2, n_3)\cdot(v_1, v_2, v_3)$. Weights are equal if and only if either (i) $v_4 = 0$, projection is parallel and $w_i = -(n_1, n_2, n_3)\cdot(v_1, v_2, v_3)$ for all i, or (ii) $v_4 \neq 0$, projection is perspective, $(n_1, n_2, n_3)\cdot\mathbf{b}_i = 0$, and the control points lie in a plane parallel to the viewplane.

Chapter 8

8.1. $\mathbf{B}_0(t) = \left(\frac{5}{2} - t + \frac{1}{2}(t-2)^2\right)\mathbf{b}_0 + \left(-\frac{3}{2} + t - (t-2)^2\right)\mathbf{b}_1 + \frac{1}{2}(t-2)^2\mathbf{b}_2$ for $t \in [2,3]$, $\mathbf{B}_1(t) = \left(\frac{7}{2} - t + \frac{1}{2}(t-3)^2\right)\mathbf{b}_1 + \left(-\frac{5}{2} + t - (t-3)^2\right)\mathbf{b}_2 + \frac{1}{2}(t-3)^2\mathbf{b}_3$ for $t \in [3,4]$.

8.2. $\mathbf{B}_0(t) = \left(\frac{5}{3} - \frac{1}{2}t + \frac{1}{2}(t-3)^2 - \frac{1}{6}(t-3)^3\right)\mathbf{b}_0 + \left(\frac{2}{3} - (t-3)^2 + \frac{1}{2}(t-3)^3\right)\mathbf{b}_1 + \left(-\frac{4}{3} + \frac{1}{2}t + \frac{1}{2}(t-3)^2 - \frac{1}{2}(t-3)^3\right)\mathbf{b}_2 + \left(\frac{1}{6}(t-3)^3\right)\mathbf{b}_3$ for $t \in [3,4]$.

8.3. $\mathbf{B}_0(t) = (t-1)^2\,\mathbf{b}_0 + \left(2t - \frac{3}{2}t^2\right)\mathbf{b}_1 + \frac{1}{2}t^2\mathbf{b}_2$,

$\mathbf{B}_1(t) = \left(\frac{3}{2} - t + \frac{1}{2}(t-1)^2\right)\mathbf{b}_1 + \left(-\frac{1}{2} + t - (t-1)^2\right)\mathbf{b}_2 + \frac{1}{2}(t-1)^2\mathbf{b}_3$,

$$\mathbf{B}_2(t) = \left(\tfrac{5}{2} - t + \tfrac{1}{2}(t-2)^2\right)\mathbf{b}_2 + \left(-\tfrac{3}{2} + t - \tfrac{3}{2}(t-2)^2\right)\mathbf{b}_3 + (t-2)^2\,\mathbf{b}_4.$$

8.5. $\mathbf{B}(2.5) = (6.25, -0.25)$, $\mathbf{B}(4.2) = (4.28, 4.0)$.

8.6. $\mathbf{b}_5(0,0)$, $\mathbf{b}_6(2,0)$, $\mathbf{b}_7(4,2)$, $t_6 = 6$, $t_7 = 7$, $t_8 = 8$, $t_9 = 9$, $t_{10} = 10$, $t_{11} = 11$.

8.12. $\mathbf{B}(2.5) = (6.25, -0.25)$, $\mathbf{B}(4.2) = (4.28, 4.0)$.

8.13. $\mathbf{B}(2.4) = (4.04, 2.48)$.

8.18. $\mathbf{B}'(2.8) = (0.8022, -3.3045)$. $\mathbf{b}_0^{(1)}(1.875, 9.375)$, $\mathbf{b}_1^{(1)}(6.0, 0.0)$, $\mathbf{b}_2^{(1)}(1.579, -3.158)$, $\mathbf{b}_3^{(1)}(-2.857, -4.286)$.

8.19. $\mathbf{b}_0^{(1)}(4/3, 2)$, $\mathbf{b}_1^{(1)}(2, -2)$, $\mathbf{b}_2^{(1)}(2, 4/3)$, knots $4, 5, 7, 8, 10$. $\mathbf{B}'(6.2) = (1.733, -0.4)$, $\mathbf{B}'(7.4) = (2, -0.667)$.

8.24. $\sum_{i=0}^{n} \frac{w_i N_{i,d}(t)}{\sum_{j=0}^{n} w_j N_{j,d}(t)} = \frac{\sum_{i=0}^{n} w_i N_{i,d}(t)}{\sum_{j=0}^{n} w_j N_{j,d}(t)} = 1.$

8.27. $\mathbf{B}'(2.2) = (1.715, -11.853)$.

8.28. $\mathbf{B}'(0.5) = (0, -4)$. $\mathbf{B}'(0.8) = (7.101, 2.959)$.

8.31. $\mathbf{B}(0.65) = (-0.6897, -0.7241)$.

Chapter 9

9.2. First derivative with respect to s:

$$\mathbf{p}_{0,0}^{(1,0)} = 2\left((4,3,1) - (2,2,0)\right) = (4,2,2),$$

$$\mathbf{p}_{1,0}^{(1,0)} = 2\left((6,2,0) - (4,3,1)\right) = (4,-2,-2),$$

$\mathbf{p}_{0,1}^{(1,0)} = 2\left((4,5,3) - (2,4,1)\right) = (4,2,4)$, similarly $\mathbf{p}_{1,1}^{(1,0)} = (4,-4,-4)$, $\mathbf{p}_{0,2}^{(1,0)} = (4,0,2)$, $\mathbf{p}_{1,2}^{(1,0)} = (4,-2,-2)$.

9.4. (a) The tangent vectors at $\mathbf{S}(0,0)$: $n(\mathbf{p}_{1,0} - \mathbf{p}_{0,0})$ and $p(\mathbf{p}_{0,1} - \mathbf{p}_{0,0})$; $\mathbf{S}(0,1)$: $n(\mathbf{p}_{0,p} - \mathbf{p}_{0,p-1})$ and $p(\mathbf{p}_{1,p} - \mathbf{p}_{0,p})$; $\mathbf{S}(1,0)$: $n(\mathbf{p}_{n,0} - \mathbf{p}_{n-1,0})$ and $p(\mathbf{p}_{n,1} - \mathbf{p}_{n,0})$; $\mathbf{S}(1,1)$: $n(\mathbf{p}_{n,p} - \mathbf{p}_{n-1,p})$ and $p(\mathbf{p}_{n,p} - \mathbf{p}_{n,p-1})$. (b) The normal at $\mathbf{S}(0,0)$ is $np(\mathbf{p}_{1,0} - \mathbf{p}_{0,0}) \times (\mathbf{p}_{0,1} - \mathbf{p}_{0,0})$;

at $\mathbf{S}(0,1)$ is $np(\mathbf{p}_{0,p} - \mathbf{p}_{0,p-1}) \times (\mathbf{p}_{1,p} - \mathbf{p}_{0,p})$ etc.

9.5. (All the points listed in the order $\mathbf{p}_{0,0}$, $\mathbf{p}_{1,0}$, etc.)

$(2,0,1)$, $(1,0,2)$, $(3,0,3)$, $(1,0,4)$, $(1,0,5)$;

$(2,2,1)$, $(1,1,2)$, $(3,3,3)$, $(1,1,4)$, $(1,1,5)$;

$(-2, 2, 1)$, $(-1, 1, 2)$, $(-3, 3, 3)$, $(-1, 1, 4)$, $(-1, 1, 5)$;

$(-2, 0, 1)$, $(-1, 0, 2)$, $(-3, 0, 3)$, $(-1, 0, 4)$, $(-1, 0, 5)$;

$(-2, -2, 1)$, $(-1, -1, 2)$, $(-3, -3, 3)$, $(-1, -1, 4)$, $(-1, -1, 5)$;

$(2, -2, 1)$, $(1, -1, 2)$, $(3, -3, 3)$, $(1, -1, 4)$, $(1, -1, 5)$;

$(2, 0, 1)$, $(1, 0, 2)$, $(3, 0, 3)$, $(1, 0, 4)$, $(1, 0, 5)$.

9.6. $\mathbf{b}_{0,0}(2, 3, 0)$, $\mathbf{b}_{1,0}(1, 5, 2)$, $\mathbf{b}_{2,0}(1, 7, -1)$, $\mathbf{b}_{3,0}(2, 9, -3)$, $\mathbf{b}_{0,1}(4, 7, -4)$,

$\mathbf{b}_{1,1}(3, 9, -2)$, $\mathbf{b}_{2,1}(3, 11, -5)$, $\mathbf{b}_{3,1}(4, 13, -7)$.

9.7. Control points as for Exercise 9.6, weights $w_{0,0} = w_{0,1} = 1$, $w_{1,0} = w_{1,1} = 2$, $w_{2,0} = w_{2,1} = 3$, $w_{3,0} = w_{3,1} = 1$.

9.9. $\mathbf{b}_{0,0}(0, 0, 0)$, $\mathbf{b}_{1,0}(0, a, 0)$, $\mathbf{b}_{2,0}(a, 2a, 0)$, $\mathbf{b}_{0,1}(0, 0, 1)$, $\mathbf{b}_{1,1}(0, a, 1)$, $\mathbf{b}_{2,1}(a, 2a, 1)$. Parabolic cylinder can be obtain by sweeping a line segment along a quadratic Bézier curve.

9.10. NURBS sphere has control points: (Listed in the order $\mathbf{p}_{0,0}$, $\mathbf{p}_{1,0}$, etc)

$(1, 0, 0)$, $(1, 0, 1)$, $(-1, 0, 1)$, $(-1, 0, 0)$, $(-1, 0, -1)$, $(1, 0, -1)$, $(1, 0, 0)$;

$(1, 1, 0)$, $(1, 1, 1)$, $(-1, -1, 1)$, $(-1, -1, 0)$, $(-1, -1, -1)$, $(1, 1, -1)$,

$(1, 1, 0)$;

$(-1, 1, 0)$, $(-1, 1, 1)$, $(1, -1, 1)$, $(1, -1, 0)$, $(1, -1, -1)$, $(-1, 1, -1)$, $(-1, 1, 0)$;

$(-1, 0, 0)$, $(-1, 0, 1)$, $(1, 0, 1)$, $(1, 0, 0)$, $(1, 0, -1)$, $(-1, 0, -1)$, $(-1, 0, 0)$;

$(-1, -1, 0)$, $(-1, -1, 1)$, $(1, 1, 1)$, $(1, 1, 0)$, $(1, 1, -1)$, $(-1, -1, -1)$, $(-1, -1, 0)$;

$(1, -1, 0)$, $(1, -1, 1)$, $(-1, 1, 1)$, $(-1, 1, 0)$, $(-1, 1, -1)$, $(1, -1, -1)$, $(1, -1, 0)$;

$(1, 0, 0)$, $(1, 0, 1)$, $(-1, 0, 1)$, $(-1, 0, 0)$, $(-1, 0, -1)$, $(1, 0, -1)$, $(1, 0, 0)$.

Weights: $w_{i,0} = \{1, 0.5, 0.5, 1, 0.5, 0.5, 1\}$,

$w_{i,1} = w_{i,2} = \{0.5, 0.25, 0.25, 0.5, 0.25, 0.25, 0.5\}$,

$w_{i,3} = \{1, 0.5, 0.5, 1, 0.5, 0.5, 1\}$,

$w_{i,4} = w_{i,5} = \{0.5, 0.25, 0.25, 0.5, 0.25, 0.25, 0.5\}$,

$w_{i,6} = \{1, 0.5, 0.5, 1, 0.5, 0.5, 1\}$.

Knots for s and t $0, 0, 0, 0.25, 0.5, 0.5, 0.75, 1, 1, 1$.

9.13. Control points: $(-1,0,1)$, $(0,0,0)$, $(1,0,1)$; $(-1,-1,1)$, $(0,0,0)$, $(1,1,1)$; $(1,-1,1)$, $(0,0,0)$, $(-1,1,1)$; $(1,0,1)$, $(0,0,0)$, $(-1,0,1)$; $(1,1,1)$, $(0,0,0)$, $(-1,-1,1)$; $(-1,1,1)$, $(0,0,0)$, $(1,-1,1)$; $(-1,0,1)$, $(0,0,0)$, $(1,0,1)$.

Weights: $w_{i,0} = w_{i,2} = \{1, 0.5, 0.5, 1, 0.5, 0.5, 1\}$,

$w_{i,1} = \{3, 1.5, 1.5, 3, 1.5, 1.5, 3\}$.

9.19.

$$\begin{aligned}&(1-s)(3t^2+4, 2t^2, -t) + s(2t, -t^4, 2t+4)\\&\quad = (4-4s+2st+3t^2-3st^2, 2t^2-2st^2-st^4, 4s-t+3st)\ .\end{aligned}$$

9.20. (a)

$$\begin{aligned}&(3(1-3s^2+2s^3)t^2+(3s^2-2s^3)(2t+10)+s-2s^2+s^3,\\&2(1-3s^2+2s^3)t^2+3(3s^2-2s^3)t+s-s^2,\\&(1-3s^2+2s^3)t+2(3s^2-2s^3)t^2)\\&= (3t^2-9t^2s^2+6t^2s^3+6ts^2+28s^2-4ts^3-19s^3+s,\\&\quad 2t^2-6t^2s^2+4t^2s^3+9ts^2-6ts^3+s-s^2,\\&\quad t-3ts^2+2ts^3+6t^2s^2-4t^2s^3)\ .\end{aligned}$$

(b)

$$\begin{aligned}&\big(3(1-3s^2+2s^3)t^2+(3s^2-2s^3)(2t+10)+(s-2s^2+s^3)s,\\&2(1-3s^2+2s^3)t^2+3(3s^2-2s^3)t-2s+4s^2-2s^3-(-s^2+s^3)s,\\&(1-3s^2+2s^3)t+2(3s^2-2s^3)t^2+(s-2s^2+s^3)s+2(-s^2+s^3)s\big)\\&= (3t^2-9t^2s^2+6t^2s^3+6ts^2+31s^2-4ts^3-22s^3+s^4,\\&\quad 2t^2-6t^2s^2+4t^2s^3+9ts^2-6ts^3-2s+4s^2-s^3-s^4,\\&\quad t-3ts^2+2ts^3+6t^2s^2-4t^2s^3+s^2-4s^3+3s^4\big)\ .\end{aligned}$$

9.21.

$$\Big(5\,(1-s)\,(1-t)-20\,(1-s)\,t+8\,s\,(1-t)-27\,st+(1-s)\,(-5+25\,t)$$
$$+\,s\left(-8\,(1-t)^2+20\,(1-t)\,t+27\,t^2\right)$$
$$+\,(1-t)\left(-5\,(1-s)^3-18\,(1-s)^2\,s-21\,(1-s)\,s^2-8\,s^3\right)$$
$$+\,t\left(20\,(1-s)^2+44\,(1-s)\,s+27\,s^2\right),$$
$$5\,(1-s)\,(1-t)-3\,(1-s)\,t-7\,st+(1-s)\,(-5+8\,t)$$
$$+\,s\left(32\,(1-t)\,t+7\,t^2\right)$$
$$+\,(1-t)\left(-5\,(1-s)^3-12\,(1-s)^2\,s-6\,(1-s)\,s^2\right)$$
$$+\,t\left(3\,(1-s)^2+10\,(1-s)\,s+7\,s^2\right),$$
$$-\,10\,s\,(1-t)-10\,st+s\left(10\,(1-t)^2+8\,(1-t)\,t+10\,t^2\right)$$
$$+\,(1-t)\left(9\,(1-s)^2\,s+21\,(1-s)\,s^2+10\,s^3\right)+t\left(8\,(1-s)\,s+10\,s^2\right)\Big)$$
$$=(-5+25\,t-3\,s+8\,st-st^2+3\,ts^2,$$
$$-\,5+8\,t+3\,s+26\,st-25\,st^2+3\,s^2-s^3-3\,ts^2+ts^3,$$
$$9\,s-13\,st+12\,st^2+3\,s^2-2\,s^3-ts^2+2\,ts^3)\,.$$

Chapter 10

10.1. (a) $\mathbf{C}'(t)=(1,\sinh(t/c))$, $\mathbf{C}''(t)=\left(0,\frac{1}{c}\cosh(t/c)\right)$,

$\mathbf{t}(t)=(1/\cosh(t/c)\,,\tanh(t/c))$, $\mathbf{n}(t)=(-\tanh(t/c)\,,1/\cosh(t/c))$.

$\kappa=1/c\cosh^2(t/c)\,.$

10.3. (a) $\theta(s)=\arcsin s$, and $x=\frac{1}{2}s\sqrt{1-s^2}+\frac{1}{2}\arcsin s$, $y=\frac{1}{2}s^2$. (b) $\theta(s)=2\sqrt{s}$, and $x=2\sqrt{s}\sin(\sqrt{s})\cos(\sqrt{s})+\cos^2(\sqrt{s})-1$, $y=\sin(\sqrt{s})\cos(\sqrt{s})-2\sqrt{s}\cos^2(\sqrt{s})+\sqrt{s}$.

(c) $\theta(s)=-\arctan(s/a)$, and $x=a\ln((s+\sqrt{a^2+s^2})/a)$, $y=a-\sqrt{a^2+s^2}$.

10.4. Reparametrize the curve so that $\mathbf{C}(s)$ is unit speed. If $\kappa(s)=0$ for all s, then $\mathbf{C}''(s)=\kappa(s)\mathbf{N}(s)$ implies that $\mathbf{C}''(s)=(0,0)$. Thus $\mathbf{C}'(s)=(a_1,a_2)$ for some constants a_1 and a_2. Finally, integrating gives $\mathbf{C}(s)=(a_1s+b_1,a_2s+b_2)$. The result can also be deduced from the fundamental theorem of plane curves: $\theta=\int\kappa(u)\,du=\int 0\,du$ etc.

10.5. In polar coordinates: $(x,y) = (r(\theta)\cos\theta, r(\theta)\sin\theta)$. Then $(x',y') = (r'(\theta)\cos\theta - r(\theta)\sin\theta, r'(\theta)\sin\theta + r(\theta)\cos\theta)$. So the arclength is

$$\int_a^b \sqrt{(r'(\theta)\cos\theta - r(\theta)\sin\theta)^2 + (r'(\theta)\sin\theta + r(\theta)\cos\theta)^2}\,d\theta$$
$$= \int \sqrt{(r(\theta)^2 + (r'(\theta))^2)}\,d\theta\,.$$

Further,

$$\begin{aligned}(x'',y'') &= (r''(\theta)\cos\theta - 2r'(\theta)\sin\theta - r(\theta)\cos\theta,\\ &\quad r''(\theta)\sin\theta + 2r'(\theta)\cos\theta - r(\theta)\sin\theta)\end{aligned}$$

and substitution into the formula for curvature gives $\kappa = \frac{2(r')^2 - rr'' + r^2}{(r^2+(r')^2)^{3/2}}$.

10.6. (a) $\kappa = \frac{1}{2(2+2\cos t)^{1/2}}$. $\mathbf{E}(t) = (t - \sin t, 3 + \cos t)$.

(b) $\dot{\mathbf{C}}(t) = (-a\sin t, b\cos t)$, $\ddot{\mathbf{C}}(t) = (-a\cos t, -b\sin t)$.

$\mathbf{n}(t) = \left(\frac{-b\cos t}{\sqrt{a^2\sin^2 t + b^2\cos^2 t}}, \frac{-a\sin t}{\sqrt{a^2\sin^2 t+b^2\cos^2 t}}\right)$,

$\kappa(t) = \frac{ab}{(a^2\sin^2 t + b^2\cos^2 t)^{3/2}}$.

Evolute is $\mathbf{E}(t) = \left(\frac{a^2-b^2}{a}\cos^3 t, -\frac{a^2-b^2}{b}\sin^3 t\right)$.

10.7. (a) Offset is $(-3\sin t, 2\cos t) + \frac{d}{(9\sin^2 t + 4\cos^2 t)^{1/2}}(-2\cos t, -3\sin t)$.

(b) $\kappa(t) = \frac{6}{(4\sin^2 t+9\cos^2 t)^{3/2}}$, $\dot\kappa(t) = \frac{90\sin t\cos t}{(4\sin^2 t + 9\cos^2 t)^{5/2}}$. Maxima and minima occur when $\dot\kappa(t) = 0$. Hence $t = 0, \pi/2, \pi, 3\pi/2$ corresponding to the points $(3,0)$, $(0,2)$, $(-3,0)$, $(0,-2)$.

(c) Maximum and minimum values of curvature are $\frac{3}{4}$ and $\frac{2}{9}$. Therefore the maximum and minimum radii of curvature are $\frac{9}{2}$ and $\frac{4}{3}$. Hence the maximum radius the ball cutter can be is $\frac{9}{2}$, otherwise the cutter will be too large to cut the ellipse at the points $(0,2)$ and $(0,-2)$.

10.10. $\dot{\mathbf{C}}(t) = (-4\sin t, -5\cos t, 3\sin t)$, $\left|\dot{\mathbf{C}}(t)\right| = 5$.

$\mathbf{t} = \left(-\frac{4}{5}\sin t, -\cos t, \frac{3}{5}\sin t\right)$. $\mathbf{n} = \left(-\frac{4}{5}\cos t, \sin t, \frac{3}{5}\cos t\right)$.

$\mathbf{b} = \left(-\frac{3}{5}, 0, -\frac{4}{5}\right)$. $\kappa = 1/5$, $\tau = 0$. Curve is a circle radius 5.

10.11. $\mathbf{t} = \left(-\frac{3}{10}\sin t - \frac{2}{5}\cos t, \frac{3}{10}\cos t - \frac{2}{5}\sin t, \frac{\sqrt{3}}{2}\right)$,

$\mathbf{n} = \left(-\frac{3}{5}\cos t + \frac{4}{5}\sin t, -\frac{3}{5}\sin t - \frac{4}{5}\cos t, 0\right)$,

$\mathbf{b} = \left(\frac{3\sqrt{3}}{10}\sin t + \frac{2\sqrt{3}}{5}\cos t, -\frac{3\sqrt{3}}{10}\cos t + \frac{2\sqrt{3}}{5}\sin t, \frac{1}{2}\right)$.

$\kappa = 1/20$, $\tau = \sqrt{3}/20$.

10.12. $\kappa = \frac{\sqrt{5}}{2}\left(2 - \cos^2 t\right)^{-3/2}$, $\tau = 0$. Curve is an ellipse.

10.13. (a) $\kappa = \tau = 1/3(1+t^2)^2$. (b) $\kappa = \left((1+\cos^2 t)+1\right)^{1/2}/(3+2\cos t)^{3/2}$, $\tau = 1/\left((1+\cos^2 t)+1\right)^{1/2}$. (c) $\kappa = \sqrt{\frac{3}{2}}t^3\left(1+t^2+t^4\right)^{-3/2}$, $\tau = 0$.

10.14. $\kappa = \tau = \frac{1}{2}\left(2-2t^2\right)^{-1/2}$.

10.17. (a) $a = \sqrt{17}$, $b = 13\sqrt{2}$, $c = 39$ so $\kappa = \frac{26}{867}\sqrt{2}\sqrt{17} = 0.175$, $\tau = \frac{1}{34}\sqrt{2}\sqrt{17} = 0.171$. (b) $\kappa = 0.185$, $\tau = 0.334$.

10.24. (a) $2(1+4u^2+4v^2)^{-1/2}$ and $2(1+4u^2+4v^2)^{-3/2}$. (c) r and $\cos u(R + r\cos u)^{-1}$.

10.25. (a) $K = -(1+u^2+v^2)^{-2}$, $H = -uv(1+u^2+v^2)^{-3/2}$.

10.26. (b) $(0,0,0)$.

10.32. $\frac{1}{\pi}\int_0^\pi \left(k_{\max}\cos^2\theta + k_{\min}\sin^2\theta\right)\, d\theta =$

$\frac{1}{\pi}\left[\frac{1}{2}(\theta+\cos\theta\sin\theta)\,k_{\max} + \frac{1}{2}(\theta-\cos\theta\sin\theta)\,k_{\min}\right]_0^\pi = \frac{1}{2}(k_{\max}+k_{\min})$.

Chapter 11

11.2. $\mathbf{N} = \frac{1}{\sqrt{1+4s^2+4t^2}}(2s,2t,1)$. So at $(0.5,0.5,-0.5)$, $\mathbf{N} = \frac{1}{\sqrt{3}}(1,1,-1)$. The incident ray has direction $(0,10,20)-(0.5,0.5,-0.5)$, giving $\mathbf{L} = \left(-\frac{\sqrt{227}}{681}, \frac{19\sqrt{227}}{681}, \frac{41\sqrt{227}}{681}\right)$. $\mathbf{R} = \left(-\frac{43\sqrt{227}}{2043}, -\frac{103\sqrt{227}}{2043}, -\frac{77\sqrt{227}}{2043}\right)$. The angle of incidence is 1.272609737 radians.

11.4. $\mathbf{V}\cdot\mathbf{N} = (0,1,0)\cdot(-t-2s, -s+t^2, 1) = -s+t^2$. So the silhouette points satisfy $s = t^2$ giving the curve $\left(t^2, t, \frac{2}{3}t^3 + t^4\right)$.

11.5. $\mathbf{V}\cdot\mathbf{N} = (1,0,0)\cdot(1-t^2-s^2, -2st, 1) = 1-s^2-t^2$. So the silhouette points satisfy $s^2+t^2 = 1$ defining the unit circle in the (s,t)-plane. The circle may be parametrized as $(s,t) = (\cos\theta, \sin\theta)$ and substituting into the parametric equation of the surface gives the silhouette curve $\left(\cos\theta, \sin\theta, \cos\theta\sin^2\theta - \cos\theta + 1/3\cos^3\theta\right)$.

11.6. $\mathbf{V}\cdot\mathbf{N} = (0,1,0)\cdot(-18st, 1-9s^2+9t^2, 1) = 1-9s^2+9t^2 = 0$. The substitution $(s,t) = (\frac{1}{3}\cosh\theta, \frac{1}{3}\sinh\theta)$ gives the silhouette curve $\left(\frac{1}{3}\cosh\theta, \frac{1}{3}\sinh\theta, \frac{1}{3}\cosh^2\theta\sinh\theta - \frac{1}{3}\sinh\theta - \frac{1}{9}\sinh^3\theta\right)$.

11.7. The plane $ax+by+cz+d=0$ has normal $\mathbf{N} = (a,b,c)$. For a parallel projection in the direction $\mathbf{V}(v_0,v_1,v_2)$, $\mathbf{V}\cdot\mathbf{N} = av_0+bv_1+cv_2$. In general, $av_0 + bv_1 + cv_2 \neq 0$ and there are no silhouette points. When $av_0 + bv_1 + cv_2 = 0$, every point is a silhouette point. (The condition corresponds to when $\mathbf{V}$ is perpendicular to $\mathbf{N}$).

For a perspective projection from the viewpoint $\mathbf{V}(v_0, v_1, v_2)$, the condition for a silhouette is $(v_0 - x, v_1 - y, v_2 - z) \cdot \mathbf{N} = av_0 + bv_1 + cv_2 - (ax + by + cz) = 0$ giving $av_0 + bv_1 + cv_2 + d = 0$. The condition is satisfied if and only if $\mathbf{V}$ is a point on the plane and every point of the plane is a silhouette point.

11.11. If the viewpoint $\mathbf{V}$ lies inside the sphere centred at $\mathbf{C}$, radius r, then $|\mathbf{V} - \mathbf{C}| < r$. Using Exercise 11.7.(a), $d > r$ and r_1, given by 11.7(b), has no solution.

11.14. The silhouettes are two lines $\left(1/\lambda, \pm\sqrt{(\lambda^1 - 1)/\lambda}, t\right)$.

11.15. $\mathbf{V} = (10, 0, 0, 1)$ and

$$\mathsf{Q} = \begin{pmatrix} 4 & 0 & -1 & 0 \\ 0 & 1 & 0 & 0 \\ -1 & 0 & -1 & 0 \\ 0 & 0 & 0 & -4 \end{pmatrix}.$$

The silhouette plane is: $40x - 10z - 4 = 0$. Solving for x and substituting into the equation of the quadric gives the conic: $-\frac{99}{25} - 5/4\, z^2 + y^2 = 0$ which can be parametrized by

$$(y, z) = \left(\pm\sqrt{99/25}\cosh t, \sqrt{396/125}\sinh t\right).$$

A parametrization for the silhouette follows.

11.17. $\mathbf{V} = (0, 0, 1)$, $\mathbf{N} = (-f(u)g'(u)\cos v, -f(u)g'(u)\sin v, f(u)f'(u))$ and $\mathbf{V} \cdot \mathbf{N} = f(u)f'(u) = 0$. Since $f(u) \neq 0$ silhouette points satisfy $f'(u) = 0$. For each u_0 such that $f'(u_0) = 0$ there is a silhouette circle given by $(f(u_0)\cos v, f(u_0)\sin v, g(u_0))$.

11.18. Since $\mathbf{VQP}^T$ is a 1×1 matrix it is equal to its own transpose, namely, $\mathbf{PQV}^T$.

11.20.

$$\mathbf{VQP}^T = \begin{pmatrix} \lambda & 0 & 0 & 1 \end{pmatrix} \begin{pmatrix} 1 & 0 & 0 & 0 \\ 0 & 1 & 0 & 0 \\ 0 & 0 & 0 & 0 \\ 0 & 0 & 0 & -1 \end{pmatrix} \begin{pmatrix} x \\ y \\ z \\ 1 \end{pmatrix} = \lambda x - 1 = 0.$$

Therefore $x = \frac{1}{\lambda}$ and $y = \pm\frac{\sqrt{\lambda^2-1}}{\lambda}$ giving two lines $(\frac{1}{\lambda}, \pm\frac{\sqrt{\lambda^2-1}}{\lambda}, z)$.

References

[1] Abhyankar, S S and Bajaj, C, 'Automatic parametrization of rational curves and surfaces I: Conics and conicoids. *Computer-Aided Design* Vol. 19, pp11–14, 1987.

[2] Bézier, P, 'Style, mathematics and NC'. *Computer-Aided Design* Vol. 22 No. 9, pp523–526, 1990.

[3] Boehm, W and Prautzsch, H, 'The insertion algorithm'. *Computer-Aided Design* Vol. 17 No. 2, pp58–59, 1985.

[4] Braid, I C, Hillyard, R C, and Stroud I A, 'Stepwise construction of polhedra in geometric modelling' in *Mathematical Methods in Computer Graphics and Design*, ed. K W Brodlie, pp123-141, Academic Press, 1980.

[5] Coolidge, J L, *A History of the Conic Sections and Quadric Surfaces.* OUP, 1945.

[6] Davis, P, 'B-splines and geometric design', *SIAM News* Vol. 29 No. 5, 1996.

[7] Dill, J. 'An application of colour graphics to the display of surface curvature'. *Computer Graphics* Vol. 15, pp153–161, 1981.

[8] Do Carmo, M P, *Differential Geometry of Curves and Surfaces.* Prentice-Hall, 1976.

[9] Farin, G, *Curves and Surfaces for Computer-Aided Geometric Design.* Third Edition. Academic Press, 1993.

[10] Forrest, A R, 'Interactive interpolation and approximation by Bézier polynomials', *Computer-Aided Design* Vol. 22 No. 9, pp527–537, 1990. Originally published in *The Computer Journal* Vol. 15 No. 1, pp71–79, 1972.

[11] Gibson, C.G., *Elementary Geometry of Algebraic Curves*. Cambridge University Press, 1998.

[12] Haralick, R M and Shapiro, L G, *Computer and Robot Vision*. Addison-Wesley, 1992.

[13] Hoschek, J and Lasser, D, *Fundamentals of Computer Aided Geometric Design*. A K Peters, 1993.

[14] Howard, T L J, Hewitt, W T, Hubbold, R J, and Wyrwas, K M, *A Practical Introduction to PHIGS and PHIGS PLUS*. Addison-Wesley, 1991.

[15] Lane, J and Riesenfeld, R, 'A geometric proof for the variation diminishing property of B-spline approximation'. *J. of Approximation Theory* Vol. 37, pp1–4, 1983.

[16] Mäntylä, M, *An Introduction to Solid Modeling*, Computer Science Press, Maryland, 1988.

[17] Munchmeyer, F, 'On surface imperfections'. In R.Martin, editor, *The Mathematics of Surfaces II*, pp459–474. OUP, 1987.

[18] Munchmeyer, F, 'Shape interrogation: a case study'. In G.Farin, editor, *Geometric Modelling: Algorithms and New Trends*, pp291–301. SIAM, Philadelphia, 1987.

[19] Phong, B-T, 'Illumination for computer-generated pictures'. Comm. ACM, Vol. 18, No. 6, pp311–317, June 1975.

[20] Piegl, L and Tiller, W, *The NURBS Book*. Springer-Verlag, 1995.

[21] Rogers, D F and Adams, J A, *Mathematical Elements for Computer Graphics*. Second Edition. McGraw-Hill, 1990.

[22] Schoenberg, I, 'Contributions to the problem of approximation of equidistant data by analytic functions', *Quart. Appl. Math.* Vol. 4, pp45–99, 1946.

[23] Sederberg, Th W, Anderson, D C and Goldman, R N, 'Implicit representation of parametric curves and surfaces'. *Computer Vision, Graphics and Image Processing* Vol. 28, pp72–84, 1984.

[24] Semple, J G and Kneebone, G T, *Algebraic Projective Geometry*. OUP, 1952.

[25] Smith, G, *Introductory Mathematics: Algebra and Analysis*. Springer-Verlag, 1998.

[26] Sommerville, D M Y, *Analytical Conics*. Bell and Sons, 1945.

[27] Spivak, M, *Calculus*. W.A.Benjamin, 1967.

Index